METABOLOMICS

Edited by **Ute Roessner**

Metabolomics
Edited by Ute Roessner

Published by InTech
Janeza Trdine 9, 51000 Rijeka, Croatia

Publishing Process Manager Igor Babic

Technical Editor Teodora Smiljanic

Cover Designer InTech Design Team

Additional hard copies can be obtained from orders@intechopen.com

Metabolomics, Edited by Ute Roessner
p. cm.
978-953-51-0046-1

For EU product safety concerns:
IN TECH d.o.o., Prolaz Marije Krucifikse Kozulić 3, 51000 Rijeka, Croatia,
info@intechopen.com or visit our website at intechopen.com.

Meet the editor

Dr Roessner obtained her PhD in Biochemistry at the MPI for Molecular Plant Physiology in Germany, where she developed novel GC-MS methods to analyse metabolites in plants. The field of metabolomics emerged together with the application of sophisticated data mining, and it is today an important tool in biological sciences, systems biology and biomarker discovery. In 2003, Dr. Roessner moved to Australia where she established a GC-MS and LC-MS based metabolomics platform at the University of Melbourne, node of the Australian Centre for Plant Functional Genomics. Since 2007, Dr. Roessner has also been involved in the setup of Metabolomics Australia (MA), an Australian federal government investment, through Bioplatforms Australia Ltd. and currently leads the MA node at the University of Melbourne.

Contents

Preface

Metabolomics is a new scientific field which has developed with an accelerating speed over the last decade as demonstrated through the increasing numbers of publications in scientific journals of any biological research field. These developments are mainly driven by increasingly robust and sensitive analytical instrumentations allowing the analysis and quantification of thousands of metabolites from any biological system. Together with the application of sophisticated computational methodology and statistical approaches the vast amount of data generated from instrumentation can be analysed and mined aiding biological and biochemical interpretation. Especially, once experimental metabolomics data can be integrated with other 'omics type data such as from genomics and proteomics analyses the path is paved for a better holistic understanding of the biological system under investigation.

This book will provide the reader with summaries of the state-of-the-art of the technologies and methodologies, especially in the data analysis and interpretation approaches as well as gives insights into exciting applications of metabolomics in human health studies, safety assessments and plant and microbial research.

Dr. Ute Roessner
School of Botany, The University of Melbourne, Victoria,
Australia

Part 1

Metabolomics of Microbes and Cell Cultures

1

Metabolomics and Mammalian Cell Culture

Kathya De la Luz-Hdez
Center of Molecular Immunology
Cuba

1. Introduction

Since the mid-1950s, when pioneering work of Earle and colleagues (1954) enable routine cell culture, mammalian cell culture has been used in the large-scale production of recombinant protein and monoclonal antibodies. Mammalian cell lines are preferred as production host for many pharmaceuticals, since complex post-translational modifications of the produced proteins (especially glycosylation) are generally not properly performed by microbial systems (Lake-Ee Quek et al., 2010).

Wagburg described that under batch conditions, mammalian cells display an inefficient metabolic phenotype characterized by high rates of glucose to lactate conversion (Warburg, 1956) together with partial oxidation of glutamine to ammonia and non-essential amino acids (Fitzpatrick et al., 1993; Jenkins et al., 1992; Ljunggren and Haggstrom, 1992; Ozturk and Palsson, 1991). The accumulation of fermentation by-products causes a reduction of the culture density and product titer that can be realized (Martinelle et al., 1998).

In order to increase the cell productivity a common optimization approach is to grow cells to moderately high density in fed-batch and the deliberately induce a prolonged, productive stationary phase. While optimization of this perturbed batch strategy is responsible for the increase of monoclonal antibodies titer seen over the past decades it has a number of short-comings, including:

a. the strategy has to be refined for each new cell line,
b. the ultimate metabolic phenotype during prolonged stationary phase varies between cell lines and it is not always possible to achieve the most productive phenotypes for a given strain
c. volumetric productivity remains relatively low due to moderate cell density (Lake Ee Quek et al., 2010).

Other approaches are related with the changes of cell phenotypes through metabolic engineering or change of culture media conditions in order to manipulate the cellular metabolic behavior.

Although transcriptomics and proteomics have been explored extensively for mammalian cell engineering (Korke et al., 2002; Seow et al., 2001; Seth et al., 2007; Smales et al., 2004; de la Luz et al., 2007, 2008) these tools fall short of generating direct measurements of the physiological state of the cell. It is essential to combine these techniques with metabolic flux analysis (MFA) a powerful method to quantify the manifestation of a phenotype: the intracellular reaction rates or the fluxome. One of the relatively new "omic" sciences is the field of metabolomics. The metabolome was first described by Oliver and collagues (1998) as being the set of all of low-molecular-mass compounds synthesized by an organism. Metabolomics is therefore the analysis of small molecules that constitute the metabolism, and it offers the closest direct measurements of a cell´s physiological activity (Beecher, 2002; Khoo and Al-Rubeai, 2007). The metabolomic analysis can be considered as "the measurement of the change in the relative concentrations of metabolites as the result of the deletion or overexpression of a gene, should allow the target of a novel gene product to be located on the metabolic map". Another definition of the metabolome states that it consists of "only those native small molecules that are participant's in general metabolic reactions and that are required for the maintenance, growth and normal function of a cell" (Khoo and Al-Rubeai, 2007).

The metabolomic as a new powerful tool to understand the complex processes of large scale mammalian cell cultures for biopharmaceutical production has not been yet embraced during process development and scale-up. This is mostly because metabolites are now not the primary focus and the relationship between metabolites and protein production in different media are not fully understood. It is for this reason that metabolomics can bridge the gap of understanding as to the dynamics of metabolism, cell growth and protein production (Khoo and Al-Rubeai, 2007). The metabolomics can be use to optimize conditions of bioreactors or the development chemically defined media. Characterizing cell lines, culture media and selection of cell lines are a vital step in the process development of biologics. In this chapter, we describe the state-of-the-art of the use of metabolomics tool in mammalian cell lines.

2. Complexity of metabolome analysis

Metabolomics requires the unbiased identification and quantification of all of the metabolites present in a specific biological sample (from an organism or *in vitro*). Metabolites are generally labile species, by their nature are chemically very diverse, and often present in a wide dynamic range. For analysis of mRNA and proteins one "only" needs to know the genome sequence of the organism and exploit this information using nucleic acid hybridization or protein separation followed by MS (although PTM are problematic). However, the analysis of metabolites is not as straightforward. In contrast to transcripts or protein identification, metabolites are not organism specific (that is to say, sequence dependent) (Hollywood et al., 2006).

In addition, their diverse chemical properties make complete metabolite analysis difficult. Genes are composed of a linear four-letter code, whereas proteins have a 20-letters code of primary amino acids. Metabolites do not have any fixed codes, and thus a general method of characterization is difficult. Present methods use the specific chemical properties of these entities to separate, identify and decipher their structures. Combinatorial approaches allow for a greater coverage. An ideal metabolic analysis should provide:

a. Give an instantaneous snapshot of all metabolites in any given system,
b. Use analytical methods that have high recovery, experimental robustness, reproducibility, high resolving power and high sensitivity (Fiehn, 2001) whilst being able to be applied universally,

c. Provide the unambiguous quantification and identification of metabolites and (d) factors to be highlighted while easily being incorporated into biochemical network models (Soo H and Al-Rubeai M, 2007).

Before any metabolome measurements are taken it is essential that metabolism is stopped as quickly as possible, especially because the enzymes are active. For animal cells liquid N_2 is used to snap freeze the sample, followed by mechanical disruption in order to release the metabolites (Viant et al., 2005). The next stage of the analysis is to extract the metabolites. There are many different methods (Tweeddale et al., 1998; Buchholz et al., 2001; Villas-Boas et al., 2005) and the most common ones are:

- Acid extraction using perchloric acid, followed by freeze thawing, then neutralization with potassium hydroxide
- Alkali extraction typically using sodium hydroxide, followed by heating (80°C)
- Ethanolic extraction by boiling the sampling in ethanol at 80°C

When the extract is finally ready, the choice of the analytical tool is based on the level of chemical information required about the metabolites, remembering that there will be a chemical bias with respect to that method, and the speed of analysis is also another consideration. The figure 1 shows different methods and approaches used for the metabolomic analysis.

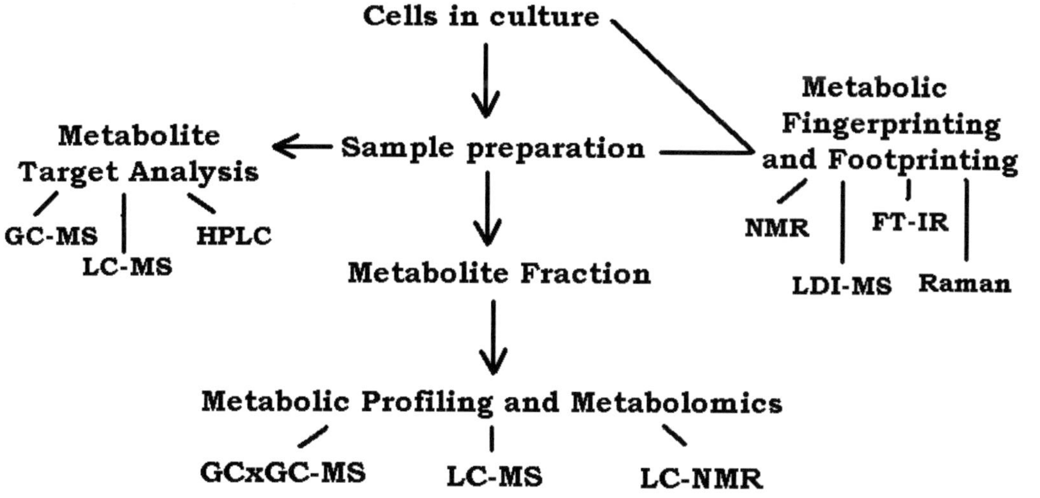

Fig. 1. Technologies for metabolome analysis. GC-MS: gas chromatography mass spectrometry, GCxGC-MS: 2 dimensional GC coupled to mass spectrometry, LC-MS: liquid chromatography mass spectrometry, HPLC: high performance liquid chromatography, LC-NMR: liquid chromatography coupled to nuclear magnetic resonance, NMR: nuclear magnetic resonance, LDI-MS: laser desorption ionization mass spectrometry, FT-IR: Fourier transform infrared spectroscopy

The most popular approaches are:

1. Metabolite target analysis: which is an approach that is restricted to metabolites for example a particular enzyme system that would be directly affected by any perturbation
2. Metabolite profiling: which is focused on a specific group of metabolites (for example lipids) or those associated with a specific pathway; within clinical and pharmaceutical analysis, this is often called metabolic profiling, which is used to trace the fate of a drug or metabolite
3. Metabolomics: is the comprehensive analysis of the entire metabolome, under a given set of conditions
4. Metabonomics: which seeks to measure the fingerprint of biochemical perturbations caused by disease, drug and toxins.
5. Metabolic fingerprinting: is used to classify samples based on provenance of either their biological relevance or origin by using a fingerprinting technology that is rapid but does not necessarily give specific metabolite information.

3. Experimental design

3.1 Cell culture growth, stimulation

The differences in optimized cell culture growth conditions present another major concern for cell line metabolomics. This is particularly an issue is studies involving comparative analysis of several different cell types, all of which might contain different levels of glucose, glutamine and lactate, as well as other nutrients and additives, which will probably lead to differences in the metabolome of the cells. If possible, it is recommended to use the same growth medium for all cell lines in the study to reduce variance in metabolic profile that can be caused by the medium (Cuperlovic-Culf et al., 2010). The standard enhancement of cell culture medium with serum of animal origin can add another level of complexity in cell growth condition optimization. Variations in serum can lead to contamination with exogenous metabolites and alterations of endogenous cell metabolite.

In order to minimize the influence of different cell culture conditions in the metabolomic final results, proper experimental designs are crucial. Nonetheless, more effort is required in the future for the determination of metabolic differences caused by various growth conditions, cell culture age and/or passage number for different cell lines.

3.2 Sample preparation and metabolite extraction

The goal of metabolomics is to analyze all or, at least, as many as possible different metabolites without selectivity for any particular molecular type and/or characteristics. The correct sample preparation is the first step in order to ensure the detection of a large number of metabolites. As metabolic processes may be rapid, varying from milliseconds to minutes (Gerdtzen et al., 2004; Taoka and Banerjee, 2002), the first necessary step is to rapidly stop any inherent enzymatic activity or any changes in the metabolite levels. The time and method of sampling are important issues to be considered to ensure reproducibility in the analytical sample, especially since a large number of biological replicates is commonly used (Khoo and Al-Rubeai, 2007).

These methods include freeze-clamping (with lower-temperature receptacles), immediate freezing in liquid nitrogen or by acidic treatments (ap Rees and Hill, 1994). Freezing in liquid nitrogen is generally considered to be the easiest way of stopping enzyme activity provided that cells or tissues are not allowed to partially thaw before extracting metabolites. In order to prevent this from happening, enzyme activity is inhibited by freezing-drying or by immediate addition of organic solvents while applying heat. Cells are subsequently disrupted, releasing the metabolites. Frozen samples may be ground down by sonication, homogenization by mechanical means in pre-chilled holders (Fiehn et al., 2000) or directly in an extraction solvent (Orth et al., 1999). The mixture of cell debris, protein, nucleotides and the desired metabolites need to be separated; this can be done by centrifugation or filtration.

For the complete analysis of a cell culture, it is important to measure both extracellular (footprint) and intracellular (fingerprint) metabolic profiles. Metabolic footprinting is technically simple because it requires only centrifugation to separate culture media and cells before the analysis. Metabolic fingerprinting, although much more technically challenging because it requires metabolite extraction from cells, provides more complete information about cellular metabolic processes (Cuperlovic-Culf et al., 2010). Recently a study related with different metabolite extraction protocols for mammalian cell culture was published (Dietmair et al., 2010). In this study, the authors compared 12 different extraction methods, according to their results; extraction in cold 50% aqueous acetonitrile was superior to other methods.

3.3 Analytical Instruments platforms

Currently, the main analytical techniques used for the analysis of the metabolome are nuclear magnetic resonance spectroscopy (NMR) and hyphenated techniques such as gas chromatography (GC) and liquid chromatography (LC) coupled to mass spectrometry (MS). In addition other combinations are possible, e.g. capillary electrophoresis (CE) coupled to MS or LC coupled to electrochemical detection. Alternatively, Fourier transform infrared spectroscopy (FTIR) and direct infusion mass spectrometry (DIMS) have been applied (Dunn and Ellis, 2005; van Greef et al., 2004; Lindon et al., 2007; Koek et al., 2010) without any prior separation, except for eventual sample preparation. NMR, FTIR and DIMS are high throughput methods and require minimal sample preparation and may be preferred techniques for metabolic fingerprint. However, the obtained spectra are composed of the signals of very many metabolites and elucidation of these complex spectra can be very complicated. In addition, detection limits for NMR and FTIR are much higher that for MS-based techniques, limiting the application range to metabolites present in higher concentrations. Therefore, GC, LC and CE coupled to MS are generally preferred in metabolomics to allow quantification and identification of as many as possible metabolites.

The general requirements for metabolomic instruments are:

- Excellent sensitivity and resolution for a wide range of molecules types
- The ability to handle a large range of concentrations (from pM to mM) for different molecular types
- The ability to identify and quantify different molecules
- Short analysis time
- To enable the measurement of many samples without sample degradation during the measurement
- Reproducible measurement across different centers and time

Several reviews have dealt with the application of NMR and MS in metabolomics (Ala-Korpela, 2008; Detmer et al., 2007; Griffin, 2003). NMR is a non-invasive, non-destructive, highly discriminatory and fast method that can analyze rather crude samples. NMR spectroscopy can be performed without extensive sample preprocessing and separation and provides several different experimental protocols optimized for mixture analysis and molecular formula or structure determination. The results of NMR measurements have proven highly replicable across centers and instruments (Viant et al., 2005). NMR can provide measurements for different types and sizes of both polar and non-polar molecules through analysis of different spectral windows. In addition, NMR instruments are highly versatile and with only minor changes in probes, users can obtain spectral information for different nuclei (^{1}H, ^{13}C, ^{15}N, and ^{32}P among others) in solvent or solid samples and even *in vivo* (Griffin, 2003). NMR is also the only method used in metabolomics that currently enables direct measurements of molecular diffusion, interactions and chemical exchange. Several databases and methods are being developed that enable metabolite identification and quantification from NMR spectra (Table 1). The major problem with NMR technology as applied to metabolomics is its low sensitivity, which limits the majority of currently available instruments to measurement of fewer than 100 metabolites.

The role of MS in metabolomic research is constantly expanding, whether the focus is on profiling (targeted analysis) or pattern-based analysis (Hollywood et al., 2006). Recent technological advances in separation science, ion sources and mass analyzers have considerably increased the sensitivity, selectivity, specificity and speed of metabolite detection and identification by MS. There are five important considerations that need to be dealt with in any global metabolite analysis by MS:

1. The efficient and unbiased extraction of metabolites from the sample matrix
2. Separation or fractionation of the analytes by chromatography
3. Ionization of the analyte metabolite
4. Detection of mass signals
5. Analyte identification

Name and availability	Instrument	Additional information
Human Metabolome Project (http:/www.hmdb.ca)	NMR, MS	Biological data; chemical and clinical data specific to humans
BMRD (http:/www.bmrb.wisc.edu)	NMR	Database search for NMR peaks assignment
Prime (http:/prime.psc.riken.jp)	MS, NMR	
Glom metabolome database (http:/csbdb.mpimp-glom.mpg.de)	MS	Specific to plants
METLIN metabolite database (http:/metlin.scrpps.edu)	MS	Drug and drug metabolites; specific to humans
NIST chemistry WebBook (http:/WebBook.nist.gov/chemistry)	NMR, MS, IR	
Madison metabolomics database (http:/mmcd.nmrfarm.wisc.edu)	MS, NMR	
NMR Lab of biomolecules (http:/spinportal.magnet.fsu.edu)	NMR	Databse search for NMR peaks assigment

Table 1. Databases of metabolomic standard data for quantification and assignment

Separation of analytes before MS detection is an important step leading to detection of more features, effectively increasing the overall "peak capacity" of the analytical platform. Separation methods include condensed-phase separation methods and gas-phase analyte separation. DIMS relies solely on the mass spectrometer to perform separation and offers an advantage in terms of speed and sample throughput. The number of identified features in a MS measurements can also be increased by changing the polarity of the ion source. Positive ion mode electrospray is optimal for basic metabolites (e.g. amines). Negative ion mode provides optimal measurement for acidic metabolites. Knowledge of the empirical formula based on exact mass can often be used to assign one or a few putative identifications that can then be used for searching metabolic or chemical databases (Table 1).

A comparative outline of the characteristics of NMR and MS methodologies as applied to metabolomics is provided in Table 2. The two methods are highly compatible and, thus ideal approach is to combine the results from NMR and MS measurements.

The most common forms of chromatography are GC and LC. GC runs are relative long, at about 60 min or more (Gummer et al., 2009); however, deconvolution software allows for the decrease in run times. In LC there is a shift from standard HPLC to UPLC (ultra-performance liquid chromatography), which can significantly increase resolution sensitivity and peak capacity (Gummer et al., 2009) due to the reduced particle size, while decreasing sample volumes and mobile phases. UPLC systems operate at high operating pressures and use sub-2-μm porous packing. Unlike pressured systems such as LC, CE (capillary electrophoresis) makes use of an electric field to move molecules towards the detector, much like gel electrophoresis. CE coupled with UV or LIF (laser-induced fluorescence) detectors are highly sensitive, but lack selectivity.

GC coupled to MS is one of the most common instrument platforms to be used in metabolomics experiments. GC-MS instruments using linear quadrupole analysers have been available for decades providing a robust technology that is amenable to automation. The identification of a wide range of primary metabolites (often after derivitisation) is greatly facilitated by the high resolution of capillary GC, the reproducible fragmentation of metabolites in the mass spectrometer and the ready availability of large mass spectral libraries (Roessner et al., 2001). Recent developments in GC-MS have resulted in improvements in both the GC and MS capabilities of this platform and a move towards the use of high mass accuracy/ high mass resolution instruments. The requirement for high throughput has led to the use of nominal GC-time-of-flight TOF-MS with much faster scan rates. High scan rates allow rapid temperature gradient programs, resulting in shorter run times and increased sensitivity (Gummer et al., 2009). Alternatively, the combination of two-dimensional GC with TOF-MS has resulted in the development of very high resolution fast MS that can be used to detect more metabolites than is possible using single queadrupole (Q), TOF and ion trap mass analyzer.

One if the drawbacks of many GC-MS metabolomics analyses is the need to derivitise metabolites before analysis. In contrast, many classes of polar metabolites can be analyzed directly by LC-MS without derivitisation. LC systems interfaced with TOF mass analyzer are now commonly used in metabolomics analyzes, delivering high throughput and high mass resolution analysis capability with mass accuracy approaching single digit ppm. Recently developed instruments also allow rapid polarity switching between positive and negative mode within a single run, reducing the need for multiple runs and cost per sample (Gummer et al., 2009).

Analysis	**NMR**	**MS**
High throughput-metabolites	No	Medium
High throughput-samples; automation	Yes	No
Quantitative	Yes	Yes
Availability in clinic	No	No
Equipment cost	High	High
Maintenance cost	Medium	High
Per sample cost	Low	High
Required technical skills	Yes	Yes
Sensitivity	Medium	High
Reproducibility	High	Low
Data analysis automation	Yes	Yes
Identification of new metabolites	Difficult	Possible
Chemical exchange analysis	Yes	No
In vivo measueremnt	Possible	Impossible

Table 2. Comparison of characteristics of major experimental methods for metabolomic analysis

LC-MS linear quadrupole, triple quadrupole (QQQ), QTrap and io trap mass analyzers have also been utilized for global and targete metabolomics, but may be limited by mass accuracy and mass resolution in identifying metabolites. However, the use of triple quadrupole and QTrap mass analyzers in various selective ion scaning modes can be used to detect specific metabolites or metabolite classes with high sensitivity and are particulary useful for targeted metabolomic analysis.

CE-MS offers a complementary approach to LC-MS for analyzing anions, cations, and neutral particles in a single run. Metabolites can be analyzed directly without derivitisation and the chromatographic resolution and sensitivity of CE is very high. However, CE is less frequently used for metabolomic analyzes tha LC-MS.

Vibrational spectroscopies are relative insensitive, but FTIR allows for high throughput screening of biological samples in an unbiased fashion. Similar to NMR, water signals pose a problem and must be subtracted electronically or attenuated total reflectance may be used. Compared with the other methods it is one of the least sensitive, but its unbiasness to compounds and ability to analyse large numbers of sample in a day makes it a plausible method for screening purposes (Khoo and Al-Rubeai, 2007).

LC-MS-based instruments can be operated in direct infusion mode with no chromatographic separation for measurement of the total mass spectrum for the mixture. The infusion can be performed with either the LC autosampler or with and offline syringe pump. Ion trap, TOF, Q-TOF, Orbitrap and FT-ICR-MS mass analyzers have been used with this mode of sample delivery. This approach relies totally on the mass analyser to resolve isobaric metabolites such as leucine or isoleucine. The key advantange of direct infusion analysis is the potential for automated high throughput sample analysis with both low and high mass resolution mass analyzers.

Another beneficial experimental method for cell culture metabolomics analysis involves stable isotope labeling followed by either MS or NMR measurement. This approach enables pathway tracing, easier metabolite assignment and metabolic flux measurements. Isotopic labeling has previously enabled detailed determination of pathways leading to the production of specific metabolites and the development of the highly accurate mathematical models of these pathways (Hollywood et al., 2006).

4. Aplications in mammalian cell culture – Study cases

4.1 Analysis of molecular mechanisms associated to the adaptation of NS0 myeloma cell line to protein-free medium

The NS0 mouse myeloma cell line has become one of the most popular systems for large-scale heterologous protein expression. For reasons of regulatory compliance, cost, batch consistency, downstream processing, and material availability, industrial applications of NS0 has moved towards serum or protein-free medium platforms (Barnes et al., 2000). For serum- or protein-free cultivation, the cell culture medium is often supplemented with lipids (derived from plant or synthetic sources) in addition to other protein supplements. The effect of lipid supplementation on the physiology of hybridomas and myelomas has been reported (Jenkins et al., 1992). NS0 cells are naturally cholesterol-dependent; not only is their growth greatly facilitated by lipid supplementation, but is also dependent on provision of cholesterol. NS0 cells capable of cholesterol-free growth can be isolated by selecting mutant clones or by adaptation. Adaptation generally involves passaging cells over a time period during which the serum concentration is decreased gradually (Sinacore et al., 2000). Eventually, the resulting population develops the capability to grow in the absence of serum. Different mechanisms underlying a cholesterol-dependent phenotype could include the absence (or mutation) of a gene or a segment of gene along the cholesterol biosynthesis pathway. There could be changes in the expression level of some proteins of the pathway due to gene regulation or other control mechanisms. In addition to the specific gene expression alterations along the cholesterol and lipid metabolism pathways, cholesterol dependence could also be the result of insufficient precursor supply (Spens and Haggstrom, 2005).

The molecular mechanisms of host and recombinant NS0 cell lines that could be related to the adaptation to protein-free medium are studied in this work. A quantitative study of proteins with differential expression levels in four conditions (host NS0 cell line adapted and non-adapted to protein-free medium, and a monoclonal antibody (Mab) transfectoma producer NS0 adapted and non-adapted to the same protein-free medium) is reported. The study is based on the use of the combination of two-dimensional electrophoresis and mass spectrometry, and a novel quantitative proteomic approach, isobaric tagging for relative and absolute quantification (iTRAQ). The metabolic study of these cell lines cultured in different nutrient conditions is also reported. Taking into account the proteomic results and metabolic analysis, a possible mechanism related to the adaptation of NS0 cell line to protein-free medium is proposed.

4.1.1 Results: Proteomic analysis

To characterize the changes associated with the adaptation to the protein-free medium 2DE gels of protein extracts from the cell line cultured in PFHM II with or without 1% (v/v) FBS were compared over a pI range of 3 to 10. Following adaptation to the protein-free medium, 78 spots changed their intensity by a factor of ≥2 from 1200 detected spots/ treatment. Interestingly, the majority of differentially expressed proteins decreased their expression in cells adapted to the protein-free medium. Fifty eight proteins were characterized by MALDI-MS and/or LC-ESI-MS-MS. The identified proteins were grouped according to their molecular function. Four major cellular pathways seem to be involved in the adaptation to the protein-free medium: i) carbohydrate metabolism and energy production, especially glycolysis and the Krebs cycle, ii) protein synthesis and folding; iii) membrane transport, and iv) cell proliferation (de la Luz et al., 2007).

In order to increase the number of proteins related with cell cycle regulation, DNA replication and lipids synthesis we used another strategy based in the isobaric labeling and the subcellular fractionation. iTRAQ reagent technology is a newly developed method for relative quantification of proteins from up to four samples. It has immense potential to improve the sensitivity and quality of mass spectrometric analysis of the proteome. We were able to identify and quantify 575 proteins simultaneously from the four states of culture. Among these 575 proteins, 43% were identified by a single significant peptide per protein; the rest were identified by at least two significant peptide of the same protein. The method used in our case to analyze the differential expression levels between two or more conditions was the locfdr. This method is a new approach to the problem of multiple comparisons and controls the number of false positive differentially expressed proteins below the user-specified threshold.

The standard deviation for each peptide value is obtained from the iTracker estimates, while in the case of proteins the variant of t-statistic suggested by Efron was used (Jung et al., 2006) to skip the inconvenience of those proteins that were identified by a single peptide and have standard deviation zero. In all analysis the condition locfdr ≥0.2 was used to select differentially expressed proteins. This threshold is a general accepted standard in fdr applications; it means that to consider a protein as differentially expressed the corresponding probability of being a false positive must be below 20%.

Following this approach we have found a set of 102 differentially expressed proteins. These proteins were classified in different functions and locations according to the KEGG database (de la Luz et al., 2008). According with the previous results four major cellular pathways seem to be involved in the adaptation to the protein-free medium (Figure 2).

4.1.2 Results: Kinetic and metabolic analysis

The host and recombinant NS0 cell line were culture in serum-supplemented and protein-free medium during 140 hours. Total cell number, viable cell number and viability were determined. The specific growth rate is different between both cells, but there was a clear decrease when both cell lines were cultured in absence of serum (Table 3). Intracellular metabolite concentrations were calculated during exponential growth phase. In contrast with previous reports, we found a lost of cholesterol auxotrophy in the host and recombinant NS0 cell line adapted to PFHM. Other metabolites such as phospholipids and fructosamine involved in specific cellular processes like membrane biogenesis and glycolipid metabolism changed their expression levels in adapted versus non-adapted cell line. In order to check if the intracellular cholesterol concentrations increase in the adapted cell line is a reversible process, cells were cultured in a medium supplemented with serum, and the initial cholesterol levels were determined (de la Luz et al., 2008).

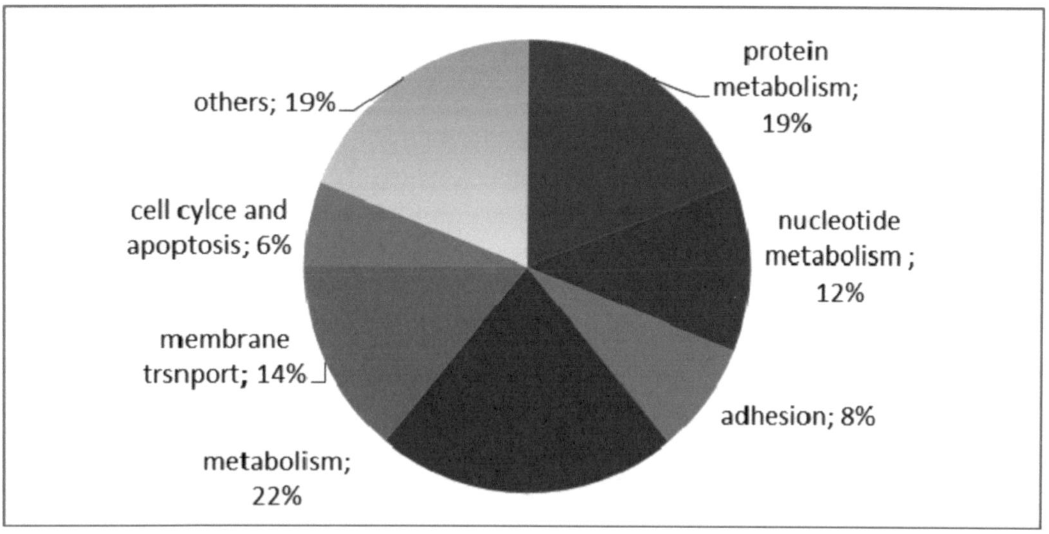

Fig. 2. Distribution of identified proteins taking in account their biological function

Cellular line	Xv_{max} (cell × mL^{-1})	Xt_{max} (cell × mL^{-1})	μ_{max} (cell × mL^{-1} × h^{-1})	IgG_{max} (µg × mL)	qIgG (µg × mL^{-1} × h^{-1})
Host NS0 (FBS)	2.14e^{+06}	2.79e^{+06}	0.034	-	-
Host NS0 (PFHM)	1.55e^{+06}	2.23e^{+06}	0.036	-	-
NS0/hR3 (FBS)	1.26e^{+06}	1.38e^{+06}	0.025	10.70	2.31e^{-07}
NS0/hR3 (PFHM)	1.51e^{+06}	1.58e^{+06}	0.037	41.50	4.73e^{-07}

Xv_{max}: maximum viable cell concentration, Xt_{max}: maximum total cell concentration, μ_{max}: maximum growth rate, IgG_{max}: maximum IgG production, q_{IgG}: specific IgG production rate.

Table 3. Cellular growth parameters from host NS0 and recombinant NS0 cell lines adapted and non-adapted to protein-free medium

Glycolysis is one of the most important metabolic pathways providing a source of precursors and energy for the cell. Previous analysis by DNA microarray studies have revealed a large number of genes involved in glycolysis, the pentose phosphate pathway and the Krebs cycle to be down-regulated in host NS0 cell line cultured in the absence of cholesterol (Seth et al., 2005). Ten proteins from glycolysis were found up-regulated in non adapted NS0 cell line with respect to adapted. This result could indicate that the glycolysis is a source of molecular precursors (cholesterol and phospholipids), especially in the adapted cell line (Figure 3). The lactate production increase after the adaptation process could be related with the higher lactate dehydrogenase enzyme activity (especific enzyme activity NS0 adapted: 20.85 U*mL^{-1}*$cell^{-1}$, non-adapted: 12.31 U*mL^{-1}*$cell^{-1}$).

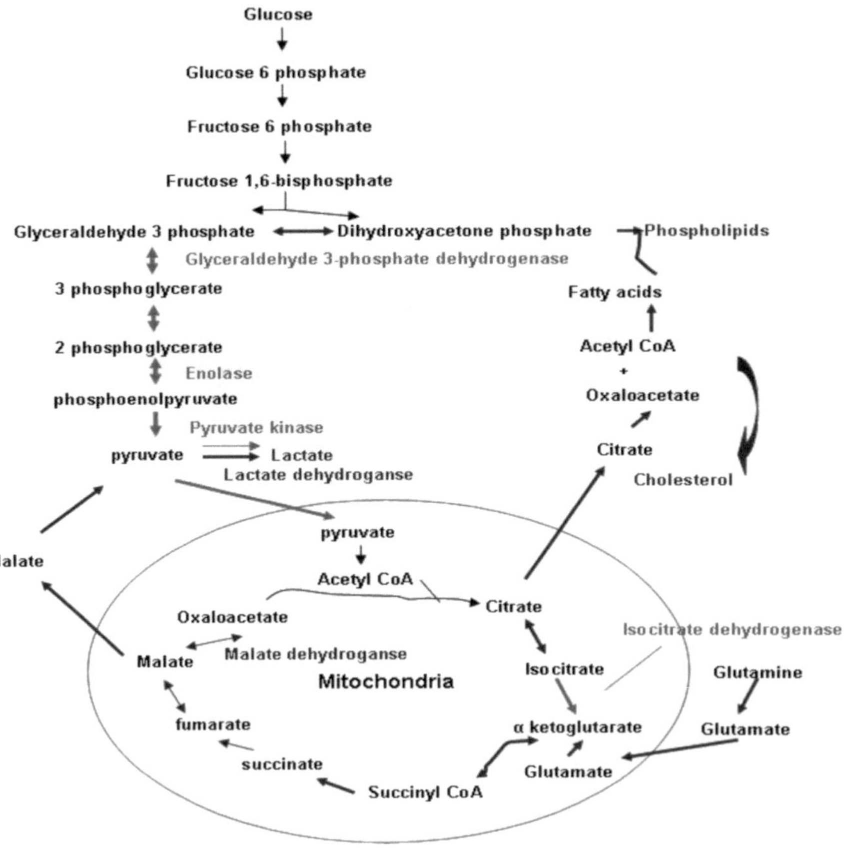

Fig. 3. Metabolic model of the adaptation of NS0 cel line to protein free-medium obtained from proteomic and metabolites analysis. Red: sub-expressed pathways. Blue: over-expressed pathways.

With the aim to calculate the relative rate of glycolysis and glutaminolysis, intracellular concentration of lactate and glucose were determined in the batch culture and the relationships between lactate production and glucose consumption (qL/qG) were calculated (figure 4). These results indicated that the lactate produced depend of the glycolysis and the glutaminolysis. Taken into account protemic and metabolic results we have proposed a metabolic mechanism where the glucose is used for the precursors synthesis. On the other hand, the cell obtain the energy from glutamine degradation.

We used the flux balance analysis (FBA) in order to compare the results obtained with an empiric metabolic network with the experimental results. In this study we used a reported metabolic network (Ma and Zeng, 2003) with changes in the cholesterol reactions, where the cholesterol synthesis pathways was eliminated in NS0 non-adapted. The comparison between adapted and non adapted metabolic network showed changes in carbohydrate and lipid metabolism, very similar with our previous experimental results. Also we analyzed the metabolites that have influence in cellular growth when they are not present in the medium. Glycine, tryptophan, phenylalanine, adenine, palmitic acid, glutamic acid, methinonine and asparagine are relatefd with the increase of cellular biomass (data not shown).

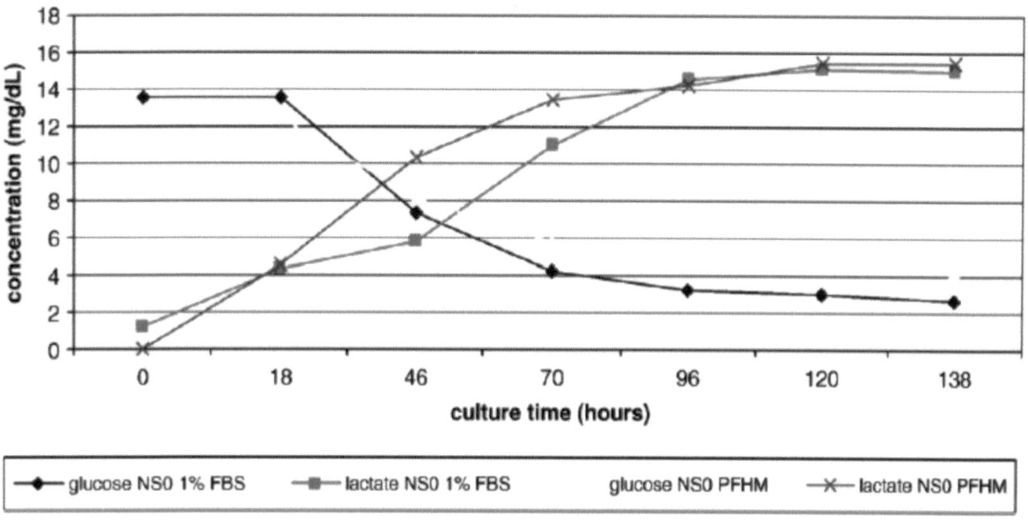

Fig. 4. Glucose consumption and lactate production during a batch culture of myeloma cell line in presence and absence of serum. During the culture period, samples were taken periodically for off-line analysis and media metabolites concentration were determined. The relation between lactate production and glucose consumption is representative of the cellular metabolic state, especially of the glycolysis efficiency

5. Conclusion

Data integration is not limited to flux data. Systems biology encompasses a holistic approach to the study of biology and the objective is to simultaneously monitor all biological processes operating as an integrated system. The use of the data obtained from studies with different "omics" techniques is not simple. In addition, a single gene may code for isoenzymes reacting with multiple metabolite substrates. The difficulty in determining the timing of different events, that it, transcription and protein activity, also contribute to the difficulty in integrating data. Hence in order for metabolomics to be used in systems biology, novel strategies will need to be created. One step forward in such an integration process is the functional assignments between protein/gene and metabolite within a system of interest. This can be done by creating models where basic biochemical pathways are modelled using static data (Khoo and Al-Rubeai, 2007). Second, time-dependent concentrations of other types of components (transcriptomics and/or proteomics) will then be incorporated followed by the reconstruction of the model with statistic data.

In contrast with previous results, changes in metabolic rates and biosynthetic machinery with respect to the presence or not of serum in the culture medium were observed in this study. The analysis was performed by two different ways. First, using iTRAQ reagents, proteins with differential expression levels in two myeloma cell lines cultured in serum-supplemented and serum-free medium were detected. These proteins belong to major pathways related with glycolysis, protein synthesis and membrane transport. These results are in accordance with previous results obtained using 2DE and the study of a revertant

NS0 cholesterol-independent (Seth et al., 2005). Second, the determination of consumption and production of different metabolites like glucose, lactate, cholesterol, phospholipids and phosphorous was performed. Differences in qL/qG were found between adapted and non-adapted cell lines, similar to the results obtained by proteomics. A significant increase was

observed in the intracellular cholesterol concentration in the adapted cell lines. However, when these cell lines adapted to PFHM were cultured in presence of serum, the intracellular cholesterol levels decreased down to the initial conditions, indicating a possible epigenetic mechanism.

6. References

Earle W.R. et al (1954) Certain factors limiting the size of the tissue culture and the development of mass cultures. *New York Academy of Sciences* 58, 1000-1011

Lake-Ee Quek et al. (2010) Metabolic flux analysis in mammalian cell culture. *Metabolic Engineering* 12, 161-171

Warburg O. (1956) On the origin of cancer cells. *Science* 123, 309-314

Fitzpatrick L. et al. (1993) Glucose and glutamine metabolism of a murine B-lymphocyte hybridoma grown in batch culture. *Applied Biochemistry and Biotechnology* 43, 93-116

Jenkins H.A. et al. (1992) Characterization of glutamine metabolism in two related murine hybridomas. *Journal of Biotechnology* 23, 167-182

Ljunggren J. and Haggstrom L. (1992) Glutamine limited fed-batch culture reduces the overflow metabolism of amino acids in myeloma cells. *Cytotechnology* 8, 45-56

Ozturk S.S. and PalssonB.O. (1991) Growth, metabolic, and antibody production kinetics of hybridoma cell culture: 1 Analysis of data from controlled batch reactors. *Biotechnology Progress* 7, 471-480

Martinelle K. et al. (1998) Elevated glutamate dehydrogenase flux in glucose-deprived hybridoma and myeloma cells: evidence from H-1/N-15 NMR. *Biotechnology and Bioengineering* 60, 508-517

Korke R. et al. (2002) Genomic and proteomic perspectives in cell culture engineering. *Journal of Biotechnology* 94, 73-92

Seow T.K. et al. (2001) Proteomic invetsigation of metabolic shift in mammalian cell culture. *Biotechnology Progress* 17, 1137-1144

Seth G. et al. (2007) Molecular portrait of high productivity in recombinant NS0 cells. *Biotechnology and Bioengineering* 97, 933-951

Smales C.M. et al. (2004) Comparative proteomic analysis of GS-NS0 murine myeloma cell lines with variyng recombinant monoclonal antibody production rate. *Biotechnology and Bioengineering* 88, 474-488

de la Luz K.R. et al. (2007) Proteomic analysis of the adaptation of the host NS0 myeloma cell line to a protein-free medium. *Biotecnología Aplicada* 24, 215-223

de la Luz K.R. et al. (2008) Metabolic and proteomic study of NS0 myeloma cell line following the adaptation to protein-free medium. *Journal of Proteomics* 71, 133-147

Oliver S.G. et al. (1998) Systematic functional analysis of the yeast genome *Trends in Biotechnology* 16, 373–378

Beecher C.W. (2002) Metabolomics: A new "omics" technology. *American Genomics - Proteomics Technology*

Khoo S.H.G and Al-Rubeai M. (2007) Metabolomics as a complementary toolo in cell culture. *Biotechnolgy and Applied Biochemistry* 47, 71-84

Hollywood K. et al. (2006) Metabolomics: current technologies and future trends. *Proteomics* 6, 4716-4723

Fiehn O. (2001) Metabolomics - the link between genotype and phenotype. *Plant Mol. Biol.* 48, 155-171

Viant M.R. et al. (2005) An NMR Metabolomic Investigation of Early Metabolic Disturbances Following Traumatic Brain Injury in a Mammalian Model. *NMR in Biomedicine* 18, 507-516

Tweeddale H. et al (1998) Effect of slow growth on metabolism of Escherichia coli, as revealed by global metabolite pool (''metabolome'') analysis. *J Bacteriol* 180, 5109–5116

Buchholz A. et al. (2001) Quantification of intracellular metabolites in Escherichia coli K12 using liquid chromatographic electrospray ionization tandem mass spectrometric techniques. *Anal Biochem* 295,129–137

Villas-Boas S.G. et al. (2005) Mass spectrometry in metabolome analysis. *Mass Spectrometry Reviews 24: 613 - 646.*

Cuperlovic-Culf M. et al. (2010) Cell culture metabolomics: applications and future directions. *Drug Discovery Today* 15, Numbers 15/16

Gerdtzen Z.P. et al. (2004) Non-linear Reduction for Kinetic Models of Metabolic Reaction Networks. *Metabolic Engineering* 6, 140-154

Taoka S. and Banerjee R.(2002) Stopped-flow kinetic analysis of the reaction catalyzed by the full-length yeast cystathionine beta-synthase. *J Biol Chem.* 277(25), 22421-22425

ap Rees T. and Hill S.A. (1994) Metabolic control analysis of plant metabolism. *Plant, Cell & Environment* 17, 587-599

Fienh O. et al. (2000) Identification of uncommon plant metabolites based on calculation of elemental compositions using gas chromatography and quadrupole mass spectrometry. *Anal Chem.* 72, 3573–3580

Orth H.C. et al. (1999) Isolation, purity analysis and stability of hyperforin as a standard material from Hypericum perforatum L. *J Pharm Pharmacol 51, 193-200*

Dietmair S. et al. (2010) Towards quantitative metabolomics of mammalian cells: development of a metabolite extraction protocol *Anal Biochem.* 404, 155-64

Dunn W.B. and Ellis D.I. (2005) Metabolomics and systems biology: making sense of the soup. *Curr. Opin. Microbiol.* 7, 296–307

van Greef J. et al. (2004) The role of analytical sciences in medical systems biology. *Curr Opin Chem Biol* 5, 559–565

Lindon J.C. et al. (2007) The handbook of metabonomics and metabolomics *Elsevier*

Koek M.M. et al. (2010) Semi-automated non-target processing in GC × GC–MS metabolomics analysis: applicability for biomedical studies *Metabolomics* 7, 1–14.

Ala-Korpela M. (2008) Critical evaluation of H NMR metabonomics of serum as a methodology for disease risk assessment and diagnostics *Clinical Chemistry and Laboratory Medicine* 46, 27–42.

Detmer K. et al. (2007) Mass spectrometry-based metabolomics *Mass Spectrometry Reviews* 26, 51-78

Griffin J.L. (2003) Metabonomics: NMR spectroscopy and pattern recognition analysis of body fluids and tissues for characterization of xenobiotic toxicity and disease diagnosis *Curr. Opin. Chem. Biol.* 7, 648-654

Barnes L. Et al. (2000) Advances in animal cell recombinant protein production: GS-NS0 expression system. *Cytotechnology* 32, 109-123

Jenkins H. et al. (1992) Characterisation of glutamine metabolism in two related murine hybridomas *Journal of Biotechnology* 23, 167-182

Gummer J. et al. (2009) Use of mass spectrometry for metabolite profiling and metabolomics *Australian Biochemist* 40, 5-8

Sinacore M. et al. (2000) Adaptation of mammalain cells to growth in serum-free media *Molecular Biotechnology* 15, 249-257

Spens E. and Haggstrom L. (2005) Defined protein and animal component-free NS0 fed-batch culture *Biotechnology and Bioengineering* 98, 6

Jung Y. et al. (2006) Identifying differentially expressed genes in meta-analysis via Bayesian model-based clustering *Biomolecules Journal* 48, 435-450

Seth G. et al. (2005) Large-scale gene expression analysis of cholesterol dependence in NS0 cells *Biotechnology and Bioengineering* 90, 552-567

Ma H. and Zeng A. (2003) Reconstruction of metabolic networks from genome data and analysis of their global structure for various organisms *Bioinformatics* 19, 270-277

2

Quantitative Metabolomics and Its Application in Metabolic Engineering of Microbial Cell Factories Exemplified by the Baker's Yeast

Mario Klimacek
Institute of Biotechnology and Biochemical Engineering, Graz University of Technology
Austria

1. Introduction

The baker's yeast *Saccharomyces cerevisiae* and its beneficial properties have been recognized very early by human beings. It has been used in the making of alcoholic beverages, bread and cake long before the term biotechnology has been coined. In addition to its great importance in food industry *S. cerevisiae* strains are nowadays applied in many other fields for example in the production of bio-fuels from corn or sugar containing crops, in the bio-sorption of heavy-metals from sewage, in pharmaceuticals or in the production of precursor compounds for the synthesis of pharmaceuticals or fine chemicals. As a consequence *S. cerevisiae* developed to one of the most important and best investigated microbial cell factories for the industrial (white) biotechnology. Furthermore *S. cerevisiae* is an important model organism used to elucidate the underlying molecular mechanistic principles that are involved in complex diseases (cancer or diabetes) and metabolic disorders (Castrillo and Oliver 2005; Castrillo and Oliver 2006; Nielsen and Jewett 2008). Other important features of *S. cerevisiae* that led to its multifaceted applicability in industry and R&D constitute its GRAS (generally recognized as safe) status and that cells are very easy to cultivate and are readily available.

The physiology of *S. cerevisiae* under various environmental conditions has been investigated intensively in the last 140 years (Racker 1974). The baker's yeast exhibits some very interesting physiological features that render it unique among all other microorganisms. It grows nearly equally fast under aerobic and anaerobic conditions with glucose as the sole carbon source (Nissen et al. 2000a; Visser et al. 1990). Under aerobic conditions and at glucose concentrations above 100 mg/L biomass formation is accompanied by the production of ethanol as a consequence of an overflow metabolism at the pyruvate node (Crabtree-effect, (Crabtree 1928)). After depletion of glucose the ethanol initially formed by the overflow metabolism is further converted into biomass under aerobic conditions (Diauxie). Under anaerobic conditions about 90% of glucose carbon is converted into ethanol and CO_2. The rate of glucose utilization and the specific ethanol yield is higher under anaerobic conditions as compared to the sugar conversion rate and ethanol yield under aerobic conditions (Pasteur-effect, (Racker 1974)). It can reduce a number of keto-compounds to the corresponding chiral alcohols that represent interesting precursors for pharmaceuticals (Csuk 1991). It can grow as a diploid as well as a haploid which highly facilitates genetic manipulation and permits high-throughput genetic engineering.

Considering the enormous early interest in studying and understanding the physiology of *S. cerevisiae* long before modern omics techniques have been developed, it is not very surprising that it was the baker's yeasts genome that was the first within the domain of eukaryotes that was completely sequenced. Genomic and biological information about *S. cerevisiae* molecular biology is comprehensively collected at the *Saccharomyces* Genome Database (SGD, http://www.yeastgenome.org/). Driven by the knowledge of the complete genomic sequence and by the steadily increasing availability of tools developed for genetic engineering, *S. cerevisiae* became a key work horse and the representative eukaryotic model organism in every modern discipline within the biosciences such as molecular and cell biology, functional genomics, systems biology or metabolic and synthetic engineering. Today's genetic work with *S. cerevisiae* cells is highly alleviated by the presence of a wide spectrum of established yeast molecular biology tool kits and availability of many wild-type and mutant strain (e.g. knock-out strains) collections as well as plasmid collections containing *S. cerevisiae* ORFs, gene deletion markers or promoter sets and many more, offered by commercial sources such as EUROSCARF (http://web.uni-frankfurt.de/fb15/mikro/euroscarf/index.html), Open biosystems (http://www.openbiosystems.com/Products/) or Addgene (http://www. addgene. org/).

The commercial establishment of genetic manipulation techniques paved the way for *S. cerevisiae* to be exploited in the field of metabolic engineering. Various novel recombinant designer strains capable of either selective formation of one desired product or of producing heterologous compounds or endogenous products from new resources (waste or renewable materials) emerged in the last decades. Metabolic engineering efforts based on *S. cerevisiae* are comprehensively summarized elsewhere and the interested reader is referred to (Bettiga et al. 2010; Nevoigt 2008). A collection of engineered substrate utilization and heterologous or homologous product formation pathways is given in Table 1.

The corresponding underlying engineering principles can be basically broken down into 4 strategies as depicted in Figure 1 panel A-D. Elucidation of the appropriate engineering approach represents the most important step in designing novel cellular properties and targets on the identification of reaction(s) or even entire pathways that are suited for the anticipated metabolic engineering objective. Relevant reaction(s) and associated gene(s) can be extracted by thorough screenings of literature data (US National Library of Medicine (http://www.ncbi.nlm.nih.gov/pubmed), SciFinder (https://scifinder.cas.org/) or Web of Knowledge (http://wokinfo.com/)) and online databases (KEGG the Kyoto Encyclopedia of Genes and Genomes (http://www. genome.jp/kegg/), the enzyme database BRENDA (http://www.brenda-enzymes.org/), or the SIB bioinformatics Resource Portal ExPASy (http://www.expasy.ch/).

To increase the probability of engineering success identified targets can be subjected to in silico modeling by employing mathematical models like restricted flux balance analysis (FBA) based on a genome-scale stoichiometric network to verify their compatibility with the underlying metabolic network (Cvijovic et al. 2010; Selvarajoo et al. 2010).

Substrate targets	Applications	References
Xylose	bio-ethanol	(Jeffries and Jin 2004; Petschacher and Nidetzky 2008; van Maris et al. 2007)
L-Arabinose	bio-ethanol	(Wisselink et al. 2007)
Lactose / whey	bio-ethanol	(Domingues et al. 2010)
Galactose	bio-ethanol	(Lee et al. 2010)
Product targets		
Insulin and insulin precursors	pharmaceuticals	(Kjeldsen 2000)
Hepatitis B antigen	pharmaceuticals	(Kuroda et al. 1993)
Cyanophycin	biopolymer	(Steinle et al. 2009)
Organic acids	chemical building blocks	(Abbott et al. 2009; Ishida et al. 2006; Raab et al. 2010)
n-Butanol	bio-fuel	(Steen et al. 2008)
Sesquiterpenes	pharmaceuticals and food	(Asadollahi et al. 2010; Jackson et al. 2003)
Carotenoids	pharmaceuticals and food	(Ukibe et al. 2009)
Diterpenoids	pharmaceutical industry	(Dejong et al. 2006)
Triterpenes	pharmaceutical industry	(Madsen et al. 2011)
Polyketides	pharmaceutical industry	(Mutka et al. 2006)
Five-carbon sugars / alcohols	food ingredients	(Toivari et al. 2007)
Ethylene	synthetic polymers	(Pirkov et al. 2008)
Flavonoids, stilbenoids	pharmaceuticals and food	(Trantas et al. 2009)
SO_2	beer flavor stability	(Yoshida et al. 2008)
Ethanol	bio-fuel	(Nissen et al. 2000b)

Table 1. Collection of metabolic engineering targets in *S. cerevisiae*

To unravel reaction(s) or pathways instead that would compromise substrate conversion and/or production of a desired product in silico modeling by for example FBA (Bro et al. 2006) or MOMA (minimization of metabolic adjustment) (Asadollahi et al. 2009) have been applied and produced potential candidate reactions which otherwise would have been often overlooked (Cvijovic et al. 2010). The open-source software platform OptFlux developed especially for in silico driven metabolic engineering is available at http://www.optflux.org/ (Rocha et al.).

Metabolic integration of a novel pathway is selected when the utilization of a new substrate (Fig. 1 A) or synthesis of a new product is intended (Fig. 1 B). New in this context means that the original, not engineered, cells are genetically not programmed to perform these reactions. Formation of an undesired side-product (Fig. 1 C) can be attributed to a split in carbon flux at the related node (C) into a productive (C→P) and undesired flux (C→D). Reasons for such flux partition compromising efficient production of the product (P) can be often traced back to the presence of an enzyme or differently regulated isoenzyme competing for the same substrate S or a promiscuous enzyme that in addition to the desired substrate is also active with other substrates (Fig. 1 D).

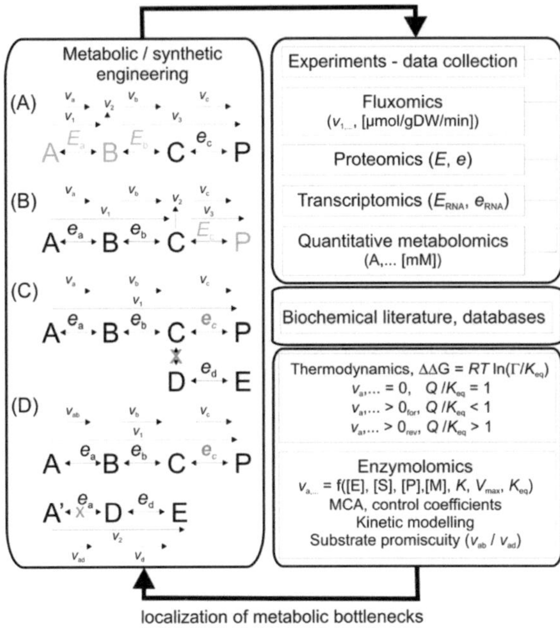

Fig. 1. Typical metabolic engineering principles based on rational design (panels A – D) are linked to a suggested experimental work-flow to unravel limiting metabolic sites. Panels A-D refer to enabling of substrate utilization (A) or product formation (B), preventing side product formation by deletion and/or overexpression of an endogenous enzyme (C), increasing selectivity of a substrate promiscuous enzyme (D); Substrate A, intermediate B, product P and enzymes *E* new to the network are indicated in grey. Overexpression of an endogenous enzyme is depicted by a grey e. Knock outs are indicated by grey x's. Subscripts of rate constants *v* given as numbers and small letters refer to fluxes based on stoichiometry (solid arrows) and individual reaction rates of enzymes (dotted arrows), respectively.

Directing the carbon flux towards P can be afforded by deletion of respective gene(s), overexpression of enzymes participating in the productive branch, or replacing the corresponding activity by a less regulated or more selective one. Furthermore unbalanced carbon usage between reaction partners participating in this pathway and/or in the recycling of, for example cofactors, can result in accumulation and release of a pathway intermediate (Krahulec et al. 2009; Krahulec et al. 2010). In this case fine-tuning of all activities involved based on for example a metabolic control analysis (MCA) or kinetic modeling analysis is required to minimize or even completely prevent by-product accumulation (Parachin et al. 2011).

Aside from rational design stochastic methods based on inverse metabolic engineering have been developed for *S. cerevisiae* to identify key target reactions and associated gene sequences enabling the desired new cellular property (Bailey et al. 2002; Bengtsson et al. 2008; Bro et al. 2005; Hong et al. 2010; Jin et al. 2005; Lee et al. 2010). Differently, methods targeting on the induction of a cellular property, such as growth, increase of substrate conversion rate or enhancing resistance to environmental stress, that is hardly to capture by in silico design because of its highly intricate metabolic relations that have to be satisfied, rely on the cellular adaptability to a certain environmental stress by evolution (Cakar et al.

2009; Cakar et al. 2005; Garcia Sanchez et al. 2010; Kuyper et al. 2005; Sonderegger and Sauer 2003; Wisselink et al. 2009).

In the course of establishing systems biology various high-throughput omics techniques such as transcriptomics, proteomics, fluxomics and others have been developed with the objective to comprehensively analyze cellular physiology at all molecular levels (DNA, RNA, protein, flux, and metabolite). Data-driven analysis is often exploited to unravel novel interrelations at the various molecular levels or to obtain a more insightful (quantitative) understanding of cellular processes. It is obvious that metabolic engineering can greatly benefit from the integration of omics techniques in the design of improved microbial cell factories (Nielsen and Jewett 2008). The various omics tools have helped to increase understanding about how cells regulate, communicate and adapt to different environmental conditions.

Depending on the metabolic engineering objective the appropriate omics tool or a combination should be selected after due consideration. For example transcriptome analysis provides a holistic image of mRNA molecule pattern and levels but do not tell us anything about metabolic fluxes. Optimization of the flux towards a specific metabolite however represents one of the major goals in metabolic engineering. Metabolic flux analyses based on stoichiometric models or ^{13}C-isotopomer analysis (provided that cells can grow under the environmental conditions used) are useful tools in this respect (Nielsen and Jewett 2008). To understand the underlying mechanistic relationships between the flux through a particular pathway and the enzymes forming the pathway, providing the relevant information for strain design, detailed knowledge about enzyme-metabolite interactions are required. Consequently quantitative information about metabolites involved together with detailed knowledge of kinetic properties of participating enzymes is mandatory. Within the omics family metabolomics represents the youngest member. This is basically due to the facts that metabolites vary greatly in their physico-chemical properties (polarity, acidity, reactivity, and stability) and are present in a large dynamic concentration range which make it almost impossible to record the entire metabolome on a single analytical platform. Another challenge represents the generation of reliable and representative metabolite data from biological samples. Cell-wall leakage, instabilities and losses of metabolites throughout the sample work-up, or strong matrix effects in the MS analysis are a few of the many causes impairing metabolite data and as a consequence adulterate molecular mechanistic interpretation. Nevertheless in the last years much progress has been made due to enormous efforts of the yeast research community to overcome these obstacles. Protocols of unbiased sample-work-up and different analytical platforms are available today that can cover more than 100 compounds quantitatively.

This review presents current accepted protocols and techniques that enable acquisition of absolute quantitative metabolite data from *S. cerevisiae* cells. The second part focuses on how quantitative metabolite data can help in the development of improved microbial cell factories.

However, before going into the details some definitions of terms used in metabolome analysis should be reminded (Nielsen 2007). Metabolite profiling targets on the qualitative or semi-quantitative analysis of specified metabolites or groups of metabolites. In contrast in metabolite target analysis selected metabolites are quantified. If the entire metabolome or a fraction of it is addressed (or as many metabolites as possible) qualitatively or quantitatively we speak of metabolomics or quantitative metabolomics.

2. Data acquisition for quantitative metabolomics in *Saccharomyces cerevisiae*

Determination of unbiased intracellular concentrations of metabolites is without doubt a prerequisite for serious interpretation of cellular properties at the molecular - kinetic level. Unbiased refers here to the physiological state, the sample work-up and preparation for metabolite analysis and metabolite quantification and calculation of intracellular molar concentration. Consequently four experimental tasks that have to be fulfilled can be formulated as follows:

- Harvesting and quenching of cells at a predefined physiological state (representative) and separation from extracellular compounds (exometabolome) without leakage of intracellular metabolites (quantitative)
- Destruction of metabolic activity and complete extraction of metabolites by maintaining the metabolite composition quantitatively unchanged, concentration and if required chemical preparation for metabolite analysis
- High-throughput quantitative analysis by employing the appropriate instrument
- Relation of molar concentrations obtained from quantitative analysis to cell specific parameters such as cell dry weight (μmol / gCDW) or cell number (μmol / cell) or if absolute molar intracellular concentrations are required to the cell (compartment) volume.

However, before we go into the details and hurdles of each task it should be emphasized here that altering the cellular network by metabolic engineering will always result in a quasi new strain with completely new properties and behaviors. In the worst case the complete sample work-up protocol established for the wild-type strain might not be applicable for the mutant strains. It is therefore strongly recommended that protocols already established for the absolute quantification of metabolites have to be re-assessed and verified for the created strains.

2.1 Representative harvesting and quantitative quenching

The concentration of a metabolite in the cell is not directly linked to genes but is determined by its formation and utilization rates. Conversion rates in turn depend on the enzymes associated catalytically with this metabolite and their kinetic parameters with respect to this metabolite as well as effectors (inhibitors, activators). In particular metabolites from catabolic reactions and reactions involved in energy metabolism display high turnover rates. For example for the frequently used metabolite ATP (~16% of all reactions present in a genome-scale stoichiometric model of *S. cerevisiae* involve ATP (Förster et al. 2003)) turnover rates of ~1.5 mM/s were reported (Rizzi et al. 1997). Hence, quenching of metabolic activity within a very short time window is required without altering the metabolomic state of the sample. Another highly desirable feature is that the extracellular environment containing substrates, products, salts and the exometabolome, that can affect subsequent metabolite analysis, is separable from the cells without losing any intracellular metabolites due to disruption of or leakage through the cell wall. Different to for example *E. coli* the cell wall of *S. cerevisiae* is less leaky and both requirements can be achieved for *S. cerevisiae* cells by spraying a defined volume of cell suspension into an appropriate quenching solution at sufficiently low temperatures. It was found that the volume ratio of sample to quenching solution affects the quenching quality and a ratio of at least 1 to 5 was suggested (Canelas et al. 2008a).

Today two methods cold methanol and cold glycerol-saline have achieved wide acceptance within the metabolomics community (Canelas et al. 2008a; Villas-Bôas and Bruheim 2007). See Fig. 2 for details. In addition to the quenching temperature and time or the volume ratio between sample and quenching solution, the time between quenching and separation by centrifugation and the centrifugation time can influence metabolic activity and metabolite leakage significantly. Harvesting by rapid sampling is often coupled to quenching. Rapid sampling is especially important in continuous steady-state cultivations and for pulse-experiments in which changes of intracellular metabolite pools induced by a certain environmental impulse are analyzed at the sub-second time scale. Ingenious devices have been developed in the last years that enable rapid sampling and quenching simultaneously at the millisecond scale. The various manual, semiautomatic and fully automatic rapid sampling techniques and their pros and cons have been comprehensively discussed and the reader is referred to (Reuss et al. 2007; van Gulik 2010; Villas-Bôas 2007a). A disadvantage in this context is that most of these devices are not available on the market and therefore not accessible to the scientific community. Commercial accessibility however would be of great importance in the context of comparability, reproducibility and standardization in quantitative metabolomics studies. In batch cultivations manual transfer of the cell suspension to the quenching solution by using a pipette or a syringe is widely accepted as environmental conditions might not change significantly during the sample transfer (3 - 6 s) (Villas-Bôas 2007b). These assumptions might hold for anaerobic or microaerobic conditions but should be reconsidered in case of aerobic cultivations for which for example the O_2/CO_2 ratio may vary and induces changes (oxygen limitation) in cell metabolism during sampling.

2.2 Quantitative metabolite extraction

The next challenge represents the quantitative extraction of all metabolites or at least of those to be of interest without reactivating any metabolic activity, enabling chemical reactions and minimizing metabolite disruption. Metabolic activity and chemical reactions can be controlled by the temperature. The addition of a denaturing (inactivating) agent bears some risks as it might be also reactive with metabolites or provides the environment (pH) for chemical reactions (Villas-Bôas 2007b). Control of metabolite degeneration or reactivity during extraction and in the following process steps represents a complex and difficult task. Furthermore metabolite degeneration is a highly specific process and strongly depends on the metabolite species, extraction solution and extraction parameters (pH, T). For example the redox cofactors NAD^+ and $NADP^+$ are stable at acid to neutral pH even at high temperatures (80-90°C) for 3 minutes (Klimacek et al. 2010). On the contrary their reduced counterparts are highly unstable under acidic conditions with NADPH generally less stable than NADH (Chaykin 1967; Lowry et al. 1961). At high pH where NAD(P)H are stable NAD(P) in turn destruct rapidly (Lowry et al. 1961). Decomposition can also be catalyzed by phosphate a prominent compound in metabolite extracts (van Eunen et al. 2010) by forming an adduct with NADH across the pyridine group (Alivisatos et al. 1964; Chaykin 1967).

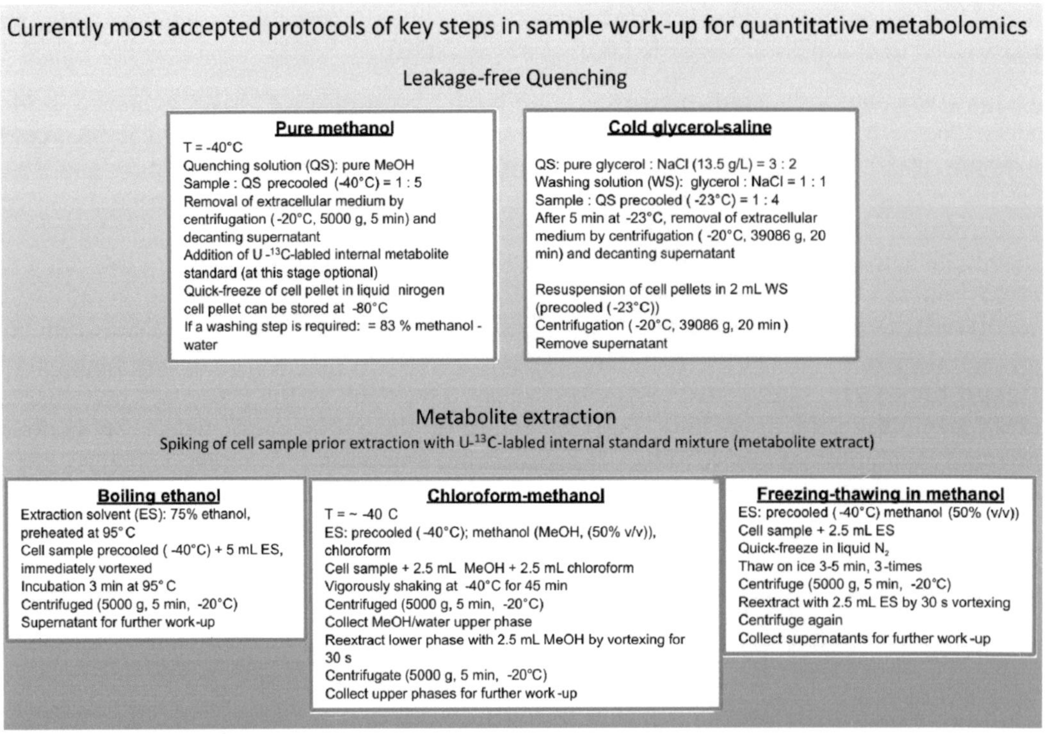

Fig. 2. Collection of state-of-the-art protocols for leakage-free quenching and quantitative extraction. Relevant literature and details can be found in the text.

It is obvious from this example considering just four metabolites that the ideal extraction procedure with which the complete metabolite consortium is extractable without any losses may not exist and calls for a compromise in the selection of conditions used for metabolite extraction. This "inconvenience" however can be circumvented, provided that subsequent metabolite detection is based on mass spectroscopy, by the addition of an aliquot of U-^{13}C-labled internal standard (IS) compounds to the biomass subsequent to quenching or prior to metabolite extraction (Büscher et al. 2009; Canelas et al. 2009; Klimacek et al. 2010; Mashego et al. 2004; Wu et al. 2005). Metabolite losses due to incomplete quantitative work-up of samples can be addressed by application of one selected IS compound. However metabolite specific instabilities, matrix effects, ion suppression, non-linear responses and day-to-day variations can be only identifed and appropriately corrected for by the addition of U-^{13}C-labled IS. A representative mixture of labeled metabolites can be easily prepared from *S. cerevisiae* wild-type and/or mutant cells cultivated in standard mineral medium supplemented with U-^{13}C-labled substrate (glucose, fructose, galactose,...) under the cultivation condition selected and by respective appropriate quenching and extraction procedures. Internal referencing by using an IS displaying a metabolite composition that is representative for the cellular state to be studied should be always taken into account because intracellular metabolite levels can vary considerably in dependence on the cultivation conditions used or on the cellular alterations introduced by pathway engineering (Klimacek et al. 2010).

Various extraction protocols with respect to extraction solvent (acids, bases, ethanol or methanol, organic solvents), buffered or non-buffered solutions, pH, temperature, etc have been tested (Villas-Bôas 2007b), evaluated and verified for *S. cerevisiae* cells in terms of metabolite coverage, efficacy and recovery (Klimacek et al. 2010) in the last years (Canelas et al. 2009; Villas-Bôas et al. 2005b). Today three extraction procedures have achieved some acceptance and are likewise used within the yeast research community (see Fig. 2 for details). That is boiling ethanol (BE; pioneered for *S. cerevisiae* cells by (Gonzalez et al. 1997)), chloroform-methanol (CM; pioneered for *S. cerevisiae* cells by (de Koning and van Dam 1992)) and to some extent freeze-thawing in methanol (FTM; pioneered for *S. cerevisiae* cells by (Villas-Bôas et al. 2005b) for which however controversial results with respect to its applicability are present in the literature. While (Villas-Bôas et al. 2005b) found extraction performance of FTM sufficient, others (Canelas et al. 2009) concluded that FTM cannot effectively prevent metabolite conversion throughout the extraction process and considered FTM therefore as not appropriate for metabolite extraction. Differences in evaluation criteria and growth conditions were used as a basis to explain the different outcomes. It should be however noted that (Canelas et al. 2009) investigated metabolite extraction performances from *S. cerevisiae* cells grown under two different physiological conditions (glucose limitation and glucose saturation; a bioreactor coupled to a rapid sampling device was used), used identical U-^{13}C-labled compounds as IS and three different analytical methods for quantification of a broad range of different compounds. Quality and metabolite recovery of FTM instead was judged by (Villas-Bôas et al. 2005b) by the application of an IS mixture composed of compounds each a representative for a substance class analyzed. The mixture was added to the quenched cells prior to extraction. Cells were cultivated under aerobic conditions in shake flasks and metabolites were quantified by established GC-MS after metabolite derivatization with methyl chloroformate. In Fig. 2 brief descriptions of respective protocols are compiled. A broad spectrum of compounds covering a wide range of different chemical properties such as acidity, polarity, size and responsiveness can be addressed with either of these extraction protocols. Details with respect to extraction method specific component coverage can be extracted from (Buescher et al. 2010; Büscher et al. 2009; Canelas et al. 2008a; Canelas et al. 2009; Klimacek et al. 2010; Villas-Bôas and Bruheim 2007; Villas-Bôas et al. 2005b).

2.3 Quantification of intracellular metabolites

Approximately 600 metabolites are present in *S. cerevisiae* cells and their concentrations range from sub-μM to mM (Nielsen 2007). Their levels vary considerably in dependence on the environmental conditions applied or mutation introduced. Hence analysis tools suited for the determination of intracellular metabolites should be able to cover quantitatively as many metabolites as possible in wide concentration ranges.

Early studies that have focused on analyzing intracellular metabolites used enzymatic assays (Ciriacy and Breitenbach 1979; de Koning and van Dam 1992; Gonzalez et al. 1997; Grosz and Stephanopoulos 1990; Kötter and Ciriacy 1993; Theobald et al. 1997) or NMR analysis (den Hollander et al. 1981; Navon et al. 1979; Shanks and Bailey 1988) for quantification of a few compounds present in the cell. To increase the coverage towards a holistic quantitative record of the endometabolome enormous efforts were put on the development of new methods and techniques in the last years. In particular mass spectrometry revealed to be excellently suited in this respect and analytics of metabolite targeting switched therefore from an enzyme assay- or NMR-based to a mass spectrometry-based technique. Today three platforms dominate the analytical part of metabolomics. That is mass spectrometric detection in the form of a single, tandem or triple quadrupole or orbitrap mass spectrometer (MS) coupled via an electrospray ionization source (ESI) to a component separation device such as gas chromatography (GC), reverse-phased ion-pairing or anion exchange liquid chromatography (LC) or capillary electrophoresis (CE). Cross-platform comparison with respect to quantitative metabolomics revealed LC as the best suited separation technique for analysis on a single platform in terms of versatility and robustness. It was suggested by the authors that it is best complemented by the use of the GC platform (Büscher et al. 2009). More than 100 metabolites could be successfully quantified by GC-MS (~100 metabolites, (Villas-Bôas and Bruheim 2007; Villas-Bôas et al. 2003; Villas-Bôas et al. 2005b; Villas-Bôas et al. 2005c)), reverse phase ion pairing LC coupled to a triple quadrupole (138 metabolites, (Buescher et al. 2010)) or an orbitrab mass spec (137 metabolites, (Lu et al. 2010)). Especially the LC-MS platform has been investigated intensively with respect to quantitative metabolite coverage, sensitivity and robustness and revealed to be very suited for the comprehensive analysis of the central carbon metabolism (Buescher et al. 2010; Büscher et al. 2009; Lu et al. 2010). Almost all metabolites involved in glycolysis, pentose phosphate pathway and TCA cycle could be addressed. In addition amino acids and their precursors, redox cofactors, nucleotides, coenzyme A esters and many more can be analyzed in one sample run in approximately half an hour (Buescher et al. 2010; Lu et al. 2010). Current LC-techniques used in metabolomics are however limited to water soluble analytes (Buescher et al. 2010).

As mentioned above absolute quantification of intracellular metabolites is indispensably linked to the use of U-^{13}C-labled internal metabolite standards. Consequently the number of compounds to be analyzed doubles which makes data analysis more demanding. Residual amounts of substrates, products and ionic components in the labeled and unlabeled metabolite extracts can significantly alter the elution profile and ionization characteristics of compounds analyzed (Buescher et al. 2010). These so called matrix effects are hardly to predict and are typically experienced as an increase of the base line signal associated with a high signal-to-noise ratio and as shifts in metabolite-specific retention times. Consequently peak-to-peak resolution and base line separation of peaks can become badly affected. Exacerbate and tedious manual peak integration is in these cases required. Alternatively complex often erroneous peak deconvolution algorithms are applied. To reduce matrix effects to a minimum demands preparation of "clean" - meaning free of disturbing media compounds - metabolite extracts for both biological samples and IS.

The recommended procedure for internal referencing by identical U-^{13}C-labled compounds involves addition of a defined volume of IS to the biological sample (prior to extraction) and to all analytical standard dilutions (minimum 6 dilutions). A standard mixture containing all the compounds at known concentrations to be analyzed is commonly used. As some metabolites are not very stable it is recommended to store a master mixture containing just the stable compounds and add the sensitive components prior to analysis. Metabolites are quantified by comparing the ratio of ^{12}C- to ^{13}C- signals with the ^{12}C/^{13}C signals of the representative standard compounds.

2.4 Calculation of molar intracellular metabolite concentrations

The ultimate goal of quantitative metabolomics is the presentation of intracellular metabolite pools in the form of absolute molar concentrations. Only the knowledge of molar concentrations enables reliable integration of metabolome data in thermodynamic analysis, application in MCA based on fundamental enzyme kinetics or kinetic modeling. In the last step of absolute quantification of intracellular metabolite pools, the unbiased molar concentrations of metabolites obtained by the methods described above have to be therefore somehow related to the cell volume. For this reason metabolite concentrations and cell volume are based on the cell dry weight (CDW) producing specific parameters for metabolite concentrations in µmol/gCDW (= [metabolite]*volume of metabolite extract / ([dry cells]*volume of biological sample)) and for the cell volume in mL/gCDW. By dividing these parameters the intracellular metabolite concentration in mM is eventually obtained. Hence accurate determination of the CDW and cell volume is mandatory for calculation of reliable intracellular metabolite data and for molecular mechanistic interpretation relying on this data. Different methods can be found in the literature. Briefly, first the physiological state has to be specified at which the CDW should be analyzed. Second an aliquot of cell suspension is separated from the medium by vacuum filtration or centrifugation. The cell pellet is washed with ice-cold water or physiological NaCl solution to remove residual medium components and subsequently dried at 100-105°C until constancy of mass is verified. One should however keep in mind that application of NaCl can compromise resulting CDW values significantly when low amounts of biomass are addressed. The volume of *S. cerevisiae* cells at a particular physiological state can be determined by applying a Coulter counter analyzer or by microscopic techniques (Lord and Wheals 1981; Tamaki et al. 2005).

The cell volume of the baker's yeast was found to depend on the growth conditions and environmental parameter settings and can vary considerably. For example, the cell volume varies by a factor of 2-3 (16 - 42 μm^3) with doubling time (Lorincz and Carter 1979; Tyson et al. 1979) and the type of substrate metabolized (Johnston 1977; Tamaki et al. 2005). The lower the doubling time the larger the cells (Tyson et al. 1979). Cells grown on glucose, the most favored carbon source, are larger than those grown on a nonfermentable carbon source (Johnston et al. 1979; Lorincz and Carter 1979; Porro et al. 2003; Tyson et al. 1979). Cells cultured in the presence of ethanol showed an enlarged size (Kubota et al. 2004). In contrast nitrogen starved cells are abnormally small (Johnston 1977). The level of repression/derepression also contribute significantly to the cell size (Mountain and Sudbery 1990). Interestingly therefore that on the basis of the more relevant parameter for quantification the specific cell volume, *S. cerevisiae* cells do not show significant variation with the growth rate (Brauer et al. 2008) and values in the range of 1.5 - 1.9 mL/gCDW can be found in the literature for the strain CEN.PK 113-7D (Canelas et al. 2011; Cipollina et al. 2008; van Eunen et al. 2009). A slightly higher value of 2.38 mL/gCDW has been reported for another strain CBS 7336 (ATCC 32167) (Ditzelmüller et al. 1983). We can conclude that although the cell volume is highly sensitive to conditions applied the specific cell volume is rather constant. Nevertheless a 1.6-fold higher specific cell volume results in likewise lower metabolite concentrations which in some cases (e.g. [substrate] < K_m) may have an influence on the data interpretation. In the light of metabolic engineering it is hard to predict whether this "constant" likewise translates to recombinant cells. For example *S. cerevisiae* cells adapted to high ethanol concentrations displayed an altered cell size (Dinh et al. 2008). Or overexpressing mannitol-1-phosphate dehydrogenase (M1PDH) in *S. cerevisiae* to produce mannitol from glucose caused a substantial increasing of the size of cells (Costenoble et al. 2003).

As for all eukaryotic organisms *S. cerevisiae* metabolism is compartmented (cytosol, mitochondrion, vacuoles) which poses a problem for the accurate determination of concentrations of relevant intracellular metabolites. Current techniques for extracting metabolites and isolating organelles do not allow for absolute separation from the cytosol without altering the respective metabolite composition and pattern. Indirect strategies based on metabolic engineering or on fundamental thermodynamic principles have been developed to address this obstacle and gave first preliminary and semi-quantitative insights into the distribution of metabolites between cytosol and mitochondrion.

Functional expression of M1PDH from *E. coli* in *S. cerevisiae* was used as indicator reaction to determine the cytosolic free NAD to NADH ratio (Canelas et al. 2008b). M1PDH catalyzes the reversible NAD(H)-dependent interconversion of fructose 6-P (F6P) and mannitol 1-P (M1P). This reaction is directly connected to the central carbon metabolism and represents a dead-end reaction in the metabolism of yeast under the conditions applied in this study. Based on the assumption that the M1PDH reaction is at equilibrium the authors were capable of calculating the NAD/NADH ratio from the equilibrium constant and the intracellular concentrations of F6P and M1P. Data were verified by thermodynamic analysis. The cytosolic ratio of NAD/NADH was found to be ~10-fold higher as compared to the same ratio when based on the whole cell. Under anaerobic conditions however mannitol is formed from M1P implying that this approach is not yet universally applicable (Costenoble et al. 2003).

A different approach based on a network-embedded thermodynamic analysis later termed anNET (Zamboni et al. 2008) was used by (Kümmel et al. 2006) to resolve intracompartmental feasible concentration ranges from cell-averaged metabolome data.

Although these first results are promising there is large open space for the development of novel strategies combined with appropriate experimental techniques that enable precise compartment-specific quantification.

3. How can metabolic engineering of *S. cerevisiae* benefit from quantitative metabolomics?

In the typical metabolic engineering approach a bunch of new recombinant strains are designed and created with respect to a particular objective (see Fig. 1) or obtained from evolutionary adaption. Their new phenotypes are tested by fermentation or conversion experiments from which the substrate uptake rate and the product pattern in the form of specific product yields are determined. Results are often applied to FBA for verification. Intracellular enzyme activities of the introduced reactions as well as of those catalyzing reactions relevant for the new phenotype are measured from cell-free extracts. This data set usually provides many valuable details about the production efficiency in terms of conversion rate (how do intracellular activities of target enzymes compare to the conversion rate measured) and product selectivity (identification of side-products and oftentimes the reactions or pathways involved).

So how do metabolomics and more specifically quantitative metabolomics come into play? As described above the composition of intracellular metabolites together with their levels represent a direct signature of the physiological state of the cells investigated. Comparing metabolite profiles of wild-type and mutant strain(s) was often used to identify target reactions limiting the conversion rate (Hasunuma et al. 2011; Kahar et al. 2011; Klimacek et al. 2010; Kötter and Ciriacy 1993; Wisselink et al. 2010; Zaldivar et al. 2002) or extract the metabolite pattern representative for the new phenotype (Canelas et al. 2008b; Devantier et al. 2005; Ding et al. 2010; Hou et al. 2009; Kamei et al. 2011; MacKenzie et al. 2008; Pereira et al. 2011; Raamsdonk et al. 2001; Ralser et al. 2007; Thorsen et al. 2007; Usaite et al. 2009; Villas-Bôas et al. 2005a; Villas-Bôas et al. 2005c; Yoshida et al. 2008). Even apparently silent phenotypes of S. *cerevisiae* single deletion mutants can be uncovered with respect to the underlying mutation based on the developed metabolome (Raamsdonk et al. 2001). The rate however at which a compound's carbon skeleton is channeled through a certain pathway is directly linked to the level of active enzymes present and their affinity to the participating reactants as well as to fundamental thermodynamic laws of the reactions involved. Consequently knowledge about intracellular concentration of metabolites and enzyme activities combined with thermodynamic and enzyme kinetic analysis can provide novel and valuable insights into the kinetic organization of the engineered pathway or even the associated metabolic network which eventually exposes key regulatory or flux limiting sites. Differently to the general holistic approach usually found in systems biology pathway analysis in metabolically engineered cells can be reduced in most cases to the components involved in the new pathway and those connecting this pathway to the central carbon metabolism (Parachin et al. 2011).

3.1 Thermodynamic pathway analysis

If we are interested in analyzing a pathway or network of pathways on the basis of thermodynamic rules with the aim to extract pathway or network relevant mechanistic relationships, knowledge of exact quantitative metabolite data of all reactants involved is mandatory. On the contrary we can also check quantitative metabolome data with respect to its thermodynamic consistency (Kümmel et al. 2006) but most importantly we can get first hits of potential candidate reactions for metabolic engineering within a metabolic network without any knowledge about enzyme activity and kinetic parameters. Consider the following reaction

$$A + B \leftrightarrow C + D \tag{1}$$

The chemical equilibrium constant (K_{eq}) associated with this reaction can be defined according to the law of mass action as

$$K_{eq} = c_C^{eq} c_D^{eq} / (c_A^{eq}\, c_B^{eq}) \tag{2}$$

The superscript ^{eq} relates to the concentrations c of reactants at the chemical equilibrium. The standard Gibbs energy of a chemical reaction ($\Delta_r G^0$, usually given in J/mol) is related to K_{eq} by the fundamental relationship

$$\Delta_r G^0 = \sum_{i=1}^{N_s} v_i \Delta_f G_i^0 = -RT \ln K_{eq} \tag{3}$$

in which v_i and $\Delta_f G_i^0$ correspond to the stoichiometric coefficient of reactant i and to the standard reaction Gibbs energy of formation of species i at a specified T, P and ionic strength, respectively. R and T denote the general gas constant (8.314 J/mol/K) and absolute temperature in Kelvin (K), respectively. The Gibbs energy of formation of a reactant i ($\Delta_f G_i$) is further defined by

$$\Delta_f G_i = \Delta_f G_i^0 + RT\ln(c_i) \tag{4}$$

c_i in Equation (4) refers to the concentration of reactants involved. The Gibbs energy of a reaction eventually is described by

$$\Delta_r G = \sum_{i=1}^{N_s} v_i \Delta_f G_i^0 + RT\sum_{i=1}^{N_s} v_i \ln c_i = \Delta_r G^0 + RT\ln Q \tag{5}$$

Q in Equation (5) indicates the reaction quotient $c_C\, c_D$ / ($c_A\, c_B$), which is also known under the term "mass action ratio" that is frequently abbreviated by Γ. Rearranging Equation (4) yields

$$\Delta_r G = RT\,\{\ln(Q/K_{eq})\} \tag{6}$$

Now the reaction in Equation (2), reading from the left to the right, takes place freely in the forward direction A,B → C,D at $\Delta_r G < 0$ ($K_{eq} > Q$), is at equilibrium and displays no net flux when $\Delta_r G = 0$ ($K_{eq} = Q$) and needs support by an external driving force when $\Delta_r G > 0$ ($K_{eq} < Q$). In other words the reverse reaction is favored under these Q-conditions implying a net flux in the back direction, C+D → A+B. Equation (6) therefore provides a very convenient and important expression that permits immediate assignment of the net flux state of a reaction within a pathway or a large metabolic network provided that K_{eq} and all reactants participating in the particular reaction are known. The value of K_{eq} is dependent on the temperature, the ionic strength and the pressure (Alberty 2003). As a consequence only those K_{eq}s should be applied that were determined under conditions representative for the cell's physiological state investigated. In particular temperature, ionic strength and pH are of considerable interest as pressure can be in most cases assumed as a constant. Thereof only the temperature during the cultivation experiment, that is usually 30°C for *S. cerevisiae,* is known. Unfortunately most K_{eq}s tabulated have been determined at 25°C. For *S. cerevisiae* cellular ionic strength and pH are often assumed to be 0.1 M and 7.0, respectively. This may be sufficient for many thermodynamic network analyses. The validity of these assumptions however should be tested in any event prior pathway or network analysis. Algorithms are available that allow to some extent compensation for ionic strength and temperature.

Equilibrium constants can be determined by three different methods based on either in vitro assay with isolated enzymes, compound-specific standard Gibbs energy of formation (see Equation (3)) or in vivo experiments that combine FBA with quantitative metabolomics.

In case of using an isolated enzyme a reaction assay is designed such that known concentrations of substrate(s) and product(s) are solved in a buffer with defined ionic strength and pH. The reaction is started by the addition of the enzyme and processed at a certain temperature as long as the equilibrium state is reached. Reactant concentrations are then analyzed by appropriate techniques and the apparent K_{eq} (K'_{eq}) at a specified pH, ionic strength and temperature is calculated with respect to the underlying stoichiometry of the reaction. Values for K'_{eq} are usually applied in the context of network analysis. Importantly all reactants and the enzyme must be of highest purity and must be stable along the time of incubation. The composition of the reaction mixture at the equilibrium must not change in the subsequent component analysis.

When the $\Delta_f G_i^0$ for all reactants participating in the investigated reaction system are known, the respective values for K_{eq} can be calculated (Alberty 1991). Values for standard Gibbs energies of formation $\Delta_f G_i^0$ have been tabulated for a number of compounds (Alberty 2003). Computer programs are available with which one can calculate transformed standard Gibbs energies of formation ($\Delta_f G_i'^0$) from $\Delta_f G_i^0$ for a specified pH and ionic strength (Alberty 2003; Zamboni et al. 2008). In addition all possible dissociation forms of a compound are also lumped into a single reactant in $\Delta_f G_i'^0$. The respective transformed Gibbs energy of formation for a reactant at a certain concentration ($\Delta_f G_i'$) is described by

$$\Delta_f G_i' = \Delta_f G_i'^0 + RT \ln(c_i) \tag{7}$$

Applying $\Delta_f G_i'$ and $\Delta_f G_i'^0$ instead of $\Delta_f G_i$ and $\Delta_f G_i^0$ in Equations (4) and (5) permit calculation of K'_{eq} of a particular reaction. Large-scale thermodynamic studies for which availability of proper K'_{eq}s are crucial often bemoan large uncertainties with which current thermodynamic data obtained from enzyme assays or based on Gibbs energies are afflicted and therefore not sufficient for data analysis.

A way to circumvent these obstacles was introduced by (Canelas et al. 2011) who developed a new method to derive apparent equilibrium constants under real in vivo conditions. The basic idea of this work was based upon that the rate v of a certain reaction is directly dependent on the maximal turnover number V_{max}, the kinetic properties (β) of the enzyme catalyzing this reaction and on the net flux state of the reaction in relation to the respective equilibrium ($1-Q/K'_{eq}$). That is $v = V_{max}\,\beta(1-Q/K'_{eq})$. *S. cerevisiae* was cultivated at 30°C in chemostats under aerobic conditions and glucose (7.5 g/L) as the sole carbon source in standard mineral medium at 32 different dilution rates spanning a wide range of specific growth rates (0.03 to 0.29 h^{-1}) and as a consequence substrate conversion rates. Extracellular and intracellular metabolites were quantified at each dilution rate. Specific substrate and product fluxes given in mmol/gCDW/h were applied to a FBA based on a stoichiometric model and stoichiometric fluxes (v) for each reaction of the model network were calculated. Mass ratio coefficients Q of 27 network reactions were calculated from quantitative metabolite data at each v. From plots of Q vs. v the authors were able to extract values for K'_{eq}s. It could be shown that Q of reactions operating close to the equilibrium state display a negative linear dependency of v. Values of K'_{eq}s representing the thermodynamic equilibrium in the cell were determined by linear extrapolation to the y-axis ($v = 0$ and $K'_{eq} = Q$) under the assumption that V_{max} and kinetic properties of enzymes involved are not dependent on v. Intracellular K'_{eq}s obtained differed in part significantly (up to a factor of 9) from their in vitro determined counterparts implying that reaction conditions in vivo (pH and ionic strength) deviate from that specified in in vitro assays or by the in silico approach. The big advantage of this approach is that in vivo K'_{eq}s can be accurately determined without knowing anything about the intracellular pH and ionic strength. However only reactions located at or close to the equilibrium state are addressable.

Combining quantitative metabolomics data with thermodynamic rules on reactions constituting an operating metabolic network enables one to map reactions according to their location with respect to K'_{eq} (Crabtree et al. 1997; Klimacek et al. 2010; Kümmel et al. 2006; Wang et al. 2004). That is close to or far away from equilibrium. A reaction reaches K'_{eq} when the activity of an enzyme catalyzing a reaction downstream is low. On the other hand reactions located far away from K'_{eq} are often catalyzed by enzymes that have regulatory functions in the cell or by flux limiting enzymes representing potential targets for strain improvement (Klimacek et al. 2010). Differently to the relative location of a particular reaction to its equilibrium, the sign (plus or minus) of $\Delta_r G'$ or whether $Q >$ or $< K'_{eq}$ gives important information about the direction of the reaction or of an entire pathway. Information about flux directions can be readily implemented as further restrictions into a stoichiometric network to increase reliability of flux distributions (Hoppe et al. 2007; Kümmel et al. 2006).

3.2 Metabolic control analysis based on quantitative metabolomics in metabolic engineering of *S. cerevisiae*

Identification of those reactions exerting significant control of flux through a particular pathway is crucial to develop strategies for flux improvement. As mentioned above reactions suspected in flux limitation can be elucidated elegantly by thermodynamic analysis. Once identified reactions can be further analyzed with respect to principles of MCA. The concept of MCA was developed and introduced by (Heinrich and Rapoport 1974; Kacser and Burns 1973) and is comprehensively described in (Stephanopoulos et al. 1998). MCA strictly applies only to steady-state conditions. To unravel whether the amount of enzyme or of a reactant participating in the reaction catalyzed by this enzyme contribute to flux limitation through this particular reaction two terms have been defined, namely the flux control coefficient (FCC) and the elasticity coefficient (ε). The FCC defines the relative change in the steady-state flux resulting from an infinitesimal change in the activity of an enzyme of the pathway divided by the relative change of the enzyme activity, represented by the relationship FCC = $E\,dJ \,/\, (J\,dE)$, in which E and J stand for the concentration or activity of the enzyme and the steady-state flux, respectively. The ε is intrinsically linked to the inherent enzyme kinetic properties and is defined by the ratio of the relative change in the reaction rate resulted by an infinitesimal change in the metabolite concentration and can be described by $\varepsilon = c_j\,\delta v_i \,/\, (v_i\,\delta c_j)$, where c_j and v_i represent the concentration of reactant j and the reaction rate of enzyme i. FCC and ε are related by the flux-control connectivity theorem (Kacser and Burns 1973).

$$\sum_{i=1}^{L} FCC_i^J \varepsilon_{c_j}^i = 0 \tag{8}$$

Equation (8) implies that large elasticities - concentration of reactant exerts large control on the enzyme reaction rate - are associated with small FCCs - the overall flux through the pathway is not very dependent on the enzyme activity - and vice versa. Consequently knowledge of either of these coefficient is only required. FCC can be determined by increasing the amount of active enzyme for example by expressing this particular enzyme at different expression levels and measuring the change in flux through the pathway. The level of enzyme present can be judged by determining its specific activity in cell-free extracts. If the specific activity (µmol/$mg_{protein}$/min) for the isolated enzyme, the cellular protein content ($mg_{protein}$/gCDW) and the specific cell volume are known then the molar concentration of this enzyme can be calculated. Enzyme levels can be also determined by proteomics techniques and to some extent extrapolated from transcriptome data. Alternatively suitable antibodies that selectively bind to the target enzymes or fusion of an indicator peptide or protein tag to the enzyme to be analyzed permitting quantification either directly in vivo (GFP) or in vitro after separation of the target enzyme by affinity chromatography (e.g. His-tag, strep-tag) can be used for determining intracellular enzyme concentrations. If rate equations and associated kinetic parameters of all enzymes involved and at least their relative activity levels are known mathematical models can be applied to estimate FCCs and ε's. Rate equations based on the steady-state or rapid equilibrium assumption for enzyme catalyzed reactions are comprehensively summarized in (Segel 1993). However kinetic parameters typically used in this approach are determined by in vitro assays and it has been shown that in vitro generated data can significantly differ from those observed in vivo (Aragón and Sánchez 1985; Mauch et al. 2000; Reuss et al. 2007). This discrepancy is most likely due to the incomplete knowledge of the cellular composition and associated enzyme metabolite interactions as well as the difficulty to analyze enzymes under in vivo-like conditions at the lab bench. A promising step forward towards generation of more reliable in vivo-like in vitro enzyme kinetic data was reported recently (van Eunen et al. 2010). The authors suggested to measure enzyme activities in a buffered reaction mixture that simulates the intracellular cellular medium composition of *S. cerevisiae*.

Using the non-linear lin-log formulation developed for metabolic network analysis by (Visser and Heijnen 2003) FCC and ε can be estimated without any prior knowledge of enzyme-specific kinetic parameters (Visser and Heijnen 2002). In this approach the reaction rate is a nonlinear function of metabolite concentrations and is proportional to enzyme levels. lin-log kinetics is suited when large perturbations on the systems are analyzed. Furthermore statistical evaluation of parameter estimates is simplified (Wu et al. 2004). Nevertheless it is an approximation that fits to fundamental enzyme-reactant properties only in a certain range of perturbation (Wu et al. 2004).

3.3 Substrate promiscuous enzymes

Enzymes that have evolved relaxed or broad substrate specificity are substrate promiscuous (Hult and Berglund 2007). The presence of substrate promiscuous enzymes in a metabolic network considerably aggravates accurate network formulation and analysis if the accurate flux partition between the individual substrates is not known. This problem is further enhanced when the utilized substrates are highly connected within the network by other (substrate promiscuous) reactions. This is possibly best exemplified by the coenzyme promiscuous enzyme xylose reductase (XR) from *Candida tenuis* (Petschacher et al. 2005). To enable metabolic integration of xylose by *S. cerevisiae* XR and xylitol dehydrogenase XDH have to be integrated (Petschacher and Nidetzky 2008). Different to XDH which is strictly dependent on NAD(H) (Nidetzky et al. 2003), XR can oxidize both NADH and NADPH. These coenzymes in turn form the redox state of a cell and are important currency metabolites in all catabolic and anabolic pathways of the cell. Hence, keeping the redox state with respect to anabolism and catabolism balanced is crucial for the cell to stay alive. It was found based on stoichiometric considerations that the NADPH usage of XR is in strong correlation with by-product formation in the form of xylitol, the reaction product of XR and the substrate of XDH in the subsequent reaction (van Maris et al. 2007). Usage of up to (0-52)% NADPH by XR was compatible with a genome-scale metabolic network (Krahulec et al. 2010). Detailed kinetic analysis from in vitro studies showed that the XR almost exclusively utilizes NADPH in terms of catalytic efficiency ($k_{cat}/K_{m,coenzyme}$) and the selectivity parameter R_{sel} ($k_{cat}/(K_{i,coenzyme}K_{m,xylose})$) which are widely used as marker parameter for coenzyme discrimination. Based on this data even a rough estimation of the coenzyme usage of XR in the cell is not possible. To solve the coenzyme usage riddle of XR we determined intracellular concentrations of NADH and NADPH (Klimacek et al. 2010) and integrated this information together with the relevant kinetic parameters (Petschacher et al. 2005) into the mechanistically appropriate enzyme kinetic rate expression (Banta et al. 2002; Petschacher and Nidetzky 2005). A balanced coenzyme usage perfectly in line with physiology observed was obtained for the XR (Klimacek et al. 2010). Information about the correct flux partition of a particular substrate promiscuous enzyme can be implemented as further restrictions into a stoichiometric network to increase reliability of flux distributions. Furthermore this approach could be successfully applied on a series of wild-type and mutant forms of XR to predict reliably formation of xylitol (Krahulec et al. 2011).

4. Conclusions

Quantitative metabolomics is especially suited to help identifying key sites limiting an engineered metabolic route either within the created pathway but also apart from it. State-of-the-art protocols for sample work-up and LC-MS and GC-MS analysis permit absolute quantification of metabolites from *S. cerevisiae* cells provided that a U-^{13}C-labled IS is applied. Quantitative data in turn are indispensible for reliable pathway and network analysis in the form of a thermodynamic analysis, MCA or kinetic modelling. In combination with other omics techniques it represents a powerful tool to create designer microbial cell factories exposing improved or novel phenotypes.

5. Acknowledgement

Financial support from the Austrian Science Fund FWF (J2698) is gratefully acknowledged.

6. References

Abbott DA, Zelle RM, Pronk JT, van Maris AJ. 2009. Metabolic engineering of *Saccharomyces cerevisiae* for production of carboxylic acids: current status and challenges. FEMS Yeast Res 9(8):1123-36.

Alberty RA. 1991. Equilibrium compositions of solutions of biochemical species and heats of biochemical reactions. Proc Natl Acad Sci U S A 88(8):3268-71.

Alberty RA. 2003. Thermodynamics of biochemical reactions. Alberty RA, editor. New Jersey: John Wiley & Sons, Inc.

Alivisatos SG, Ungar F, Abraham G. 1964. Non-Enzymatic Interactions of Reduced Coenzyme I with Inorganic Phosphate and Certain Other Anions. Nature 203:973-5.

Aragón JJ, Sánchez V. 1985. Enzyme concentration affects the allosteric behavior of yeast phosphofructokinase. Biochem Biophys Res Commun 131(2):849-55.

Asadollahi MA, Maury J, Patil KR, Schalk M, Clark A, Nielsen J. 2009. Enhancing sesquiterpene production in Saccharomyces cerevisiae through in silico driven metabolic engineering. Metab Eng 11(6):328-34.

Asadollahi MA, Maury J, Schalk M, Clark A, Nielsen J. 2010. Enhancement of farnesyl diphosphate pool as direct precursor of sesquiterpenes through metabolic engineering of the mevalonate pathway in *Saccharomyces cerevisiae*. Biotechnol Bioeng 106(1):86-96.

Bailey JE, Sburlati A, Hatzimanikatis V, Lee K, Renner WA, Tsai PS. 2002. Inverse metabolic engineering: a strategy for directed genetic engineering of useful phenotypes. Biotechnol Bioeng 79(5):568-79.

Banta S, Boston M, Jarnagin A, Anderson S. 2002. Mathematical modeling of in vitro enzymatic production of 2-Keto-L-gulonic acid using NAD(H) or NADP(H) as cofactors. Metab Eng 4(4):273-84.

Bengtsson O, Jeppsson M, Sonderegger M, Parachin NS, Sauer U, Hahn-Hagerdal B, Gorwa-Grauslund MF. 2008. Identification of common traits in improved xylose-growing *Saccharomyces cerevisiae* for inverse metabolic engineering. Yeast 25(11):835-47.

Bettiga M, Gorwa-Grauslund MF, Hahn-Hägerdal B. 2010. Metabolic engineering in yeast. In: Smolke CD, editor. The metabolic pathway engineering handbook. Fundamentals. Boca Raton: CRC Press. p 1 - 48.

Brauer MJ, Huttenhower C, Airoldi EM, Rosenstein R, Matese JC, Gresham D, Boer VM, Troyanskaya OG, Botstein D. 2008. Coordination of growth rate, cell cycle, stress response, and metabolic activity in yeast. Mol Biol Cell 19(1):352-67.

Bro C, Knudsen S, Regenberg B, Olsson L, Nielsen J. 2005. Improvement of galactose uptake in *Saccharomyces cerevisiae* through overexpression of phosphoglucomutase: example of transcript analysis as a tool in inverse metabolic engineering. Appl Environ Microbiol 71(11):6465-72.

Bro C, Regenberg B, Förster J, Nielsen J. 2006. In silico aided metabolic engineering of *Saccharomyces cerevisiae* for improved bioethanol production. Metab Eng 8(2):102-11.

Buescher JM, Moco S, Sauer U, Zamboni N. 2010. Ultrahigh performance liquid chromatography-tandem mass spectrometry method for fast and robust quantification of anionic and aromatic metabolites. Anal Chem 82(11):4403-12.

Büscher JM, Czernik D, Ewald JC, Sauer U, Zamboni N. 2009. Cross-platform comparison of methods for quantitative metabolomics of primary metabolism. Anal Chem 81(6):2135-43.

Cakar ZP, Alkim C, Turanli B, Tokman N, Akman S, Sarikaya M, Tamerler C, Benbadis L, François JM. 2009. Isolation of cobalt hyper-resistant mutants of *Saccharomyces cerevisiae* by in vivo evolutionary engineering approach. J Biotechnol 143(2):130-8.

Cakar ZP, Seker UO, Tamerler C, Sonderegger M, Sauer U. 2005. Evolutionary engineering of multiple-stress resistant *Saccharomyces cerevisiae*. FEMS Yeast Res 5(6-7):569-78.

Canelas AB, Ras C, Ten Pierick A, van Dam JC, Heijnen JJ, van Gulik W. 2008a. Leakage-free rapid quenching technique for yeast metabolomics. Metabolomics 4:226-239.

Canelas AB, Ras C, ten Pierick A, van Gulik WM, Heijnen JJ. 2011. An in vivo data-driven framework for classification and quantification of enzyme kinetics and determination of apparent thermodynamic data. Metab Eng 13(3):294-306.

Canelas AB, ten Pierick A, Ras C, Seifar RM, van Dam JC, van Gulik WM, Heijnen JJ. 2009. Quantitative evaluation of intracellular metabolite extraction techniques for yeast metabolomics. Anal Chem 81(17):7379-89.

Canelas AB, van Gulik WM, Heijnen JJ. 2008b. Determination of the cytosolic free NAD/NADH ratio in Saccharomyces cerevisiae under steady-state and highly dynamic conditions. Biotechnol Bioeng 100(4):734-43.

Castrillo JI, Oliver SG. 2005. Towards integrative functional genomics using yeast as a reference model. In: Vaidyanathan S, Harrigan GG, Goodacre R, editors. Metabolome analysis: Strategies for systems biology. Berlin: Springer-Verlag. p 9 - 30.

Castrillo JI, Oliver SG. 2006. Metabolomics and systems biology in Saccharomyces cerevisiae. In: Brown AJP, editor. Fungal Genomics. Berlin: Springer-Verlag. p 3 - 7.

Chaykin S. 1967. Nicotinamide coenzymes. Annu Rev Biochem 36:149-70.

Cipollina C, van den Brink J, Daran-Lapujade P, Pronk JT, Vai M, de Winde JH. 2008. Revisiting the role of yeast Sfp1 in ribosome biogenesis and cell size control: a chemostat study. Microbiology 154(Pt 1):337-46.

Ciriacy M, Breitenbach I. 1979. Physiological effects of seven different blocks in glycolysis in *Saccharomyces cerevisiae*. J Bacteriol 139(1):152-60.

Costenoble R, Adler L, Niklasson C, Lidén G. 2003. Engineering of the metabolism of *Saccharomyces cerevisiae* for anaerobic production of mannitol. FEMS Yeast Res 3(1):17-25.

Crabtree B, Newsholme EA, Reppas NB. 1997. Principles of regulation and control in biochemistry: a paradigmatic, flux-oriented approach. In: Hoffman JF, Jamieson JD, editors. Handbook of Physiology. New York, U.S.A.: Oxford University Press. p 117-180.

Crabtree HG. 1928. The carbohydrate metabolism of certain pathological overgrowths. Biochem J 22(5):1289-98.

Csuk R. 1991. Baker's yeast mediated transformations on organic chemistry. Chem Rev 91:49 - 97.

Cvijovic M, Bordel S, Nielsen J. 2010. Mathematical models of cell factories: moving towards the core of industrial biotechnology. Microb Biotechnol 4(5):572-84.

de Koning W, van Dam K. 1992. A method for the determination of changes of glycolytic metabolites in yeast on a subsecond time scale using extraction at neutral pH. Anal Biochem 204(1):118-23.

Dejong JM, Liu Y, Bollon AP, Long RM, Jennewein S, Williams D, Croteau RB. 2006. Genetic engineering of taxol biosynthetic genes in *Saccharomyces cerevisiae*. Biotechnol Bioeng 93(2):212-24.

den Hollander JA, Ugurbil K, Brown TR, Shulman RG. 1981. Phosphorus-31 nuclear magnetic resonance studies of the effect of oxygen upon glycolysis in yeast. Biochemistry 20(20):5871-80.

Devantier R, Scheithauer B, Villas-Bôas SG, Pedersen S, Olsson L. 2005. Metabolite profiling for analysis of yeast stress response during very high gravity ethanol fermentations. Biotechnol Bioeng 90(6):703-14.

Ding M-Z, Zhou X, Yuan Y-J. 2010. Metabolome profiling reveals adaptive evolution of *Saccharomyces cerevisiae* during repeated vacuum fermentations. Metabolomics 6:42 - 55.

Dinh TN, Nagahisa K, Hirasawa T, Furusawa C, Shimizu H. 2008. Adaptation of *Saccharomyces cerevisiae* cells to high ethanol concentration and changes in fatty acid composition of membrane and cell size. PLoS One 3(7):e2623.

Ditzelmüller G, Wöhrer W, Kubicek CP, Röhr M. 1983. Nucleotide pools of growing, synchronized and stressed cultures of *Saccharomyces cerevisiae*. Arch Microbiol 135(1):63-7.

Domingues L, Guimarães PM, Oliveira C. 2010. Metabolic engineering of *Saccharomyces cerevisiae* for lactose/whey fermentation. Bioeng Bugs 1(3):164-171.

Förster J, Famili I, Fu P, Palsson BØ, Nielsen J. 2003. Genome-scale reconstruction of the *Saccharomyces cerevisiae* metabolic network. Genome Res 13(2):244-53.

Garcia Sanchez R, Karhumaa K, Fonseca C, Sànchez Nogué V, Almeida JR, Larsson CU, Bengtsson O, Bettiga M, Hahn-Hägerdal B, Gorwa-Grauslund MF. 2010. Improved xylose and arabinose utilization by an industrial recombinant *Saccharomyces cerevisiae* strain using evolutionary engineering. Biotechnol Biofuels 3:13.

Gonzalez B, François J, Renaud M. 1997. A rapid and reliable method for metabolite extraction in yeast using boiling buffered ethanol. Yeast 13(14):1347-55.

Grosz R, Stephanopoulos G. 1990. Physiological, biochemical, and mathematical studies of micro-aerobic continuous ethanol fermentation by Saccharomyces cerevisiae. II: intracellular metabolite and enzyme assays at steady state chemostat cultures. Biotechnol Bioeng 36(10):1020-9.

Hasunuma T, Sanda T, Yamada R, Yoshimura K, Ishii J, Kondo A. 2011. Metabolic pathway engineering based on metabolomics confers acetic and formic acid tolerance to a recombinant xylose-fermenting strain of *Saccharomyces cerevisiae*. Microb Cell Fact 10(1):2.

Heinrich R, Rapoport TA. 1974. A linear steady-state treatment of enzymatic chains. General properties, control and effector strength. Eur J Biochem 42(1):89-95.

Hong ME, Lee KS, Yu BJ, Sung YJ, Park SM, Koo HM, Kweon DH, Park JC, Jin YS. 2010. Identification of gene targets eliciting improved alcohol tolerance in Saccharomyces cerevisiae through inverse metabolic engineering. J Biotechnol 149(1-2):52-9.

Hoppe A, Hoffmann S, Holzhutter HG. 2007. Including metabolite concentrations into flux balance analysis: thermodynamic realizability as a constraint on flux distributions in metabolic networks. BMC Syst Biol 1:23.

Hou J, Lages NF, Oldiges M, Vemuri GN. 2009. Metabolic impact of redox cofactor perturbations in *Saccharomyces cerevisiae*. Metab Eng 11(4-5):253-61.

Hult K, Berglund P. 2007. Enzyme promiscuity: mechanism and applications. Trends Biotechnol 25(5):231-8.

Ishida N, Saitoh S, Ohnishi T, Tokuhiro K, Nagamori E, Kitamoto K, Takahashi H. 2006. Metabolic engineering of *Saccharomyces cerevisiae* for efficient production of pure L-(+)-lactic acid. Appl Biochem Biotechnol 129-132:795-807.

Jackson BE, Hart-Wells EA, Matsuda SP. 2003. Metabolic engineering to produce sesquiterpenes in yeast. Org Lett 5(10):1629-32.

Jeffries TW, Jin YS. 2004. Metabolic engineering for improved fermentation of pentoses by yeasts. Appl Microbiol Biotechnol 63(5):495-509.

Jin YS, Alper H, Yang YT, Stephanopoulos G. 2005. Improvement of xylose uptake and ethanol production in recombinant *Saccharomyces cerevisiae* through an inverse metabolic engineering approach. Appl Environ Microbiol 71(12):8249-56.

Johnston GC. 1977. Cell size and budding during starvation of the yeast *Saccharomyces cerevisiae*. J Bacteriol 132(2):738-9.

Johnston GC, Ehrhardt CW, Lorincz A, Carter BL. 1979. Regulation of cell size in the yeast *Saccharomyces cerevisiae*. J Bacteriol 137(1):1-5.

Kacser H, Burns JA. 1973. The control of flux. Symp Soc Exp Biol 27:65 - 104.

Kahar P, Taku K, Tanaka S. 2011. Enhancement of xylose uptake in 2-deoxyglucose tolerant mutant of *Saccharomyces cerevisiae*. J Biosci Bioeng 111(5):557 - 563.

Kamei Y, Tamura T, Yoshida R, Ohta S, Fukusaki E, Mukai Y. 2011. GABA metabolism pathway genes, *UGA1* and *GAD1*, regulate replicative lifespan in *Saccharomyces cerevisiae*. Biochem Biophys Res Commun 407(1):185-90.

Kjeldsen T. 2000. Yeast secretory expression of insulin precursors. Appl Microbiol Biotechnol 54(3):277-86.

Klimacek M, Krahulec S, Sauer U, Nidetzky B. 2010. Limitations in xylose-fermenting *Saccharomyces cerevisiae*, made evident through comprehensive metabolite profiling and thermodynamic analysis. Appl Environ Microbiol 76(22):7566-74.

Kötter P, Ciriacy M. 1993. Xylose fermentation by *Saccharomyces cerevisiae*. Appl Microbiol Biotechnol 38:776 - 783.

Krahulec S, Klimacek M, Nidetzky B. 2009. Engineering of a matched pair of xylose reductase and xylitol dehydrogenase for xylose fermentation by *Saccharomyces cerevisiae*. Biotechnol J 4(5):684-94.

Krahulec S, Klimacek M, Nidetzky B. 2011. Analysis and prediction of the physiological effects of altered coenzyme specificity in xylose reductase and xylitol dehydrogenase during xylose fermentation by *Saccharomyces cerevisiae*. J Biotechnol in press.

Krahulec S, Petschacher B, Wallner M, Longus K, Klimacek M, Nidetzky B. 2010. Fermentation of mixed glucose-xylose substrates by engineered strains of *Saccharomyces cerevisiae*: role of the coenzyme specificity of xylose reductase, and effect of glucose on xylose utilization. Microb Cell Fact 9:16.

Kubota S, Takeo I, Kume K, Kanai M, Shitamukai A, Mizunuma M, Miyakawa T, Shimoi H, Iefuji H, Hirata D. 2004. Effect of ethanol on cell growth of budding yeast: genes that are important for cell growth in the presence of ethanol. Biosci Biotechnol Biochem 68(4):968-72.

Kümmel A, Panke S, Heinemann M. 2006. Putative regulatory sites unraveled by network-embedded thermodynamic analysis of metabolome data. Mol Syst Biol 2:2006 0034.

Kuroda S, Miyazaki T, Otaka S, Fujisawa Y. 1993. *Saccharomyces cerevisiae* can release hepatitis B virus surface antigen (HBsAg) particles into the medium by its secretory apparatus. Appl Microbiol Biotechnol 40(2-3):333-40.

Kuyper M, Toirkens MJ, Diderich JA, Winkler AA, van Dijken JP, Pronk JT. 2005. Evolutionary engineering of mixed-sugar utilization by a xylose-fermenting *Saccharomyces cerevisiae* strain. FEMS Yeast Res 5(10):925-34.

Lee KS, Hong ME, Jung SC, Ha SJ, Yu BJ, Koo HM, Park SM, Seo JH, Kweon DH, Park JC and others. 2010. Improved galactose fermentation of *Saccharomyces cerevisiae* through inverse metabolic engineering. Biotechnol Bioeng 108(3):621-31.

Lord PG, Wheals AE. 1981. Variability in individual cell cycles of *Saccharomyces cerevisiae*. J Cell Sci 50:361-76.

Lorincz A, Carter BL. 1979. Control of cell size at bud initiation in *Saccharomyces cerevisiae*. J Gen Microbiol 113:287 - 295.

Lowry OH, Passonneau JV, Rock MK. 1961. The stability of pyridine nucleotides. J Biol Chem 236:2756-9.

Lu W, Clasquin MF, Melamud E, Amador-Noguez D, Caudy AA, Rabinowitz JD. 2010. Metabolomic analysis via reversed-phase ion-pairing liquid chromatography coupled to a stand alone orbitrap mass spectrometer. Anal Chem 82(8):3212-21.

MacKenzie DA, Defernez M, Dunn WB, Brown M, Fuller LJ, de Herrera SR, Günther A, James SA, Eagles J, Philo M and others. 2008. Relatedness of medically important strains of *Saccharomyces cerevisiae* as revealed by phylogenetics and metabolomics. Yeast 25(7):501-12.

Madsen KM, Udatha GD, Semba S, Otero JM, Koetter P, Nielsen J, Ebizuka Y, Kushiro T, Panagiotou G. 2011. Linking genotype and phenotype of *Saccharomyces cerevisiae* strains reveals metabolic engineering targets and leads to triterpene hyper-producers. PLoS One 6(3):e14763.

Mashego MR, Wu L, Van Dam JC, Ras C, Vinke JL, Van Winden WA, Van Gulik WM, Heijnen JJ. 2004. MIRACLE: mass isotopomer ratio analysis of U-^{13}C-labeled extracts. A new method for accurate quantification of changes in concentrations of intracellular metabolites. Biotechnol Bioeng 85(6):620-8.

Mauch K, Vaseghi S, Reuss M. 2000. Quantitative analysis of metabolic signalling pathways in Saccharomyces cerevisiae. In: Schügerl K, Bellgardt KH, editors. Bioreaction engineering. Berlin: Springer Verlag. p 435 - 477.

Mountain HA, Sudbery PE. 1990. The relationship of growth rate and catabolite repression with WHI2 expression and cell size in *Saccharomyces cerevisiae*. J Gen Microbiol 136(4):733-7.

Mutka SC, Bondi SM, Carney JR, Da Silva NA, Kealey JT. 2006. Metabolic pathway engineering for complex polyketide biosynthesis in *Saccharomyces cerevisiae*. FEMS Yeast Res 6(1):40-7.

Navon G, Shulman RG, Yamane T, Eccleshall TR, Lam KB, Baronofsky JJ, Marmur J. 1979. Phosphorus-31 nuclear magnetic resonance studies of wild-type and glycolytic pathway mutants of *Saccharomyces cerevisiae*. Biochemistry 18(21):4487-99.

Nevoigt E. 2008. Progress in metabolic engineering of *Saccharomyces cerevisiae*. Microbiol Mol Biol Rev 72(3):379-412.

Nidetzky B, Helmer H, Klimacek M, Lunzer R, Mayer G. 2003. Characterization of recombinant xylitol dehydrogenase from *Galactocandida mastotermitis* expressed in *Escherichia coli*. Chem. Biol. Interact. 143-144:533-542.

Nielsen J. 2007. Metabolomics in functional genomics and systems biology. In: Villas-Bôas SG, Roessner U, Hansen MAE, Smedsgaard J, Nielsen J, editors. Metabolome analysis: An Introduction. Hoboken: John Wiley & Sons, Inc. p 3 - 14.

Nielsen J, Jewett MC. 2008. Impact of systems biology on metabolic engineering of Saccharomyces cerevisiae. FEMS Yeast Res 8(1):122-31.

Nissen TL, Hamann CW, Kielland-Brandt MC, Nielsen J, Villadsen J. 2000a. Anaerobic and aerobic batch cultivations of *Saccharomyces cerevisiae* mutants impaired in glycerol synthesis. Yeast 16(5):463-74.

Nissen TL, Kielland-Brandt MC, Nielsen J, Villadsen J. 2000b. Optimization of ethanol production in *Saccharomyces cerevisiae* by metabolic engineering of the ammonium assimilation. Metab Eng 2(1):69-77.

Parachin NS, Bergdahl B, van Niel EW, Gorwa-Grauslund MF. 2011. Kinetic modelling reveals current limitations in the production of ethanol from xylose by recombinant *Saccharomyces cerevisiae*. Metab Eng in press.

Pereira FB, Guimarães PMR, Teixeira JA, Domingues L. 2011. Robust industrial *Saccharomyces cerevisiae* strains for very high gravity bio-ethanol fermentations. J Biosci Bioeng 112(2):130 - 136.

Petschacher B, Leitgeb S, Kavanagh KL, Wilson DK, Nidetzky B. 2005. The coenzyme specificity of *Candida tenuis* xylose reductase (AKR2B5) explored by site-directed mutagenesis and X-ray crystallography. Biochem. J. 385(Pt 1):75-83.

Petschacher B, Nidetzky B. 2005. Engineering *Candida tenuis* Xylose reductase for improved utilization of NADH: antagonistic effects of multiple side chain replacements and performance of site-directed mutants under simulated *in vivo* conditions. Appl. Environ. Microbiol. 71(10):6390-6393.

Petschacher B, Nidetzky B. 2008. Altering the coenzyme preference of xylose reductase to favor utilization of NADH enhances ethanol yield from xylose in a metabolically engineered strain of *Saccharomyces cerevisiae*. Microb Cell Fact 7:9.

Pirkov I, Albers E, Norbeck J, Larsson C. 2008. Ethylene production by metabolic engineering of the yeast *Saccharomyces cerevisiae*. Metab Eng 10(5):276-80.

Porro D, Brambilla L, Alberghina L. 2003. Glucose metabolism and cell size in continuous cultures of *Saccharomyces cerevisiae*. FEMS Microbiol Lett 229(2):165-71.

Raab AM, Gebhardt G, Bolotina N, Weuster-Botz D, Lang C. 2010. Metabolic engineering of *Saccharomyces cerevisiae* for the biotechnological production of succinic acid. Metab Eng 12(6):518-25.

Raamsdonk LM, Teusink B, Broadhurst D, Zhang N, Hayes A, Walsh MC, Berden JA, Brindle KM, Kell DB, Rowland JJ and others. 2001. A functional genomics strategy that uses metabolome data to reveal the phenotype of silent mutations. Nat Biotechnol 19(1):45-50.

Racker E. 1974. History of the Pasteur effect and its pathobiology. Mol Cell Biochem 5(1-2):17-23.

Ralser M, Wamelink MM, Kowald A, Gerisch B, Heeren G, Struys EA, Klipp E, Jakobs C, Breitenbach M, Lehrach H and others. 2007. Dynamic rerouting of the carbohydrate flux is key to counteracting oxidative stress. J Biol 6(4):10.

Reuss M, Aguilera-Vázques L, Mauch K. 2007. Reconstruction of dynamic network models from metabolite measurements. In: Nielsen J, Jewett MC, editors. Metabolomics. Berlin: Springer-Verlag. p 97 - 127.

Rizzi M, Baltes M, Theobald U, Reuss M. 1997. In vivo analysis of metabolic dynamics in *Saccharomyces cerevisiae*: II. Mathematical model. Biotechnol Bioeng 55(4):592-608.

Rocha I, Maia P, Evangelista P, Vilaca P, Soares S, Pinto JP, Nielsen J, Patil KR, Ferreira EC, Rocha M. OptFlux: an open-source software platform for in silico metabolic engineering. BMC Syst Biol 4:45.

Segel IH. 1993. Enzyme Kinetics - Behavior and analysis of rapid equilibrium and steady-state enzyme systems. New York: Wiley interscience.

Selvarajoo K, Arjunan SNV, Tomita M. 2010. In silico models for metabolic systems engineering. In: Smolke CD, editor. The metabolic pathway engineering handbook. Tools and applications. Boca Raton: CRC Press. p 1 - 22.

Shanks JV, Bailey JE. 1988. Estimation of intracellular sugar phosphate concentrations in *Saccharomyces cerevisiae* using ^{31}P nuclear magnetic resonance spectroscopy. Biotechnol Bioeng 32(9):1138-52.

Sonderegger M, Sauer U. 2003. Evolutionary engineering of *Saccharomyces cerevisiae* for anaerobic growth on xylose. Appl Environ Microbiol 69(4):1990-8.

Steen EJ, Chan R, Prasad N, Myers S, Petzold CJ, Redding A, Ouellet M, Keasling JD. 2008. Metabolic engineering of *Saccharomyces cerevisiae* for the production of n-butanol. Microb Cell Fact 7:36.

Steinle A, Bergander K, Steinbüchel A. 2009. Metabolic engineering of *Saccharomyces cerevisiae* for production of novel cyanophycins with an extended range of constituent amino acids. Appl Environ Microbiol 75(11):3437-46.

Stephanopoulos GN, Aristidou A, Nielsen J. 1998. Metabolic engineering, principles and methodologies. San Diego: Academic Press.

Tamaki H, Yun CW, Mizutani T, Tsuzuki T, Takagi Y, Shinozaki M, Kodama Y, Shirahige K, Kumagai H. 2005. Glucose-dependent cell size is regulated by a G protein-coupled receptor system in yeast *Saccharomyces cerevisiae*. Genes Cells 10(3):193-206.

Theobald U, Mailinger W, Baltes M, Rizzi M, Reuss M. 1997. In vivo analysis of metabolic dynamics in Saccharomyces cerevisiae : I. Experimental observations. Biotechnol Bioeng 55(2):305-16.

Thorsen M, Lagniel G, Kristiansson E, Junot C, Nerman O, Labarre J, Tamás MJ. 2007. Quantitative transcriptome, proteome, and sulfur metabolite profiling of the *Saccharomyces cerevisiae* response to arsenite. Physiol Genomics 30(1):35-43.

Toivari MH, Ruohonen L, Miasnikov AN, Richard P, Penttilä M. 2007. Metabolic engineering of *Saccharomyces cerevisiae* for conversion of D-glucose to xylitol and other five-carbon sugars and sugar alcohols. Appl Environ Microbiol 73(17):5471-6.

Trantas E, Panopoulos N, Ververidis F. 2009. Metabolic engineering of the complete pathway leading to heterologous biosynthesis of various flavonoids and stilbenoids in *Saccharomyces cerevisiae*. Metab Eng 11(6):355-66.

Tyson CB, Lord PG, Wheals AE. 1979. Dependency of size of *Saccharomyces cerevisiae* cells on growth rate. J Bacteriol 138(1):92-8.

Ukibe K, Hashida K, Yoshida N, Takagi H. 2009. Metabolic engineering of *Saccharomyces cerevisiae* for astaxanthin production and oxidative stress tolerance. Appl Environ Microbiol 75(22):7205-11.

Usaite R, Jewett MC, Oliveira AP, Yates JR, 3rd, Olsson L, Nielsen J. 2009. Reconstruction of the yeast Snf1 kinase regulatory network reveals its role as a global energy regulator. Mol Syst Biol 5:319.

van Eunen K, Bouwman J, Daran-Lapujade P, Postmus J, Canelas AB, Mensonides FI, Orij R, Tuzun I, van den Brink J, Smits GJ and others. 2010. Measuring enzyme activities under standardized in vivo-like conditions for systems biology. Febs J 277(3):749-60.

van Eunen K, Bouwman J, Lindenbergh A, Westerhoff HV, Bakker BM. 2009. Time-dependent regulation analysis dissects shifts between metabolic and gene-expression regulation during nitrogen starvation in baker's yeast. Febs J 276(19):5521-36.

van Gulik WM. 2010. Fast sampling for quantitative microbial metabolomics. Curr Opin Biotechnol 21(1):27-34.

van Maris AJ, Winkler AA, Kuyper M, de Laat WT, van Dijken JP, Pronk JT. 2007. Development of efficient xylose fermentation in *Saccharomyces cerevisiae*: xylose isomerase as a key component. Adv Biochem Eng Biotechnol 108:179-204.

Villas-Bôas SG. 2007a. Microbial metabolomics: Rapid sampling techniques to investigate intracellular metabolite dynamics - an overview. In: Villas-Bôas SG, Roessner U, Hansen MAE, Smedsgaard J, Nielsen J, editors. Metabolome analysis: An introduction: John Wiley & Sons, Inc. p 203 - 214.

Villas-Bôas SG. 2007b. Sampling and sample preparation. In: Villas-Bôas SG, Roessner U, Hansen MAE, Smedsgaard J, Nielsen J, editors. Metabolome analysis: An introduction: John Wiley & Sons, Inc. p 39 - 82.

Villas-Bôas SG, Åkesson M, Nielsen J. 2005a. Biosynthesis of glyoxylate from glycine in *Saccharomyces cerevisiae*. FEMS Yeast Res 5(8):703-9.

Villas-Bôas SG, Bruheim P. 2007. Cold glycerol-saline: the promising quenching solution for accurate intracellular metabolite analysis of microbial cells. Anal Biochem 370(1):87-97.

Villas-Bôas SG, Delicado DG, Åkesson M, Nielsen J. 2003. Simultaneous analysis of amino and nonamino organic acids as methyl chloroformate derivatives using gas chromatography-mass spectrometry. Anal Biochem 322(1):134-8.

Villas-Bôas SG, Hojer-Pedersen J, Åkesson M, Smedsgaard J, Nielsen J. 2005b. Global metabolite analysis of yeast: evaluation of sample preparation methods. Yeast 22(14):1155-69.

Villas-Bôas SG, Moxley JF, Åkesson M, Stephanopoulos G, Nielsen J. 2005c. High-throughput metabolic state analysis: the missing link in integrated functional genomics of yeasts. Biochem J 388(Pt 2):669-77.

Visser D, Heijnen JJ. 2002. The mathematics of metabolic control analysis revisited. Metab Eng 4(2):114-23.

Visser D, Heijnen JJ. 2003. Dynamic simulation and metabolic re-design of a branched pathway using linlog kinetics. Metab Eng 5(3):164-76.

Visser W, Scheffers WA, Batenburg-van der Vegte WH, van Dijken JP. 1990. Oxygen requirements of yeasts. Appl Environ Microbiol 56(12):3785-92.

Wang L, Birol I, Hatzimanikatis V. 2004. Metabolic control analysis under uncertainty: framework development and case studies. Biophys J 87(6):3750-63.

Wisselink HW, Cipollina C, Oud B, Crimi B, Heijnen JJ, Pronk JT, van Maris AJ. 2010. Metabolome, transcriptome and metabolic flux analysis of arabinose fermentation by engineered Saccharomyces cerevisiae. Metab Eng 12(6):537-51.

Wisselink HW, Toirkens MJ, del Rosario Franco Berriel M, Winkler AA, van Dijken JP, Pronk JT, van Maris AJ. 2007. Engineering of *Saccharomyces cerevisiae* for efficient anaerobic alcoholic fermentation of L-arabinose. Appl Environ Microbiol 73(15):4881-91.

Wisselink HW, Toirkens MJ, Wu Q, Pronk JT, van Maris AJ. 2009. Novel evolutionary engineering approach for accelerated utilization of glucose, xylose, and arabinose mixtures by engineered *Saccharomyces cerevisiae* strains. Appl Environ Microbiol 75(4):907-14.

Wu L, Mashego MR, van Dam JC, Proell AM, Vinke JL, Ras C, van Winden WA, van Gulik WM, Heijnen JJ. 2005. Quantitative analysis of the microbial metabolome by isotope dilution mass spectrometry using uniformly 13C-labeled cell extracts as internal standards. Anal Biochem 336(2):164-71.

Wu L, Wang W, van Winden WA, van Gulik WM, Heijnen JJ. 2004. A new framework for the estimation of control parameters in metabolic pathways using lin-log kinetics. Eur J Biochem 271(16):3348-59.

Yoshida S, Imoto J, Minato T, Oouchi R, Sugihara M, Imai T, Ishiguro T, Mizutani S, Tomita M, Soga T and others. 2008. Development of bottom-fermenting *Saccharomyces* strains that produce high SO_2 levels, using integrated metabolome and transcriptome analysis. Appl Environ Microbiol 74(9):2787-96.

Zaldivar J, Borges A, Johansson B, Smits HP, Villas-Boas SG, Nielsen J, Olsson L. 2002. Fermentation performance and intracellular metabolite patterns in laboratory and industrial xylose-fermenting Saccharomyces cerevisiae. Appl Microbiol Biotechnol 59(4-5):436-42.

Zamboni N, Kümmel A, Heinemann M. 2008. anNET: a tool for network-embedded thermodynamic analysis of quantitative metabolome data. BMC Bioinformatics 9:199.

Part 2

Data Analysis and Integration

3

Online Metabolomics Databases and Pipelines

Adam J. Carroll
The Australian National University
Australia

1. Introduction

As metabolomics becomes an increasingly major component of modern biological research, steps must be taken to preserve and make maximal use of the ever increasing torrents of new data entering the public domain. While this task is by no means unique to the field of metabolomics, the complexity, heterogeneity and large sizes of metabolomics datasets make the development of effective metabolomics bioinformatics tools particularly challenging. Despite these challenges, metabolomics specialists have recently been making rapid progress in this area. A wide range of powerful web-based tools designed to facilitate the systematic online storage, processing, dissemination and biological interpretation of technically and biologically diverse metabolomics datasets have now emerged and are rapidly becoming cornerstones of advancement in biological science.

Web-based tools for metabolomics perform a wide variety of functions. These can be divided into several broad categories, including:

1. Storage and dissemination of technical, biological, and physicochemical reference data for metabolites
2. Processing of raw instrument data to generate [metabolite x sample] data matrices suitable for statistical and multivariate data-analysis
3. Database storage and querying of pre-processed relative and/or absolute metabolite level data
4. Statistical and multivariate analysis of pre-processed relative and/or absolute metabolite level data
5. Aiding biological interpretation of metabolomics results by integration of biological knowledge such as known biomarkers or metabolic pathway information.

While some tools are broader in scope than others and some tools can essentially fully service the data-processing requirements of certain metabolomics approaches, it is important to note that no single tool is currently capable of fulfilling every requirement of every metabolomics researcher. This chapter will review the current state of development in the area of web-based informatics tools for metabolomics and explain how currently available tools can be used to accelerate scientific discovery. It will then attempt to predict future developments in the area of metabolomics web-tool development and advise new metabolomics researchers on strategies to maximise their own benefit from these developments.

2. Information about metabolites: Biological cheminformatics

2.1 Background

One of the fundamental questions of metabolomics is "how many metabolites occur in nature, what are their structures, what are their physical, chemical and biological properties and how are they distributed amongst species?". Large-scale efforts to build comprehensive databases of metabolite-related knowledge are beginning to provide at least approximate answers to these questions. Defining "the metabolome" of an organism in qualitative sense, by building well-annotated catalogues of metabolites and their properties, is analogous to sequencing and mapping the genome of an organism. That is, it provides a crucial foundation for the development of analytical approaches and experiments, aids in the interpretation of analytical results and provides an important scaffold upon which to attach new information as it becomes available.

Because metabolites are small molecule chemicals of biological origin, organisation of metabolite information lies at the interface between *bioinformatics* (the management of biological information) and *cheminformatics* (the management of chemical information). While metabolomics researchers will find useful information about metabolites in broad-scoped, general cheminformatic databases, a new generation of biology-focused cheminformatic databases are making it easier for biologists to find cheminformatic data specifically related to biology. This section will guide the reader towards online sources of metabolite information and explain how these information sources can be used to aid metabolomics research.

2.2 Molecular semantics: The metabolite naming issue

One of the challenges associated with finding online information about metabolites can be figuring out what text to enter into search engines. Metabolites can be named in many different ways in many different places online and searching with one name will generally only retrieve resources tagged with that particular name. Moreover, in cutting-edge metabolomics research, it is frequently the case that one is searching for information about a poorly-known or even completely hypothetical metabolite for which one has a structure in mind but for which its common name, if indeed it has one, is unknown. Fortunately, there are ways around these problems, thanks to the thoughtful design of cheminformatic databases. These will be explained below.

For well-known metabolites, finding detailed information is particularly easy. Most metabolite information databases annotate each metabolite entry with a large set of 'synonyms' – a range of different names commonly used to refer a given metabolite. As a result, if one uses a common name to search those databases for information about a well-known metabolite, one will usually find the information they need. In those cases, the key thing is to know which databases to search (these will be outlined shortly).

Finding information on well-known metabolites is relatively easy. However, metabolomics researchers are often interested in discovering new metabolites or learning what little is known about more about poorly-known metabolites. Often, a researcher may know the structure of a theoretical metabolite but have no idea whether it has been observed in nature before let alone what its common name might be. Indeed, such 'theoretical' metabolites often *have* been observed in nature before and have a common name, but finding this out can be challenging if one does not know where to start. This is where *InChI* codes and comprehensive InChI-enabled cheminformatic databases become *indispensible* (Wohlgemuth et al., 2010).

"InChI" is an abbreviation for "**In**ternational **Ch**emical **I**dentifier", a system of expressing chemical structures as compact strings of text suitable for efficiently and unambiguously conveying chemical structures across text-based systems such as web search engines. The InChI system was developed by the International Union of Pure and Applied Chemistry (IUPAC) and the National Institute of Standards and Technology (NIST). Each unique chemical structure can be converted into its own unique InChI code and vice-versa[1]. There are a range of freely-available software tools that allow one to draw a chemical structure and obtain its InChI code or enter an InChI code and have its structure drawn automatically (see Table 1 for examples). All the major metabolite information databases tag their entries with InChI codes, so if one is uncertain of the name of a target metabolite, the best approach is to generate its InChI code and search with that. Some cheminformatic databases provide web-based structure drawing tools allowing users to effectively generate an InChI code and search with it in a single step. One of the advantages of using an unambiguous structural identifier such as InChI to search a database is that if no hits are obtained, one may fairly safely conclude that the target molecule was not in the database[2]. When a hit is obtained, however, the returned information may include common name(s) for the molecule that can aid in subsequent literature searches. For anyone building metabolite databases or supplying supplementary tables of metabolite data for publication, annotation of these data with InChI codes is highly-recommended (Wohlgemuth et al., 2010). Online tools for generating InChI codes from structures or other identifiers are listed in Table 1. A particularly useful tool for metabolomics researchers is the Chemical Translation Service provided by the lab of Oliver Fiehn (Wohlgemuth et al., 2010) since this tool is capable of batch translations of miscellaneous metabolite identifiers and synonyms to standard InChI codes and other common identifiers.

2.3 Chemical ontologies: Organising metabolites into useful categories

In scientific communication, biologists frequently refer to broad 'classes' of metabolites using terms related to their functional groups (eg. 'alcohols'), their chemical properties (eg. 'organic acids') or biological roles (eg. 'hormones'). Moreover, researchers are often interested in obtaining lists of metabolites that fall in a particular class. For example, a researcher may want to identify metabolites in an organism that contain a particular functional group and will therefore be expected to undergo certain chemical reactions. Potential classes range in scope from very broad (eg. 'organic') to moderately specific (eg. 'alkaloids') to even more specific (eg. 'monoterpenoid indole alkaloids') and so on. While "metabolite classes" like these appear throughout the scientific literature, formalising them

[1] There is one caveat to this statement. The only truly non-ambiguous InChI codes are called "Standard" InChI (often abbreviated to "StdInChI" - these always begin with the string "InChI=1S/"). If building a metabolomics database, it is advisable to use only standard InChI codes.

[2] Some metabolite databases were built prior to the release of Standard InChI and have been annotated using non-standard InChI codes (always beginning with "InChI=1/"). It is always a good idea to check which InChI type a database uses before searching it with an InChI code.

into accurately and systematically defined and 'chemical ontologies' that can be used in practically useful ways is a non-trivial task. Despite this, a number of metabolite-related databases have begun developing and/or employing hierarchical systems of compound classification, allowing users to browse lists of metabolites via classification trees (ontologies). Examples of databases employing compound ontologies or hierarchical compound taxonomies for annotation of metabolite information include PubChem, ChEBI (Degtyarenko et al., 2008), the BioCyc family of metabolic pathway databases (Caspi et al., 2010), the Human Metabolome Database (HMDB) (Wishart et al., 2007) and MetabolomeExpress (Carroll et al., 2010). The ChEBI compound ontology is by far the most advanced and comprehensive ontology for biological small molecules and is downloadable in open formats from the ChEBI website. Its adoption is recommended in the development of new metabolomics databases.

Tool (URL)	Features
ChEBI Advanced Search (http://www.ebi.ac.uk/chebi/advancedSearchForward.do#)	**Input:** Structure drawing tool **Output:** StdInChI, SMILES **Search capabilities:** substructure, similarity, identical structure
PubChem Structure Search (http://pubchem.ncbi.nlm.nih.gov/search/search.cgi)	**Input:** Structure drawing tool **Output:** StdInChI, StdInChIKey, SMILES, SMARTS, Formula **Search capabilities:** substructure, superstructure, similarity, identical structure
Fiehn Lab Chemical Translation Service (http://uranus.fiehnlab.ucdavis.edu:8080/cts/homePage)	**Input:** Any major database ID, common synonym or structure identifier **Output:** Any major database ID, common synonym or structure identifier **Search capabilities:** simple search by any major database ID, common synonym or structure identifier

Table 1. Recommended online tools for generating unambiguous InChI structural identifier strings from structures, names or other identifiers

2.4 Physicochemical information

Physicochemical information about metabolites includes information about their physical and chemical properties such as their structures, molecular formulas, molecular weights, melting and boiling points, solubilities in different solvents at different temperatures, polarities, pKa, light absorbance and fluorescence properties, energy contents, refractive indices and other similar types of basic empirical information. This kind of information can be extremely useful when designing extraction, sample clean-up or analyte-enrichment protocols, for example.

2.5 Recommended sources of general metabolite information

Many online databases offer information about metabolites. These have varying scales and scopes of content, query tools and modes of access. In these aspects, several databases stand out from all others and these are described below.

2.5.1 ChemSpider

Description: A freely-accessible collection of compound data from across the web with a very versatile search engine.

Scope: all chemicals - not just metabolites

Semantic content: *Many* synonyms, identifiers and external database IDs and link-outs

Physicochemical content: Masses, formula, experimental melting point, physical state, appearance, stability, storage compatibility, safety. A substantial amount of additional predicted data.

Biological content: Links to MeSH

Analytical content: Some compounds have spectra

Noteworthy tools: Search by physicochemical properties

Modes of access: search (by synonym, InChI, SMILES, CAS, structure), API. Limited to 5000 structures per day.

Strengths: Enormous index of chemicals that is widely linked to external online resources - a good starting point if looking for information on a particular chemical.

Limitations: Broad focus means extracting desired subsets of information can be difficult. Cannot be downloaded. Results only returned in HTML - not spread sheet format. Limited to 5000 structures per day.

URL: http://www.chemspider.com/

2.5.2 Human Metabolome Database (HMDB)

Description: A comprehensive, freely-available knowledgebase of human metabolite information.

Scope: Human (*Homo sapiens*) metabolites

Semantic content: metabolite names, formulas, masses, structures, InChI, SMILES, external database IDs and link-outs; Chemical taxonomy

Physicochemical content: Masses, formula, water solubility, hydrophobicity, melting point, physiological charge, physical state

Biological content: Presence in human cellular compartments and biofluids; Measured concentrations in biofluids; reactions; enzymes; enzyme genes; disease associations; descriptions of biological roles

Analytical content: many compounds have LC/MS, GC/MS and/or NMR spectra obtained under standardised conditions

Noteworthy tools: Versatile 'Data Extractor'. Searching based on spectral properties

Modes of access: browse, search, complex query with data extractor, download

Strengths: Comprehensive, may be freely downloaded in entirety. Human focus is good for human metabolomics researchers.

Limitations: Important fields are empty for some very common metabolites. Being limited to human metabolites limits utility for other research areas. Downloadable flat-file format requires parsing in order to be usable in spread sheets or local databases.

Reference: (Wishart et al., 2007)

URL: http://www.hmdb.ca/

2.5.3 Chemical Entities of Biological Interest (ChEBI)

Description: A freely-available dictionary of small molecule chemicals of interest to biologists.

Scope: Small molecules of biological interest (endogenous biochemicals and exogenous bioactive compounds)

Semantic content: metabolite names, formulas, masses, structures, InChI, SMILES, external database IDs and link-outs; ChEBI Chemical Ontology

Physicochemical content: Formal charge

Biological content: None

Analytical content: None

Noteworthy tools: Structure based search, Versatile advanced query, Chemical and functional ontology-based browsing

Modes of access: browse, search, complex query, FTP download

Strengths: May be downloaded in bulk. Versatile advanced query tool. Query results downloadable in useful formats. Well-designed ontology.

Limitations: Human-centric. Far from comprehensive, particularly for non-human-related information. References to supporting literature are not provided with biofunctional ontology assignments. No species occurrence information.

Reference: (Degtyarenko et al., 2008)

URL: http://www.ebi.ac.uk/chebi

2.5.4 PubChem

Description: A freely available general dictionary of chemicals.

Scope: Any small molecules

Semantic content: metabolite names, formulas, masses, structures, InChI, SMILES, external database IDs and link-outs; MeSH chemical classification

Physicochemical content: Formal charge, partition coefficient, H-bonding donor and acceptor counts

Biological content: Bioactivity, bioassay results, safety and toxicology, associations with metabolic pathways in KEGG

Analytical content: None

Noteworthy tools: Structure and structural similarity-based search, versatile advanced query, chemical and biomedical ontology-based browsing, chemical structure clustering

Modes of access: browse, search, complex query, FTP download

Strengths: Huge number of compounds. Highly annotated. Some information is available for bulk download. Versatile advanced query tool. Extensive link-outs.

Limitations: Much compound information displayed on the website is not provided for bulk download. No species occurrence information.

URL: http://pubchem.ncbi.nlm.nih.gov

2.6 Metabolic pathway databases

A wide range of biological information about metabolites is available online. Utilising this information can aid in the development of hypotheses, the design of experiments and the biological interpretation of metabolomics results. For this purpose, among the most useful types of database are metabolic pathway databases. These play a crucial role in metabolomics research by systematically capturing and providing a close representation of current knowledge about: a) which metabolites occur in particular biological systems; b) the enzymatic and non-enzymatic reactions that link different metabolites together into metabolic pathways; c) the enzymes that carry out these reactions and the genes that encode them; and d) the allosteric interactions and signalling networks that regulate these genes and gene products. Another highly useful function that some metabolic pathway databases carry out is to visually overlay metabolomic datasets over pathway diagrams to provide biological contexts aiding the biological interpretation of results. Some of most useful metabolic pathway databases are described below.

2.6.1 Kyoto Encyclopedia of Genes and Genomes (KEGG)

Description: A knowledgebase of genomes, genes, gene-products their properties and the metabolic and regulatory pathways they form.

Species: many species from many different classes

Metabolic pathway content: metabolite names, formulas, masses, structures and external database IDs; reactions; reactant-product atom mappings; pathways; enzymes; enzyme genes; orthologies; bioactivities; allosteric interactions / regulatory pathways; pathway, compound, taxonomy and biological process ontologies

Noteworthy features: Structural similarity search

Modes of access: browse, search, API, FTP download (requires subscription)

Strengths: Enormous amount of information. The largest source of atom-mapped reactions available.

Limitations: Broad focus means extracting desired subsets of information can be challenging. Query tools are limited.

Reference: (Ogata et al., 1999)

URL: http://www.genome.jp/kegg/

2.6.2 BioCyc and the "Cyc" family of metabolic pathway databases

Description: Similar to KEGG. A collection of Pathway / Genome Databases (PGDBs) built using software that predicts metabolic pathways from genome sequences and subsequently refined by varying degrees of expert curation.

Species: BioCyc itself includes highly-curated PGDBs for 3 organisms: *Escherichia coli* (EcoCyc), *Arabidopsis thaliana* (AraCyc), *Saccharomyces cerevisiae* (YeastCyc). Another highly-curated PGDB called MetaCyc compiles pathway and enzyme information from >1900 organisms (mainly single-cell organisms) into a single reference database. See also the separate HumanCyc, PlantCyc and many other "Cyc" databases.

Metabolic pathway content: metabolite names, formulas, masses, structures, InChI, SMILES, Gibbs free energies and external database IDs; reactions; pathways; enzymes; enzyme genes; allosteric interactions / regulatory pathways; compound, pathway, gene and enzyme ontologies; links to literature supporting pathways

Noteworthy features: Interactive cellular overview pathway display, Regulatory overview, Genome browser, Advanced query tool, Powerful API, Omics viewer

Modes of access: browse, search, API, FTP

Strengths: Enormous amount of information. Powerful and intuitive query tools and API make extraction of data subsets easy. Returns results in simple tables and open XML pathway exchange formats. Omics viewer allows overlaying of omics (including metabolomics) data onto pathway views.

Limitations: Some useful and easily-fillable fields are empty for some metabolites. The Cyc databases often refer to generic entities such as 'a fatty acid' – this can limit their utility when researchers are interested in modelling connections between certain specific entities.

Reference: (Caspi et al., 2010)

URLs: http://biocyc.org, http://humancyc.org, http://www.plantcyc.org and others

2.6.3 Reactome

Description: An interactive collection of curated, peer-reviewed metabolic pathways with cross-referencing of reactions and pathways between organisms. Pathways are displayed via an intuitive GUI but may be downloaded in a variety of open formats.

Species: A variety of species. Most comprehensive for human.

Metabolic pathway content: hierarchically organised curated and peer-reviewed metabolic pathways; reactions; reaction-gene associations

Noteworthy features: Interactive pathway viewer

Modes of access: browse, search and download

Strengths: Peer-reviewed, user-friendly, different subcellular metabolite pools are treated as separate entities

Limitations: Reaction-centric. Not much information about metabolites and does not provide any tools for overlaying metabolite expression data.

Reference: (Croft et al., 2011)

URL: http://www.reactome.org

2.6.4 KappaView

Description: A web-based tool allowing users to overlay metabolite- and gene-expression responses and correlations onto custom pathway diagrams or onto a collection of neat, simple and interactive metabolic pathway diagrams.

Species: A variety of species.

Metabolism-related content: hierarchically organised curated metabolic pathways; reactions; reaction-gene associations

Noteworthy features: Gene and metabolite expression overlay

Modes of access: browse, search and download

Strengths: User-friendly; neat/simple diagrams; may be integrated into third party websites using a flexible API; can also overlay metabolite-metabolite, gene-gene and metabolite-gene correlations

Limitations: Does not support InChI

Reference: (Sakurai et al., 2010)

URL: http://kpv.kazusa.or.jp/kpv4/

2.6.5 Human Metabolome Database (HMDB) Pathways

Description: A comprehensive, freely-available knowledgebase of human metabolite information.

Species: Human (*Homo sapiens*)

Metabolism-related content: A set of 'textbook' style metabolic pathway diagrams with metabolites hyperlinked to HMDB metabolite information pages and enzymes hyperlinked to UniProt database.

Noteworthy features: None

Modes of access: browse, search

Strengths: Easy to understand

Limitations: Not downloadable. No documented API.

Reference: (Wishart et al., 2007)

URL: http://www.hmdb.ca

2.6.6 KNApSAcK

Description: A comprehensive species-metabolite relationship database for plants. Although not strictly a metabolic pathway database, this database is useful for identifying plant species that contain a certain chemical or identifying chemicals that have been reported in a particular plant species or higher level taxon.

Species: Plants

Metabolism-related content: References to literature reporting the presence of compounds in different plant species. Chemical structures. Masses.

Noteworthy features: References to literature.

Modes of access: browse, search

Strengths: Contains information on many plant-specific specialised metabolites.

Limitations: Data itself is not downloadable.

Reference: (Shinbo et al., 2006)

URL: http://kanaya.naist.jp/KNApSAcK/

3. Online analytical reference spectra for metabolomics

3.1 The roles of analytical reference libraries in metabolomics research

The first online metabolomics databases to store and disseminate actual instrument data for metabolites generally provided spectral reference libraries. These spectral libraries provide reference signals for authentic standard compounds and sometimes also for 'unknown' metabolites obtained through the analysis of standards and biological materials under controlled conditions. The de-novo construction of large analytical reference libraries requires expertise in chemistry, is time consuming and expensive. Centralization of spectral reference data in expert-curated public repositories helps the metabolomics community by: 1) making it easier and cheaper for new labs to build their own data processing pipelines; 2) reducing the probability of metabolite misidentification by non-specialists; and 3) promoting efficient communication about 'unknown' metabolites that are recognisable on the basis of their analytical properties but for which no structural information is available.

3.2 Types of analytical reference spectra available online

Reference spectra are available from a number of online sources. Types of reference data available include downloadable mass-spectral and retention-index (MSRI) libraries for gas chromatography / mass spectrometry (GC/MS) (Kopka et al., 2005; Schauer et al., 2005; Carroll et al., 2010), searchable but not-downloadable MSRI data (Skogerson et al., 2011), NMR spectra collected under standardized conditions (Wishart et al., 2007; Cui et al., 2008; Ulrich et al., 2008) and MS and MS/MS spectra from a wide range of platforms including accurate mass instruments (Smith et al., 2005; Horai et al., 2010). In addition, most cheminformatic and metabolic pathway databases provide accurate monoisotopic mass information for metabolites which can help provide candidate identities for accurate-mass LC/MS and direct-infusion (DI)/MS peaks. These data sources are described in detail later.

3.3 Reference data for Nuclear Magnetic Resonance (NMR)

One of the great advantages that NMR has over mass-spectrometry is that chemical shifts and coupling constants - unlike mass-spectral fragmentation patterns - are, under readily controllable conditions, absolute physical constants that may be readily and accurately reproduced between different makes and model of instrument. Reference libraries of NMR spectra of metabolites, acquired under standardized conditions, are therefore of broad utility by the metabolomics research community. The major sources of standardized NMR spectra for metabolomics are the Madison Metabolomics Consortium Database (Cui et al., 2008), the Biological Magnetic Resonance Bank (Ulrich et al., 2008) and the HMDB (Wishart et al., 2007). These are detailed shortly.

3.4 Reference data for Gas-Chromatography / Mass-Spectrometry (GC/MS)

The most useful reference data for GC/MS are downloadable MSRI libraries. These are libraries of mass-spectra and retention indices for peaks observed in GC/MS chromatograms obtained by the GC/MS analysis of pure compounds and biological samples under standardised conditions (Kopka et al., 2005; Schauer et al., 2005). When the same standardized conditions are employed for GC/MS analysis in different laboratories, a single common MSRI library can be used for the high-confidence identification of common metabolite signals in those different labs (Schauer et al., 2005). Researchers setting up new GC/MS metabolomics platforms are advised to consider adopting a standardised GC/MS protocol already supported by a publicly-available MSRI library such as those available from the Golm Metabolome Database (Kopka et al., 2005) or MetabolomeExpress (Carroll et al., 2010) since this will enable them to share MSRI libraries with those labs and benefit from ongoing efforts to extend those libraries and annotate the large number of 'unknown' metabolites detected in GC/MS chromatograms of biological samples.

3.5 Reference data for liquid chromatography-MS, MS/MS and MS^n

While the low-cost and operational simplicity of GC/MS has led it to become the most widely employed analytical platform in metabolomics, an increasing number of laboratories are adopting complementary techniques based on liquid chromatography (LC)- and direct infusion (DI)/MS methods that employ different ionisation techniques and more advanced mass-spectrometers capable of MS, MS/MS, MS^3 and MS^n modes of analysis together with much higher mass accuracy and resolution than is provided by most standard GC-MS systems. In the paragraphs below, the various types of non GC/MS, MS-based metabolomics techniques such as LC/MS, DI/MS and capillary electrophoresis (CE)/MS including tandem MS and MS^n methods will be referred to collectively as "LC/MS" techniques.

While GC/MS metabolomics is dominated almost entirely by electron impact ionisation (EI) methods using the industry-standardised ionisation energy of 70eV, yielding highly-reproducible fragmentation spectra between different GC/MS instruments, such broad standardisation has not occurred for LC/MS. For LC/MS, the enormous diversity of mass-spectrometer types, combined with a lack of highly-developed LC 'retention-index' systems present significant challenges towards the creation of standardized MSRI reference libraries, analogous to those available for GC/MS, capable of unambiguous cross-laboratory peak identification for LC/MS.

The simplest type of online reference data for LC/MS metabolomics are the accurate, monoisotopic masses and molecular formulas of metabolites and, in some cases, their stable-isotope-labelled isotopomers. The data-processing packages provided with MS instruments capable of high-accuracy mass measurements generally allow users to create custom libraries of accurate masses and/or molecular formulas (for improved match scoring based on the shapes of isotopic envelopes) for target analytes to assist with peak identification. Although accurate masses or molecular formulas alone are not sufficient to unambiguously identify metabolite signals (due to the high frequency of structural isomers across nature), using these data in a rational manner can often provide valuable clues about the possible identities of peaks.

A good way of reducing (*but not eliminating*) ambiguity in accurate mass-based assignments is to build a separate accurate mass library for each biological system under investigation and to include in each library only those metabolites for which literature evidence exists to support their presence in that organism. An easy way of doing this is to use the advanced query tool provided with each of the BioCyc family of metabolic pathway databases (of which there are many). While the metabolite sets thus obtained may not be complete, this is a fast way of obtaining a good quality starting set.

Another approach for reducing ambiguity in LC/MS peak identifications is to use MS/MS spectral similarity as a scoring parameter to complement accurate-mass MS based assignments (see (Matsuda et al., 2009; Matsuda et al., 2010) for good examples). The major online sources of MS/MS spectra for metabolites are MassBank (Horai et al., 2010), METLIN (Smith et al., 2005), ReSpect for Phytochemicals (http://spectra.psc.riken.jp/menta.cgi/index) and the HMDB (Wishart et al., 2007). These databases each have different strengths and limitations which will be outlined shortly. With the notable exception of ReSpect for Phytochemicals, a drawback that these databases share is a lack of support for bulk downloading of spectra. That said, MassBank does provide a powerful API to partially overcome the need for bulk download while the METLIN website currently reports that an API is in development.

3.6 The need for chromatographic retention data in LC/MS reference databases

It is important to note that, for high-confidence peak identifications that meet minimum reporting standards outlined by the Metabolomics Standards Initiative (MSI) (Sansone et al., 2007), it is necessary to support peak identifications with an additional, orthogonal identification parameter. In the case of LC/MS, where chromatography is used, this parameter is generally retention time or relative retention time agreement with an authentic standard. Unfortunately, there appear to be few if any LC/MS reference databases that provide retention time or relative retention time information. Absolute retention times vary from instrument to instrument and from column to column (even between columns of the same make and model), and are therefore considered to be of limited use for high-confidence inter-laboratory peak identification. However, *relative* retention times (or *retention indices*), where the retention time of each peak is expressed relative to one or two other peaks in the same chromatogram, are far more stable (Tarasova et al., 2009) and may provide an avenue to the compilation of LC-MS reference libraries capable of providing MSI-compliant peak identifications by combining accurate mass MS or MS/MS spectra with meaningful and highly reproducible retention index (RI) properties. Complementary to this approach would be the further development of RI-prediction models that can accurately predict the LC retention indices of metabolites based on their structures (Hagiwara et al., 2010).

It is important to note that sufficient RI reproducibility may only be achievable with certain simple types of stationary and mobile phase combinations whereby a single stationary phase interaction mechanism (eg. hydrophobic interactions in C18 reversed-phase chromatography or hydrogen-bonding interactions in silanol based normal phase chromatography) applies to all analytes. In separations over mixed-mode stationary phases where multiple interaction mechanisms occur, there is more potential for variations in chromatographic conditions to differentially affect different peaks, thus changing their relative retention times. Public databases of "Accurate Mass / retention Time (AMT) tags" are playing increasingly important roles in peptide identification in LC-MS proteomics (Hagiwara et al., 2010). A similar trend is to be expected in metabolomics.

3.7 Major online sources of analytical reference spectra for metabolomics

3.7.1 Madison Metabolomics Consortium Database (MMCD)

Description: An analytical reference database and signal-matching tool for metabolomics.

Species: Not species-constrained

Reference data: Standardized NMR spectra for 791 different metabolites (^{1}H, ^{13}C, DEPT90, DEPT135, [^{1}H, ^{1}H] TOCSY and [^{1}H, ^{13}C] HSQC). General information on >20 000 metabolites.

Noteworthy features: NMR spectrum-based search. Batch search capability.

Modes of access: browse, search and download individual spectra via web interface. Bulk FTP download of raw spectra via BMRB FTP site.

Strengths: Enormous resource for NMR metabolomics. Includes a wide range of metabolites including those that don't occur in humans (eg. plant-specific metabolites). Spectral matching tools provide batch-processing capability.

Limitations: No support for bulk download of metabolite information based on complex query

Reference: (Cui et al., 2008)

URL: http://mmcd.nmrfam.wisc.edu

3.7.2 Human Metabolome Database (HMDB)

Description: A comprehensive, freely-available knowledgebase of human metabolite information.

Species: Human (*Homo sapiens*)

Reference data: Standardized MS/MS and NMR spectra (^{1}H, ^{13}C, ^{13}C HSQC, TOCSY) for >780 metabolites. GC/MS MSRI reference data on ~300 metabolites.

Noteworthy features: NMR, MS/MS and GC/MS spectrum-based search

Modes of access: browse, search and bulk download (bulk download of MS/MS spectra only provides *images* of spectra).

Strengths: A large set of standardized NMR and GC/MS spectra help new labs to quickly set up metabolite profiling platforms.

Limitations: No support for bulk download of metabolite information based on complex query. No batch-processing capabilities for spectral matching. No API for integration with other web tools.

Reference: (Wishart et al., 2007)

URL: http://www.hmdb.ca

3.7.3 METLIN

Description: A repository for metabolite information and tandem mass spectrometry data.

Species: Not formally species-constrained but is fairly human-centric

Reference data: Accurate masses of >44000 metabolites. >28000 high-resolution Quadrupole/Time-Of-Flight (Q/TOF) MS/MS spectra for ~5000 metabolites. Multiple collision energies.

Noteworthy features: Batch searching of mzXML MS/MS files against the database. Integration with XCMS LC/MS data-processing pipeline. Neutral loss search.

Modes of access: Search only. API in development

Strengths: A large set of standardized NMR and GC/MS spectra help new labs to quickly set up metabolite profiling platforms.

Limitations: No bulk-download (must be purchased from instrument manufacturer).

Reference: (Smith et al., 2005)

URL: http://metlin.scripps.edu

3.7.4 MassBank

Description: A repository for mass-spectra of pure compounds. Features a unique design involving a centralised interface but a distributed network of data servers providing the mass-spectra.

Species: Not species constrained. Not limited to biological metabolites.

Reference data: >29000 mass spectra from a wide range of instrument types including, but not limited to, GC/MS, LC/MS and LC-MS/MS.

Noteworthy features: Batch searching of MS/MS files against the database. Neutral loss search. Most sophisticated and powerful spectral search and visualisation capabilities of all available mass-spectral repositories.

Modes of access: Search, browse and API.

Strengths: Many spectra, powerful search capabilities.

Limitations: No bulk-download. However, individual spectra may be downloaded in text format.

Reference: (Horai et al., 2010)

URL: http://www.massbank.jp/

3.7.5 ReSpect for Phytochemicals

Description: An interactive collection of MS^n spectra of plant metabolites, collected by the LC/MS metabolomics group of the RIKEN Plant Science Center.

Species: Plant species.

Reference data: A total of >8500 MS/MS spectra including >3000 spectra from the literature, >4000 triple quadrupole MS/MS spectra corresponding to >861 standard compounds and >1000 Q/TOF spectra corresponding to >550 standard compounds. Includes both +ve and -ve ionization modes.

Noteworthy features: Spectral search online using cosine method

Modes of access: Search, browse and complete download.

Strengths: Contains many plant-specific spectra not available elsewhere. Free for bulk download.

Limitations: No API. No batch search capability.

URL: http://spectra.psc.riken.jp/menta.cgi/index

3.7.6 MS-MS Fragment Viewer 1.0

Description: A database of Liquid Chromatography Fourier Transfer Ion Cyclotron Resonance Mass Spectroscopy (LC-FT/ICR-MS), Ion-Trap Tandem Mass Spectroscopy (IT-MS/MS), Fourier Transform Tandem Mass Spectroscopy (FT-MS/MS) and photodiode array (PDA) spectra with predicted structures of fragment ions observed in LC-FT/ICR-MS.

Species: Plant species.

Reference data: Spectral data for 116 different flavonoids.

Noteworthy features: Predicted structures of fragment ions observed in LC-FT/ICR-MS.

Modes of access: Search only.

Strengths: Ultra-high mass accuracy of FT/ICR-MS.

Limitations: Limited range of spectra. Search only. No browse capability means it is impossible to know what to search for. No spectral-based searching.

URL: http://webs2.kazusa.or.jp/msmsfragmentviewer/

3.7.7 MoTo DB

Description: A liquid chromatography-mass spectrometry-based metabolome database for tomato

Species: Tomato (*Solanum lycopersicum*)

Reference data: Masses, retention times, UV/Vis properties and MS/MS fragment information for a range of metabolites reported to occur in tomato plants.

Noteworthy features: Includes retention times.

Modes of access: Search only.

Strengths: Provides literature references to support peak annotations.

Limitations: Very limited search capability. No browse capability. No download.

Reference: (Moco et al., 2006)

URL: http://appliedbioinformatics.wur.nl/moto/

3.7.8 The Golm Metabolome Database (GMD)

Description: An interactive and downloadable database of electron impact (EI) ionization mass-spectra and associated retention indices of metabolite peaks detected by GC-EI-Quadrupole (GC-EI-Q-MS) and GC-EI-Time Of Flight (GC-EI-TOF-MS) instruments operated under standardized conditions.

Species: Not formally species-constrained but is plant-centric.

Reference data: Contains MSRI data for ~4500 analytes (different chemical derivatives) corresponding to ~1500 metabolites

Noteworthy features: Decision-tree based substructure prediction.

Modes of access: Search, browse and API

Strengths: Very comprehensive. Free for download. Well curated and supported.

Limitations: Does not provide innate support for sharing of MSRI libraries by arbitrary users.

Reference: (Kopka et al., 2005)

URL: http://gmd.mpimp-golm.mpg.de/Default.aspx

3.7.9 MetabolomeExpress

Description: An interactive database of downloadable MSRI libraries, raw and processed GC/MS metabolite profiling datasets and a database of metabolic phenotypes observed in any organism using any analytical technique. Includes a complete GC/MS data processing pipeline and cross-study data mining tools.

Species: Not formally species-constrained but current content is plant-centric.

Reference data: A number of GC/MS MSRI libraries are downloadable from the website. Golm Metabolome Database MSRI libraries are provided for use within the data processing pipeline.

Noteworthy features: Members may independently upload their own MSRI libraries for interactive dissemination and use within the GC/MS data-processing pipeline.

Modes of access: browse and FTP

Strengths: Libraries free for download. Provides a built-in GC/MS data processing pipeline.

Limitations: No API. No search.

Reference: (Carroll et al., 2010)

URL: http://www.metabolome-express.org

4. Web-based data processing pipelines for metabolomics

4.1 Background

Less than a decade ago, software packages enabling processing and analysis of metabolomics datasets were restricted to a limited range of desktop software programs. Would-be metabolomics researchers would have to download or purchase and install software on local computers, set up local reference libraries for peak identification and sometimes develop custom in-house computer scripts to adapt the outputs of various programs into the formats required by programs used for downstream analysis. These challenges were compounded by the fact that available programs often lacked the kinds of specialised, biology-related features desirable for metabolomics research. However, the understandable widespread dissatisfaction of metabolomics researchers with this situation has, over the last decade, driven rapid development of powerful online, platform-independent data processing pipelines tailored towards the needs of metabolomics research. Thanks to the availability of these packages and the availability of standardised analytical reference libraries, it is now quite feasible for researchers with limited experience to conduct detailed processing and analysis of their instrumental datasets with little more than a fast internet connection, an up-to-date web-browser and, in some cases, an FTP-client program for uploading data. This section will provide an overview of the types of data processing pipelines that are currently accessible online and compare the most powerful examples in more detail.

4.2 Functions carried out by online data-processing tools

Any ideal metabolomics data-processing pipeline, whether online or offline, should be able to: a) identify and quantify biologically-relevant signals from raw instrument files and distinguish them from biologically irrelevant signals; b) identify non-redundant metabolite signals and, where possible, annotate them with their molecular identities; c) assemble a [metabolite x sample] data matrix appropriately normalised to sample volumes, internal standards and/or other useful normalisation factors; d) facilitate determination and statistical analysis of relative metabolite levels between sample classes; e) carry out multivariate analyses such as principal components analysis (PCA), hierarchical clustering analysis (HCA) and partial least squares discriminant analysis (PLS-DA); and f) provide facilities to assist biological interpretation of results (eg. mapping of detected metabolite responses onto metabolic pathways, over-representation analysis and biomarker detection). While the vast majority of online metabolomics data-processing tools carry out only one or a few of these functions, there are systems capable of carrying out all of these functions. The functionalities of a variety of web-based data processing tools for metabolomics are summarised in Table 2.

5. Online metabolomics data repositories

5.1 Background

The long-standing scientific tradition of openly disclosing supporting primary data whenever scientific claims are made has been a fundamental factor underlying the credibility of science. However, in more recent years, the scale and complexity of primary datasets has risen dramatically, presenting ever-new challenges to this tradition with the widespread emergence of high-throughput metabolomics technologies in bioscience being a good example.

In this author's view, it is absolutely crucial that the culture of open primary data disclosure is maintained, and that "challenges" should not become "excuses". Even in the extreme case of next-generation DNA sequencing where the sizes of typical primary datasets (after parsing of raw image data) are typically measured in the 10's of gigabytes (at least 10 times larger than typical metabolomics datasets), scientists have risen to the challenge by providing online storage space and developing specialised data repository systems capable of systematically archiving and effectively disseminating these data (Kaminuma et al., 2010; Cochrane et al., 2011; Leinonen et al., 2011).

Given the relatively small sizes of metabolomics datasets and the fact that metabolomics techniques predate next-generation sequencing by a considerable number of years, it is difficult to think of a satisfactory justification for the number of scientific claims that have been made on the basis of metabolomics datasets that have not, at the very least, been made freely available for download from a publicly accessible web site. That said, recent years

Feature / Tool Name:	**Metabolome Express**	**XCMS Online**	**Metabo Analyst**	**MeltDB**	**metDAT 2.0**
Reference:	(Carroll et al., 2010)	(Smith et al., 2006)	(Xia et al., 2009)	(Neuweger et al., 2008)	(Biswas et al., 2010)
Raw data processing	GC/MS	LC/MS	GC/MS, LC/MS, NMR	GC/MS, LC/MS	DI/MS
Raw data visualization	Interactive	Static		Static	Static
Peak detection	+	+	+	+	binning
Peak ID method	MSRI	accurate mass		MS	accurate mass
Peak ID ambiguity	Unambiguous	Ambiguous		Ambiguous	Ambiguous
Peak alignment	identification-based	COW	COW (XCMS)	COW (XCMS)	+
Processed data handling					
Data matrix construction	+	+	+	+	+
Normalisation	+		+	+	+
Fold change calculation	+	+	+	+	+
Univariate statistics	t-test	t-test	t-test, ANOVA	+	+
Correlation analysis	+		+	+	+
Multivariate analysis					
PCA	+		+	+	+
PLS-DA			+	+	+
Cluster analysis	+		+	+	+
Classification and feature analysis			+	+	
Biological interpretation					
Pathway mapping	In development			+	+
Phenotype recognition	+				
ORA	In development		+		

Table 2. Comparison of features of major web-based data processing pipelines for metabolomics. Only tools capable of some level of raw data processing have been included. A '+' indicates the presence of a feature. COW = Correlation Optimized Warping (the algorithm used by XCMS); MSRI = Mass Spectral and Retention Index; PCA = Principal Components Analysis; PLS-DA=Partial Least Squares-Discriminant Analysis; ORA=Over-representation Analysis; DI/MS = Direct Infusion / Mass Spectrometry

have seen a strong increase in the number of metabolomics labs sharing primary datasets from their own websites and even the emergence of centralized metabolomics data repositories allowing arbitrary labs to share their datasets publicly without even having to set up their own website. These groups that have been voluntarily driving the free and open dissemination of primary metabolomics data should be commended! The following sections will highlight the data sharing efforts that have been made by individual groups within the metabolomics community and describe the centralized metabolomics data repositories that are currently in operation and/or development.

5.2 Online databases sharing raw and/or partially-processed experimental datasets

5.2.1 DROP met: Data resources of plant metabolomics

Description: A part of the PRIMe (Platform for RIKEN Metabolomics) website.

Species: Plant species

Reference data: Provides a simple download page allowing free download of raw and/or processed LC/MS and GC/MS datasets and metadata from 8 different peer-reviewed publications emerging from the RIKEN Plant Science Center.

Noteworthy features: Metadata for each raw data file is provided in a systematic, MSI-compliant format.

Modes of access: browse

Strengths: Data are easy to find and well annotated.

Limitations: Metabolic phenotypes are not stored in a database. There is no way of querying the data without downloading, extracting biological information and importing into a local database.

URL: http://prime.psc.riken.jp/?action=drop_index

5.2.2 KomicMarket (Kazusa omics data market)

Description: A freely accessible database of annotations of metabolite peaks from FT-ICR-MS analysis of standard compounds and plant samples.

Species: Plant species

Reference data: Metabolites detected in tomato fruits by FT-ICR-MS. 215 standard compounds detected by FT-ICR-MS.

Noteworthy features: None

Modes of access: Search, browse and API.

Strengths: Good collection of high mass-accuracy flavonoid spectra. API makes download of spectra and associated annotations relatively easy.

Limitations: No bulk download of spectra needlessly makes access to spectra more challenging.

Reference: (Iijima et al., 2008)

URL: http://webs2.kazusa.or.jp/komics/index.php

5.2.3 MassBase 1.0

Description: A mass-spectral tag archive for metabolomics.

Species: Plant species

Reference data: Provides raw and processed GC/MS, LC/MS and CE/MS data for download.

Noteworthy features: None

Modes of access: Search, browse and download.

Strengths: One of very few sites to archive and disseminate raw chromatograms.

Limitations: No bulk-download of data – data files must be downloaded one at a time via the web interface. Chromatograms provided in proprietary binary formats. Limited metadata provided. Peak annotations are not provided.

URL: http://webs2.kazusa.or.jp/massbase/index.php/

5.2.4 SetupX

Description: A study design database for GC/MS metabolomics experiments.

Species: Plant species

Reference data: Provides raw and processed GC/MS data for download together with metadata.

Noteworthy features: Metabolite detections are searchable by species and species are searchable by metabolite detections.

Modes of access: Search, browse and download.

Strengths: One of very few sites to archive and disseminate raw chromatograms. Experimental datasets may be downloaded as single zipped files.

Limitations: Enormous sizes of zipped experimental dataset files means that download errors frequently occur during long downloads. No quantitative information is provided with metabolite detections and there is no way to compare the results of different experiments.

Reference: (Scholz and Fiehn, 2007)

URL: http://fiehnlab.ucdavis.edu:8080/m1/

5.2.5 PlantMetabolomics.org

Description: A database of processed, large-scale metabolic phenotype information obtained from an array of different *Arabidopsis thaliana* T-DNA insertion mutants.

Species: Arabidopsis thaliana

Reference data: Provides relative metabolite levels of a large number of metabolites in a large number of *Arabidosis thaliana* mutants.

Noteworthy features: None.

Modes of access: Search, browse and download.

Strengths: Data on a very wide range of metabolites. Incorporates phenotypic notes on mutants.

Limitations: Important metadata fields are frequently left empty. Raw data files are not provided. Origins of processed results are not transparent. There is no way to align and compare global phenotypes of mutants.

Reference: (Bais et al., 2010)

URL: http://plantmetabolomics.vrac.iastate.edu/ver2/index.php

5.2.6 Mery-B

Description: A repository for plant metabolomics datasets including experimental metadata processed data and raw data for NMR experiments.

Species: Plants.

Reference data: Provides NMR-based metabolite quantification data for a variety of tissues from a variety of species grown under a variety of conditions. Based on ~1000 spectra. Chemical shift peak assignment information is provided.

Noteworthy features: Interactive raw data viewers for 1D NMR and GC/MS data.

Modes of access: Search and browse.

Strengths: Contains data from a range of peer-reviewed publications and references to literature are clearly presented. Raw NMR spectra and GC chromatograms are available for visualisation. All experimental protocols are provided.

Limitations: Tools for statistical analysis are not yet functional. Data are not downloadable for offline analysis. Analytical reference libraries are not provided. Peak assignments are not seamlessly integrated into the raw data viewer. No direct links between statistical results and raw data vizualisation. Interface is not very intuitive.

Reference: (Ferry-Dumazet et al., 2011)

URL: http://www.cbib.u-bordeaux2.fr/MERYB/home/home.php

5.2.7 MetabolomeExpress

Description: An interactive, centralized metabolomics data repository for metabolomics data from all organisms and all analytical platforms that provides a variety of cross-study data-mining tools for analysis of metabolic phenotypes. Processed data may be uploaded in a simple tab-delimited format. Alternatively, raw GC/MS data may be uploaded and processing online using the integrated data-processing pipeline before being imported into the data repository.

Species: Not formally species-constrained but current content is plant-centric. Data from other systems is currently being gathered from the literature.

Reference data: MSRI libraries, GC/MS chromatograms, processed results, metadata in systematic formats. Database currently includes >12000 publicly available metabolite response statistics representing >100 metabolic phenotypes from 8 species under 22 different experiments in 16 different peer-reviewed publications.

Noteworthy features: Members may independently upload their own MSRI libraries for interactive dissemination and use within the GC/MS data-processing pipeline. Provides tools for cross-study meta-analysis and database-driven phenotype recognition by pattern matching.

Modes of access: browse and FTP

Strengths: All public data free for download. Provides a built-in GC/MS data processing pipeline. Allows cross-study analysis. Processed metabolite response statistics are transparently linked to underlying raw data in an interactive raw data viewer.

Limitations: No API. No search. Raw data processing pipeline needs to be extended to support analytical platforms other than GC/MS. Does not provide as many multivariate analysis and classification tools as other web-based metabolomics data-processing systems.

Reference: (Carroll et al., 2010)

URL: http://www.metabolome-express.org

6. Conclusion

The field of metabolomics informatics development is moving very rapidly. New data-processing tools and new data repositories will continue to emerge. As they do, an increasingly important area to make progress in will be in the standardization of universal data exchange formats that allow free flow of data between compliant databases. Similarly important will be the development of user-friendly metadata capture tools that make systematic annotation of their datasets as painless as possible for biologists. These developments will require the development of new ontologies and/or the extension of existing ontologies that do not cover all of the terms required to describe metabolomics experiments. The efficient sharing and mining of well-annotated and well-quality-controlled metabolomics data across the internet will undoubtedly lead to many important discoveries in the future.

7. References

Bais P, Moon SM, He K, Leitao R, Dreher K, Walk T, Sucaet Y, Barkan L, Wohlgemuth G, Roth MR, Wurtele ES, Dixon P, Fiehn O, Lange BM, Shulaev V, Sumner LW, Welti R, Nikolau BJ, Rhee SY, Dickerson JA (2010) PlantMetabolomics.org: a web portal for plant metabolomics experiments. Plant physiology 152: 1807-1816

Biswas A, Mynampati KC, Umashankar S, Reuben S, Parab G, Rao R, Kannan VS, Swarup S (2010) MetDAT: a modular and workflow-based free online pipeline for mass spectrometry data processing, analysis and interpretation. Bioinformatics 26: 2639-2640

Carroll AJ, Badger MR, Harvey Millar A (2010) The MetabolomeExpress Project: enabling web-based processing, analysis and transparent dissemination of GC/MS metabolomics datasets. BMC Bioinformatics 11: 376

Caspi R, Altman T, Dale JM, Dreher K, Fulcher CA, Gilham F, Kaipa P, Karthikeyan AS, Kothari A, Krummenacker M, Latendresse M, Mueller LA, Paley S, Popescu L, Pujar A, Shearer AG, Zhang P, Karp PD (2010) The MetaCyc database of metabolic pathways and enzymes and the BioCyc collection of pathway/genome databases. Nucleic acids research 38: D473-479

Cochrane G, Karsch-Mizrachi I, Nakamura Y (2011) The International Nucleotide Sequence Database Collaboration. Nucleic acids research 39: D15-D18

Croft D, O'Kelly G, Wu G, Haw R, Gillespie M, Matthews L, Caudy M, Garapati P, Gopinath G, Jassal B, Jupe S, Kalatskaya I, Mahajan S, May B, Ndegwa N, Schmidt E, Shamovsky V, Yung C, Birney E, Hermjakob H, D'Eustachio P, Stein L (2011) Reactome: a database of reactions, pathways and biological processes. Nucleic acids research 39: D691-697

Cui Q, Lewis IA, Hegeman AD, Anderson ME, Li J, Schulte CF, Westler WM, Eghbalnia HR, Sussman MR, Markley JL (2008) Metabolite identification via the Madison Metabolomics Consortium Database. Nature biotechnology 26: 162-164

Degtyarenko K, de Matos P, Ennis M, Hastings J, Zbinden M, McNaught A, Alcantara R, Darsow M, Guedj M, Ashburner M (2008) ChEBI: a database and ontology for chemical entities of biological interest. Nucleic acids research 36: D344-350

Ferry-Dumazet H, Gil L, Deborde C, Moing A, Bernillon S, Rolin D, Nikolski M, de Daruvar A, Jacob D (2011) MeRy-B: a web knowledgebase for the storage, visualization, analysis and annotation of plant NMR metabolomic profiles. BMC plant biology 11: 104

Hagiwara T, Saito S, Ujiie Y, Imai K, Kakuta M, Kadota K, Terada T, Sumikoshi K, Shimizu K, Nishi T (2010) HPLC Retention time prediction for metabolome analysi. Bioinformation 5: 255-258

Horai H, Arita M, Kanaya S, Nihei Y, Ikeda T, Suwa K, Ojima Y, Tanaka K, Tanaka S, Aoshima K, Oda Y, Kakazu Y, Kusano M, Tohge T, Matsuda F, Sawada Y, Hirai MY, Nakanishi H, Ikeda K, Akimoto N, Maoka T, Takahashi H, Ara T, Sakurai N, Suzuki H, Shibata D, Neumann S, Iida T, Funatsu K, Matsuura F, Soga T, Taguchi R, Saito K, Nishioka T (2010) MassBank: a public repository for sharing mass spectral data for life sciences. Journal of mass spectrometry : JMS 45: 703-714

Iijima Y, Nakamura Y, Ogata Y, Tanaka Ki, Sakurai N, Suda K, Suzuki T, Suzuki H, Okazaki K, Kitayama M, Kanaya S, Aoki K, Shibata D (2008) Metabolite annotations based on the integration of mass spectral information. The Plant Journal 54: 949-962

Kaminuma E, Mashima J, Kodama Y, Gojobori T, Ogasawara O, Okubo K, Takagi T, Nakamura Y (2010) DDBJ launches a new archive database with analytical tools for next-generation sequence data. Nucleic acids research 38: D33-D38

Kopka J, Schauer N, Krueger S, Birkemeyer C, Usadel B, Bergmuller E, Dormann P, Weckwerth W, Gibon Y, Stitt M, Willmitzer L, Fernie AR, Steinhauser D (2005) GMD@CSB.DB: the Golm Metabolome Database. Bioinformatics 21: 1635-1638

Leinonen R, Sugawara H, Shumway M (2011) The Sequence Read Archive. Nucleic acids research 39: D19-D21

Matsuda F, Hirai MY, Sasaki E, Akiyama K, Yonekura-Sakakibara K, Provart NJ, Sakurai T, Shimada Y, Saito K (2010) AtMetExpress Development: A Phytochemical Atlas of Arabidopsis Development. Plant Physiology 152: 566-578

Matsuda F, Yonekura-Sakakibara K, Niida R, Kuromori T, Shinozaki K, Saito K (2009) MS/MS spectral tag-based annotation of non-targeted profile of plant secondary metabolites. Plant J 57: 555-577

Moco S, Bino RJ, Vorst O, Verhoeven HA, de Groot J, van Beek TA, Vervoort J, de Vos CH (2006) A liquid chromatography-mass spectrometry-based metabolome database for tomato. Plant Physiology 141: 1205-1218

Neuweger H, Albaum SP, Dondrup M, Persicke M, Watt T, Niehaus K, Stoye J, Goesmann A (2008) MeltDB: a software platform for the analysis and integration of metabolomics experiment data. Bioinformatics 24: 2726-2732

Ogata H, Goto S, Sato K, Fujibuchi W, Bono H, Kanehisa M (1999) KEGG: Kyoto Encyclopedia of Genes and Genomes. Nucleic acids research 27: 29-34

Sakurai N, Ara T, Ogata Y, Sano R, Ohno T, Sugiyama K, Hiruta A, Yamazaki K, Yano K, Aoki K, Aharoni A, Hamada K, Yokoyama K, Kawamura S, Otsuka H, Tokimatsu T, Kanehisa M, Suzuki H, Saito K, Shibata D (2010) KaPPA-View4: a metabolic pathway database for representation and analysis of correlation networks of gene co-expression and metabolite co-accumulation and omics data. Nucleic acids research

Sansone SA, Fan T, Goodacre R, Griffin JL, Hardy NW, Kaddurah-Daouk R, Kristal BS, Lindon J, Mendes P, Morrison N, Nikolau B, Robertson D, Sumner LW, Taylor C, van der Werf M, van Ommen B, Fiehn O (2007) The metabolomics standards initiative. Nature biotechnology 25: 846-848

Schauer N, Steinhauser D, Strelkov S, Schomburg D, Allison G, Moritz T, Lundgren K, Roessner-Tunali U, Forbes MG, Willmitzer L, Fernie AR, Kopka J (2005) GC-MS libraries for the rapid identification of metabolites in complex biological samples. FEBS Lett 579: 1332-1337

Scholz M, Fiehn O (2007) SetupX--a public study design database for metabolomic projects. Pacific Symposium on Biocomputing. Pacific Symposium on Biocomputing: 169-180

Shinbo Y, Nakamura Y, Altaf-Ul-Amin M, Asahi H, Kurokawa K, Arita M, Saito K, Ohta D, Shibata D, Kanaya S (2006) KNApSAcK: A Comprehensive Species-Metabolite Relationship Database. *In* K Saito, RA Dixon, L Willmitzer, eds, Plant Metabolomics, Vol 57. Springer Berlin Heidelberg, pp 165-181

Skogerson K, Wohlgemuth G, Barupal DK, Fiehn O (2011) The volatile compound BinBase mass spectral database. BMC Bioinformatics 12: 321

Smith CA, O'Maille G, Want EJ, Qin C, Trauger SA, Brandon TR, Custodio DE, Abagyan R, Siuzdak G (2005) METLIN: a metabolite mass spectral database. Therapeutic drug monitoring 27: 747-751

Smith CA, Want EJ, O'Maille G, Abagyan R, Siuzdak G (2006) XCMS: processing mass spectrometry data for metabolite profiling using nonlinear peak alignment, matching, and identification. Analytical chemistry 78: 779-787

Tarasova IA, Guryca V, Pridatchenko ML, Gorshkov AV, Kieffer-Jaquinod S, Evreinov VV, Masselon CD, Gorshkov MV (2009) Standardization of retention time data for AMT tag proteomics database generation. Journal of Chromatography B 877: 433-440

Ulrich EL, Akutsu H, Doreleijers JF, Harano Y, Ioannidis YE, Lin J, Livny M, Mading S, Maziuk D, Miller Z, Nakatani E, Schulte CF, Tolmie DE, Kent Wenger R, Yao H, Markley JL (2008) BioMagResBank. Nucleic acids research 36: D402-408

Wishart DS, Tzur D, Knox C, Eisner R, Guo AC, Young N, Cheng D, Jewell K, Arndt D, Sawhney S, Fung C, Nikolai L, Lewis M, Coutouly MA, Forsythe I, Tang P, Shrivastava S, Jeroncic K, Stothard P, Amegbey G, Block D, Hau DD, Wagner J, Miniaci J, Clements M, Gebremedhin M, Guo N, Zhang Y, Duggan GE, Macinnis GD, Weljie AM, Dowlatabadi R, Bamforth F, Clive D, Greiner R, Li L, Marrie T, Sykes BD, Vogel HJ, Querengesser L (2007) HMDB: the Human Metabolome Database. Nucleic acids research 35: D521-526

Wohlgemuth G, Haldiya PK, Willighagen E, Kind T, Fiehn O (2010) The Chemical Translation Service--a web-based tool to improve standardization of metabolomic reports. Bioinformatics 26: 2647-2648

Xia J, Psychogios N, Young N, Wishart DS (2009) MetaboAnalyst: a web server for metabolomic data analysis and interpretation. Nucleic acids research 37: W652-660

4

Generic Software Frameworks for GC-MS Based Metabolomics

Nils Hoffmann and Jens Stoye
Genome Informatics, Faculty of Technology, Bielefeld University
Germany

1. Introduction

Metabolomics has seen a rapid development of new technologies, methodologies, and data analysis procedures during the past decade. The development of fast gas- and liquid-chromatography devices coupled to sensitive mass-spectrometers, supplemented by the unprecedented precision of nuclear magnetic resonance for structure elucidation of small molecules, together with the public availability of database resources associated to metabolites and metabolic pathways, has enabled researchers to approach the metabolome of organisms in a high-throughput fashion. Other "omics" technologies have a longer history in high-throughput, such as next generation sequencing for genomics, RNA microarrays for transcriptomics, and mass spectrometry methods for proteomics. All of these together give researchers a unique opportunity to study and combine multi-omics aspects, forming the discipline of "Systems Biology" in order to study organisms at multiple scales simultaneously.

Like all other "omics" technologies, metabolomics data acquisition is becoming more reliable and less costly, while at the same time throughput is increased. Modern time-of-flight (TOF) mass spectrometers are capable of acquiring full scan mass spectra at a rate of 500Hz from 50 to 750 m/z and with a mass accuracy <5 ppm with external calibration (Neumann & Böcker, 2010). At the opposite extreme of machinery, Fourier-transform ion-cyclotron-resonance (FTICR) mass spectrometers coupled to liquid chromatography for sample separation reach an unprecedented mass accuracy of <1 ppm m/z and very high mass resolution (Miura et al., 2010). These features are key requirements for successful and unique identification of metabolites. Coupled to chromatographic separation devices, these machines create datasets ranging in size from a few hundred megabytes to several gigabytes per run. While this is not a severe limitation for small scale experiments, it may pose a significant burden on projects that aim at studying the metabolome or specific metabolites of a large number of specimens and replicates, for example in medical research studies or in routine diagnostics applications tailored to the metabolome of a specific species (Wishart et al., 2009).

Thus, there is a need for sophisticated methods that can treat these datasets efficiently in terms of computational resources and which are able to extract, process, and compare the relevant information from these datasets. Many such methods have been published, however there is a high degree of fragmentation concerning the availability and accessibility of these methods, which makes it hard to integrate them into a lab's workflow.

The aim of this work is to discuss the necessary and desirable features of a software framework for metabolomics data preprocessing based on gas-chromatography (GC) and comprehensive two-dimensional gas-chromatography (GCxGC) coupled to single-dimension detectors (flame/photo ionization, FID/PID) or multi-dimension detectors (mass spectrometry, MS). We compare the features of publicly available Open Source frameworks that usually have a steep learning curve for end-users and bioinformaticians alike, owing to their inherent complexity. Many users will thus be appaled by the effort it takes to get used to a framework. Thus, the main audience of this work are bioinformaticians and users willing to invest some time in learning to use and/or program in these frameworks in order to set up a lab specific analytical platform. For a review of LC-MS based metabolomics data preprocessing consider (Castillo, Mattila, Miettinen, Orešič & Hyötyläinen, 2011).

Before we actually compare the capabilities of these different frameworks, we will first define a typical workflow for automatic data processing of metabolomics experiments and will discuss available methods within each of the workflow's steps.

We will concentrate on frameworks available under an Open Source license, thus allowing researchers to examine their actual implementation details. This distinguishes these frameworks from applications that are only provided on explicit request, under limited terms of use, or that are not published together with their source code (Lommen, 2009; Stein, 1999), which is still often the case in metabolomics and may hamper comparability and reuse of existing solutions. Additionally, all frameworks compared in this work are available for all major operating systems such as Microsoft Windows, Linux, and Apple Mac OSx as standalone applications or libraries.

Web-based methods are not compared within this work as they most often require a complex infrastructure to be set up and maintained. However, we will give a short overview of recent publications on this topic and provide short links to the parts of the metabolomics pipeline that we discuss in the following section. A survey of web-based methods is provided by Tohge & Fernie (2009). More recent web-based applications for metabolomics include the retention time alignment methods Warp2D (Ahmad et al., 2011) and ChromA (Hoffmann & Stoye, 2009), which are applicable to GC-MS or LC-MS data, and Chromaligner (Wang et al., 2010), which aligns GC and LC data with single-dimension detectors like FID.

Tools for statistical analysis of multiple sample groups and with different phenotypes have been reported by Kastenmüller et al. (2011). However, other tools aim to integrate a more complete metabolomics workflow including preprocessing, peakfinding, alignment and statistical analysis combined with pathway mapping information like MetaboAnalyst (Xia & Wishart, 2011), MetabolomeExpress (Carroll et al., 2010), or MeltDB (Neuweger et al., 2008). These larger web-based frameworks integrate other functionality for time-course analysis (Xia et al., 2011), pathway mapping (Neuweger et al., 2009; Xia & Wishart, 2010a) and metabolite set enrichment analysis (Kankainen et al., 2011; Xia & Wishart, 2010b).

In the Application section, we will exemplarily describe two pipelines for metabolomics analyses based on our own Open Source framework Maltcms: ChromA, which is applicable to GC-MS, and ChromA4D, which is applicable to data from comprehensive GCxGC-MS experiments. We show how to set up, configure and execute each pipeline using instructional datasets. These two workflows include the typical steps of raw-data preprocessing in metabolomics, including peak-finding and integration, peak-matching among multiple replicate groups and tentative identification using mass-spectral databases, as well as visualizations of raw and processed data. We will describe the individual steps of the workflows of the two application pipelines to give the reader a thorough understanding of the methods used by ChromA and ChromA4D.

Finally, we discuss the current state of the presented Open Source frameworks and give an outlook into the future of software frameworks and data standards for metabolomics.

2. A typical workflow for a metabolomics experiment

Metabolomics can be defined as the study of the metabolic state of an organism or its response to direct or indirect perturbation. In order to find differences between two or more states, for example before treatment with a drug and after, and among one or multiple specimens, the actual hypothesis for the experiment needs to be defined. Based on this hypothesis, a design for the structure of the experiments and their subsequent analysis can be derived. This involves, among many necessary biological or medical considerations, the choice of sample extraction procedures and preparation methods, as well as the choice of the analytical methods used for downstream sample analysis.

Preprocessing of the data from those experiments begins after the samples have been acquired using the chosen analytical method, such as GC-MS or LC-MS. Owing to the increasing amount of data produced by high-throughput metabolomics experiments, with large sample numbers and high-accuracy/high-speed analytical devices, it is a key requirement that the resulting data is processed with very high level of automation. It is then that the following typical workflow is applied in some variation, as illustrated in Figure 1.

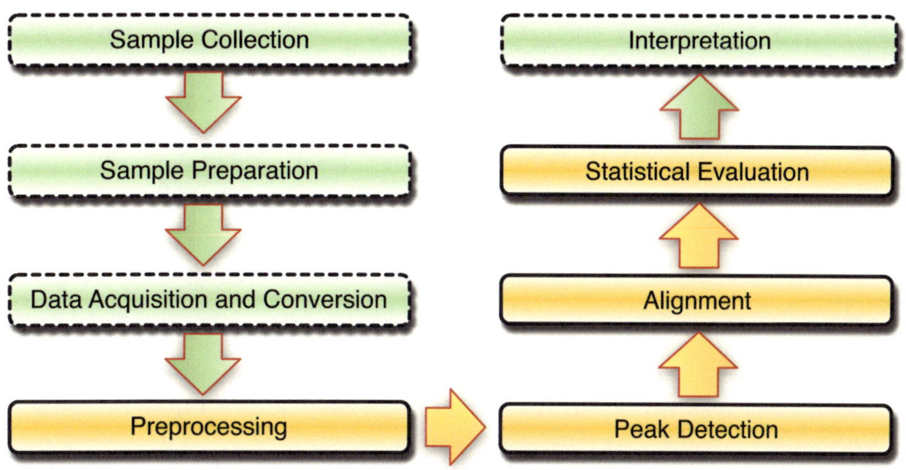

Fig. 1. A typical workflow for a metabolomics experiment. Steps shown in orange (solid border) are usually handled within the bioinformatics domain, while the steps shown in green (dashed border) often involve co-work with scientists from other disciplines.

2.1 Data acquisition and conversion

The most common formats exported from GC-MS and LC-MS machines today are NetCDF (Rew & Davis, 1990), based on the specifications in the ASTM/AIA standard ANDI-MS (Matthews, 2000), mzXML (Oliver et al., 2004), mzData (Orchard et al., 2005), and more recently as the successor to the latter two, mzML (Deutsch, 2008; Martens et al., 2010). All of these formats include well-defined data structures for meta-information necessary to interpret data in the right context, such as detector type, chromatographic protocol, detector potential and other details about the separation and acquisition of the data. Furthermore, they explicitly model chromatograms and mass spectra, with varying degrees of detail.

NetCDF is the oldest and probably most widely used format today. It is routinely exported even by older machinery, which offers backwards compatibility to those. It is a general-purpose binary format, with a header that describes the structure of the data contained in the file, grouped into variables and indexed by dimensions. In recent years, efforts were made to establish open formats for data exchange based on a defined grammar in extensible markup language (*XML*) with extendable controlled vocabularies, to allow new technologies to be easily incorporated into the file format without breaking backwards compatibility. Additionally, XML formats are human readable which narrows the technology gap. mzXML was the first approach to establish such a format. It has been superseded by mzData and, more recently, mzML was designed as a super-set of both, incorporating extensibility through the use of an indexed controlled vocabulary. This allows mzML to be adapted to technologies like GCxGC-MS without having to change its definition, although its origins are in the proteomics domain. One drawback of XML-based formats is often claimed to be their considerably larger space requirements when compared to the supposedly more compact binary data representations. Recent advances in mzML approach this issue by compressing spectral data using gzip compression.

The data is continuously stored in a vendor-dependent native format during sample processing on a GC-MS machine. Along with the mass spectral information, like ion mass (or equivalents) and abundance, the acquisition time of each mass spectrum is recorded. Usually, the vendor software includes methods for data conversion into one of the aforementioned formats. However, especially when a high degreee of automation is desired, it may be beneficial to directly access the data in their native format. This avoids the need to run the vendor's proprietary software manually for every data conversion task. Both the ProteoWizard framework (Kessner et al., 2008) and the Trans Proteomic Pipeline (Deutsch et al., 2010) include multiple vendor-specific libraries for that use case.

2.2 Preprocessing

Raw mass specrometry data is usually represented in sparse formats, only recording those masses whose intensities exceed a user-defined threshold. This thresholding is usually applied within the vendor's proprietary software and may lead to artificial *gaps* within the data. Thus, the first step in preprocessing involves the binning of mass spectra over time into bins of defined size in the m/z dimension, followed by interpolation of missing values. After binning, the data is stored as a rectangular array of values, with the first dimension representing time, the second dimension representing the approximate bin mass values, and the third dimension representing the intensity corresponding to each measured ion. This process is also often described as resampling (Lange et al., 2007).

Depending on various instrumental parameters, the raw exported data may require additional processing. The most commonly reported methods for smoothing are the Savitzky-Golay filter (Savitzky & Golay, 1964), LOESS regression (Smith et al., 2006) and variants of local averaging, for example by a windowed moving average filter. These methods can also be applied to interpolate values where gaps are present in the original data. The top-hat filter (Bertsch et al., 2008; Lange et al., 2007) is used to remove a varying baseline from the signal. More refined methods use signal decomposition and reconstruction methods, such as Fourier transform and continuous wavelet transform (CWT) (Du et al., 2006; Fredriksson et al., 2009; Tautenhahn et al., 2008) in order to remove noise and baseline contributions from the signal and simultaneously find peaks.

2.3 Peak detection

Often the process of peak detection is decoupled from the actual preprocessing of the data. XCMS (Smith et al., 2006), for example, uses a Gaussian second derivative peak model with a fixed kernel width and signal-to-noise threshold to find peaks along the chromatographic domain of each ion bin. Other methods extend this approach to use a multi-scale continuous wavelet transform using such a kernel over various widths, tracking the response of the transformed signal in order to locate peak apex positions in scale-space before estimating the true peak widths based on the kernel scale with maximum response (Fredriksson et al., 2009; Tautenhahn et al., 2008). However, these methods usually allow only a small number of co-eluting peaks in different mass-bins, since they were initially designed to work with LC-MS data mainly, where only one parent ion and a limited number of accompanying adduct ions are expected. In GC-MS, electron-ionization creates rich fragmentation mass spectra, which pose additional challenges to deconvolution of co-eluting ions and subsequent association to peak groups. Even though its source code is not publicly available, the method used by AMDIS (Stein, 1999) has seen wide practical application and is well accepted as a reference by the metabolomics and analytical chemistry communities.

2.4 Alignment

The alignment problem in metabolomics and proteomics stems from the analytical methods used. These produce sampled sensor readings acquired over time in fixed or programmed intervals, usually called chromatograms. The sensor readings can be one- or multidimensional. In the first case, detectors like ultra violet and visible light absorbance detectors (UV/VIS) or flame ionization detectors (FID) measure the signal response as one-dimensional features, e.g. as the absorbance spectrum or electrical potential, respectively. Multi-dimensional detectors like mass spectrometers record a large number of features simultaneously, e.g. mass and ion count. The task is then to find corresponding and non-corresponding features between different sample acquisitions. This *correspondence problem* is a term used by Åberg et al. (2009) which describes the actual purpose of alignment, namely to find *true* correspondences between related analytical signals over a number of sample acquisitions. For GC-MS- and LC-MS-based data, a number of different methods have been developed, some of which are described in more detail by Castillo, Gopalacharyulu, Yetukuri & Orešič (2011) and Åberg et al. (2009). Here, we will concentrate on those methods that have been reported to be applicable to GC-MS. In principle, alignment algorithms can be classified into two main categories: peak- and signal-based methods. Methods of the first type start with a defined set of peaks, which are present in most or all samples that are to be aligned before determining the best correspondences of the peaks between samples in order to then derive a time correction function. Krebs et al. (2006) locate *landmark* peaks in the TIC and then select pairs of those peaks with a high correlation between their mass spectra in order to fit an interpolating spline between a reference chromatogram and the to-be-aligned one. The method of Robinson et al. (2007) is inspired by multiple sequence alignment algorithms and uses dynamic programming to progressively align peak lists without requiring an explicit reference chromatogram. Other methods, like that of Chae et al. (2008) perform piecewise, block-oriented matching of peaks, either on the TIC, on selected masses, or on the complete mass spectra. Time correction is applied after the peak assignments between the reference chromatogram and the others have been calculated. Signal-based methods include recent variants of correlation optimized warping (Smilde & Horvatovich, 2008), parametric time warping (Christin et al., 2010) and dynamic time warping (Christin et al., 2010; Clifford et al., 2009; Hoffmann & Stoye, 2009; Prince & Marcotte, 2006) and usually consider the complete chromatogram for comparison. However, attempts are made to reduce the computational burden associated with a complete pairwise comparison of mass spectra by partitioning the chromatograms into similar regions (Hoffmann & Stoye, 2009), or by selecting a representative subset of mass traces (Christin et al., 2010). Another distinction in alignment algorithms is the requirement of an explicit reference for alignment. Some methods apply clustering techniques to select one chromatogram that is most similar to all others (Hoffmann & Stoye, 2009; Smilde & Horvatovich, 2008), while other methods choose such a reference based on the number of features contained in a chromatogram (Lange et al., 2007) or by manual user choice (Chae et al., 2008; Clifford et al., 2009). For high-throughput applications, alignments should be fast to calculate and reference selection should be automatic. Thus, a sampling method for time correction has recently been reported by Pluskal et al. (2010) for LC-MS. A comparison of these methods is given in the same publication.

2.5 Statistical evaluation

After peaks have been located and integrated for all samples, and their correspondence has been established, peak report tables can be generated, containing peak information for each sample and peak, with associated corrected retention times and peak areas. Additionally, peaks may have been putatively identified by searching against a database, such as the GMD (Hummel et al., 2007) or the NIST mass-spectral database (Babushok et al., 2007).

These peak tables can then be analyzed with further methods, in order to detect e.g. systematic differences between different sample groups. Prior to such an analysis, the peak areas need to be normalized. This is usually done by using a spiked-in compound which is not expected to occur naturally as a reference. The normalization compound is supposed to have the same concentration in all samples. The compound's peak area can then be used to normalize all peak areas of a sample with respect to it (Doebbe et al., 2010).

Different experimental designs allow to analyze correlations of metabolite levels for the same subjects under different conditions (paired), or within and between groups of subjects. For simple paired settings, multiple t-tests with corrections for multiple testing can be applied (Berk et al., 2011), while for comparisons between groups of subjects, Fisher's F-Statistic (Pierce et al., 2006) and various analysis of variance (ANOVA), principal component analysis (PCA) and partial least squares (PLS) methods are applied (Kastenmüller et al., 2011; Wiklund et al., 2008; Xia et al., 2011).

2.6 Evaluation of hypothesis

Finally, after peak areas have been normalized and differences have been found between sample groups, the actual results need to be put into context and be interpreted in their biological context. This task is usually not handled by the frameworks described in this chapter. Many web-based analysis tools allow to put the data into a larger context, by providing name- or id-based mapping of the experimentally determined metabolite concentrations onto biochemical pathways like MetaboAnalyst (Xia & Wishart, 2011), MetabolomeExpress (Carroll et al., 2010), or MeltDB (Neuweger et al., 2008). The latter allows association of the metabolomics data with other results for the same subjects under study or with results from other "omics" experiments on the same target subjects, but this is beyond the scope of the frameworks presented herein.

3. Frameworks for GC-MS analysis

A number of Open Source frameworks have been developed for LC-MS based proteomics frameworks like OpenMS (Bertsch et al., 2008), ProteoWizard (Kessner et al., 2008), and most notably the TransProteomicPipeline (Deutsch et al., 2010). Even though many of the steps required for proteomics apply similarily to metabolomics applications, there are still some essential differences due to the different analytical setups and technologies (e.g. matrix assisted laser desorption ionization mass spectrometry, MALDI-MS) used in the two fields. XCMS (Smith et al., 2006) was among the first frameworks to offer support for data preprocessing in LC-MS based metabolomics. Later, MZmine2 (Pluskal et al., 2010) offered an alternative with a user-friendly interface and easy extendability. Lately, Scheltema et al. (2011) published their PeakML format and mzMatch framework also for LC-MS applications. As of now, there seem to be only a few frameworks available for GC-MS based metabolomics that offer similar methods, namely PyMS (Callaghan et al., 2010; Isaac et al., 2009) and Maltcms/ChromA (Hoffmann & Stoye, 2009; *Maltcms*, 2011) . These will be presented in more detail in this section. A compact overview of the Open Source frameworks discussed herein is given in Table 1. A detailed feature comparison can be found in Table 2.

3.1 XCMS

XCMS (Smith et al., 2006) is a very mature framework and has seen constant development during the last five years. It is mainly designed for LC-MS applications, however its binning, peak finding and alignment are also applicable to GC-MS data. XCMS is implemented in the *GNU R* programming language, the de-facto standard for Open Source statistics. Since *GNU R* is an interpreted scripting language, it is easy to write custom scripts that realize additional functionality of the typical GC-MS workflow described above. XCMS is part of the Bioconductor package collection, which offers many computational methods for various "omics" technologies. Further statistical methods are available from *GNU R*.

XCMS supports input in NetCDF, mzXML, mzData and, more recently, mzML format. This allows XCMS to be used with virtually any chromatography-mass spectrometry data, since vendor software supports conversion to at least one of those formats. XCMS uses the *xcmsRaw* object as its primary tabular data structure for each binned data file. The *xcmsSet* object is then used to represent peaks and peak groups and is used by its peak alignment and *diffreport* features.

The peak finding methods in XCMS are quite different from each other. For data with normal or low mass resolution and accuracy, the matched filter peak finder (Smith et al., 2006) is usually sensitive enough. It uses a Gaussian peak template function with user defined width and signal-to-noise critera to locate peaks on individual binned extracted ion current

(EIC) traces over the complete time range of the binned chromatogram. The other method, CentWave (Tautenhahn et al., 2008) is based on a continuous wavelet transform on areas of interest within the raw data matrix. Both peak finding methods report peak boundaries and integrated areas for raw data and for the data reconstructed from the peak finder's signal response values.

Initially designed for LC-MS, XCMS does not have a method to group co-eluting peaks into peak groups, as is a requirement in GC-MS methods using electron ionization. However, CAMERA (Tautenhahn et al., 2007) shows how XCMS can be used as a basis in order to create a derived application, in this case for ion annotation between samples.

Peak alignment in XCMS is performed using local LOESS regression between peak groups with very similar m/z and retention time behaviour and good support within each sample group. This allows a simultaneous alignment and retention time correction of all peaks. The other available method is based on the Obi-Warp dynamic time warping (Prince & Marcotte, 2006) algorithm and is capable of correcting large non-linear retention time distortions. It uses the peak set with the highest number of features as alignment reference, which is comparable to the approach used by Lange et al. (2007). However, it is much more computationally demanding then the LOESS-based alignment.

XCMS's *diffreport* generates a summary report of significant analyte differences between two sample sets. It uses Welch's two-sample t-statistic to calculate p-values for each analyte group. ANOVA may be used for more than two sample sets.

A number of different visualizations are also available, both for raw and processed data. These include TIC plots, EIC plots, analyte group plots for grouped features, and chromatogram (rt, m/z, intensity) surface plots.

XCMS can use GNU R's Rmpi infrastructure to execute arbitary function calls, such as profile generation and peak finding, in parallel on a local cluster of computers.

3.2 PyMS

PyMS (Callaghan et al., 2010; Isaac et al., 2009) is a programming framework for GC-MS metabolomics based on the *Python* programming language. It can therefore use a large number of scientific libraries which are accessible via the SciPy and NumPy packages (*SciPy*, 2011). Since Python is a scripting language, it allows to do rapid prototyping, comparable to GNU R. However, Python's syntax may be more familiar for programmers with a background in object-oriented programming languages.

The downloadable version of PyMS currently only supports NetCDF among the more recent open data exchange formats. Nonetheless, it is the only framework in this comparison with support for the JCAMP GC-MS file format.

PyMS provides dedicated data structures for chromatograms, allowing efficient access to EICs, mass spectra, and peak data.

In order to find peaks, PyMS also builds a rectangular profile matrix with the dimensions time, m/z and intensity. Through the use of slightly shifted binning boundaries, they reduce the chance of false assignments of ion signals to neighboring bins, when binning is performed with unit precision (bin width of 1 m/z). PyMS offers the moving average and the Savitzky-Golay (Savitzky & Golay, 1964) filters for signal smoothing of EICs within the profile matrix. Baseline correction can be performed by the top-hat filter (Lange et al., 2007). The actual peak finding is based on the method described by Biller & Biemann (1974) and involves the matching of local peak maxima co-eluting within a defined window. Peaks are integrated for all co-eluting masses, starting from a peak apex to both sides and ending if the increase in area falls below a given threshold.

Peak alignment in PyMS is realized by the method introduced by Robinson et al. (2007). It is related to progressive multiple sequence alignment methods and is based on a generic dynamic programming algorithm for peak lists. It proceeds by first aligning peak lists within sample groups, before aligning the aligned peak lists of different groups, until all groups have been aligned.

Visualizations of chromatogram TICs, EICs, peaks and mass spectra are available and are displayed to the user in an interactive plot panel.

For high-throughput applications, PyMS can be used together with MPI to parallelize tasks within a local cluster of computers.

3.3 Maltcms

The framework *Maltcms* allows to set up and configure individual processing components for various types of computational analyses of metabolomics data. The framework is implemented in *JAVA* and is modular using the service provider pattern for maximal decoupling of interface and implementation, so that it can be extended in functionality at runtime.

Maltcms can read data from files in NetCDF, mzXML, mzData or mzML format. It uses a pipeline paradigm to model the typical preprocessing workflow in metabolomics, where each processing step can define dependencies on previous steps. This allows automatic pipeline validation and ensures that a user can not define an invalid pipeline. The workflow itself is serialized to XML format, keeping track of all resources created during pipeline execution. Using a custom post-processor, users can define which results of the pipeline should be archived.

Maltcms uses a generalization of the ANDI-MS data schema internally and a data provider interface with corresponding implementations to perform the mapping from any proprietary data format to an internal data object model. This allows efficient access to individual mass spectra and other data available in the raw-data files. Additionally, developers need no special knowledge of any supported file format, since all data can be accessed generically. Results from previous processing steps are referenced in the data model to allow both shadowing of data, e.g. creating a processing result variable with the same name as an already existing variable, and aggregation of processing results. Thus, all previous processing results are transparently accessible for downstream elements of a processing pipeline, unless they have been shadowed.

Primary storage of processing results is performed on a per-chromatogram basis in the binary NetCDF file format. Since metabolomics experiments create large amounts of data, a focus is put on efficient data structures, data access, and scalability of the framework.

Embedding Maltcms in existing workflows or interfacing with other software is also possible, as alignments, peak-lists and other feature data can be exported as comma separated value files or in specific xml-based formats, which are well-defined by custom schemas.

To exploit the potential of modern multi-core CPUs and distributed computing networks, Maltcms supports multi-threaded execution on a local machine or within a grid of connected computers using an OpenGrid infrastructure (e.g. Oracle Grid Engine or Globus Toolkit (Foster, 2005)) or a manually connected network of machines via remote method invocation (RMI).

The framework is accompanied by many libraries for different purposes, such as the *JFreeChart* library for 2D-plotting or, for BLAS compatible linear algebra, math and statistics implementations, the *Colt* and *commons-math* libraries. Building upon the base library *Cross*, which defines the commonly available interfaces and default implementations, Maltcms provides the domain dependent data structures and specializations for processing of chromatographic data.

Name	Version	Analytical method	Software license	Programming language
XCMS	1.26.1[a]	LC-MS/GC-MS	GNU GPL v2	*GNU R* 2.13/C++
PyMS	r371	GC-MS	GNU GPL v2	*Python* 2.5
Maltcms/ChromA	1.1	GC-MS	GNU L-GPL v3	*JAVA* 6

Table 1. Overview of available Open Source software frameworks for GC-MS based metabolomics. a: Part of Bioconductor 2.8

Feature (GC-MS pipeline)	XCMS	PyMS	ChromA
Data formats	A, B, C, D	A, E	A, B, C, D
Signal preprocessing	MM	SG, TH	MA, MM, TH
Peak detection	MF, CWT	BB	MAX
Multiple peak alignment	LOESS, DTW	PROGDP	DTW, CLIQUE
Visualization	TIC, EIC, SURF	TIC, EIC	TIC, EIC, SURF
DB search	no (LC-MS only)	no	MSP
Normalization	no	no	RP, EV
Statistical evaluation	TT	no	FT

Table 2. Feature comparison of Open Source software frameworks for preprocessing of GC-MS based metabolomics data. Keys to abbreviations: **Data formats** A: NetCDF, B: mzXML, C: mzData, D: mzML, E: JCAMP GC-MS. **Signal preprocessing** MM: moving median, SG: Savitzky-Golay filter, TH: top-hat filter, MA: moving average. **Peak detection** MF: matched Gaussian filter, CWT: continuous wavelet transform, BB: Biller-Biemann, MAX: TIC local maxima. **Multiple peak alignment** LOESS: LOESS regression, DTW: dynamic time warping, PROGDP: progressive using dynamic programming, CLIQUE: progressive clique-based. **Visualization** (of unaligned and aligned data) TIC: plots of total ion chromatogram/peaks, EIC: plots of extracted ion chromatograms/peaks, SURF: surface plots of profile matrix (rt x m/z x I). **DB search** MSP: msp-format, compatible with AMDIS and GMD format. **Normalization** RP: reference peak area, EV: external value, e.g. dry weight. **Statistical evaluation** TT: groupwise t-test, multiple testing correction, FT: F-test, between group vs. within group variance

3.3.1 ChromA

ChromA is a configuration of Maltcms that includes preprocessing, in the form of mass binning, time-scale alignment and annotation of signal peaks found within the data, as well as visualizations of unaligned and aligned data from GC-MS and LC-MS experiments. The user may supply mandatory alignment anchors as CSV files to the pipeline and a database location for tentative metabolite identification. Further downstream processing can be performed either on the retention time-corrected chromatograms in NetCDF format, or on the corresponding peak tables in either CSV format or XML format.

Peaks can either be imported from other tools, by providing them in CSV format to ChromA, giving at least the scan index of each peak in a file per row. Alternatively, ChromA has a fast peak finder that locates peaks based on derivatives of the smoothed and baseline-corrected TIC, using a moving average filter followed by top-hat filter baseline-substraction, with a predefined minimum peak-width. Peak alignment is based on a star-wise or tree-based application of an enhanced variant of pairwise dynamic time warping (DTW) (Hoffmann & Stoye, 2009). To reduce both runtime and space requirements, conserved signals throughout the data are identified, constraining the search space of DTW to a precomputed closed polygon. The alignment anchors can be augmented or overwritten by user-defined anchors, such as previously identified compounds, characteristic mass or MS/MS identifications. Then, the candidates are paired by means of a bidirectional best-hits (BBH) criterion, which can compare different aspects of the candidates for similarity. Paired anchors are extended to k-cliques with configurable k, which help to determine the conservation or absence of signals across measurements, especially with respect to replicate groups. Tentative identification of peaks against a database using their mass spectra is possible using the MetaboliteDB module. This module provides access to mass-spectral databases in msp-compatible format, for example the Golm Metabolite Database or the NIST EI-MS database.

ChromA visualizes alignment results including paired anchors in birds-eye view or as a simultaneous overlay plot of the TIC. Additionally, absolute and relative differential charts are provided, which allow easy spotting of quantitative differences.

Peak tables are exported in CSV format, including peak apex positions, area under curve, peak intensity and possibly tentative database identifications. Additionally, information about the matched and aligned peak groups is saved in CSV format.

4. Frameworks for GCxGC-MS analysis

The automatic and routine analysis of comprehensive GCxGC-MS data is yet to be established. GCxGC-MS couples a second chromatographic column to the first one, thereby achieving a much higher peak capacity and thus a better separation of closely co-eluting analytes (Castillo, Mattila, Miettinen, Orešič & Hyötyläinen, 2011). Usually, for a one-hour run, the raw data file size exceeds a few Gigabytes. Quite a number of algorithms have been published on alignment of peaks in such four-dimensional (first column retention time, second column retention time, mass, and intensity values) data (Kim et al., 2011; Oh et al., 2008; Pierce et al., 2005; Vial et al., 2009; Zhang, 2010), however only a few methods are available for a more complete typical preprocessing workflow. A compact overview of the available frameworks, their licenses and programming languages is given in Table 3. Table 4 gives a more detailed feature matrix of these frameworks. The remainder of this section gives a concise overview of the frameworks Guineu (Castillo, Mattila, Miettinen, Orešič & Hyötyläinen, 2011) and ChromA4D (*Maltcms*, 2011).

Name	Version	Supported methods	Software license	Programming language
Guineu	0.8.2	GCxGC-MS (LC-MS)	GNU GPL v2	*JAVA* 6
Maltcms/ChromA4D	1.1	GCxGC-MS	GNU L-GPL v3	*JAVA* 6

Table 3. Feature comparison of Open Source software frameworks for GCxGC-MS based metabolomics

4.1 Guineu

Guineu is a recently published graphical user interface and application for the comparative analysis of GCxGC-MS data (Castillo, Mattila, Miettinen, Orešič & Hyötyläinen, 2011). It currently reads LECO ChromaTOF software's peak list output after smoothing, baseline correction, peak finding, deconvolution, database search and retention index (RI) calculation have been performed within ChromaTOF.

The peak lists are aligned pairwise using the score alignment algorithm, which requires user-defined retention time windows for both separation dimensions. Additionally, the one-dimensional retention index (RI) of each peak is used within the score calculation. Finally,

Feature (GCxGC-MS pipeline)	Guineu	ChromA4D
Data formats	G	A,H
Signal preprocessing	no	MA, MM, TH, CV
Peak detection	no	MAX-SRG
Multiple peak alignment	SCORE	CLIQUE
Visualization	STATS	STATS, TIC, EIC, TIC2D
DB search	GMD, PUBCHEM, KEGG	MSP (GMD)
Normalization	RP	RP, EV
Statistical evaluation	CV, FLT, TT, PCA, CDA, SP, ANOVA	FT

Table 4. Feature comparison of Open Source software frameworks for preprocessing of GCxGC-MS based metabolomics data. Key to abbreviations: **Data formats** A: NetCDF, G: ChromaTOF peak lists, H: CSV peak lists. **Signal preprocessing** MA: moving average, MM: moving median, TH: top-hat filter, CV: coefficient of variation threshold. **Peak detection** MAX-SRG: TIC local maxima, seeded region growing based on ms similarity. **Multiple peak alignment** SCORE: parallel iterative score-based, CLIQUE: progressive clique-based.**Visualization** (of unaligned and aligned data) TIC: plots of total ion chromatogram/peaks, EIC: plots of extracted ion chromatograms/peaks, SURF: surface plots of profile matrix (rt x m/z x I), STATS: visualization of statistical values. **DB search** GMD: Golm metabolite database webservice, PUBCHEM: pubchem database webservice, KEGG: kegg metabolite database, MSP: msp-format, compatible with AMDIS and GMD format. **Normalization** RP: reference peak area, EV: external value, e.g. dry weight. **Statistical evaluation** CV: coefficient of variation, FLT: fold-test, TT: groupwise t-test, PCA: principal components analysis, CDA: curvilinear distance analysis, SP: Sammon's projection, ANOVA: analysis of variance, FT: F-test, between group vs. within group variance.

a threshold for mass spectral similarity is needed in order to create *putative* peak groups. Additional peak lists are added incrementally to an already aligned *path*, based on the individual peaks' score against those peaks that are already contained within the path.

Guineu provides different filters to remove peaks by name, group occurrence count, or other features from the ChromaTOF peak table. In order to identify compound classes, the Golm metabolite database (GMD) substructure search is used. Peak areas can be extracted from ChromaTOF using the TIC, or using extracted, informative or unique masses. Peak area normalization is available relative to multiple user-defined standard compounds.

After peak list processing, Guineu produces an output table containing information for all aligned peaks, containing information on the original analyte annotation as given by ChromaTOF, peak areas, average retention times in both dimensions together with the average RI and further chemical information on the functional group and substructure prediction as given by the GMD. It is also possible to link the peak data to KEGG and Pubchem via the CAS annotation, if it is available for the reported analyte.

For statistical analysis of the peak data, Guineu provides fold change- and t-tests, principal component analysis (PCA), analysis of variance (ANOVA) and other methods.

Guineu's statistical analysis methods provide different plots of the data sets, e.g. for showing the principal components of variation within the data sets after analysis with PCA.

4.2 ChromA4D

For the comparison of comprehensive two-dimensional gas chromatography-mass spectrometry (GCxGC-MS) data, ChromA4D accepts NetCDF files as input. Additionally, the user needs to provide the total runtime on the second orthogonal column (modulation time) to calculate the second retention dimension information from the raw data files. For tentative metabolite identification, the location of a database can be given by the user. ChromA4D reports the located peaks, their respective integrated TIC areas, their best matching corresponding peaks in other chromatograms, as well as a tentative identification for each peak. Furthermore, all peaks are exported together with their mass spectra to MSP format, which allows for downstream processing and re-analysis with AMDIS and other tools. The exported MSP files may be used to define a custom database of reference spectra for subsequent analyses.

Peak areas are found by a modified seeded region growing algorithm. All local maxima of the TIC representation that exceed a threshold are selected as initial seeds. Then, the peak area is determined by using the distance of the seed mass spectrum to all neighbor mass spectra as a measure of the peak's coherence. The area is extended until the distance exceeds a given threshold. No information about the expected peak shape is needed. The peak integration is based on the sum of TICs of the peak area. An identification of the area's average or apex mass spectrum or the seed mass spectrum is again possible using the MetaboliteDB module.

To represent the similarities and differences between different chromatograms, bidirectional best hits are used to find co-occurring peaks. These are located by using a distance that exponentially penalizes differences in the first and second retention times of the peaks to be compared. To avoid a full computation of all pairs of peaks, only those peaks within a defined window of retention times based on the standard deviation of the exponential time penalty function are evaluated.

ChromA4D's visualizations represent aligned chromatograms as color overlay images, similar to those used in differential proteomics. This allows a direct visual comparison of signals present in one sample, but not present in another sample.

ChromA4D creates peak report tables in CSV format, which include peak apex positions in both chromatographic dimensions, area under curve, peak intensity and possibly tentative database identifications. Additionally, information about the matched and aligned peak groups is saved in CSV format.

5. Application examples

The following examples for GC-MS and GCxGC-MS are based on the Maltcms framework, using the ChromA and ChromA4D configurations described in the previous sections. In order to run them, the recent version of *Maltcms* needs to be downloaded and unzipped to a local folder on a computer. Additionally, Maltcms requires a *JAVA* runtime environment version 6 or newer to be installed. If these requirements are met, one needs to start a command prompt and change to the folder containing the unzipped Maltcms.

5.1 An example workflow for GC-MS

The experiment used to illustrate an example workflow for one-dimensional GC-MS consists of two samples of standard compounds, which contain mainly sugars, amino acids, other organic acids and nucleosides, measured after manual (MD) and after automatic derivatization (AD) with the derivatization protocol and substances given below. Group AD consists of a sample of n-alkanes standard and two replicates of mix1, namely mix1-1 and mix1-2. We will show how ChromA can be used to find and integrate peaks, as well as compare and align the peaks between the samples, and finally how the alignment results can be used for quality control.

5.1.1 Sample preparation

20 μL of each sample were incubated with 60 μL methoxylamine hydrochloride (Sigma Aldrich) in pyridine (20 mg/ml) for 90 min at 60°C before 100 μL of N-Methyl-N-(trimethylsilyl)-trifluoroacetamide (MSTFA) (Macherey & Nagel) were added for 60 min at 37°C.

5.1.2 Acquisition and data processing

The samples were acquired on an Agilent GC 7890N with MSD 5975C triple axis detector. An Agilent HP5ms column with a length of 30 m, a diameter of 0.25 mm, and a film thickness of 0.25 μm (Agilent, Santa Clara CA, USA) was used for the gas-chromatographic separation, followed by a deactivated restriction capillary with 50 cm length and a diameter of 0.18 mm. Per sample, 1 μL was injected onto the column in pulsed splitless mode (30 psi for 2 min). The flow rate was set to 1.5 mL/min of Helium. The linear temperature ramp started at 50 °C for 2 min until it reached its maximum of 325 °C at a rate of 10 °C/min. The raw data were exported to NetCDF format using the Agilent ChemStation software v.B.04.01 (Agilent, Santa Clara CA, USA) with default parameters and without additional preprocessing applied.

A sample containing n-alkanes was measured as an external standard for manual (MD) and automatic derivatization (AD) in order to be able to later determine retention indices for

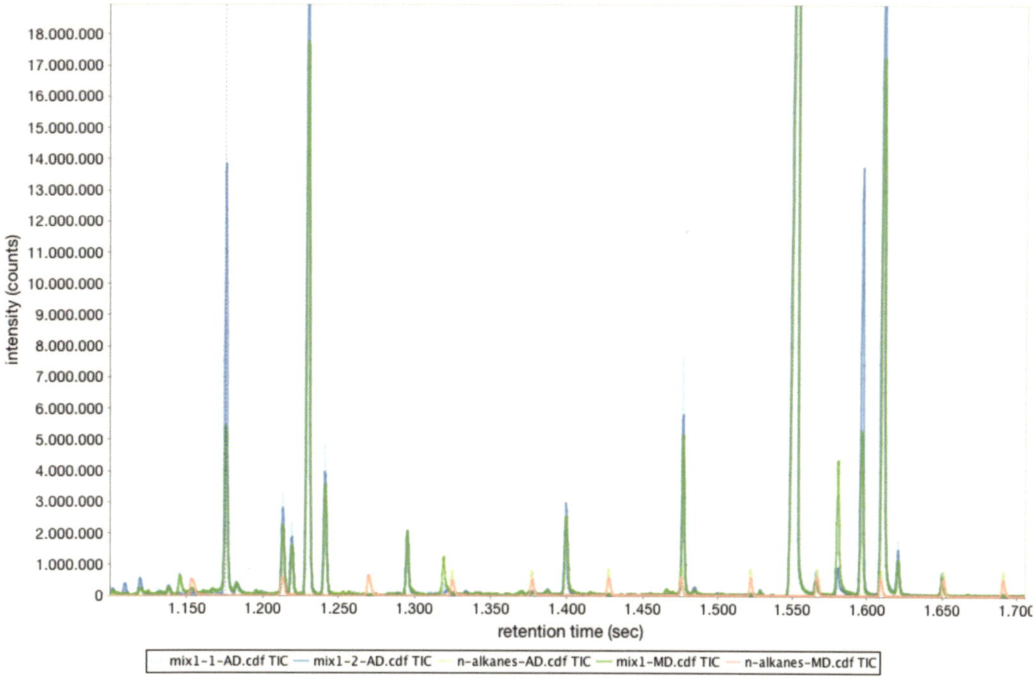

(a) Overlay of unaligned data sets, extracted from middle section within a time range of 1100 to 1700 seconds.

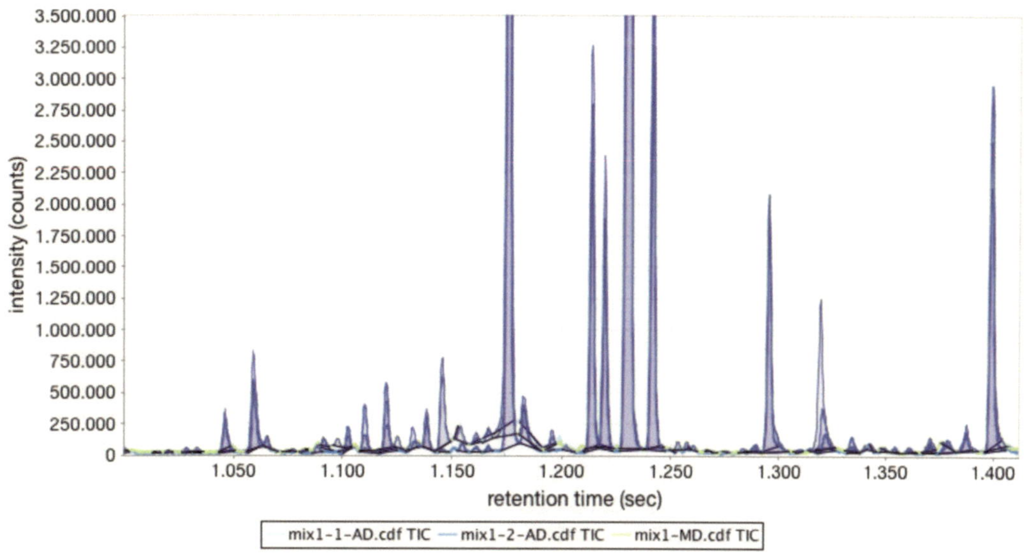

(b) Overlay with highlighted peak areas (without n-alkanes) after peak finding and integration. Zoomed in to provide more detail.

Fig. 2. TIC overlay plots of the raw GC-MS data sets.

the other samples. The acquired data were exported to ANDI-MS (NetCDF) format before ChromA was applied. The default ChromA pipeline `chroma.properties` was run from the unzipped Maltcms directory with the following command (issued on a single line of input):

```
> java -Xmx1G -jar maltcms.jar -i ../data/ -o ../output/ -f *.CDF  \
 -c cfg/chroma.properties
```

`-i` points to the directory containing the input data, `-o` points to the directory where output should be placed, `-f` can be a comma separated list of filenames or, as in this case, a wildcard expression, matching all files in the input directory having a file name ending with `.CDF`. The final argument indicated by `-c` is the path to the configuration file used for definition of the pipeline and its commands. An overlay of the raw TICs of the samples is depicted in Figure 2(a). The default ChromA pipeline configuration creates a profile matrix with nominal mass bin width. Then, the TIC peaks are located separately within each sample data file and are integrated (Figure 2(b)). The peak apex mass spectra are then used in the next step in order to build a multiple peak alignment between all peaks of all samples by finding large cliques, or clusters of peaks exhibiting similar retention time behaviour and having highly similar mass spectra. This coarse alignment could already be used to calculate a polynomial fit, correcting retention time shift for all peaks. However, the ChromA pipeline uses the peak clusters in order to constrain a dynamic time warping (DTW) alignment in the next step, which is calculated between all pairs of samples. The resulting distances are used to determine the reference sample with the lowest sum of distances to all remaining samples. Those are then aligned to the reference using the warp map obtained from the pairwise DTW calculations. The pairwise DTW distances can easily be used for a hierarchical cluster analysis. Similar samples should be grouped into the same cluster, while dissimilar samples should be grouped into different clusters. Figure 3 shows the results of applying a complete linkage clustering algorithm provided by *GNU R* to the pairwise distance matrix. It is clearly visible that the samples are grouped correctly, without incorporation of any external group assignment. Thus, this method can be used for quality control of multiple sample acquisitions, when the clustering results are compared against a pre-defined number of sample groups.

5.2 An example workflow for GCxGC-MS

The instructional samples presented in this section were preprocessed according to the protocol given by Doebbe et al. (2010). The description of the protocol has been adapted from that reference where necessary.

5.2.1 Sample preparation

The samples were incubated with 100 μl methoxylamine hydrochloride (Sigma Aldrich) in pyridine (20 mg/ml) for 90 min at 37°C while stirring. N-Methyl-N-(trimethylsilyl)-trifluoroacetamide (MSTFA) (Macherey & Nagel) was then added and incubated for another 30 min at 37°C with constant stirring.

5.2.2 Acquisition and data processing

The sample acquisition was performed on a LECO Pegasus 4D TOF-MS (LECO, St. Joseph, MI, USA). The Pegasus 4D system was equipped with an Agilent 6890 gas chromatograph

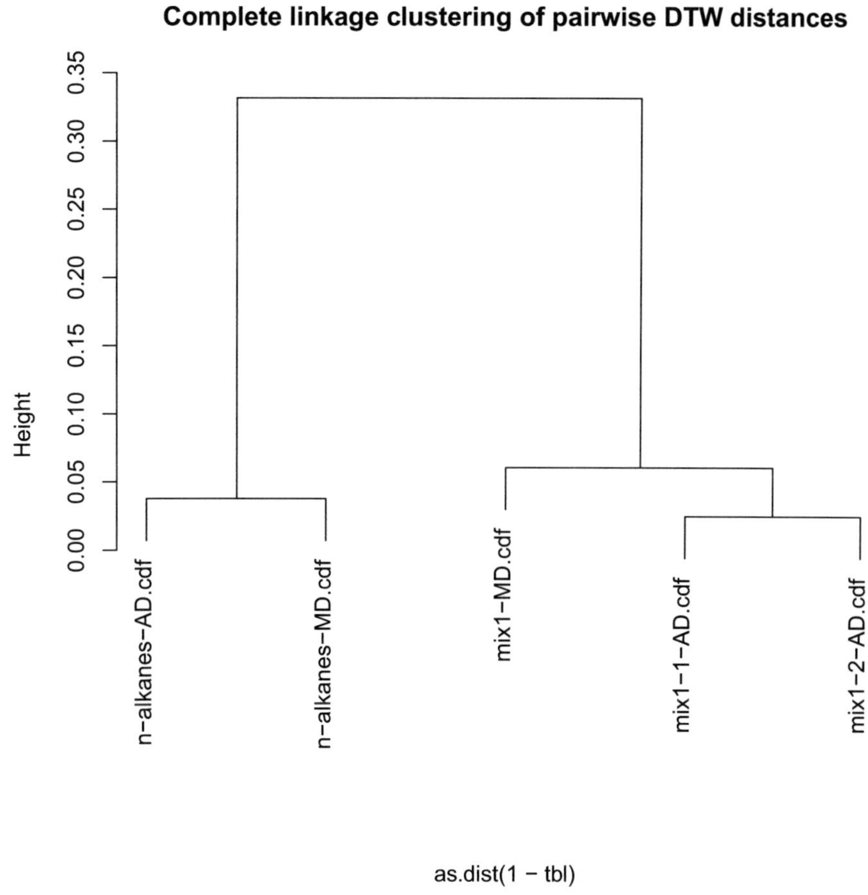

Fig. 3. Clustering of GC-MS samples based on pairwise DTW similarities transformed to distances. The samples are clearly separated into two clusters, one containing the n-alkane standard samples, the other one containing the mix1 samples.

(Agilent, Santa Clara, CA, USA). The inlet temperature was set to 275°C. An Rtx-5ms (Restek, Bellefonte, PA, USA) capillary column was used with a length of 30 m, 0.25 mm diameter and 0.25 μm film thickness as the primary column. The secondary column was a BPX-50 (SGE, Ringwood, Victoria, Australia) capillary column with a length of 2 m, a diameter of 0.1 mm and 0.1 μm film thickness. The temperature program of the primary oven was set to the following conditions: 70°C for 2 min, 4°C/min to 180°C, 2°C/min to 230°C, 4°C/min to 325°C hold 3 min. This program resulted in a total runtime of about 70 min for each sample. The secondary oven was programmed with an offset of 15°C to the primary oven temperature. The thermal modulator was set 30°C relative to the primary oven and to a modulation time of 5 seconds with a hot pulse time of 0.4 seconds. The mass spectrometer ion source temperature was set to 200°C and the ionization was performed at -70eV. The detector voltage was set to 1600V and the stored mass range was 50-750 m/z with an acquisition rate of 200 spectra/second.

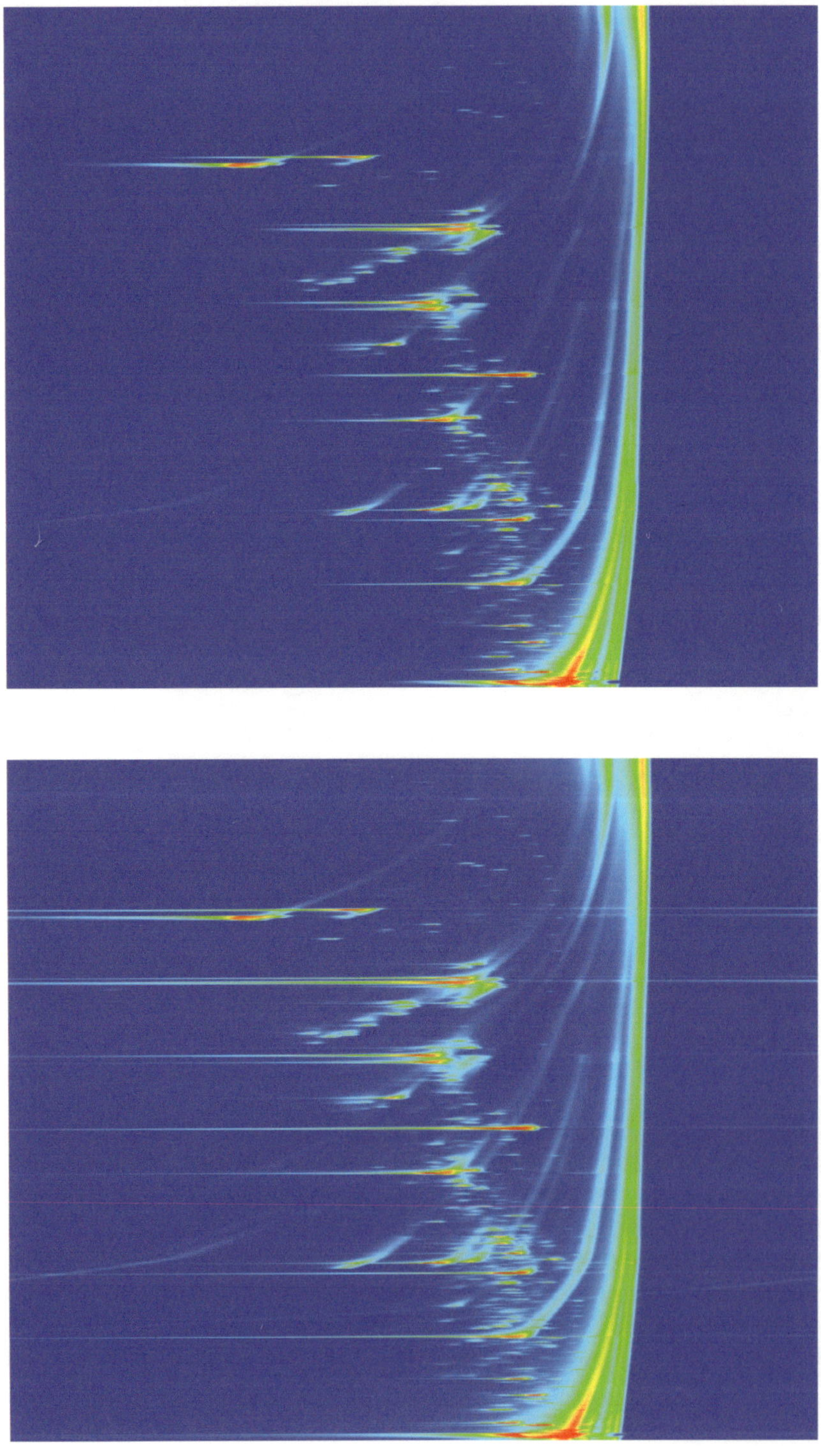

(a) 2D-TIC plot before filters were applied. Long tailing peaks are visible within the vertical dimension. Additionally, high frequency noise is present in the raw exported data, which is barely visible at this resolution.

(b) 2D-TIC plot after application of a moving median filter with window size 3 for smoothing of high-frequency noise and successive application of a top hat filter with a window size of 301 for baseline removal in order to reduce false positive peak finding results.

Fig. 4. Visualizations of Standard-Mix1-1 before and after signal filtering with the ChromA4D processing pipeline.

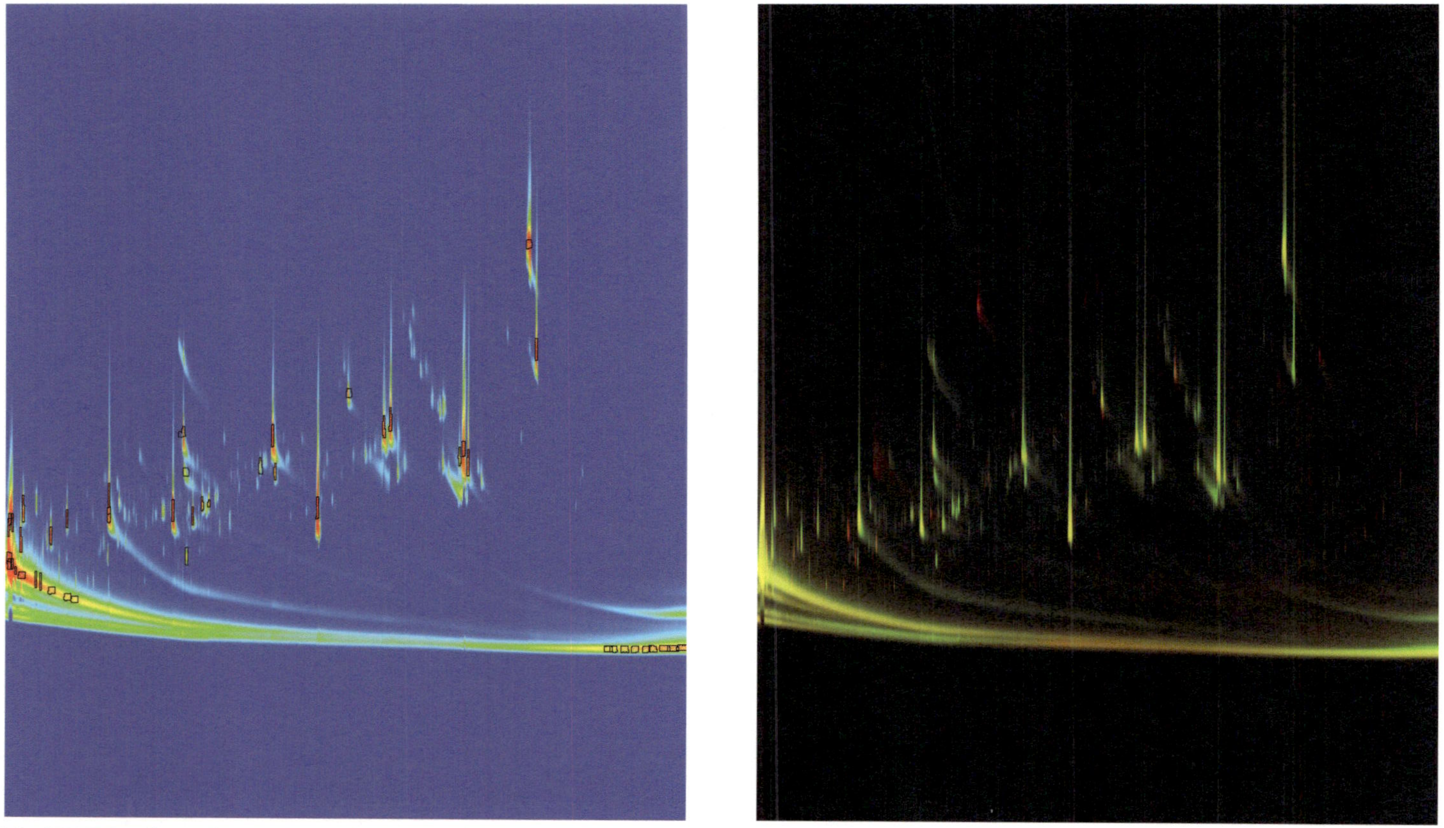

(a) 2D-TIC plot of Standard-Mix1-1 after peak finding and integration with seeded region growing based on the cosine mass spectral similarity with a fusion threshold of 0.99. Peak areas were limited to contain at most 100 points.

(b) Differential plot of the two Standard-Mix1 samples after DTW alignment based on vertical TIC slices. Yellow color indicates similar amounts of total ion intensity in both samples. Green shows a surplus in Standard-Mix1-1, while red shows a surplus in Standard-Mix1-2.

Fig. 5. Visualizations of Standard-Mix1-1 after peak finding and of Standard-Mix1-1 and Standard-Mix1-2 after alignment with DTW.

The raw acquired samples in LECO's proprietary ELU format were exported to NetCDF format using the LECO ChromaTOF® software v.4.22 (LECO, St. Joseph, MI, USA). Initial attempts to export the full, raw data failed with a crash beyond a NetCDF file size of 4GBytes. Thus, we resampled the data with ChromaTOF to 100 Hz (resampling factor 2) and exported with automatic signal smoothing and baseline offset correction value of 1 which resulted in file sizes around 3GBytes per sample. The samples presented in this section are named "Standard-Mix1-1" and "Standard-Mix1-2" and were measured on different days (Nov. 29th, 2008 and Dec. 12th, 2008).

The default ChromA4D pipeline for peak finding was called from within the unzipped Maltcms directory (issued on a single line of input):

```
> java -Xmx2G -jar maltcms.jar -i ../data/ -o ../output/ \
 -f *.cdf -c cfg/4Dpeakfinding.properties
```

The pipeline first preprocesses the data by applying a median filter followed by a top hat filter in order to remove high- and low-frequency noise contributions (Figures 4(a) and 4(b)). ChromA4D then uses a variant of seeded region growing in order to extend peak seeds, which are found as local maxima of the 2D-TIC. These initial seeds are then extended until the mass spectral similarity of the seed and the next evaluated candidate drops below a user-defined threshold, or until the peak area reaches its maximum, pre-defined size (Figure 5(a)). After peak area integration, the pipeline clusters peaks between samples based on their mass spectral similarity and retention time behaviour in both dimensions to form peak cliques (not shown) as multiple peak alignments, which are then exported into CSV format for further downstream processing. Another possible application shown in Figure 5(b) is the visualization of pairwise GCxGC-MS alignments using DTW on the vertical 2D-TIC slices, which can be useful for qualitative comparisons.

6. Summary and outlook

The present state of Open Source frameworks for metabolomics is very diverse. A number of tools have seen steady development and improvement over the last years, such as XCMS, MZmine, and PyMS, while others are still being developed, such as mzMatch, Guineu, and Maltcms. There is currently no framework available that covers every aspect of metabolomics data preprocessing. Most of the frameworks concentrate on one or a few analytical technologies with the largest distinction being between GC-MS and LC-MS. GCxGC-MS raw data processing is currently only handled by Maltcms' ChromA4D pipeline, while Guineu processes peak lists exported from LECO's ChromaTOF software and offers statistical methods for sample comparison together with a user-friendly graphical interface.

We showed two instructive examples on setting up and running the basic processing pipelines ChromA and ChromA4D for GC-MS and GCxGC-MS raw data. The general structure of these pipelines would be slightly different for each of the Open Source frameworks presented in this chapter, however, the basic concepts behind the processing steps are the same for all tools. Since metabolomics is an evolving field of research, no framework captures all possible use-cases, but it will be interesting to see which frameworks will be flexible and extendable enough to be adapted to new requirements in the near future.

In order to combine experiments from multiple "omics" experiments, another level of abstraction on top of local or web-service based tools for data processing, fusion, and integration of metabolomics experiments is a necessary future requirement. Generic workflow systems like Taverna (Hull et al., 2006) or Conveyor (Linke et al., 2011) offer integration of such resources, augmented with graphical editors for *point-and-click* user interaction. However, due to their generic nature these systems are far away from being as user-friendly as applications designed for a specific data analysis task and require some expert knowledge when assembling task-specific processing graphs.

One point that requires further attention is the definition and controlled evolution of peak data formats for metabolomics, along with other formats for easier exchange of secondary data between applications and frameworks. A first step in this direction has been taken by Scheltema et al. (2011) by defining the PeakML format. However, it is important that such formats are curated and evolved, possibly by a larger non-profit organization like the HUPO within its proteomics standards initiative *HUPO PSI*. Primary data is already acessible in a variety of different, defined formats, the most recent addition being mzML (Martens et al., 2010) which is curated by the PSI. Such standardization attempts can however only be successful and gain the required momentum if also the manufacturers of analytical machinery support the formats with their proprietary software within a short time frame after the specification and see a benefit in offering such functionality due to the expressed demand of scientists working in the field as in case of NetCDF, mzData, or mzML.

7. Acknowledgements

We would like to thank Manuela Meyer at the Center for Biotechnology (CeBiTec), Bielefeld University, Germany, for the preparation of the GC-MS samples used in this work. We furthermore thank Denise Schöfbeck and Rainer Schumacher, Center for Analytical Chemistry, IFA Tulln, University of Natural Resources and Life Sciences, Vienna, Austria, for measuring those samples and for kindly providing us with the datasets. Furthermore, we would like to express our gratitude to Anja Döbbe and Olaf Kruse, Algae BioTech Group at the CeBiTec, for providing the GCxGC-MS samples and datasets. We finally thank our students Mathias Wilhelm for his work on ChromA4D and Kai-Bernd Stadermann for his work on the RMI-based remote execution framework.

8. References

Åberg, K., Alm, E. & Torgrip, R. (2009). The correspondence problem for metabonomics datasets, *Analytical and Bioanalytical Chemistry* 394(1): 151–162.

Ahmad, I., Suits, F., Hoekman, B., Swertz, M. A., Byelas, H., Dijkstra, M., Hooft, R., Katsubo, D., van Breukelen, B., Bischoff, R. & Horvatovich, P. (2011). A high-throughput processing service for retention time alignment of complex proteomics and metabolomics LC-MS data, *Bioinformatics* 27(8): 1176–1178.

Babushok, V. I., Linstrom, P. J., Reed, J. J., Zenkevich, I. G., Brown, R. L., Mallard, W. G. & Stein, S. E. (2007). Development of a database of gas chromatographic retention properties of organic compounds., *Journal of Chromatography A* 1157(1-2): 414–421.

Berk, M., Ebbels, T. & Montana, G. (2011). A statistical framework for biomarker discovery in metabolomic time course data, *Bioinformatics* 27(14): 1979–1985.

Bertsch, A., Hildebrandt, A., Hussong, R. & Zerck, A. (2008). OpenMS - An open-source software framework for mass spectrometry., *BMC Bioinformatics* 9(1): 163.

Biller, J. E. & Biemann, K. (1974). Reconstructed Mass Spectra, A Novel Approach for the Utilization of Gas Chromatograph—Mass Spectrometer Data, *Analytical Letters* 7(7): 515–528.

Callaghan, S., De Souza, D., Tull, D., Roessner, U., Bacic, A., McConville, M. & Likić, V. (2010). Application and comparative study of PyMS Python toolkit for processing of gas chromatography-mass spectrometry (GC-MS) data, *2nd Australasian Symposium on Metabolomics*, Melbourne 2010.

Carroll, A. J., Badger, M. R. & Harvey Millar, A. (2010). The MetabolomeExpress Project: enabling web-based processing, analysis and transparent dissemination of GC/MS metabolomics datasets, *BMC Bioinformatics* 11(1): 376.

Castillo, S., Gopalacharyulu, P., Yetukuri, L. & Orešič, M. (2011). Algorithms and tools for the preprocessing of LC–MS metabolomics data, *Chemometrics and Intelligent Laboratory Systems* 108(1): 23–32.

Castillo, S., Mattila, I., Miettinen, J., Orešič, M. & Hyötyläinen, T. (2011). Data Analysis Tool for Comprehensive Two-Dimensional Gas Chromatography/Time-of-Flight Mass Spectrometry, *Analytical Chemistry* 83(8): 3058–3067.

Chae, M., Reis, R. & Thaden, J. J. (2008). An iterative block-shifting approach to retention time alignment that preserves the shape and area of gas chromatography-mass spectrometry peaks, *BMC Bioinformatics* 9(Suppl 9): S15.

Christin, C., Hoefsloot, H. C. J., Smilde, A. K., Suits, F., Bischoff, R. & Horvatovich, P. L. (2010). Time Alignment Algorithms Based on Selected Mass Traces for Complex LC-MS Data, *Journal of Proteome Research* 9(3): 1483–1495.

Clifford, D., Stone, G., Montoliu, I., Rezzi, S., Martin, F.-P., Guy, P., Bruce, S. & Kochhar, S. (2009). Alignment Using Variable Penalty Dynamic Time Warping, *Analytical Chemistry* 81(3): 1000–1007.

Deutsch, E. (2008). mzML: a single, unifying data format for mass spectrometer output., *Proteomics* 8(14): 2776–2777.

Deutsch, E. W., Mendoza, L., Shteynberg, D., Farrah, T., Lam, H., Tasman, N., Sun, Z., Nilsson, E., Pratt, B., Prazen, B., Eng, J. K., Martin, D. B., Nesvizhskii, A. I. & Aebersold, R. (2010). A guided tour of the Trans-Proteomic Pipeline, *Proteomics* 10(6): 1150–1159.

Doebbe, A., Keck, M., Russa, M. L., Mussgnug, J. H., Hankamer, B., Tekce, E., Niehaus, K. & Kruse, O. (2010). The Interplay of Proton, Electron, and Metabolite Supply for Photosynthetic H2 Production in Chlamydomonas reinhardtii, *Journal of Biological Chemistry* 285(39): 30247–30260.

Du, P., Kibbe, W. A. & Lin, S. M. (2006). Improved peak detection in mass spectrum by incorporating continuous wavelet transform-based pattern matching, *Bioinformatics* 22(17): 2059–2065.

Foster, I. T. (2005). Globus Toolkit Version 4: Software for Service-Oriented Systems., *in* H. Jin, D. A. Reed & W. Jiang (eds), *IFIP International Conference on Network and Parallel Computing*, Springer, pp. 2–13.

Fredriksson, M. J., Petersson, P., Axelsson, B.-O. & Bylund, D. (2009). An automatic peak finding method for LC-MS data using Gaussian second derivative filtering., *Journal of Separation Science* 32(22): 3906–3918.

GNU R (2011).
URL: *http://www.r-project.org/*

Hoffmann, N. & Stoye, J. (2009). ChromA: signal-based retention time alignment for chromatography-mass spectrometry data, *Bioinformatics* 25(16): 2080–2081.

Hull, D., Wolstencroft, K., Stevens, R., Goble, C., Pocock, M. R., Li, P. & Oinn, T. (2006). Taverna: a tool for building and running workflows of services, *Nucleic Acids Research* 34(suppl 2): W729–W732.

Hummel, J., Selbig, J., Walther, D. & Kopka, J. (2007). The Golm Metabolome Database: a database for GC-MS based metabolite profiling, *in* J. Nielsen & M. Jewett (eds), *Metabolomics*, Springer Berlin / Heidelberg, pp. 75–95.

HUPO PSI (2011).
URL: *http://www.psidev.info/*

Isaac, A., Lee, L., Keen, W., Erwin, T., Wang, Q., De Souza, D., Roessner, U., Pyke, J., Kotagiri, R., Wettenhall, R., McConville, M., Bacic, A. & Likić, V. (2009). PyMS: A Python toolkit for processing of gas chromatography-mass spectrometry data, *Bioinformatics Australia Conference*, Melbourne 2009.

JAVA (2011).
URL: *http://www.java.com/en/*

Kankainen, M., Gopalacharyulu, P., Holm, L. & Orešič, M. (2011). MPEA–metabolite pathway enrichment analysis, *Bioinformatics* 27(13): 1878–1879.

Kastenmüller, G., Römisch-Margl, W., Wägele, B., Altmaier, E. & Suhre, K. (2011). metaP-Server: A Web-Based Metabolomics Data Analysis Tool, *Journal of Biomedicine and Biotechnology* 2011: 1–8.

Kessner, D., Chambers, M., Burke, R., Agus, D. & Mallick, P. (2008). ProteoWizard: open source software for rapid proteomics tools development., *Bioinformatics* 24(21): 2534–2536.

Kim, S., Fang, A., Wang, B., Jeong, J. & Zhang, X. (2011). An Optimal Peak Alignment For Comprehensive Two-Dimensional Gas Chromatography Mass Spectrometry Using Mixture Similarity Measure, *Bioinformatics* 27(12): 1660–1666.

Krebs, M. D., Tingley, R. D., Zeskind, J. E., Holmboe, M. E., Kang, J.-M. & Davis, C. E. (2006). Alignment of gas chromatography-mass spectrometry data by landmark selection from complex chemical mixtures, *Chemometrics and Intelligent Laboratory Systems* 81(1): 74–81.

Lange, E., Gropl, C., Schulz-Trieglaff, O., Huber, C. & Reinert, K. (2007). A geometric approach for the alignment of liquid chromatography mass spectrometry data, *Bioinformatics* 23(13): i273–i281.

Linke, B., Giegerich, R. & Goesmann, A. (2011). Conveyor: a workflow engine for bioinformatic analyses, *Bioinformatics* 27(7): 903–911.

Lommen, A. (2009). MetAlign: Interface-Driven, Versatile Metabolomics Tool for Hyphenated Full-Scan Mass Spectrometry Data Preprocessing, *Analytical Chemistry* 81(8): 3079–3086.

Maltcms (2011).
URL: *http://maltcms.sourceforge.net*

Martens, L., Chambers, M., Sturm, M., Kessner, D., Levander, F., Shofstahl, J., Tang, W. H., Rompp, A., Neumann, S., Pizarro, A. D., Montecchi-Palazzi, L., Tasman, N., Coleman, M., Reisinger, F., Souda, P., Hermjakob, H., Binz, P. A. & Deutsch, E. W. (2010). mzML–a Community Standard for Mass Spectrometry Data, *Molecular and Cellular Proteomics* 10(1): R110.000133–R110.000133.

Matthews, L. (2000). ASTM Protocols for Analytical Data Interchange, 5(5): 60–61.

Miura, D., Tsuji, Y., Takahashi, K., Wariishi, H. & Saito, K. (2010). A strategy for the determination of the elemental composition by fourier transform ion cyclotron resonance mass spectrometry based on isotopic peak ratios., *Technical Report 13*, Innovation Center for Medical Redox Navigation, Kyushu University, 3-1-1 Maidashi, Higashi-ku, Fukuoka 12-8582, Japan.

Neumann, S. & Böcker, S. (2010). Computational mass spectrometry for metabolomics: Identification of metabolites and small molecules, *Analytical and Bioanalytical Chemistry* 398(7-8): 2779–2788.

Neuweger, H., Albaum, S. P., Niehaus, K., Stoye, J. & Goesmann, A. (2008). MeltDB: a software platform for the analysis and integration of metabolomics experiment data, *Bioinformatics* 24(23): 2726–2732.

Neuweger, H., Persicke, M., Albaum, S. P., Bekel, T., Dondrup, M., Hüser, A. T., Winnebald, J., Schneider, J., Kalinowski, J. & Goesmann, A. (2009). Visualizing post genomics data-sets on customized pathway maps by ProMeTra-aeration-dependent gene expression and metabolism of Corynebacterium glutamicum as an example., *BMC Systems Biology* 3: 82.

Oh, C., Huang, X., Regnier, F. E., Buck, C. & Zhang, X. (2008). Comprehensive two-dimensional gas chromatography/time-of-flight mass spectrometry peak sorting algorithm, *Journal of Chromatography A* 1179(2): 205–215.

Oliver, S. G., Paton, N. W. & Taylor, C. F. (2004). A common open representation of mass spectrometry data and its application to proteomics research, *Nature Biotechnology* 22(11): 1459–1466. 10.1038/nbt1031.

Orchard, S., Hermjakob, H., Taylor, C. F., Potthast, F., Jones, P., Zhu, W., Julian, R. K. & Apweiler, R. (2005). Second proteomics standards initiative spring workshop., *Expert review of proteomics*, EMBL Outstation - European Bioinformatics Institute, Wellcome Trust Genome Campus, Hinxton, Cambridge CB10 1SD, UK. pp. 287–289.

Pierce, K. M., Hoggard, J. C., Hope, J. L., Rainey, P. M., Hoofnagle, A. N., Jack, R. M., Wright, B. W. & Synovec, R. E. (2006). Fisher Ratio Method Applied to Third-Order Separation Data To Identify Significant Chemical Components of Metabolite Extracts, *Analytical Chemistry* 78(14): 5068–5075.

Pierce, K. M., Wood, L. F., Wright, B. W. & Synovec, R. E. (2005). A comprehensive two-dimensional retention time alignment algorithm to enhance chemometric analysis of comprehensive two-dimensional separation data, *Analytical Chemistry* 77(23): 7735–7743.

Pluskal, T., Castillo, S., Villar-Briones, A. & Orešič, M. (2010). MZmine 2: Modular framework for processing, visualizing, and analyzing mass spectrometry-based molecular profile data, *BMC Bioinformatics* 11(1): 395.

Prince, J. & Marcotte, E. (2006). Chromatographic alignment of ESI-LC-MS proteomics data sets by ordered bijective interpolated warping, *Analytical Chemistry* 78(17): 6140–6152.

Python (2008).
URL: *http://www.python.org/download/releases/2.5.2/*

Rew, R. & Davis, G. (1990). NetCDF: an interface for scientific data access, *Computer Graphics and Applications, IEEE* 10(4): 76–82.

Robinson, M. D., De Souza, D. P., Saunders, E. C., Mcconville, M. J., Speed, T. P. & Likić, V. A. (2007). A dynamic programming approach for the alignment of signal peaks in multiple gas chromatography-mass spectrometry experiments, *BMC Bioinformatics* 8(1): 419.

Savitzky, A. & Golay, M. J. E. (1964). Smoothing and Differentiation of Data by Simplified Least Squares Procedures., *Analytical Chemistry* 36(8): 1627–1639.

Scheltema, R. A., Jankevics, A., Jansen, R. C., Swertz, M. A. & Breitling, R. (2011). PeakML/mzMatch: A File Format, Java Library, R Library, and Tool-Chain for Mass Spectrometry Data Analysis, *Analytical Chemistry* 83(7): 2786–2793.

SciPy (2011).
URL: *http://www.scipy.org/*

Smilde, A. K. & Horvatovich, P. L. (2008). Optimized time alignment algorithm for LC-MS data: correlation optimized warping using component detection algorithm-selected mass chromatograms, *Analytical Chemistry* 80(18): 7012–7021.

Smith, C. A., Want, E. J., O'Maille, G., Abagyan, R. & Siuzdak, G. (2006). XCMS: Processing Mass Spectrometry Data for Metabolite Profiling Using Nonlinear Peak Alignment, Matching, and Identification, *Analytical Chemistry* 78(3): 779–787.

Stein, S. (1999). An integrated method for spectrum extraction and compound identification from gas chromatography/mass spectrometry data, *Journal of the American Society for Mass Spectrometry* 10(8): 770–781.

Tautenhahn, R., Böttcher, C. & Neumann, S. (2007). Annotation of LC/ESI-MS Mass Signals, *in* S. Hochreiter & R. Wagner (eds), *Bioinformatics Research and Development*, Springer Berlin / Heidelberg, pp. 371–380. 10.1007/978-3-540-71233-6_29.

Tautenhahn, R., Böttcher, C. & Neumann, S. (2008). Highly sensitive feature detection for high resolution LC/MS, *BMC Bioinformatics* 9: 504.

Tohge, T. & Fernie, A. R. (2009). Web-based resources for mass-spectrometry-based metabolomics: A user's guide, *Phytochemistry* 70(4): 450–456.

Vial, J., Noçairi, H., Sassiat, P., Mallipatu, S., Cognon, G., Thiébaut, D., Teillet, B. & Rutledge, D. N. (2009). Combination of dynamic time warping and multivariate analysis for the comparison of comprehensive two-dimensional gas chromatograms: application to plant extracts, *Journal of Chromatography A* 1216(14): 2866–2872.

Wang, S. Y., Ho, T. J., Kuo, C. H. & Tseng, Y. J. (2010). Chromaligner: a web server for chromatogram alignment, *Bioinformatics* 26(18): 2338–2339.

Wiklund, S., Johansson, E., Sjöström, L., Mellerowicz, E. J., Edlund, U., Shockcor, J. P., Gottfries, J., Moritz, T. & Trygg, J. (2008). Visualization of GC/TOF-MS-based metabolomics data for identification of biochemically interesting compounds using OPLS class models, *Analytical Chemistry* 80(1): 115–122.

Wishart, D. S., Knox, C., Guo, A. C., Eisner, R., Young, N., Gautam, B., Hau, D. D., Psychogios, N., Dong, E., Bouatra, S., Mandal, R., Sinelnikov, I., Xia, J., Jia, L., Cruz, J. A., Lim, E., Sobsey, C. A., Shrivastava, S., Huang, P., Liu, P., Fang, L., Peng, J., Fradette, R., Cheng, D., Tzur, D., Clements, M., Lewis, A., De Souza, A., Zuniga, A., Dawe, M., Xiong, Y., Clive, D., Greiner, R., Nazyrova, A., Shaykhutdinov, R., Li, L., Vogel, H. J. & Forsythe, I. (2009). HMDB: a knowledgebase for the human metabolome, *Nucleic Acids Research* 37(Database): D603–D610.

Xia, J., Sinelnikov, I. V. & Wishart, D. S. (2011). MetATT: a web-based metabolomics tool for analyzing time-series and two-factor data sets, *Bioinformatics* 27(17): 2455–2456.

Xia, J. & Wishart, D. S. (2010a). MetPA: a web-based metabolomics tool for pathway analysis and visualization, *Bioinformatics* 26(18): 2342–2344.

Xia, J. & Wishart, D. S. (2010b). MSEA: a web-based tool to identify biologically meaningful patterns in quantitative metabolomic data, *Nucleic Acids Research* 38(suppl 2): W71–W77.

Xia, J. & Wishart, D. S. (2011). Web-based inference of biological patterns, functions and pathways from metabolomic data using MetaboAnalyst, *Nature Protocols* 6(6): 743–760.

XML (2008).
URL: *http://www.w3.org/TR/REC-xml/*

Zhang, X. (2010). DISCO: distance and spectrum correlation optimization alignment for two-dimensional gas chromatography time-of-flight mass spectrometry-based metabolomics, *Analytical Chemistry* 82(12): 5069–5081.

5

Computational Methods to Interpret and Integrate Metabolomic Data

Feng Li[1], Jiangxin Wang[2], Lei Nie[3] and Weiwen Zhang[2]
[1]*Department of Mathematics and Statistics, University of Maryland, Baltimore County, Baltimore, MD,*
[2]*School of Chemical Engineering and Technology, Tianjin University, Tianjin,*
[3]*Division of Biostatistics, Department of Epidemiology and Preventive Medicine, University of Maryland Baltimore, Baltimore, MD,*
[2]*P.R. China*
[1,3]*U.S.A.*

1. Introduction

Revolutionary improvements in high-throughput DNA sequencing technologies have made it possible to measure gene, mRNA, proteins and metabolites, as well as their interaction at global level. In the past decades, significant efforts in improving analytical technologies pertaining to measuring mRNA, proteins and metabolites have been made. These efforts have led to the generation of several new 'omics' research fields: transcriptomics, proteomics, metabolomics, interactomics and so on (Singh & Nagaraj, 2006; Fiehn 2007; Lin & Qian, 2007; Kandpal et al., 2009; Ishii & Tomita, 2009). Among them, metabolomics is an approach to obtain a comprehensive evaluation of metabolites in cells. Compared with transcriptomics and proteomics approaches, metabolomics can achieve large-scale quantitative and qualitative measurements of cellular metabolites, which can thus generate a high-resolution biochemical and functional information of an organism.

Due to the chemical complexity of cellular metabolites, it is generally accepted that no single analytical technique can provide a comprehensive visualization of all metabolites, so multiple technologies are generally employed (Dunn & Ellis, 2005; Villas-Boas Silas et al., 2005; Hollywood et al., 2006; Dettmer et al., 2007; Lenz & Wilson, 2007; Seger & Sturm, 2007). The selection of the most suitable technology is typically a compromise between speed, chemical selectivity, and instrumental sensitivity. Tools such as nuclear magnetic resonance spectroscopy (NMR) are rapid, highly selective and non-destructive, but have relatively lower sensitivities. Other tools such as capillary electrophoresis (CE) coupled to laser-induced fluorescence detection are highly sensitive, but have limited chemical selectivity (Ramautar et al., 2006). So far mass spectrometry (MS) measurement following chromatographic separation offers the best combination of sensitivity and selectivity (Dunn & Ellis, 2005; Bedair & Sumner, 2008). Mass-selective detection provides highly specific chemical information including molecular mass and/or characteristic fragment-ion that is directly related to chemical structure of molecules. This information can be utilized for compound identification through spectral matching with data compiled in libraries for authentic compounds or used for *de novo* structural elucidation. Further, chemically selective MS information can be obtained from extremely small quantities of metabolites in the *p*mole and *f*mole level for many primary and secondary metabolites. Different technologies, either individual or integrated, could be employed for different study aims, based on metabolite identification, detection speed, high throughput and sensitivity. In the chapter, we will first review current MS technologies that have been incorporated into many metabolomics research programs as well as some of the emerging MS technologies that hold additional promise for the future advancement of metabolomics.

In contrast to classical biochemical approaches that typically focus on a single metabolite, single metabolic reaction or their kinetic properties, metabolomics involves collection of large amount of quantitative data on a broad series of metabolites in an attempt to gain an overall understanding of metabolism and/or metabolic dynamics associated with conditions of interest, such as disease or drug exposure. Generally speaking, metabolomics data share a great deal of similarity with transcriptomics data: both types of data matrices are large, feature rich, and challenged with issues of dealing with a limited sample size and a high-dimensional feature space. Thus in many cases the robust data processing algorithms originally developed for transcriptomic analysis have been adapted directly for metabolomic analysis. However, the challenges with metabolomic data can be unique, and may require new methodologies supported with a detailed knowledge of cheminformatics, bioinformatics, optimization, dynamic system theory and statistics. In recent years, many computational methods have been developed specifically for metabolomic data, varying from metabolic network analysis oriented or feature selection/data mining oriented. In the chapter, we will introduce each of these methods and their relevant applications, and will also discuss all the computational challenges associated.

2. Major analytical technologies of metabolomics

2.1 Mass spectrometry (MS)-based metabolomics

Several MS-based metabolomics technologies have been developed in the past decades, a brief introduction was presented here. Much detailed information regarding these technologies can be found from several recent excellent reviews (Dunn & Ellis, 2005; Hollywood et al., 2006; Dettmer et al., 2007; Bedair & Sumner, 2008).

Direct infusion mass spectrometry (DIMS) is a method for direct analysis of complex metabolic extracts without extraction or separation *via* electrospray ionization (ESI) MS, which provides a sensitive, high-throughput method to make it possible for several hundred samples per day. In terms of disadvantages, DIMS analysis is susceptible to ionization suppression due to competitive ionization with other components in the matrix (*e.g.*, salts and other ionic compounds, organic acids/bases, and hydrophobic compounds), although ionization suppression effects could be reduced by nano electrospray (nano-DIMS) with increased ionization efficiency (Southam et al., 2007). In addition, typical DIMS is not able to discriminate isomeric compounds; however, when coupled with tandem MS (DIMSMS) or Fourier transform ion cyclotron resonance (FT-ICR) spectrometers can trap and accumulate fragment ions that often enable to determine different isomeric structures (Aharoni et al., 2002). A comparison between direct infusion negative-ESI iontrap MS and GC-quadrupole MS analysis for the metabolic fingerprinting of five yeast mutants was reported recently

(Mas et al., 2007). Negative ESI LCQ ion-trap MS was reported as an effective method for the characterization of plant extracts with well-defined clusters in comparison to positive-ion ESI and 1H-NMR profiling (Mattoli et al., 2006).

Gas chromatography-mass spectrometry (GC-MS) has been a very useful technology for volatile and thermally stable polar and nonpolar metabolites (Tanaka et al., 1980). Metabolite identification or confirmation is performed by retention time or index comparisons with pure compounds and mass spectral interpretation or comparison using retention index/mass spectral library databases (Wagner et al., 2003). Metabolites can be classified into two classes: volatile metabolites not requiring chemical derivatisation (Yassaa et al., 2001; Mallouchos et al., 2002; Deng et al., 2004) and non-volatile metabolites requiring chemical derivatisation (Roessner et al., 2000). GC-MS based metabolic profiling has been used to compare four *Arabidopsis* genotypes and showed each genotype exhibited a different metabolite profile (Birkemeyer et al., 2003), and to compare transgenic tomato plants over expressing hexokinase (Roessner-Tunali et al., 2003). Silent phenotypes of potatoes have been distinguished from their parental background by employing metabolic profiling (Weckwerth et al., 2004). The same approach has recently been employed in microbial metabolomics to study the effect of different growth conditions on *Corynebacterium glutamicum* (Strelkov et al., 2004).

The application of Liquid Chromatography coupled to Mass Spectrometry (LC-MS) in metabolomics has been growing over the past few years (Wittmann et al., 2004). As a universal separation technique that can be tailored for the targeted analysis of specific metabolite groups or utilized in a broader non-targeted manner, LC offers additional benefits of analyte recovery by fraction collection and/or concentration, which has been difficult for GC separation. In addition, LC-MS operates at lower analysis temperatures than GC-MS, which enables the analysis of heatlabile metabolites. LC-MS analysis does not involve sample derivatization, which simplifies the sample-preparation and improves the identification of metabolites. The major disadvantage of LC-MS relative to GC-MS is the lack of transferable LC-MS libraries for metabolite identifications, although some efforts have been initiated to construct in-house LC-MS or LC-MS-MS libraries for automated metabolite identifications (Noteborn et al., 2000). Two-dimensional LC has also been utilized to increase the peak separation capacity (Aharoni et al., 2002). Recent LC-MS metabolite-profiling examples include the identification of flavonoids and isoflavonoids in *Medicago truncatula* (Daykin et al., 2002), the revelation of novel pathways by studying the differential and elicitor-specific responses in phenylpropanoid and isoflavonoid biosynthesis in *Medicago truncatula* cell cultures (Farag et al., 2008), and the investigation of small polar-metabolite responses to salt stress in *Arabidopsis thaliana* (Lindon et al., 2000). LC-MS has also been used in the non-targeted analysis of endogenous metabolites in an unbiased manner (Rashed et al., 1997; De Vos et al., 2007).

Capillary electrophoresis mass spectrometry (CE-MS) is a powerful separation technique for charged metabolites (Ramautar et al., 2006; Monton et al., 2007). CE has superior separation efficiencies compared to LC due to the plug-flow profile generated by the electroosmotic flow (EOF) as compared to the parabolic flow in LC. Capillary zone electrophoresis (CZE) has been the major CE mode used for CE-MS analysis of metabolites, due to the simplicity of the running buffer. Simultaneous separation of charged and neutral metabolites can be achieved using other CE modes (*e.g.*, micellar electrokinetic chromatography (MEKC) or capillary electrochromatography (CEC)). Cationic and anionic CE-MS analysis of *Bacillus subtilis* extracts detected 1692 metabolite features of which 150 were identified (Soga et al., 2003). The same analytical procedure was recently used to study the alteration of metabolic pathways in transgenic rice lines that over-express dihydroflavonol-4-reductase (Sato et al., 2004). Non-aqueous CE-ESI ion-trap MSn was utilized for quantitative and qualitative profiling of isoquinoline alkaloids in single-plant tubers of four central European Corydalis species (Sturm et al., 2007).

2.2 Other emerging mass spectrometry technologies

Matrix assisted laser desorption ionization mass spectrometry (MALDI-MS) is a popular analytical technique for biopolymer analysis. It has a high throughput capacity and a higher tolerance for salts than ESI. In metabolomics, MALDI has largely been confined to the targeted analysis of high-molecular-weight metabolites due to the substantial chemical-background signals generated by the matrix in the low-molecular-weight region (<1,000 *m/z*) (e.g., the analysis of phospholipids in mammalian tissues (Jones et al., 2006), plant carotenoids (Fraser et al., 2007), and plant cell wall xyloglycans (Lerouxel et al., 2002). MALDI has also been used for imaging MS (IMS) of proteins and small molecules in tissues (Reyzer & Caprioli, 2007). Whole organisms or selected tissue sections are analyzed through an array of spots in which MS spectra are acquired at spatial intervals that define the image resolution. The *m/z* intensities of the acquired spectra are then plotted in the x and y coordinates to form a 2D image of the *m/z* values, which represents the spatial distribution of that metabolite/ion in the tissue (Rubakhin et al., 2005). MALDI-TOF IMS has been applied successfully for the study of drug and metabolite distributions in rat-brain tissues (Hsieh et al., 2006) and whole rat body (Khatib-Shahidi et al., 2006).

Desorption electrospray ionization (DESI) is a new, ambient, soft-ionization technique that combines features from both ESI and desorption ionization (DI) methods (Takats et al, 2004; Cooks et al., 2006). In DESI, an electrospray emitter is used to generate a spray of charged micro droplets that is directed towards an ambient sample surface. There is virtually no sample preparation required for DESI, thus allowing the direct analysis of animal and plant tissues. The application of DESI in metabolomics is relatively new, but its ambient DI properties as well as its high-throughput capacity make it an attractive tool for metabolomics. One promising area for DESI is *in vivo* metabolomics, which was demonstrated through the direct profiling of alkaloids from plant tissues of Conium maculatum without sample preparation while still identifying all of its previously reported alkaloids using tandem MS (Talaty et al., 2005). Although this technique can be incorporated into an IMS configuration, the spatial resolution of the DESI source (0.5–1.0 mm) is currently less than that of MALDI ion imaging (50–100 µm) (Wiseman et al., 2005).

Extractive electrospray ionization (EESI) is another new ESI technique that uses two separate sprayers. Although the exact sample-ionization mechanism is still unclear, the ionization process depends on liquid–liquid extraction between the colliding micro-droplets of the sample spray and the charged reagent-solvent spray (Chen et al., 2006). The advantage of EESI is its ability to analyze complex biological samples, such as urine and serum, directly with minimum or no sample preparation for an extended period of time. EESI-MS along with ^{1}H NMR was recently used to monitor the effect of diet on the metabolites founds in rate urine (Gu et al., 2007).

2.3 Nuclear magnetic resonance (NMR) spectroscopy

NMR spectroscopy is a technique that exploits the magnetic properties of certain atomic nuclei to determine physical and chemical properties of atoms or the molecules in which they are contained. It relies on the phenomenon of nuclear magnetic resonance and can provide detailed information about the structure, dynamics, reaction state, and chemical environment of molecules. Most frequently, NMR spectroscopy is used to investigate the properties of organic molecules, though it is applicable to any nucleus possessing spin. This can range from small compounds analyzed with 1-dimensional proton or carbon-13 NMR to large proteins or nucleic acids using 3 or 4-dimensional techniques (Grivet & Delort, 2009). NMR presents an unbiased technique for metabolite fingerprinting that is quantitative even in complex mixtures. Nicholson and co-workers have pioneered the application of NMR for metabolite fingerprinting (Lindon et al., 2003). The analysis of six yeast knock-out strains proved to classify and relate the genotypes by multivariate statistics, which potentially can be applied for functional genomics (Raamsdonk et al., 2001). NMR is non-destructive and therefore *in vivo* analysis is also possible (Gmati et al., 2005). Thus, NMR provides a powerful method for accessing metabolite complement (metabolome) and metabolic fluxes (fluxome) at a fine scale (metabolite identification) and a global scale (metabolomics).

2.4 Vibrational spectroscopy

Vibrational spectroscopy is one of the oldest spectroscopic methods. The vibrational states of a molecule can be probed in a variety of ways. The most direct way is through infrared spectroscopy (IRS), as vibrational transitions typically require an amount of energy that corresponds to the infrared region of the spectrum. Raman spectroscopy (RS), which typically uses visible light, can also be used to measure vibration frequencies directly. RS has been used for the identification of microorganisms of medical relevance (Dunn et al., 2005); however, its application for complex biological systems outside the area of microbiology is still in its infancy, although the potential of using 1064 nm excitation has been demonstrated in studies of the biochemical analysis of honey (de Oliveira et al., 2002) and in the analysis of plant pigments and essential oils (Schrader et al., 1999). In contrast, IRS has been applied for diagnostics, characterisation of microorganism and plant, adulteration and quality assurance, biomarker discovery and biochemical responses (Dunn et al., 2005).

2.5 Single-cell metabolomics

Single cell analysis is the new frontier in "OMICS" (Wang & Bodoritz, 2010). Most current metabolomic technologies only collect data averaged over thousands or millions of cells. However, cellular heterogeneity within a cell population is a widespread event (Irish et al., 2006; Graf & Stadtfeld, 2008). Since the metabolome provides biological processes occurring in the cells, it will be imperative to establish a reliable metabolomic method to measure at a single-cell level.

The challenges for single-cell metabolomics include: *i*) tiny quantities of metabolites from a single cell. A type single cell is about 1-500 *f*L in volume (from *E. coli* to large mammalian), and with metabolites as low as *a*mole to *f*mole (Schmid A et al., 2010), 10^6 times lower than a typical population-based metabolomics. While amplification of DNA/RNA and highly sensitive fluorescence measurements could be employed in single cell genomics, transcriptomics and proteomics, no similar technique is available for single-cell metabolomics. *ii*) Sample processing for a single cell is extremely challenging. Even though detection limits for metabolites using MS can be as low as *f*moles to *a*mole range (Amantonico et al., 2008); however, transferring of a cell or cell content to mass spectrometer, conserving the original metabolome, and separating metabolites from cell debris, proteins and salts, would be critical.

In recent years, several approaches have been established for MS-based single-cell metabolomics (Figure 1) (see review by Heinemann & Zenobi, 2011). *i*) Sampling the cell contents with a micropipette, followed by injection into a mass spectrometer using a nano-electrospray ionization (nano-ESI) source (Masujima 2009). This approach, probably only suitable for very large cells, can only measure a few cells per hour; *ii*) Sample preparation on a microfluidic chip, followed by deposition on a sample plate for (matrix-assisted) laser desorption/ionization (MALDI or LDI) mass spectrometry (Lu et al., 2006; Mellors et al., 2008; Amantonico et al., 2008, 2010; Holmes et al., 2009). Once a complete setup is realized, it has the potential to generate high throughput data in an automated way; *iii*) Cell arraying, single cells are deposited on a sample plate for LDI or MALDI covered by a solvent-repelling, application of a MALDI matrix in an organic solvent will then lyse the cells and extract the compounds of interest for analysis by MALDI. This approach is a true high-throughput operation because the sample arraying can be automated, and thus the speed of MS instrument is the only limited factor (1000s of cells/hour) (Urban et al., 2010); *iv*) Imaging mass spectrometry, many modern mass spectrometers have imaging capabilities, with a spatial resolution of typically ~50 μm (MALDI or LDI), and ~1 μm (secondary ion mass spectrometry, SIMS), at relatively fast acquisition speed (Fletcher, 2009). With SIMS, the distribution of ions such as Na^+, K^+, Ca^{2+}, as well as cationized cholesterol, lipids present at cell surfaces can be imaged (Fletcher 2009). However, so far the data generated through this approach has been less quantitative (Heinemann & Zenobi, 2011).

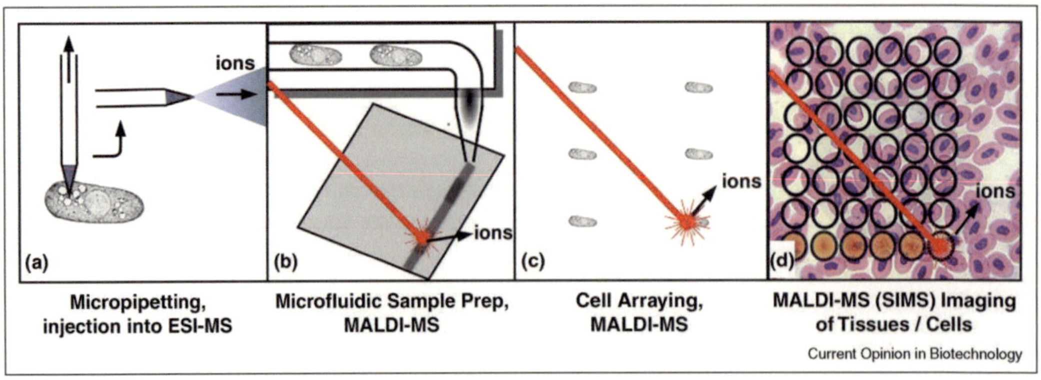

Fig. 1. Schematics of the four MS-based approaches for single cell metabolomics. (Heinemann & Zenobi, 2011)

Only very few metabolites can be analyzed directly in single cells by autofluorescence (Amantonico et al., 2010); however, by incorporating fluorescent tags or probes, researchers have been able to detect more metabolites. For example, Fehr et al. (2002) developed a protein based nanosensor for detection of maltose uptake by living yeast cells. In another study, a genetically encoded fluorescent sensor was expressed in living cells for detecting adenylate nucleotides (Berg et al., 2009). However, it is still arguable whether the formation of these foreign complexes between sensor and metabolites in cells will cause damages and lead to alteration of physiological status of cells.

The high sensitivity of electrochemical detectors to electroactive species makes them suitable for targeted studies of metabolites in single-cell analysis. A range of microscale electrochemical methods have been introduced to monitor various physiological processes (Huang & Kennedy, 1995). For example, release of metabolites, such as catecholamines and oxygen, can be readily measured electrochemically (Cannon et al., 2000). Specific methods targeting particular metabolic pathways in single cells have been in use for a long time, including autoradiography of cells preincubated with radioisotopically labeled compounds (Fliermans & Schmidt, 1975).

Other methods used in single cells including single-cell spectroscopy in conjunction with image analysis for glycogen metabolism in yeast cells (Cahill et al., 2000), enzyme-catalyzed luminescence method for dopamine release from a mammalian nerve cell (Shinohara & Wang, 2007), synchrotron Fourier transform infrared spectromicroscopy for ethanol formation in single living cells of unicellular algae (Goff et al., 2009), raman spectroscopy (Schuster et al., 2000; Buckmaster et al., 2009; Hermelink et al., 2009) for detecting nucleic bases and amino acids in single cells, and nuclear magnetic resonance (NMR) for structural characterization of organic compounds, including metabolites (Beckonert et al., 2007; Motta et al., 2010). Brief summary and comparison for various metabolomics techniques discussed above are listed in Table 1.

3. Computational methods for analyzing metabolomic data

Several computational methods have been developed in recent years to analyze metabolomic data. The overview of the metabolomic data processing is shown in Figure 2. These computational methods can be divided into two major categories: methods for data pre-processing (low-level, such as noise reduction) or methods for interpretation (high-level, such as feature selection). Pre-processing methods concern the improvement and the enhancement of raw signals, which typically include noise reduction, peak detection, baseline correction, peak alignment and normalization. Pre-processing methods for metabolomic data has been reviewed in details by researchers from different perspectives (Jewett et al., 2007; Enot et al., 2011). In this chapter, we will focus on the most widely used or recently developed high-level methods for interpreting metabolomic data. Some computational methods may also depend on the platform or instruments used. The reader should refer Section 2 of the chapter for detailed explanation on popular platforms such as GC-MS, LC-MS and NMR etc. Some software packages have been developed for interpreting metabolomic data in recent years, although review of the software tools are beyond the scope of this chapter, most of the software utilize the statistical methods discussed here.

Techniques	Advantages	Disadvatanges
DIMS	high-throughput, simple sample preparation	nondiscrimination of isomeric compounds, susceptilble to ionization
GC-MS	high sensitivity, ideal to complex samples, versatile	derivation needed for non-volatile metabolites
LC-MS	high sensitivity, average/high resolution	limited structural information, matrix effects
CE-MS	high sensitivity, quantification, label free, small volume	no robust, destructive
MALDI-MS	high sensitivity, detection of a wide range of molecule, label free	no quantification, destructive
DESI	direct analysis, high-throughput, ambient depsorption ionization	optimization required for each sample, low resolution
EESI	high-throughput, simple sample preparation	destructive
NMR	structural information, qualitative and quantitative, versatile	expensive, time consuming, difficult interpration
IR	versatile, easy to identify functional groups	poor structural information, difficult sample preparation, destructive
Raman	label free	low selectivity, poor sturctural information
FM	high sensitivity, imaging capabilities, dynamic	no structural information, targeted analysis only, difficult labelling
EC	high sensitivity, quantification, label free	not comprehensive, no structural information, vulnerable to interferences

FM: fluorescence microscopy
EC: electrochemical

Table 1. Summary for metabolomics techniques: characteristic of the main techniques considered for application in metabolomics

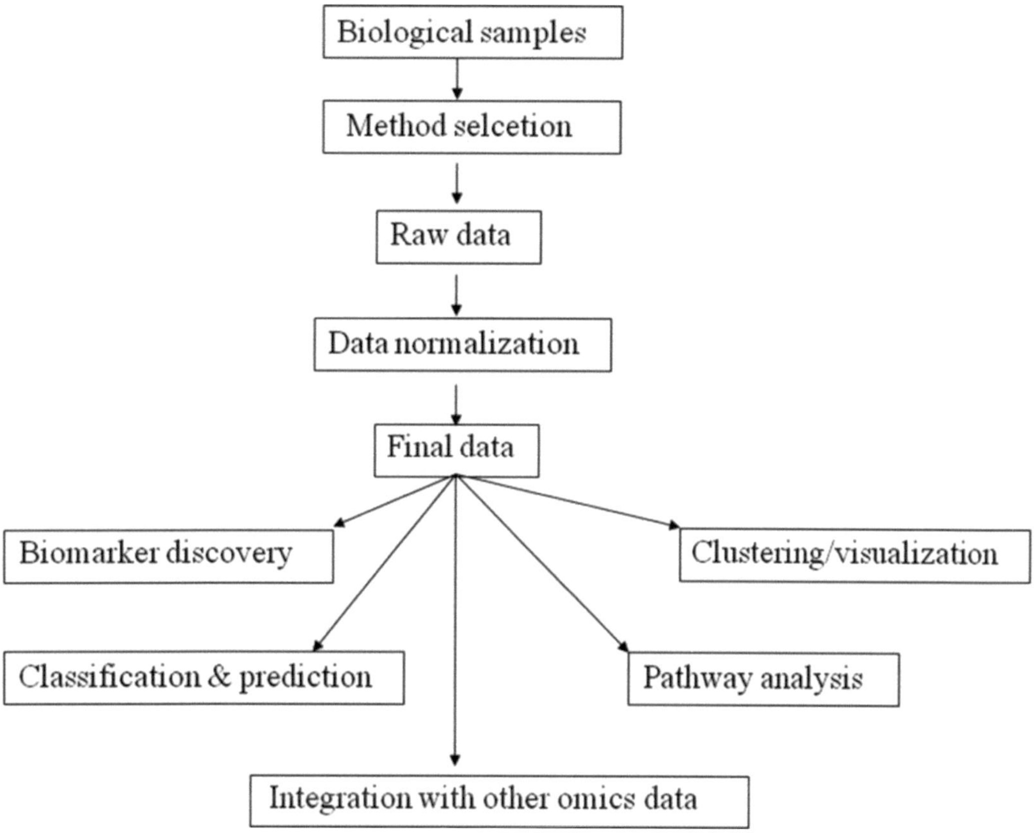

Fig. 2. An overview of metabolomic processing pipeline.

3.1 Computational methods for interpreting metabolomics

The computational methods to interpreting metabolomics data should be selected according to the aim of a study. If the aim is for sample classification and *prior* information about the sample identity is unknown, unsupervised methods such as hierarchical clustering analysis (HCA), principal component analysis (PCA), or self-organizing map (SOM) are typically used. On the other hand, in the case that sample identity is known and the aim of a study is to discover characteristic biomarkers (*e.g.*, search for biomarkers of a disease comparing samples from healthy and diseased subjects), supervised methods such as linear discriminant analysis, artificial neural networks (ANN) or support vector machine (SVM) can be used. The supervised methods use *prior* information about sample class and perform better in biomarker discovery. However, if the aim of the study is a mere biomarker discovery in samples of known classes, common statistical methods such as ANOVA with multiplicity control can also be used (Jonsson et al., 2005). Below, we categorize and discuss different computational methods according to their usage in practice. We will not specifically differentiate them by the mathematical forms such as univariate or multivariate models. It should be noted that most of the methods are multivariate and are often used in combination in practical application.

Biomarker discovery: Like other "omics" studies, the primary objective of many metabolomics studies is to find biomarkers that are discriminative between matched "case" and "control" samples, *i.e.,* which metabolites are apparently altered under different physiological conditions. In pharmaceutical research, metabolomics study has been used for biomarker discovery for different diseases, safety markers, or drug mechanism research. However, given the large number of metabolites studied simultaneously with usually small sample size, it is very common to find metabolites that appear persuasive but in fact spurious. Thus, it is of crucial importance to control the rate of false positive (Broadhurst & Kell, 2006). To tackle this problem, many statistical methods have been developed under the term of large scale hypothesis testing (Benjamini & Hochberg, 1995; Storey 2002; Efron 2003, 2004a, 2004b, 2007a, 2007b, 2008; Storey & Tibshirani, 2003; Reiner et al., 2003; Xie et al., 2005). In classical hypothesis testing, the fundamental problem is to control type I error, the probability that a non-trivial finding is declared while it actually happens by chance. Type I error increases with the number of independent hypothesis considered simultaneously. The well-known and widely used strategy to control the overall type I error rate is Bonferroni-correction, in which the critical value for individual hypothesis testing is obtained by dividing the significance level by the number of hypothesis considered. For example, in metabolomics, if the search for discriminating biomarkers is performed using 500 metabolites and an acceptable chance to reject one true hypothesis is 0.05, and then the Bonferroni-corrected critical value for rejecting an individual null hypothesis for a metabolite should be 0.05/500 = 0.0001). Berferroni-correction is conservative in the sense that it excludes type I errors at the cost of increasing the potential for type II errors (false negatives) (Broadhurst & Kell, 2006).

A widely accepted error measure in microarray literature for large-scale hypothesis testing is the false discovery rate (FDR), the proportion of false positives among all the discoveries. The procedure controlling false discovery rate proposed by Benjamini and Hochberg (1995) has been recognized as a breakthrough and widely applied or adapted by statistical researchers (Efron, 2004a). Most of the literatures assume that the theoretical null hypothesis is known in advance. However, Efron argued that in large scale hypothesis testing, like in 'omics' studies, the theoretical null often fails for reasons like correlations among proteins or genes, unknown confounding factors, or systematic bias (Efron, 2004b, 2007a, 2008). Thus, it will be appropriate to estimate the distribution of the null statistics from the data in order to have a more meaningful discovery. Translating to the setting of a metabolomics study, Efron's concept aims to find a subset of metabolites that behave very differently from the majority of the metabolites. Efron's creative idea has received significant attention from the research field. Applications to metabolite biomarker finding has not been reported but certainly expected.

Metabolomics-based biomarker discovery have been reported. In one example for invasive ovarian carcinomas and ovarian borderline tumors, a differential analysis of 291 detected metabolites in sixty-six invasive ovarian carcinomas and nine borderline tumors of the ovary revealed 51 metabolites that were significantly different between borderline tumors and carcinoma with a FDR controlled at 7.8% (Denkert et al., 2006). For *Onchocera volvulus*, analysis of an African sample set of 73 serum and plasm samples based on LC-MS revealed a set of 14 biomarkers that showed excellent discrimination between *Onchocera volvulus*-positive and negative individuals (Denery et al., 2010). Controlling FDR at 54% using

Storey's q-value approach (Storey, 2002) resulted in 194 features selected from a total of 2350 mass features. Among the 194 features, the authors selected the top 14 feature for investigation.

Data clustering and visualization: Clustering or unsupervised modeling is useful for class discovery and provides information on data similarity: metabolomic samples clustered or grouped together can be objectively considered to be similar.

Principal components analysis (PCA) is probably the most widely used unsupervised approach to data mining or visualization. PCA is a multivariate technique that transforms the data into a coordinate system where each new projection (also called principal components (PC)) is a linear combination of the original variables. PCs are orthogonal so that each dimension is related to different data characteristics and source of variability in a mathematical sense (Enot et al., 2008). As a dimension reduction tool, PCA is very useful for metabolomics data visualization and further data clustering. However, PCA may not work well for metabolomics data where the differences between groups are minor and obscured by other covariates.

Self-organizing map (SOM) is another visualization tool for high-dimensional data (Kohonen, 1998, 2001). The SOM describes a mapping from a higher dimensional input space to a lower dimensional map space. The procedure for placing a vector from data space onto the map is to find the node with the closest weight vector to the vector taken from data space. Once the closest node is located, it is assigned the values from the vector taken from the data space. The SOM places similar input data in adjacent nodes. Therefore, SOM forms a semantic map where similar samples are mapped close together and dissimilar apart. One disadvantage of SOM is that the final map solution is dependent on the order of the presentation of the training data. The batching-learning version of the algorithm (Kohonen, 2001) overcame this problem and yields reproducible maps. SOM has been applied to metabolic profiling for clustering blood plasma (Kaartinen et al., 1998), and NMR spectra of breast cancer tissues (Beckonet et al., 2003b). More recently, Kouskoumvekaki et al. (2008) applied SOM to identify similarities among the metabolic profiles of different filamentous fungi. Meinicke et al. (2008) proposed one-dimensional SOM for metabolite-based clustering and visualization of marker candidates. In a case study on the wound response of *Arabidopsis thaliana*, they showed how the clustering and visualization capabilities of SOM can be utilized to identify relevant groups of biomarkers.

As a popular unsupervised learning method, Hierarchical cluster analysis (HCA) clusters the data to form a tree diagram or dendrogram which shows the relationships between samples (Ebbels, 2007). The algorithm begins by computing the distances between all pairs of samples. Initially each cluster consists of a single sample. The algorithm proceeds iteratively until all samples are members of a single cluster. The final structure of the resulting clusters depends on the choice of distance function or "linkage" between two clusters as well as a similarity cut-off. The most popular choices of linkage are centroidal, average, single (nearest neighbor) and complete (farthest neighbor) linkages. The centroidal linkage defines the inter-cluster distance as the distance between the centroids. To determine the cluster membership, one must decide on a similarity cut-off which breaks the dendrogram into a number of separate clusters. As an example, Beckobert et al. (2003) used the HCA method to explore a set of toxicology studies. HCA allowed interpretation of the data in terms of the magnitude and site of toxicological effect, and helped to explain misclassifications by other methods.

The *k*-means clustering is a method of cluster analysis that aims to partition *n* observations into *k* clusters in which each observation belongs to the cluster with the nearest mean. In *k*-means clustering, the Euclidean distance is used as a distance metric and variance is used as a measure of cluster scatter. The number of clusters *k* is an input parameter. When performing *k*-means, it is important to run diagnostic checks for determining the number of clusters in the data set. Thus, *k*-means is often used in conjunction with other clustering and visualization methods (Ebbels, 2007).

Fuzzy *k*-means (also called Fuzzy c-means) is an extension of *k*-means clustering technique based on Fuzzy logic. While *k*-means discovers hard clusters (a point belong to only one cluster), Fuzzy *k*-means is a more statistically formalized method and discovers soft clusters where a particular point can belong to more than one cluster with certain probability. In Fuzzy *k*-means algorithm, one sample can be assigned to more than one class instead of only one. The membership of each sample is calculated and then represented by a membership value between 0 and 1, instead of just 0 and 1 in the hard clustering. Cuperlovic-Culf et al. (2009) presented the application of fuzzy *k*-means clustering method for the classification of samples based on metabolomics 1D ^{1}H-NMR fingerprints. The sample classification was performed on NMR spectra of cancer cell line extracts and of urine samples of type 2 diabetes patients and animal models. The fuzzy *k*-means clustering method allowed more accurate sample classification in both datasets relative to the other tested methods including PCA, HCA and *k*-means clustering. Li et al. (2009) applied fuzzy *k*-means to cluster three gene types of *Escherichia coli* on the basis of their metabolic profiles and delivered better results than PCA. On the basis of the optimized parameters, the fuzzy *k*-means was able to reveal main phenotype changes and individual characters of three gene types of *E. coli*, while PCA failed to model the metabolite data.

Clustering of metabolomics data can be hampered by noise originating from biological variation, physical sampling error and analytical error. Bootstrap aggregating (bagging) is a resampling technique that can deal with noise and improves accuracy. Hageman et al. (2006) demonstrated the application of bagged clustering to metabolomics data. It was argued that the bagged *k*-means should be favored against ordinary *k*-means clustering when dealing with noisy metabolomics data.

In practice, it is common to combine dimensionality reduction and clustering methods. For example, first, a sample-based principal component analysis (PCA) is performed to compute a subset of principal components. Then the metabolite-specific PCA loadings of these components are used for metabolite-based clustering using *k*-means or hierarchical methods (Pohjanen et al., 2006).

Classification and prediction: While the purpose of clustering is to group similar data together, classification aims at finding a rule to discriminate the classes in an optimal way as well as selecting the subset of features that are most discriminative or predicative. In contrast to clustering applications, the class label and the number of classes are known for a subset of data (training samples) in *a priori* in classification problem. Once the rule or classifies are determined using training dataset, it can be used to predict the class label (such as diseased or not) of a test sample.

The *k*-Nearest Neighbour (*k*NN) rule for classification may be the simplest of all supervised classification approaches (Ebbels, 2007). Different from other supervised learning methods, the training phase of *k*NN consists of only storing the training samples and the corresponding class labels. In the classification (prediction) phase, a test sample is classified by assigning the label which is most frequent among the *k* nearest training samples. The method requires only the choice of *k*, the number of neighbors to be considered when making the classification. Greater values of *k* reduce the effect of noise on the classification, but make boundaries between classes less distinct. A limitation of this classification method is that the classes with the more frequent examples tend to dominate the prediction of a test sample. Usually, *k* is chosen through a cross validation procedure. *k*NN has been often used as a comparator for other methods in literature (Beckonert et al., 2003; Baumgartner et al., 2004).

Partial least square for discriminant analysis (PLS-DA) is a regression extension of PCA that takes advantage of class label information to maximize separation between groups of observations. PLS-DA models the relationship between the class affiliation matrix (Y) and feature matrix (X), which is a generalization of multiple linear regressions. It determines a set of latent variables explaining as much as possible of the covariance between the two matrices. PLS-DA can deal with uncompleted dataset and multicollinearity problem. The output of PLS-DA is the score matrix that can be plotted similarly as in PCA and the predictor matrix containing estimated class affiliation (Ciosek et al., 2005; Trygg & Lundstedt, 2007). The ortogonal-PLS (OPLS) method (Cloarec et al., 2005; Trygg & Lundstedt, 2007) is a recent modification of the PLS method. The main idea of OPLS is to separate the systematic variation in X into two parts, one that is linearly related to Y and one that is orthogonal to Y. The OPLS method provides a prediction similar to that of PLS. However, the interpretation of the models is improved because the structured noise is modeled separately from the variation common in X and Y. Analogous to PLS, when the Y matrix is class affiliation, the corresponding analysis is named OPLS-DA. Cloarec et al. (2005) illustrated the applicability of the method in combination with statistical total correlation spectroscopy to ^{1}H NMR spectra of urine from a metabonomic study of a model of insulin resistance based on the administration of a carbohydrate diet to three different mice strains. Tapp and Kemsley (2009) recently discussed similarities and differences between PLS-DA and OPLS-DA with a focus on the usage of OPLS in the analytical chemistry literature. They concluded that the two methods are very similar and no one outperforms the other, and the reported discrepancies in the literature must be due to differences in the implementation details, or some otherwise "unfair" comparison between the methods

An Artificial neural network (ANN) is a widely used non-linear data modeling tool (Bishop, 1995). An ANN is a computational model that is inspired by the structure and functional aspects of biological neural network. An ANN consists of a layered network of nodes with simple linear or sigmoid activation function. The most widely used type of ANN is the multilayer perceptron (MLP), which has at least three layers including the input layer, one or more hidden layers and output layer. The most attractive feature of a MLP is its capacity to approximate any continuous function in arbitrary precision given enough number of nodes in the hidden layers with sigmoid-type activation functions. ANN has been applied to classification of tumor cells by different researchers (Maxwell et al., 1998; Ott et al., 2003).

One disadvantage of ANN is that non-linear optimization algorithm is needed to train a neural network. It is easy for the optimization procedure to be stuck in local minima while finding the optimal parameters. Thus, global optimal solution is not guaranteed. Moreover, it is difficult to interpret the connection weights to gain some biological insight for feature ranking in applications.

Support vector machines (SVMs) is a supervised learning technique for classification and regression (Cristianini & Shawe-Taylor, 2000). A support vector machine constructs an optimal hyperplane for classification. The hyperplane is constructed in such a way that it has the largest distance to the nearest training data points of any class. In contrast to ANN, SVMs is trained by using convex programming such as quadratic programming. Thus, it can find the global optimal solution efficiently. The computation complexity of SVMs depends on the number of samples instead of the dimension of each sample. Thus there is no curse of dimensionality problem. When cases are not linearly separable, appropriate kernel functions can be chosen to transform the original data into high dimensional feature space. SVMs is well known for its good generalization capacity. For nearly a decade, SVMs has been used in the field of bioinformatics for classifying and evaluating gene expression microarray data (Furey et al., 2000; Guyon et al., 2002). Mahadevan et al. (2008) compared the performance of PLS-DA multivariate analysis to SVMs and showed that SVMs were superior to PLS-DA in terms of predictive accuracy with the least number of features. With fewer features, SVMs were able to give better predictive model when compared to that of PLS-DA.

Computation methods for identification of metabolites: Metabolomics studies are targeted at identifying and quantifying all metabolites in a given biological context. One central problem is the translation of the measured mass into molecular formulae. Given the observed mass and atoms, the first problem is to find all the possible elemental compositions by solving the Diophantine equation:

$$\sum_i n_i \cdot m_i = M ,$$

where M is the observed mass and m_i is the mass of the ith atom. Diophantine equation is the basis for much of the mass spectrometry software to obtain compositions. Usually, there are many integer solutions to it mathematically. Among all the mathematical solutions, we then seek all of the integers n_i that are chemically feasible considering some chemical contextual information such as the valence rules, double-bond equivalents or exact mass (Meija, 2006). Even with very high mass accuracy (<1 ppm), many chemically possible formulae can be obtained in higher mass regions. To further reduce the number of potential elemental compositions, it is necessary to utilize isotope abundance pattern (Kind & Fiehn, 2006).

The identification of small metabolites has been seen as one of the bottlenecks in interpreting metabolomics data. Neumann & Bocker (2010) provided a review focusing on the computational methods for electronspary ionization (ESI) mass spectrometry. One of the most common methods for the identification of compounds using mass spectrometry is the comparison with spectra of authentic standards. The Metabolomics Standards Initiative

(MSI) has defined several confidence levels for the identification of non-novel chemical compounds, ranging from level 1 for a rigorous identification to unidentified signals at level 4 (Sumner et al., 2007). The difference between level 1 and 2 is that the former requires the comparison with authentic standards based on in-house data measured under identical analytical conditions, whereas the latter allows one to use literature or external databases. Level 1 or level 2 identifications are based on a comparison of "exact mass and isotope pattern". Even with the most exact mass and isotope pattern the identification will be limited to the elemental composition. Many compounds share the same sum formula for known metabolites in databases such as KEGG or PubChem. For all other MSI levels, the "identification" usually reduces to an annotation with lower levels of confidence (Neumann & Bocker, 2010).

When a reference spectrum is used, a similarity or distance function is needed for selecting database entries. The most basic similarity functions are those based on counting the number of matching peaks between a query spectrum and each of the database spectra. For this, both spectra can be considered as binary vectors with 0's and 1's for "peak absent" and "peak present", respectively (Neumann & Bocker, 2010). Common distance functions on binary vectors are the Hamming distance (counting any difference) or the Jaccard coefficient (the fraction of matching peaks). Besides counting matches, other measures also consider their actual mass and intensity, such as the Euclidean distance, the probability-based matching (PBM), the normalized dot product (NDP), and a modified cosine distance for the database search of EI spectra (Stein, 1994).

Oberacher et al. (2009) proposed and optimized a search function for tandem mass spectrometry (MS/MS-spectra) based on a combination of relative and absolute match probabilities, which combines the principle of peak counting and summed intensities of matching peaks. The X-Rank algorithm (Mylonas et al., 2009) for MS/MS-spectra match is based on probability calculations. It sorts peak intensities of a spectrum and then establishes a correlation between two sorted spectra. X-Rank computes the probability that a rank from an experimental spectrum matches a rank from a reference library spectrum. The solution requires training on a representative dataset. In a training step, characteristic parameter values are generated for a given data set. Identification of small compounds is still challenging, especially for compounds that have not been recorded in any library or structure database. Methods for these tasks are highly sought.

3.2 Computational methods for metabolic pathway analysis using metabolomic data

Metabolomics data provides a series of snapshots of cellular metabolism, which can be combined with metabolic flux data for further analysis. Metabolic pathways are the true functional units of metabolic systems (Schilling et al., 2000b). Finding biochemically plausible pathways between metabolites in a metabolic network is a central problem in computational metabolic modeling. Mathematical modeling approaches to metabolic regulation analysis involve different levels of details and complexities ranging from detailed kinetic models, stoichiometric analysis, structural kinetic models, to large scale topological network analysis (Steuer, 2007).

Detailed kinetic models of metabolic pathways, based on explicit enzymekinetic rate equations, is a bottom-up approach towards more comprehensive large-scale dynamic models. It allows for the most detailed quantitative evaluation of the dynamics of metabolic systems, which is very important for improving the understanding of metabolic regulation and control. The metabolic control analysis (MCA) is the culminating mathematical theory from kinetic models, which describes the control and regulatory properties of metabolic systems (Heinrich & Schuster, 1996; Fell, 1997).

A metabolic network is a collection of enzyme-catalyzed reactions and transport processes that serve to dissipate substrate metabolites and generate final metabolites. The dynamics of a metabolic system can be described by a set of ordinary differential equations:

$$\frac{dX_i}{dt} = \sum_j S_{ij} v_j(X,k) = S \cdot v(X,k), \tag{1}$$

where X_i represents the concentration of the metabolite and S_{ij} stands for the stoichiometric coefficient for the reactant i in the jth reaction. $v_j(X,k)$ corresponds to the flux through the jth reaction. The vector $v(X,k)$ consists of nonlinear enzyme-kinetic rate functions, which depends on the concentration X and a set of kinetic parameters k. Given the $v(X,k)$ function form, the set of kinetic parameters k and an initial state $X(0)$, the differential equations can be solved numerically to obtain the time-dependent behavior of all metabolites under consideration. The stoichiometric matrix S is an m by n matrix where m corresponds to the number of metabolites and n is the total number of metabolites and n is the total number of fluxes taking place in the network.

The stoichiometric analysis approach takes advantage of the structure nature of metabolic system. Knowledge of the stoichiometry puts constraints on the feasible flux distributions, which can be utilized to model the functional capabilities of metabolic networks (Varma & Palsson 1994; Edwards & Palsson, 2000; Stelling et al., 2002; Famili et al., 2003; Price et al., 2003).

The pathway structure should be an invariant property of the network along with stoichiometry. Under steady-state, the set of ordinary equations reduce to linear equations:

$$0 = S \cdot v(X^0,k) = S \cdot v^0, \tag{2}$$

which is typically underdetermined as the number of reactions exceeds the number of participating metabolites. The set of all solutions v^0 to the steady state equation (2) is called null space. A set of basis vectors corresponding to a steady-state biochemical pathway can be selected to describe the null space. Additional constraints from biological insights are necessary to determine the system completely (Schilling et al., 2000b), which generally results in a set of linear equalities and inequalities. The set of feasible solutions under bio-chemical constraints form a convex steady-state flux cone. The convex analysis of biochemical networks was founded by Clarke (Clarke, 1980, 1988). The study of convex flux cones utilizes methods and concepts rooting in linear algebra and optimization and forms the underlying mathematical structure for metabolic pathway analysis.

Flux balance analysis (FBA) is a computational approach to reduction of the admissible flux space (Schilling et al., 1999; Edwards & Palsson, 2000). FBA optimizes an objective function such as maximal biomass yield or maximal energy production through the steady-state flux space has resulted in many applications (Papin et al., 2004; Almaas et al., 2004; Stephanopoulos et al., 2004). Using FBA, *in silico* studies on the systemic properties of the *Haemophilus influenzae* and *E. coli* (Edwards & Palsson, 2000) metabolic networks have been completed. Under various substrate conditions, Schilling et al. (2000a) explored the metabolic capabilities and predicted functions of a sub-system of the *E. coli* using FBA.

A metabolic network can be decomposed into distinct pathways, termed elementary flux modes (EFM). An EFM is the minimal set of reactions capable of working together in a steady state, which is unique for a given metabolic network. Another closely related concept is extreme pathways, which are a subset of elementary modes (Klamt & Stelling, 2003). All feasible flux vectors can be described as linear combinations of EFMs. The concept of EFM has resulted in a vast number of applications for metabolic network analysis (Stelling et al., 2002; Schuster et al., 2002; Klamt & Schuster, 2002; Klamt & Gilles, 2004; Klamt et al., 2006). For medium-sized metabolic networks, software packages have been developed for the computation of elementary flux modes (Hoops et al., 2006; Klamt et al., 2007). Owing to a combinatorial explosion of the number of elementary vectors, this approach becomes computationally intractable for genome scale networks. To develop an analysis approach computationally feasible even for genome scale networks, Urbanczik & Wagner (2005) proposed to focus on conversion cone, the projection of the flux cone, which describes the interaction of the metabolism with its external chemical environment. The method for calculating the elementary vectors of this cone was applied to study the metabolism of *Saccharomyces cerevisiae*.

Stoichiometric analysis does not incorporate dynamic properties into the description of the system. Steuer (2006, 2007) proposed a structure kinetic modeling approach to augmenting the stoichiometric analysis with kinetic properties. The idea of the proposed approach is to use a local linear approximation to explicit kinetic model to capture the dynamic response to perturbations, the stability of a metabolic state, as well as the transition to oscillatory behavior. The local linear approximation is obtained from a Taylor series expansion of the metabolic system. The linear term of the expansion is the derivative of the kinetic rate equations with respect to the metabolic concentration X at a given state, which usually requires knowledge of the enzyme-kinetic rate equations. Even in the absence of enzyme-kinetic information, it is still possible to specify the structure of the linear term. Structure kinetic modeling approach allows quantitative conclusions about the possible dynamics of the system, based on only a minimal amount of additional information.

The extension of the detailed kinetic models to whole cell models is faced with some fundamental difficulties including the absence of comprehensive measured kinetic parameter values, and the observed inconsistency in the available kinetic data, and the computational complexity of such models. Traditionally, kinetic models are constructed using rate equations derived to describe conditions in vitro and thus rely on the use of in vitro measured kinetic parameters. However, the conditions at which *in vitro* experiments are performed are often very different from those inside the cell. Thus, *in vitro* kinetic rates and *in vitro* kinetic parameters describe enzymatic behaviors that may not truly represent the observed physiological kinetic behavior in the cell. Several methods have been proposed to address this issue by incorporating *in vivo* measurements in constructing kinetic models (Visser & Heijnen, 2003) or estimating kinetic parameters in biochemical networks using measured variables (Lei & Jorgensen, 2001; Moles et al., 2003; Segre et al., 2003). These methods require considerable mathematical efforts or utilize nonlinear optimization techniques. (Famili et al., 2003) proposed an approach for incorporating steady-state *in vivo* data with constraint based modeling approach to determine all candidate numerical values of kinetic constants. The kinetic solution space, termed *k*-cone, contains all the allowable numerical values of the kinetic constants. The *k*-cone is obtained by approximating the nonlinear kinetic rate laws as a linear or bilinear function of the kinetic constants. The *k*-cone approach can be used to determine consistency between *in vitro* measured kinetic values and *in vivo* concentration and flux measurements when used in a network-scale kinetic model. To calculate the relationship between kinetic parameters measured in vitro and the *k*-cone, optimization methods were essential. It was successful in determining whether *in vitro* measured kinetic values used in the reconstruction of a kinetic-based model of *Saccharomyces cerevisiae* central metabolism could reproduce *in vivo* measurements.

Despite the number of research published, information gained from theoretical or experimental metabolic network has not fully enabled probing biochemical pathway structure with the aim at detecting novel metabolic routes (Fiehn, 2007).

4. Computational methods to integrate metabolomic data with other "omics" datasets

It is also becoming clear that any single "omics" approach may not be sufficient to characterize the complexity of biological systems and an integrated "omics" approach may be a key to decipher complex biological systems (Gygi et al., 1999; Zhang et al., 2010). In general, integrated analysis of metabolomics dataset with other types of "omics" datasets can increase both dimension of information sources and statistical power in order to generate a conclusion with high confidence. In recent years, several computational methods have been applied to integrate metabolomics with other "omics" datasets, and the results demonstrated that better pattern recognition and association identification can be achieved when proper mathematical, statistical and other computational tools were applied.

The simple correlation analysis, *Pearson* or *Spearman*'s correlation has been used to assess degree of association between metabolomic and transcriptomic data. For example, it has been used to distinguish different potato tuber systems, and to determine the relationship between genes and their paired metabolites (Urbanczyk-Wochniak et al., 2003, 2007), through integrating metabolomic and transcriptomic datasets.

Because of the high dimensionality involved in the metabolomic data and other "omics" data, the dimension reduction tool such as PCA can demonstrate its power in the integrated analysis. It has been used in two different but related scenarios: *i*) PCA applies to each "omics" data to reduce the dimensionality of each omics data so that the "omics" data are ready for integration; *ii*) PCA applies to the integrated metabolomics and other "omics" data directly to identify a particular pattern. In scenario 1, metabolomic data or other "omics" data are not directly used in any integrated analysis. Rather, PCA was used to reduce the dimension of each data, so that researchers can focus on the most important components of metabolomic data or other "omics" data (Urbanczyk-Wochniak et al., 2003; Rubingh et al.,

2009; van den Berg et al., 2009). In scenario 2, PCA was directly used to identify pattern of the integrated metabolomic data with other "omics" data. For example, it was used to a integrated metabolomic data and a proteomic data to reveal clustering of the two genotypes (Weckwerth et al., 2004). While PCA can also be used for exploring polynomial relationships and for multivariate outlier detection, this method is restricted to linear relationships.

In both correlation and PCA analysis, the roles of all variables are the same and they are interchangeable. They are used to explore associations between factors. In some other analyses, some factors (independent variables *X*) are used to explain or predict the variable of main interest (dependent variable *Y*). For example, PLS is a statistical method that models *Y* over *X* through a linear relationship. Rather than considering all dependent variables as regressors in a multivariate regression analysis, PLS regresses *Y* over principal components resulted from a principal components analysis (Garthwaite 1994). The method was in fact previously used to model metabolomic variables as a function of the transcriptome profiles (Pir et al., 2006). The analysis allowed the discrimination between the effects that the growth media, dilution rate and deletion of specific genes on the transcriptomic and metabolomic profiles (Pir et al. 2006). The method was also used to relate quantifiable phenotypes of interest such as protease activity or productivity, to concentrations of each of the metabolites determined (Braaksma et al., 2011). The analysis revealed various sugar derivatives correlated with glucoamylase activity.

As an extension of PLS, Le Cao and colleagues proposed a sparse PLS approach to combine integration and simultaneous variable (*e.g.*, gene) selection in one step (Le Cao et al., 2008, 2009). In the approach, the PLS was penalized by the sum of the absolute values of the coefficients through least absolute shrinkage and selection operator (LASSO) (Tibshirani 1996), therefore automatically eliminating variables (e.g., genes) with negligible effects. The model selection approach, together with the smoothly clipped absolute deviation approach (Fan & Li, 2001) is effective in analyzing data with sparsity (*e.g.*, only a few genes have significant effects).

The methods previously discussed in this section, including *Perason* correlation, PCA, and PLS, are all methods to explore linear relationship. On the other hand, the kinetic model and artificial network could be more sensible when nonlinearity occurs. In an analysis to integrate metabolomics and pharmacokinetics (or nutrikinetics), Van Velzen et al. (2009) presented a one-compartment nutrikinetic model with first-order excretion, a lag time, and a baseline function was fitted to the time courses of these selected biomarkers based on metabolomic data. A kietic model was also used to model the relationship between enzyme kinetics and intracellular metabolites through a two-substrate Michaelis–Menten equation with competitive substrate inhibition or competitive product inhibition (Schroer et al., 2009). The kinetic constants were estimated by nonlinear regression of initial rate measurements. Martense & Vanrolleghem (2010) summarized a few other modeling approaches. Compared to data driving unsupervised analysis, the mathematical modeling may provide a meaning relationship for a better understanding. However, because the modeling is generally based on approximation under some restricted assumptions, the simple model may not precisely describe the complex biology system.

Artificial network is another method to cope with the nonlinearity. A batch-learning self-organizing network was utilized to classify the metabolomes and the transcriptomes according to their time-dependent pattern of changes (Kanaya et al., 2001); the results showed that the metabolomes and transcriptomes regulated by the same mechanism tended to be clustered together (Hirai et al., 2004, 2005). The A batch-learning self-organizing network is artificial neural network that is trained using unsupervised learning to produce a low-dimensional representation of the input space of the training samples. A Network-embedded thermodynamic analysis (NET analysis) is presented as a framework for mechanistic and model-based analysis of metabolite data. By coupling the data to an operating metabolic network via the second law of thermodynamics and the metabolites' Gibbs energies of formation, NET analysis allows inferring functional principles from quantitative metabolite data; for example it identifies reactions that are subject to active allosteric or genetic regulation as exemplified with quantitative metabolomic data from *E. coli* and *S. cerevisiae* (Kummel et al., 2006). The network typically creates a graphic representing the global relationship. In a review article, Feist et al. (2006) classified studies using network into three categories: studies that use a reconstruction to examine topological network properties, studies that use a reconstruction in constraint-based modeling for quantitative or qualitative analyses, and studies that are purely data driven. Some of the networks' mathematical frameworks are graph theory. It provides a visual presentation of the complex biology system. However, when it involves more features, the network approach often become too complicated to provide a clear clue.

Bayesian graphical modeling approaches infer biological regulatory networks by integrating expression levels of different types. Specific sequence/structure information will be incorporated into the *prior* probability models (Webb-Robertson et al., 2009) presented a Bayesian approach to integration that uses posterior probabilities to assign class memberships to samples using individual and multiple data sources; these probabilities are based on lower-level likelihood functions derived from standard statistical learning algorithms. The approach was demonstrated by integrating two proteomic datasets and one metabolic dataset from microbial infections of mice; the results showed that integration of the different datasets improved classification accuracy to ~89% from the best individual dataset at ~83 %.

Integrative interpretations of data from different "omics" including metabolomics, are still in it early development stage. More thoughtful interpretation methods that are capable to reveal biology at a system level are yet to come. The collaboration between mathematicians, statisticians, bioinformaticians and experimental biologists will be the key to the success of these efforts.

5. Final remarks

Although comprehensive coverage of metabolome in cells is not yet possible, significant advancements in the large-scale profiling of metabolites have been achieved in recent years and these analyses have offered unique insight into the metabolic and regulatory networks of cells. In this chapter, we first reviewed some of the widely used and emerging technologies for metabolomics analysis, and then focus on recent progress in developing computational methodologies to improve biological interpretation of high throughput metabolomic data. In addition, we present some mathematical, statistical and bioinformatics methods that have been utilized for the integration of metabolomics data with other type of "omics" datasets and how this integrative analysis has improved our interpretation of biological systems.

6. Acknowledgements

The research of J. Wang and W. Zhang of Tianjin University was supported by National Basic Research Program of China (National "973" program, project No. 2011CBA00803 and No. 2012CB721101), and National Natural Science Foundation of China (Project No. 31170043).

7. Disclaimer

Feng Li is currently working for Office of Biostatistics, Food and Drug Administration. Views expressed in this paper are the author's professional opinions and do not necessarily represent the official positions of the U.S. Food and Drug Administration

8. References

Aharoni, A., C. H. R. De Vos, H. A. Verhoeven, C. A. Maliepaard, G. Kruppa, R. Bino & D. B. Goodenowe (2002) Nontargeted metabolome analysis by use of fourier transform ion cyclotron mass spectrometry. *OMICS A Journal of Integrative Biology*, 6, 217-234.

Almaas, E., B. Kovacs, T. Vicsek, Z. N. Oltvai & A. L. Barabasi (2004) Global organization of metabolic fluxes in the bacterium *Escherichia coli*. *Nature*, 427, 839-843.

Amantonico, A., P. L. Urban, J. Y. Oh & R. Zenobi (2009) Interfacing microfluidics and laser desorption/Ionization mass spectrometry by continuous deposition for application in single cell analysis. *Chimia*, 63, 185-188.

Amantonico, A., J. Y. Oh, J. Sobek, M. Heinemann & R. Zenobi (2008) Mass spectrometric method for analyzing metabolites in yeast with single cell sensitivity. *Angewandte Chemie-International Edition*, 47, 5382-5385.

Amantonico, A., P. L. Urban & R. Zenobi (2010) Analytical techniques for single-cell metabolomics: state of the art and trends. *Analytical and Bioanalytical Chemistry*, 398, 2493-2504.

Anderle, M., S. Roy, H. Lin, C. Becker & K. Joho (2004) Quantifying reproducibility for differential proteomics: noise analysis for protein liquid chromatography-mass spectrometry of human serum. *Bioinformatics*, 20, 3575-3582.

Baumgartner, C., C. Bohm, D. Baumgartner, G. Marini, K. Weinberger, B. Olgemoller, B. Liebl & A. A. Roscher. 2004. Supervised machine learning techniques for the classification of metabolic disorders in newborns. In *Bioinformatics*, 2985-96. England.

Beckonert, O., M. E. Bollard, T. M. D. Ebbels, H. C. Keun, H. Antti, E. Holmes, J. C. Lindon & J. K. Nicholson (2003) NMR-based metabonomic toxicity classification: hierarchical cluster analysis and k-nearest-neighbour approaches. *Analytica Chimica Acta*, 490, 3-15.

Beckonert, O., H. C. Keun, T. M. D. Ebbels, J. G. Bundy, E. Holmes, J. C. Lindon & J. K. Nicholson (2007) Metabolic profiling, metabolomic and metabonomic procedures for NMR spectroscopy of urine, plasma, serum and tissue extracts. *Nature Protocols*, 2, 2692-2703.

Bedair, M. & L. W. Sumner (2008) Current and emerging mass-spectrometry technologies for metabolomics. *Trac-Trends in Analytical Chemistry*, 27, 238-250.

Benjamini, Y. & Y. Hochberg (1995) Controlling the False Discovery Rate: A Practical and Powerful Approach to Multiple Testing. *Journal of the Royal Statistical Society. Series B (Methodological)*, 57, 289-300.

Berg, J., Y. P. Hung & G. Yellen (2009) A genetically encoded fluorescent reporter of ATP:ADP ratio. *Nature Methods*, 6, 161-166.

Birkemeyer, C., A. Kolasa & J. Kopka (2003) Comprehensive chemical derivatization for gas chromatography-mass spectrometry-based multi-targeted profiling of the major phytohormones. *Journal of Chromatography A*, 993, 89-102.

Bishop, C. M. 1995. *Neural Networks for Pattern Recognition*. Oxford, UK: Clarendon Press.

Braaksma, M., S. Bijlsma, L. Coulier, P. J. Punt & M. J. van der Werf (2011) Metabolomics as a tool for target identification in strain improvement: the influence of phenotype definition. *Microbiology*, 157, 147-59.

Breitling, R., A. R. Pitt & M. P. Barrett (2006) Precision mapping of the metabolome. *Trends in Biotechnology*, 24, 543-548.

Broadhurst, D. & D. Kell (2006) Statistical strategies for avoiding false discoveries in metabolomics and related experiments. *Metabolomics*, 2, 171-196.

Buckmaster, R., F. Asphahani, M. Thein, J. Xu & M. Zhang (2009) Detection of drug-induced cellular changes using confocal Raman spectroscopy on patterned single-cell biosensors. *Analyst*, 134, 1440-1446.

Cahill, G., P. K. Walsh & D. Donnelly (2000) Determination of yeast glycogen content by individual cell spectroscopy using image analysis. *Biotechnology and Bioengineering*, 69, 312-322.

Cannon, D. M., N. Winograd & A. G. Ewing (2000) Quantitative chemical analysis of single cells. *Annual Review of Biophysics and Biomolecular Structure*, 29, 239-263.

Chen, H. W., A. Venter & R. G. Cooks (2006) Extractive electrospray ionization for direct analysis of undiluted urine, milk and other complex mixtures without sample preparation. *Chemical Communications*, 2042-2044.

Ciosek, P., Z. Brzozka, W. Wroblewski, E. Martinelli, C. Di Natale & A. D'Amico. 2005. Direct and two-stage data analysis procedures based on PCA, PLS-DA and ANN for ISE-based electronic tongue-Effect of supervised feature extraction. In *Talanta*, 590-6. England.

Clarke, B. L. 1980. Stability of Complex Reaction Networks. In *Advances in Chemical Physics*, 1-215. John Wiley & Sons, Inc.

--- (1988) Stoichiometric network analysis. *Cell Biophysics*, 12, 237-53.

Cloarec, O., M. E. Dumas, A. Craig, R. H. Barton, J. Trygg, J. Hudson, C. Blancher, D. Gauguier, J. C. Lindon, E. Holmes & J. Nicholson (2005) Statistical total correlation spectroscopy: an exploratory approach for latent biomarker identification from metabolic 1H NMR data sets. *Analytical Chemistry*, 77, 1282-9.

Cooks, R. G., Z. Ouyang, Z. Takats & J. M. Wiseman (2006) Ambient mass spectrometry. *Science*, 311, 1566-1570.

Cristianini, N. & J. Shawe-Taylor. 2000. *n Introduction to Support Vector Machines*. Cambridge University Press.

Cuperlovic-Culf, M., N. Belacel, A. S. Culf, I. C. Chute, R. J. Ouellette, I. W. Burton, T. K. Karakach & J. A. Walter (2009) NMR metabolic analysis of samples using fuzzy K-means clustering. *Magnetic Resonance in Chemistry*, 47 Suppl 1, S96-104.

Daykin, C. A., P. J. D. Foxall, S. C. Connor, J. C. Lindon & J. K. Nicholson (2002) The comparison of plasma deproteinization methods for the detection of low-molecular-weight metabolites by H-1 nuclear magnetic resonance spectroscopy. *Analytical Biochemistry,* 304, 220-230.

Denery, J., A. Nunes, M. Hixon, T. Dickerson, K. Janda (2010) Metabololics-based discovery of diagnostic biomarkers for onchoceriasis. PLoS Neglected Tropical Diseases, 4(10), e384.

Deng, C. H., X. M. Zhang & N. Li (2004) Investigation of volatile biomarkers in lung cancer blood using solid-phase microextraction and capillary gas chromatography-mass spectrometry. *Journal of Chromatography B-Analytical Technologies in the Biomedical and Life Sciences,* 808, 269-277.

Denkert, C., J. Budczies, T. Kind, W. Weichert, P. Tablack, J. Sehouli, S. Niesporek, D., Konsgen, M. Dietel, and O. Fiehn (2006) Mass Spectrometry–Based Metabolic Profiling Reveals Different Metabolite Patterns in Invasive Ovarian Carcinomas and Ovarian Borderline TumorsSpec. Cancer Res, 66: (22).

de Oliveira, L. F. C., R. Colombara & H. G. M. Edwards (2002) Fourier transform Raman spectroscopy of honey. *Applied Spectroscopy,* 56, 306-311.

Dettmer, K., P. A. Aronov & B. D. Hammock (2007) Mass spectrometry-based metabolomics. *Mass Spectrometry Reviews,* 26, 51-78.

De Vos, R. C. H., S. Moco, A. Lommen, J. J. B. Keurentjes, R. J. Bino & R. D. Hall (2007) Untargeted large-scale plant metabolomics using liquid chromatography coupled to mass spectrometry. *Nature Protocols,* 2, 778-791.

Dieterle, F., A. Ross, G. Schlotterbeck & H. Senn (2006) Metabolite projection analysis for fast identification of metabolites in metabonomics. Application in an amiodarone study. *Analytical Chemistry,* 78, 3551-61.

Dunn, W. B., N. J. C. Bailey & H. E. Johnson (2005) Measuring the metabolome: current analytical technologies. *Analyst,* 130, 606-625.

Dunn, W. B. & D. I. Ellis (2005) Metabolomics: current analytical platforms and methodologies. *Trac-Trends in Analytical Chemistry,* 24, 285-294.

Ebbels, T. M. D. 2007. Non-linear Methods for the Analysis of Metabolic Profiles. In *The Handbook of Metabonomics and Metabolomics,* eds. J. C. Lindon, J. K. Nicholson & E. Holmes, 201--226. London: Elsevier.

Edwards, J. S. & B. O. Palsson. 2000. The Escherichia coli MG1655 in silico metabolic genotype: its definition, characteristics, and capabilities. In *Proceedings of the National Academy of Sciences of the United States of America,* 5528-33. United States.

Efron, B. (2003) Robbins, Empirical Bayes and Microarrays. *The Annals of Statistics,* 31, 366-378.

--- (2004a) Large-Scale Simultaneous Hypothesis Testing. *Journal of the American Statistical Association,* 99, 96-104.

--- (2004b) Large-Scale Simultaneous Hypothesis Testing: The Choice of a Null Hypothesis. *Journal of the American Statistical Association,* 99, 96-104.

--- (2007a) Correlation and Large-Scale Simultaneous Significance Testing. *Journal of the American Statistical Association,* 102, 93-103.

--- (2007b) Size, power and false discovery rates. *The Annals of Statistics,* 35, 1351-1377.

--- (2008) Microarrays, Empirical Bayes and the Two-Groups Model. *Statistical Science,* 23, 1-22.

Enot, D. P., B. Haas & K. M. Weinberger (2011) Bioinformatics for mass spectrometry-based metabolomics. *Methods in Molecular Biology,* 719, 351-75.

Enot, D. P., W. Lin, M. Beckmann, D. Parker, D. P. Overy & J. Draper (2008) Preprocessing, classification modeling and feature selection using flow injection electrospray mass spectrometry metabolite fingerprint data. *Nature Protocols,* 3, 446-70.

Famili, I., J. Forster, J. Nielsen & B. O. Palsson. 2003. *Saccharomyces cerevisiae* phenotypes can be predicted by using constraint-based analysis of a genome-scale reconstructed metabolic network. In *Proceedings of the National Academy of Sciences of the United States of America,* 13134-9. United States.

Fan, J. Q. & R. Z. Li (2001) Variable selection via nonconcave penalized likelihood and its oracle properties. *Journal of the American Statistical Association,* 96, 1348-1360.

Farag, M. A., D. V. Huhman, R. A. Dixon & L. W. Sumner (2008) Metabolomics reveals novel pathways and differential mechanistic and elicitor-specific responses in phenylpropanoid and isoflavonoid biosynthesis in Medicago truncatula cell cultures. *Plant Physiology,* 146, 387-402.

Fehr, M., W. B. Frommer & S. Lalonde (2002) Visualization of maltose uptake in living yeast cells by fluorescent nanosensors. *Proceedings of the National Academy of Sciences of the United States of America,* 99, 9846-9851.

Feist, A. M., J. C. Scholten, B. O. Palsson, F. J. Brockman & T. Ideker (2006) Modeling methanogenesis with a genome-scale metabolic reconstruction of *Methanosarcina barkeri. Molecular Systems Biology,* 2, 2006 0004.

Fell, D. 1997. *Understanding the Control of Metabolism.* Potland Press.

Fiehn, O. (2001) Combining genomics, metabolome analysis, and biochemical modelling to understand metabolic networks. *Comparative and Functional Genomics,* 2, 155-168.

Fiehn, O. 2007. Cellular Metabolomics: The Quest for Pathway Structure. In *The Handbook of Metabonomics and Metabolomics,* eds. J. C. Lindon, J. K. Nicholson & E. Holmes, 35-54. Elsevier.

Fletcher, J. S. (2009) Cellular imaging with secondary ion mass spectrometry. *Analyst,* 134, 2204-2215.

Fliermans, C. B. & E. L. Schmidt (1975) Authoradiography and immunofluorescence combined for autecological study of single cell activity with nitrobacter as a model system. *Applied Microbiology,* 30, 676-684.

Fraser, P. D., E. M. A. Enfissi, M. Goodfellow, T. Eguchi & P. M. Bramley (2007) Metabolite profiling of plant carotenoids using the matrix-assisted laser desorption ionization time-of-flight mass spectrometry. *Plant Journal,* 49, 552-564.

Furey, T. S., N. Cristianini, N. Duffy, D. W. Bednarski, M. Schummer & D. Haussler (2000). Support vector machine classification and validation of cancer tissue samples using microarray expression data. *Bioinformatics,* 16, 906-914.

Garthwaite, P. H. (1994) An Interpretation of Partial Least-Squares. *Journal of the American Statistical Association,* 89, 122-127.

Gmati, D., J. K. Chen & M. Jolicoeur (2005) Development of a small-scale bioreactor: Application to *in vivo* NMR measurement. *Biotechnology and Bioengineering,* 89, 138-147.

Goff, K. L., L. Quaroni & K. E. Wilson (2009) Measurement of metabolite formation in single living cells of *Chlamydomonas reinhardtii* using synchrotron Fourier-Transform Infrared spectromicroscopy. *Analyst,* 134, 2216-2219.

Graf, T. & M. Stadtfeld (2008) Heterogeneity of Embryonic and Adult Stem Cells. *Cell Stem Cell,* 3, 480-483.

Grivet, J.-P. & A.-M. Delort (2009) NMR for microbiology: In vivo and in situ applications. *Progress in Nuclear Magnetic Resonance Spectroscopy,* 54, 1-53.

Gu, H., H. Chen, Z. Pan, A. U. Jackson, N. Talaty, B. Xi, C. Kissinger, C. Duda, D. Mann, D. Raftery & R. G. Cooks (2007) Monitoring diet effects via biofluids and their implications for metabolomics studies. *Analytical Chemistry,* 79, 89-97.

Guyon, I., J. Weston, S. Barnhill & V. Vapnik (2002) Gene Selection for Cancer Classification using Support Vector Machines. *Machine Learning,* 46, 389-422.

Gygi, S. P., Y. Rochon, B. R. Franza & R. Aebersold (1999) Correlation between protein and mRNA abundance in yeast. *Molecular and Cellular Biology,* 19, 1720-30.

Hageman, J. A., R. A. van den Berg, J. A. Westerhuis, H. C. J. Hoefsloot & A. K. Smilde (2006) Bagged K-Means Clustering of Metabolome Data. *Critical Reviews in Analytical Chemistry,* 36, 211-220.

Heinemann, M. & R. Zenobi (2011) Single cell metabolomics. *Current Opinion in Biotechnology.* 22, 26-31.

Heinrich, R. & S. Schuster. 1996. *The Regulation of Cellular Systems*. Chapman and Hall.

Hermelink, A., A. Brauer, P. Lasch & D. Naumann (2009) Phenotypic heterogeneity within microbial populations at the single-cell level investigated by confocal Raman microspectroscopy. *Analyst,* 134, 1149-1153.

Hirai, M. Y., M. Klein, Y. Fujikawa, M. Yano, D. B. Goodenowe, Y. Yamazaki, S. Kanaya, Y. Nakamura, M. Kitayama, H. Suzuki, N. Sakurai, D. Shibata, J. Tokuhisa, M. Reichelt, J. Gershenzon, J. Papenbrock & K. Saito (2005) Elucidation of gene-to-gene and metabolite-to-gene networks in arabidopsis by integration of metabolomics and transcriptomics. *Journal of Biological Chemistry,* 280, 25590-5.

Hirai, M. Y., M. Yano, D. B. Goodenowe, S. Kanaya, T. Kimura, M. Awazuhara, M. Arita, T. Fujiwara & K. Saito (2004) Integration of transcriptomics and metabolomics for understanding of global responses to nutritional stresses in Arabidopsis thaliana. *Proceedings of the National Academy of Sciences of the United States of America,* 101, 10205-10.

Hollywood, K., D. R. Brison & R. Goodacre (2006) Metabolomics: Current technologies and future trends. *Proteomics,* 6, 4716-4723.

Holmes, D., D. Pettigrew, C. H. Reccius, J. D. Gwyer, C. van Berkel, J. Holloway, D. E. Davies & H. Morgan (2009) Leukocyte analysis and differentiation using high speed microfluidic single cell impedance cytometry. *Lab on a Chip,* 9, 2881-2889.

Hoops, S., S. Sahle, R. Gauges, C. Lee, J. Pahle, N. Simus, M. Singhal, L. Xu, P. Mendes & U. Kummer. 2006. COPASI--a COmplex PAthway SImulator. In *Bioinformatics,* 3067-74. England.

Hsieh, Y., R. Casale, E. Fukuda, J. W. Chen, I. Knemeyer, J. Wingate, R. Morrison & W. Korfmacher (2006) Matrix-assisted laser desorption/ionization imaging mass spectrometry for direct measurement of clozapine in rat brain tissue. *Rapid Communications in Mass Spectrometry,* 20, 965-972.

Huang, L. & R. T. Kennedy (1995) Exploring single-cell dynamics using chemically-modified microelectrodes. *Trends in Analytical Chemistry,* 14, 158-164.

Irish, J. M., N. Kotecha & G. P. Nolan (2006) Innovation - Mapping normal and cancer cell signalling networks: towards single-cell proteomics. *Nature Reviews Cancer,* 6, 146-155.

Ishii, N. & M. Tomita. 2009. *Multi-Omics Data-Driven Systems Biology of E. coli.*

Jones, J. J., S. Borgmann, C. L. Wilkins & R. M. O'Brien (2006) Characterizing the phospholipid profiles in mammalian tissues by MALDI FTMS. *Analytical Chemistry,* 78, 3062-3071.

Jewett, M., M. Hansen & J. Nielsen. 2007a. Data acquisition, analysis, and mining: Integrative tools for discerning. In *Metabolomics: A Powerful Tool in Systems Biology,* eds. J. Nielsen & M. C. Jewett, 159-187. Springer.

---. 2007b. Data acquisition, analysis, and mining: Integrative tools for discerning. In *Metabolomics, A Powerful Tool in Systems Biology,* eds. J. Nielsen & M. C. Jewett, 159-187. Springer.

Jonsson, P., S. J. Bruce, T. Moritz, J. Trygg, M. Sjostrom, R. Plumb, J. Granger, E. Maibaum, J. K. Nicholson, E. Holmes & H. Antti (2005) Extraction, interpretation and validation of information for comparing samples in metabolic LC/MS data sets. *Analyst,* 130, 701-7.

Kaartinen, J., Y. Hiltunen, P. T. Kovanen & M. Ala-Korpela (1998) Application of self-organizing maps for the detection and classification of human blood plasma lipoprotein lipid profiles on the basis of 1H NMR spectroscopy data. *NMR in Biomedicine,* 11, 168-176.

Kandpal, R. P., B. Saviola & J. Felton (2009) The era of 'omics unlimited. *Biotechniques,* 46, 351.

Kanaya, S., M. Kinouchi, T. Abe, Y. Kudo, Y. Yamada, T. Nishi, H. Mori & T. Ikemura (2001) Analysis of codon usage diversity of bacterial genes with a self-organizing map (SOM): characterization of horizontally transferred genes with emphasis on the *E. coli* O157 genome. *Gene,* 276, 89-99.

Khatib-Shahidi, S., M. Andersson, J. L. Herman, T. A. Gillespie & R. M. Caprioli (2006) Direct molecular analysis of whole-body animal tissue sections by imaging MALDI mass spectrometry. *Analytical Chemistry,* 78, 6448-6456.

Kind, T. & O. Fiehn. 2006. Metabolomic database annotations via query of elemental compositions: mass accuracy is insufficient even at less than 1 ppm. In *BMC Bioinformatics*, 234 7:234. England.

Klamt, S. & E. D. Gilles (2004) Minimal cut sets in biochemical reaction networks. *Bioinformatics,* 20, 226-34.

Klamt, S., J. Saez-Rodriguez & E. D. Gilles (2007) Structural and functional analysis of cellular networks with CellNetAnalyzer. *BMC Systems Biology,* 1, 2.

Klamt, S., J. Saez-Rodriguez, J. A. Lindquist, L. Simeoni & E. D. Gilles (2006) A methodology for the structural and functional analysis of signaling and regulatory networks. *BMC Bioinformatics,* 7, 56.

Klamt, S. & S. Schuster (2002) Calculating as many fluxes as possible in underdetermined metabolic networks. *Mol Biol Rep,* 29, 243-8.

Klamt, S. & J. Stelling (2003) Two approaches for metabolic pathway analysis? *Trends in Biotechnology,* 21, 64-9.

Kohonen, T. (1998) The self-organizing map. *Neurocomputing,* 21, 1-6.

---. 2001. *Self-Organizing Maps.* Springer.

Kouskoumvekaki, I., Z. Yang, S. O. Jonsdottir, L. Olsson & G. Panagiotou. 2008. Identification of biomarkers for genotyping Aspergilli using non-linear methods for clustering and classification. In *BMC Bioinformatics,* 59. England.

Kummel, A., S. Panke & M. Heinemann (2006) Putative regulatory sites unraveled by network-embedded thermodynamic analysis of metabolome data. *Molecular Systems Biology,* 2, 2006 0034.

Le Cao, K. A., P. G. Martin, C. Robert-Granie & P. Besse (2009) Sparse canonical methods for biological data integration: application to a cross-platform study. *BMC Bioinformatics,* 10, 34.

Le Cao, K. A., D. Rossouw, C. Robert-Granie & P. Besse (2008) A sparse PLS for variable selection when integrating omics data. *Statistical and Applied Genetics and Molecular Biology,* 7, Article 35.

Lenz, E. M. & I. D. Wilson (2007) Analytical strategies in metabonomics. *Journal of Proteome Research,* 6, 443-458.

Lei, F. & S. B. Jorgensen. 2001. Estimation of kinetic parameters in a structured yeast model using regularisation. In *Journal of Biotechnology,* 223-37. Netherlands.

Lerouxel, O., T. S. Choo, M. Seveno, B. Usadel, L. Faye, P. Lerouge & M. Pauly (2002) Rapid structural phenotyping of plant cell wall mutants by enzymatic oligosaccharide fingerprinting. *Plant Physiology,* 130, 1754-1763.

Li, X., X. Lu, J. Tian, P. Gao, H. Kong & G. Xu (2009) Application of fuzzy c-means clustering in data analysis of metabolomics. *Analytical Chemistry,* 81, 4468-75.

Lin, J. & J. Qian (2007) Systems biology approach to integrative comparative genomics. *Expert Review of Proteomics,* 4, 107-119.

Lindon, J. C., E. Holmes & J. K. Nicholson (2003) So whats the deal with metabonomics? Metabonomics measures the fingerprint of biochemical perturbations caused by disease, drugs, and toxins. *Analytical Chemistry,* 75, 384A-391A.

Lindon, J. C., J. K. Nicholson & I. D. Wilson (2000) Directly coupled HPLC-NMR and HPLC-NMR-MS in pharmaceutical research and development. *Journal of Chromatography B,* 748, 233-258.

Listgarten, J. & A. Emili. 2005. Statistical and computational methods for comparative proteomic profiling using liquid chromatography-tandem mass spectrometry. In *Molecular and Cellular Proteomics,* 419-34. United States.

Lu, K.-Y., A. M. Wo, Y.-J. Lo, K.-C. Chen, C.-M. Lin & C.-R. Yang (2006) Three dimensional electrode array for cell lysis via electroporation. *Biosensors & Bioelectronics,* 22, 568-574.

Mahadevan, S., S. L. Shah, T. J. Marrie & C. M. Slupsky (2008) Analysis of metabolomic data using support vector machines. *Analytical Chemistry,* 80, 7562-70.

Mallouchos, A., M. Komaitis, A. Koutinas & M. Kanellaki (2002) Investigation of volatiles evolution during the alcoholic fermentation of grape must using free and immobilized cells with the help of solid phase microextraction (SPME) headspace sampling. *Journal of Agricultural and Food Chemistry,* 50, 3840-3848.

Mas, S., S. G. Villas-Boas, M. E. Hansen, M. Akesson & J. Nielsen (2007) A comparison of direct infusion MS and GC-MS for metabolic footprinting of yeast mutants. *Biotechnology and Bioengineering,* 96, 1014-1022.

Masujima, T. (2009) Live single-cell mass spectrometry. *Analytical Sciences,* 25, 953-960.

Mattoli, L., F. Cangi, A. Maidecchi, C. Ghiara, E. Ragazzi, M. Tubaro, L. Stella, F. Tisato & P. Traldi (2006) Metabolomic fingerprinting of plant extracts. *Journal of Mass Spectrometry,* 41, 1534-1545.

Maxwell, R. J., I. Martinez-Perez, S. Cerdan, M. E. Cabanas, C. Arus, A. Moreno, A. Capdevila, E. Ferrer, F. Bartomeus, A. Aparicio, G. Conesa, J. M. Roda, F. Carceller, J. M. Pascual, S. L. Howells, R. Mazucco & J. R. Griffiths (1998) Pattern recognition analysis of 1H NMR spectra from perchloric acid extracts of human brain tumor biopsies. *Magnetic Resonance in Medicine,* 39, 869-77.

Mellors, J. S., V. Gorbounov, R. S. Ramsey & J. M. Ramsey (2008) Fully integrated glass microfluidic device for performing high-efficiency capillary electrophoresis and electrospray ionization mass spectrometry. *Analytical Chemistry,* 80, 6881-6887.

Meija, J. (2006) Mathematical tools in analytical mass spectrometry. *Analytical and Bioanalytical Chemistry*, 486-499.

Meinicke, P., T. Lingner, A. Kaever, K. Feussner, C. Gobel, I. Feussner, P. Karlovsky & B. Morgenstern. 2008. Metabolite-based clustering and visualization of mass spectrometry data using one-dimensional self-organizing maps. In *Algorithms Mol Biol*, 9. England.

Moles, C. G., P. Mendes & J. R. Banga (2003) Parameter Estimation in Biochemical Pathways: A Comparison of Global Optimization Methods. *Genome Research*, 13, 2467-2474.

Monton, M. R. N. & T. Soga (2007) Metabolome analysis by capillary electrophoresis-mass spectrometry. *Journal of Chromatography A*, 1168, 237-246.

Motta, A., D. Paris & D. Melck (2010) Monitoring real-time metabolism of living cells by fast two-dimensional NMR spectroscopy. *Analytical Chemistry*, 82, 2405-2411.

Mylonas, R., Y. Mauron, A. Masselot, P. A. Binz, N. Budin, M. Fathi, V. Viette, D. F. Hochstrasser & F. Lisacek (2009) X-Rank: a robust algorithm for small molecule identification using tandem mass spectrometry. *Analytical Chemistry*, 81, 7604-10.

Neumann, S. & S. Bocker (2010) Computational mass spectrometry for metabolomics: Identification of metabolites and small molecules. *Analytical and Bioanalytical Chemistry*, 2779-2788.

Noteborn, H., A. Lommen, R. C. van der Jagt & J. M. Weseman (2000) Chemical fingerprinting for the evaluation of unintended secondary metabolic changes in transgenic food crops. *Journal of Biotechnology*, 77, 103-114.

Oberacher, H., M. Pavlic, K. Libiseller, B. Schubert, M. Sulyok, R. Schuhmacher, E. Csaszar & H. C. Kofeler (2009) On the inter-instrument and the inter-laboratory transferability of a tandem mass spectral reference library: 2. Optimization and characterization of the search algorithm. *Journal of Mass Spectrometry*, 44, 494-502.

Ott, K. H., N. Aranibar, B. Singh & G. W. Stockton. 2003. Metabonomics classifies pathways affected by bioactive compounds. Artificial neural network classification of NMR spectra of plant extracts. In *Phytochemistry*, 971-85. United States.

Papin, J. A., J. Stelling, N. D. Price, S. Klamt, S. Schuster & B. O. Palsson. 2004. Comparison of network-based pathway analysis methods. In *Trends in Biotechnology*, 400-5. England.

Pir, P., B. Kirdar, A. Hayes, Z. Y. Onsan, K. O. Ulgen & S. G. Oliver (2006) Integrative investigation of metabolic and transcriptomic data. *BMC Bioinformatics*, 7, 203.

Podwojski, K., A. Fritsch, D. C. Chamrad, W. Paul, B. Sitek, P. Mutzel, C. Stephan, H. E. Meyer, W. Urfer, K. Ickstadt & J. Rahnenführer (2009) Retention Time Alignment Algorithms for LC/MS Data must consider Nonlinear Shifts. *Bioinformatics*. 25: 758-64.

Pohjanen, E., E. Thysell, J. Lindberg, I. Schuppe-Koistinen, T. Moritz, P. Jonsson & H. Antti (2006) Statistical multivariate metabolite profiling for aiding biomarker pattern detection and mechanistic interpretations in GC/MS based metabolomics. *Metabolomics*, 2, 257-268-268.

Price, N. D., J. A. Papin, C. H. Schilling & B. O. Palsson (2003) Genome-scale microbial in silico models: the constraints-based approach. *Trends in Biotechnology*, 21, 162-9.

Raamsdonk, L. M., B. Teusink, D. Broadhurst, N. S. Zhang, A. Hayes, M. C. Walsh, J. A. Berden, K. M. Brindle, D. B. Kell, J. J. Rowland, H. V. Westerhoff, K. van Dam & S. G. Oliver (2001) A functional genomics strategy that uses metabolome data to reveal the phenotype of silent mutations. *Nature Biotechnology*, 19, 45-50.

Ramautar, R., A. Demirci & G. J. de Jong (2006) Capillary electrophoresis in metabolomics. *Trac-Trends in Analytical Chemistry*, 25, 455-466.

Rashed, M. S., M. P. Bucknall, D. Little, A. Awad, M. Jacob, M. Alamoudi, M. Alwattar & P. T. Ozand (1997) Screening blood spots for inborn errors of metabolism by electrospray tandem mass spectrometry with a microplate batch process and a computer algorithm for automated flagging of abnormal profiles. *Clinical Chemistry*, 43, 1129-1141.

Reiner, A., D. Yekutieli & Y. Benjamini (2003) Identifying differentially expressed genes using false discovery rate controlling procedures. *Bioinformatics*, 19, 368-75.

Reyzer, M. L. & R. M. Caprioli (2007) MALDI-MS-based imaging of small molecules and proteins in tissues. *Current Opinion in Chemical Biology*, 11, 29-35.

Roessner, U., C. Wagner, J. Kopka, R. N. Trethewey & L. Willmitzer (2000) Simultaneous analysis of metabolites in potato tuber by gas chromatography-mass spectrometry. *Plant Journal*, 23, 131-142.

Roessner-Tunali, U., B. Hegemann, A. Lytovchenko, F. Carrari, C. Bruedigam, D. Granot & A. R. Fernie (2003) Metabolic profiling of transgenic tomato plants overexpressing hexokinase reveals that the influence of hexose phosphorylation diminishes during fruit development. *Plant Physiology*, 133, 84-99.

Rubakhin, S. S., J. C. Jurchen, E. B. Monroe & J. V. Sweedler (2005) Imaging mass spectrometry: fundamentals and applications to drug discovery. *Drug Discovery Today*, 10, 823-837.

Rubingh, C. M., S. Bijlsma, R. H. Jellema, K. M. Overkamp, M. J. van der Werf & A. K. Smilde (2009) Analyzing longitudinal microbial metabolomics data. *Journal of Proteome Research*, 8, 4319-27.

Sato, S., T. Soga, T. Nishioka & M. Tomita (2004) Simultaneous determination of the main metabolites in rice leaves using capillary electrophoresis mass spectrometry and capillary electrophoresis diode array detection. *Plant Journal*, 40, 151-163.

Schmid, A., H. Kortmann, P. S. Dittrich & L. M. Blank (2010) Chemical and biological single cell analysis. *Current Opinion in Biotechnology*, 21, 12-20.

Schilling, C. H., J. S. Edwards, D. Letscher & B. Palsson (2000a) Combining pathway analysis with flux balance analysis for the comprehensive study of metabolic systems. *Biotechnology & Bioengineering*, 71, 286-306.

Schilling, C. H., D. Letscher & B. O. Palsson (2000b) Theory for the systemic definition of metabolic pathways and their use in interpreting metabolic function from a pathway-oriented perspective. *Journal of Theoretical Biology*, 203, 229-48.

Schilling, C. H., S. Schuster, B. O. Palsson & R. Heinrich (1999) Metabolic pathway analysis: basic concepts and scientific applications in the post-genomic era. *Biotechnology Progress*, 15, 296-303.

Schrader, B., H. H. Klump, K. Schenzel & H. Schulz (1999) Non-destructive NIR FT Raman analysis of plants. *Journal of Molecular Structure*, 509, 201-212.

Schroer, K., B. Zelic, M. Oldiges & S. Lutz (2009) Metabolomics for biotransformations: Intracellular redox cofactor analysis and enzyme kinetics offer insight into whole cell processes. *Biotechnology & Bioengineering*, 104, 251-60.

Schuster, K. C., I. Reese, E. Urlaub, J. R. Gapes & B. Lendl (2000) Multidimensional information on the chemical composition of single bacterial cells by confocal Raman microspectroscopy. *Analytical Chemistry*, 72, 5529-5534.

Schuster, S., T. Pfeiffer, F. Moldenhauer, I. Koch & T. Dandekar (2002) Exploring the pathway structure of metabolism: decomposition into subnetworks and application to Mycoplasma pneumoniae. *Bioinformatics*, 18, 351-361.

Seger, C. & S. Sturm (2007) Analytical aspects of plant metabolite profiling platforms: Current standings and future aims. *Journal of Proteome Research*, 6, 480-497.

Segre, D., J. Zucker, J. Katz, X. Lin, P. D'Haeseleer, W. P. Rindone, P. Kharchenko, D. H. Nguyen, M. A. Wright & G. M. Church (2003) From annotated genomes to metabolic flux models and kinetic parameter fitting. *OMICS*, 7, 301-16.

Shinohara, H. & F. Wang (2007) Real-time detection of dopamine released from a nerve model cell by an enzyme-catalyzed luminescence method and its application to drug assessment. *Analytical Sciences*, 23, 81-84.

Singh, O. V. & N. S. Nagaraj (2006) Transcriptomics, proteomics and interactomics: unique approaches to track the insights of bioremediation. *Briefings in Functional Genomics & Proteomics*, 4, 355-362.

Soga, T., Y. Ohashi, Y. Ueno, H. Naraoka, M. Tomita & T. Nishioka (2003) Quantitative metabolome analysis using capillary electrophoresis mass spectrometry. *Journal of Proteome Research*, 2, 488-494.

Southam, A. D., T. G. Payne, H. J. Cooper, T. N. Arvanitis & M. R. Viant (2007) Dynamic range and mass accuracy of wide-scan direct infusion nanoelectrospray Fourier transform ion cyclotron resonance mass spectrometry-based metabolomics increased by the spectral stitching method. *Analytical Chemistry*, 79, 4595-4602.

Stein, S. E. (1994) Estimating probabilities of correct identification from results of mass spectral library searches. *Journal of the American Society for Mass Spectrometry*, 5, 316-323.

Stelling, J., S. Klamt, K. Bettenbrock, S. Schuster & E. D. Gilles (2002) Metabolic network structure determines key aspects of functionality and regulation. *Nature*, 420, 190-193.

Stephanopoulos, G., H. Alper & J. Moxley (2004) Exploiting biological complexity for strain improvement through systems biology. *Nature Biotechnology*, 22, 1261-1267.

Steuer, R. 2006. Review: on the analysis and interpretation of correlations in metabolomic data. In *Briefing in Bioinformatics*, 151-8. England.

---. 2007. Computational approaches to the topology, stability and dynamics of metabolic networks. In *Phytochemistry*, 2139-51. United States.

Storey, J. D. (2002) A direct approach to false discovery rates. *Journal of the Royal Statistical Society: Series B (Statistical Methodology)*, 64, 479-498.

Storey, J. D. & R. Tibshirani. 2003. Statistical significance for genomewide studies. In *Proceedings of the National Academy of Sciences of the United States of America*, 9440-5. United States.

Strelkov, S., M. von Elstermann & D. Schomburg (2004) Comprehensive analysis of metabolites in *Corynebacterium glutamicum* by gas chromatography/mass spectrometry. *Biological Chemistry*, 385, 853-861.

Sturm, S., C. Seger & H. Stuppner (2007) Analysis of Central European Corydalis species by nonaqueous capillary electrophoresis-electrospray ion trap mass spectrometry. *Journal of Chromatography A*, 1159, 42-50.

Suits, F., J. Lepre, P. Du, R. Bischoff & P. Horvatovich (2008) Two-dimensional method for time aligning liquid chromatography-mass spectrometry data. *Analytical Chemistry*, 80, 3095-104.

Sumner, L., A. Amberg, D. Barrett, M. Beale, R. Beger, C. Daykin, T. Fan, O. Fiehn, R. Goodacre, J. Griffin, T. Hankemeier, N. Hardy, J. Harnly, R. Higashi, J. Kopka, A. Lane, J. Lindon, P. Marriott, A. Nicholls, M. Reily, J. Thaden & M. Viant (2007) Proposed minimum reporting standards for chemical analysis. *Metabolomics*, 211-221.

Sysi-Aho, M., M. Katajamaa, L. Yetukuri & M. Oresic. 2007. Normalization method for metabolomics data using optimal selection of multiple internal standards. In *BMC Bioinformatics*, 8: 93. England.

Takats, Z., J. M. Wiseman, B. Gologan & R. G. Cooks (2004) Mass spectrometry sampling under ambient conditions with desorption electrospray ionization. *Science*, 306, 471-473.

Talaty, N., Z. Takats & R. G. Cooks (2005) Rapid in situ detection of alkaloids in plant tissue under ambient conditions using desorption electrospray ionization. *Analyst*, 130, 1624-1633.

Tanaka, K., A. Westdull, D. G. Hine, T. B. Lynn & T. Lowe (1980) Gas-chromategraphic method of analysis for urinary organic-acids. 2. Description of the procedure, and its application to diagnosis of patients with organic acidurias. *Clinical Chemistry*, 26, 1847-1853.

Tapp, H. S. & E. K. Kemsley (2009) Notes on the practical utility of OPLS. *TrAC Trends in Analytical Chemistry*, 28, 1322-1327.

Tibshirani, R. (1996) Regression shrinkage and selection via the Lasso. *Journal of the Royal Statistical Society Series B-Methodological*, 58, 267-288.

Trygg, J. & T. Lundstedt. 2007. Chemometrics Techniques for Metabonomics. In *The Handbook of Metabonomics and Metabolomics*, eds. J. C. Lindon, J. K. Nicholson & E. Holmes. London: Elseivier.

Urban, P. L., K. Jefimovs, A. Amantonico, S. R. Fagerer, T. Schmid, S. Maedler, J. Puigmarti-Luis, N. Goedecke & R. Zenobi (2010) High-density micro-arrays for mass spectrometry. *Lab on a Chip*, 10, 3206-3209.

Urbanczik, R. & C. Wagner (2005) An improved algorithm for stoichiometric network analysis: theory and applications. *Bioinformatics*, 21, 1203-10.

Urbanczyk-Wochniak, E., A. Luedemann, J. Kopka, J. Selbig, U. Roessner-Tunali, L. Willmitzer & A. R. Fernie (2003) Parallel analysis of transcript and metabolic profiles: a new approach in systems biology. *EMBO Report*, 4, 989-93.

Urbanczyk-Wochniak, E., L. Willmitzer & A. R. Fernie (2007) Integrating profiling data: using linear correlation to reveal coregulation of transcript and metabolites. *Methods in Molecular Biology*, 358, 77-85.

van den Berg, R. A., C. M. Rubingh, J. A. Westerhuis, M. J. van der Werf & A. K. Smilde (2009) Metabolomics data exploration guided by prior knowledge. *Anal Chim Acta*, 651, 173-81.

van Nederkassel, A. M., M. Daszykowski, P. H. Eilers & Y. V. Heyden. 2006. A comparison of three algorithms for chromatograms alignment. In *J Chromatogr A*, 199-210. Netherlands.

van Velzen, E. J., J. A. Westerhuis, J. P. van Duynhoven, F. A. van Dorsten, C. H. Grun, D. M. Jacobs, G. S. Duchateau, D. J. Vis & A. K. Smilde (2009) Phenotyping tea consumers by nutrikinetic analysis of polyphenolic end-metabolites. *Journal of Proteome Research*, 8, 3317-30.

Vandenbogaert, M., S. Li-Thiao-Te, H. M. Kaltenbach, R. Zhang, T. Aittokallio & B. Schwikowski (2008) Alignment of LC-MS images, with applications to biomarker discovery and protein identification. *Proteomics,* 8, 650-72.

Varma, A. & B. O. Palsson (1994) Metabolic Flux Balancing: Basic Concepts, Scientific and Practical Use. *Nature Biotechnology,* 12, 994-998.

Villas-Boas, S. G., S. Mas, M. Akesson, J. Smedsgaard & J. Nielsen (2005) Mass spectrometry in metabolome analysis. *Mass Spectrometry Reviews,* 24, 613-646.

Visser, D. & J. J. Heijnen. 2003. Dynamic simulation and metabolic re-design of a branched pathway using linlog kinetics. In *Metabolic Engineering,* 164-76. United States.

Wagner, C., M. Sefkow & J. Kopka (2003) Construction and application of a mass spectral and retention time index database generated from plant GC/EI-TOF-MS metabolite profiles. *Phytochemistry,* 62, 887-900.

Wang, D. & S. Bodovitz (2010) Single cell analysis: the new frontier in 'omics'. *Trends in Biotechnology,* 28, 281-290.

Warrack, B. M., S. Hnatyshyn, K. H. Ott, M. D. Reily, M. Sanders, H. Zhang & D. M. Drexler. 2009. Normalization strategies for metabonomic analysis of urine samples. In *J Chromatogr B Analyt Technol Biomed Life Sci,* 547-52. Netherlands.

Webb-Robertson, B. J., L. A. McCue, N. Beagley, J. E. McDermott, D. S. Wunschel, S. M. Varnum, J. Z. Hu, N. G. Isern, G. W. Buchko, K. McAteer, J. G. Pounds, S. J. Skerrett, D. Liggitt & C. W. Frevert (2009) A Bayesian integration model of high-throughput proteomics and metabolomics data for improved early detection of microbial infections. *Pac Symp Biocomput,* 451-63.

Weckwerth, W., M. E. Loureiro, K. Wenzel & O. Fiehn (2004) Differential metabolic networks unravel the effects of silent plant phenotypes. *Proceedings of the National Academy of Sciences of the United States of America,* 101, 7809-7814.

Weckwerth, W., K. Wenzel & O. Fiehn (2004) Process for the integrated extraction, identification and quantification of metabolites, proteins and RNA to reveal their co-regulation in biochemical networks. *Proteomics,* 4, 78-83.

Wiseman, J. M., S. M. Puolitaival, Z. Takats, R. G. Cooks & R. M. Caprioli (2005) Mass spectrometric profiling of intact biological tissue by using desorption electrospray ionization. *Angewandte Chemie-International Edition,* 44, 7094-7097.

Wittmann, C., J. O. Kromer, P. Kiefer, T. Binz & E. Heinzle (2004) Impact of the cold shock phenomenon on quantification of intracellular metabolites in bacteria. *Analytical Biochemistry,* 327, 135-139.

Xie, Y., W. Pan & A. B. Khodursky (2005) A note on using permutation-based false discovery rate estimates to compare different analysis methods for microarray data. *Bioinformatics,* 21, 4280-4288.

Yassaa, N., E. Brancaleoni, M. Frattoni & P. Ciccioli (2001) Trace level determination of enantiomeric monoterpenes in terrestrial plant emission and in the atmosphere using a beta-cyclodextrin capillary column coupled with thermal desorption and mass spectrometry. *Journal of Chromatography A,* 915, 185-197.

Zhang, J., E. Gonzalez, T. Hestilow, W. Haskins & Y. Huang (2009) Review of peak detection algorithms in liquid-chromatography-mass spectrometry. *Current Genomics,* 10, 388-401.

Zhang, W., F. Li & L. Nie (2010) Integrating multiple 'omics' analysis for microbial biology: application and methodologies. *Microbiology,* 156, 287-301.

6

Metabotype Concept: Flexibility, Usefulness and Meaning in Different Biological Populations

Nabil Semmar[1, 2]
[1]*Plateau BioMeT, UMR INRA 1260, Medical School of Marseilles,Marseille,*
[2]*Université de Tunis El Manar, Institut Supérieur des Sciences Biologiques Appliquées de Tunis (ISSBAT), Rue Zouhair Essafi, Tunis,*
[1]*France*
[2]*Tunisia*

1. Introduction

Metabolism represents a junction system in biological body receiving cumulated signals from upstream (genome, transcriptome, proteome) and downstream (environment) systems. This median position of the metabolic system makes it to be very sensitive toward internal and external signals resulting in its regulatory role in physiological homeostasy and adaptive responses to endogenous and exogenous factors. These factors have initiation, modulation or pressure effects on the biological organisms and species which adapt, resist or react through different types of metabolisms based on different synthesis and regulation levels of metabolites (Wilson, 2009).

These characteristics give to metabolism a flexibility that may be described by means of four variability criteria (Fig. 1a): presence-absence of metabolites, concentration levels, relative levels or ratios between metabolites, and metabolic profiles characterizing different structural and functional states in biosystems through different metabolites' levels. Presence-absence of metabolites is a qualitative criterion that concerns metabolites that are stimulated by particular internal biological states (species physiology, disease, stress, etc.) or external governing factors (climate, threat, etc.). Beyond this binary aspect of metabolic responses, increase or decrease in concentration levels of some metabolites can be sensitive responses to the degree of a governing implicit factor (e.g. wounding, toxin, pollutant exposure levels, etc.). More precision on association between biological state and governing factor can be extracted from concentration ratios between sensitive metabolites. Complex situations integrating many interactive factors can be characterized by metabolic profiles in which several metabolites levels increase or decrease compared with control conditions or neutral situations.

The four metabolic variability criteria can be used separately or in association to provide reliable pictures on different metabolic phenotypes of a biological system; such pictures characterizing the metabolic phenotypes are called metabotypes (or chemotypes) (Fig. 1a). Identifications of relationships between metabotypes and quantitative or qualitative control factors lead to the concept of metabolic markers. Metabolic markers can be used to anticipate (predict), alert and control responses and states of different components in biological systems. These components include cells, biofluids, tissues, organisms and biological populations or species.

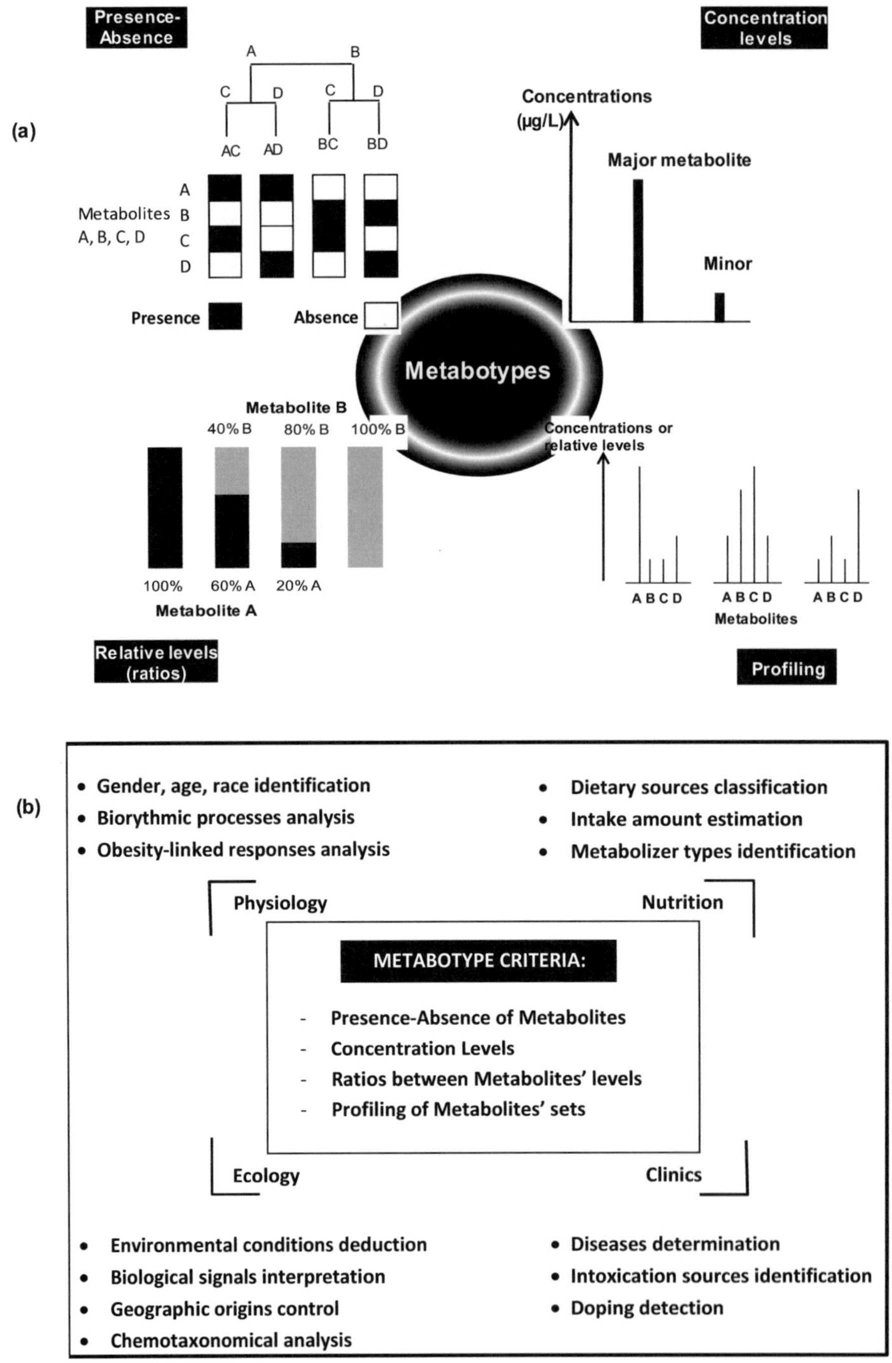

Fig. 1. (a) Schematic representations of four metabotype definitions based on occurrences, concentration levels, ratios and profiles of metabolites in biological matrices. (b) Interest of the four metabotype criteria to develop biomarkers in different biological fields (physiology, nutrition, clinics, ecology).

This chapter illustrates the usefulness of metabotype concept as flexible biomarkers in different biological fields: physiology, clinics, nutrition and ecology (Fig. 1b): in physiology, metabotypes can be markers of biological varieties, gender, aging, biorhythms, etc. (Bell et al., 1991; Bollard et al., 2001; Duncan et al., 2007 ; Holmes et al., 2008). In nutrition, different metabotypes can be correlated to different diets; secondary metabolites, particularly produced by plants, represent strong biomarkers of consumed fruits, herbals and legumes (Kaufman et al., 1997; Yanez, 2008). In clinical field, different metabotypes can be indicator of different disease types, intoxication levels or doping sources (Timbrell, 1998; Kintz et al., 1999; Barderas et al., 2011). In ecology, different plant metabotypes based on secondary metabolites are indicative of different physicochemical conditions (temperature, light, humidity, oxidative stress, etc.) or biological relationships (parasite or herbivore attacks, pollination states, etc.) submitted by the plant (Semmar et al., 2008). Also, metabotype (or chemotype) concept is helpfully used in chemotaxonomy to highlight chemical polymorphism such as chemical varieties of floral colors (Torskangerpoll et al., 2005).

2. Physiological metabotypes

2.1 Metabotypes based on metabolites' occurrences

Presence-absence of metabolites is an efficient metabolomic parameter to characterize different biological species or varieties known to represent different physiological systems (Fig. 2): for instance, benzoic acid is metabolized almost entirely to hippuric acid by primates, rodents and rabbits (mammals) (Fig. 2a). However, it is excreted unchanged and as glucuronide by insects, birds and reptiles (Jones, 1982). Similarly, the excretion of phenylacetic acid (Fig. 2b) as the parent compound or as glutamine, glycine or taurine conjugates is species-dependent (Robertson et al., 2002): for instance, it is excreted as phenylacetyl glutamine in humans and phenylacetylglycine in rats.

Comparison of ^{1}H NMR spectra from control B6C3F1 mouse urines with those of control SD rat urines revealed the presence of guanidinoacetic acid and trimethylamine in mouse but their absence in rat (Bollard et al., 2005).

2.2 Metabotypes based on concentration levels

In animals, several metabolites have been found to quantitatively vary in relation to gender, age, body weight, etc. :

Plasma NMR profiles showed reduced concentration of the triglyceride resonances in female rats compared with male counterparts (Fig.3a) (Stanley, 2002).

In humans, urinary citrate levels have been found to be generally greater in females (Fig.3b) (Hodgkinson, 1962). Moreover, serum cholesterol levels are different between sexes at matched ages, with in general, pre-menopausal females being less susceptible to high cholesterol levels than males (Fig.3c) (Joossen, 1988).

Oestrogenic hormones are known to exhibit some control over the kidney: in humans, glycosuria (glucose in urine) is common in pregnant females (Fig.3d) (Davison and Hytten, 1975). In the rat, pregnancy is associated with an increased glucose filtration rate and decreased urine flow rate (Bishop and Green, 1980).

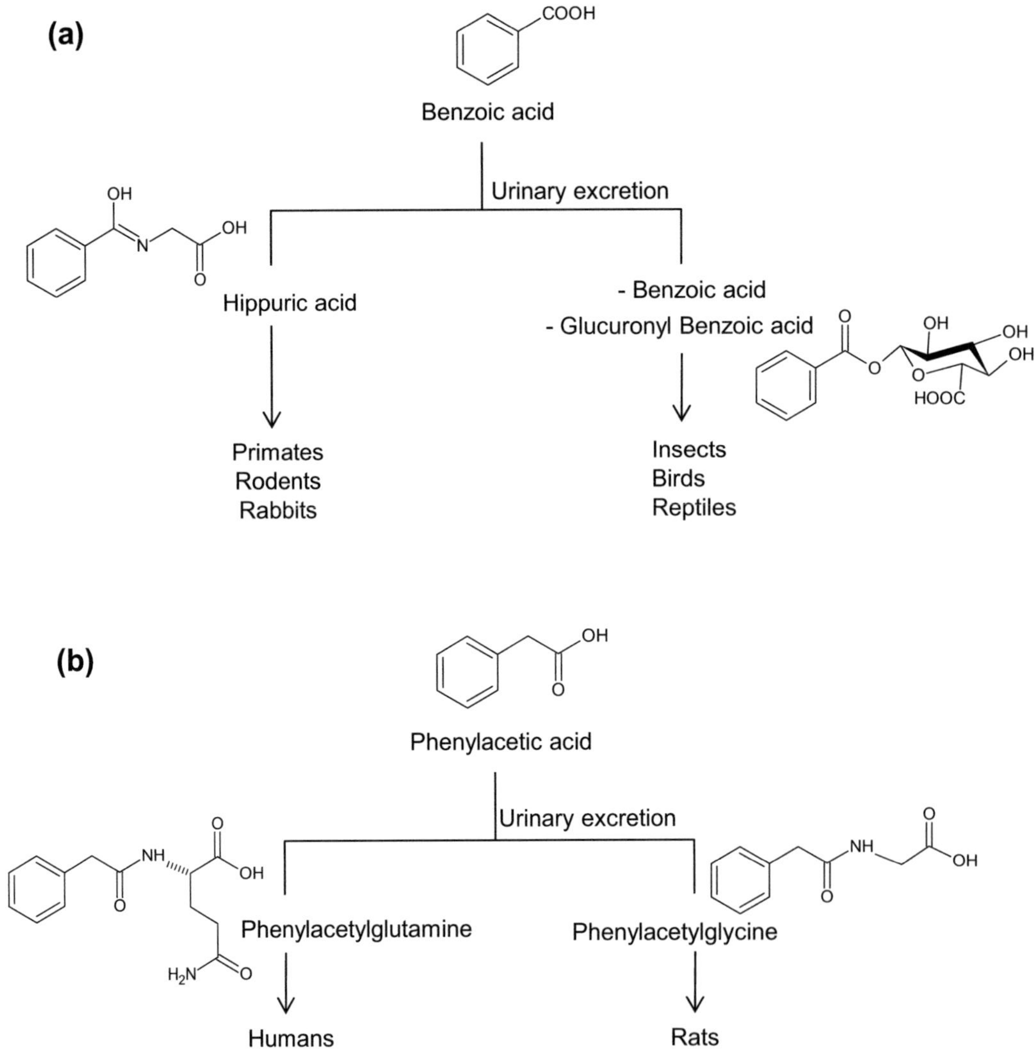

Fig. 2. Metabolic derivatives of benzoic acid (a) and phenylacetic acid (b) excreted in urine of human and animal species and from which specific metabotypes can be defined.

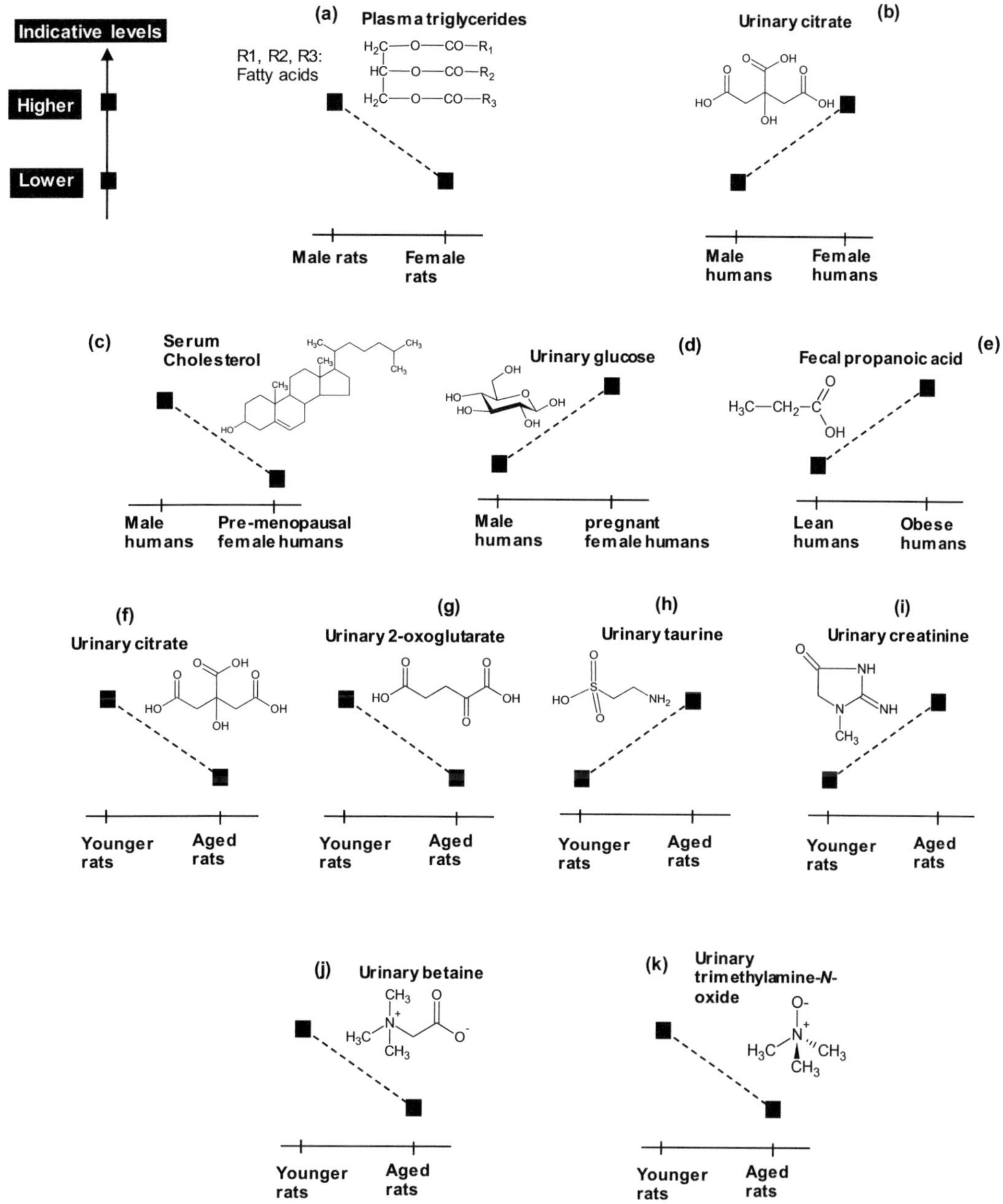

Fig. 3. Quantitative variations of different metabolites in different biological matrices (serum, plasma, urine) in relation to different physiological factors (gender, age, sexual maturity degree, body weight).

Differences in the concentrations of stool short-chain fatty acids (SCFA) between the lean, overweight and obese human subjects have been shown to be important (Schwiertz et al., 2010): the mean of total SCFA concentration in fecal samples of obese volunteers was by more than 20% higher in total than of lean volunteers. The highest increase was seen for propionate (Fig. 3e) with 41%.

The age is known to result in numerous physiological changes that are reflected by physical and metabolic changes:

Previous studies have shown that ageing rats decrease their excretion of citrate and 2-oxoglutarate versus increase in their taurine and creatinine output (Fig.3f-i) (Bell et al., 1991). The increase in creatinine with age (up to 6 months) may be associated with the increase output from the muscle of larger rats as well as an age-related increase in the glomerular filtration rate (the rat kidney becomes fully functional at 3 months).

Moreover, in young rats (1 month or less), the urinary excretion of trimethylglycine (betaine) and trimethylamine-*N*-oxide was higher than in older rats (Fig. 3j, k). The increased level of betaine and trimethylamine-*N*-oxide was higher than in older rats. The increased level of betaine in the urine of young rats may be due to high choline levels (Bell et al., 1991). Betaine (or *N*-trimethylglycine) results from oxidation of choline; it represents a reservoir of methyl groups and plays the role of methyl donor in the synthesis of methionine in mammals.

In plant world, secondary metabolites showed significant variations in relation to species, age and maturity level: for instance, among two birch species, Silver Birch (*Betula pendula*) does not emit sesquiterpenes (SQT), while Downy Birch (*B. pubescens*) does (Hakola et al., 2001). Moreover, older trees *B. pubescens* emitted greater quantities and higher proportions of SQT than younger ones. In the common snapdragon, *Antirrhinum majus*, the emission of methyl benzoate (MeBA) is increased in pollination period (Dudareva et al., 2000); this may serve as guide for bees to find their way inside the flower. After pollination, emission of MeBA decreases dramatically.

SCFA are produced by the intestinal microbiota which represent a large part of bacteria belonging to the phyla of Firmicus, Bacteroides, Actinobacteria, Proeobacteria and Verrumicrobia (Zoetendal et al., 2008). Some phyla were characterized by high level production of some metabolites: for instance, Bacteroides phylum produces high levels of acetate and propionate, whereas several members of the Firmicus phylum produce high amount of butyrates (Maslowski et al., 2009).

2.3 Metabotypes based on ratios between concentration levels

Metabolic ratios between concentration levels of structurally close metabolites have been used to characterize different biological states in human populations:

Menstruation affects the *N*-oxidation of trimethylamine resulting in a fall in the ratio of trimethylamine-*N*-oxide on trimethylamine in the urine (Fig. 4a) (Zhang et al., 1996).

A metabolomic study on different body weight human subjects showed that leaner people had higher ratios of acetate to butyrate and propionate (Fig. 4b) (Schwiertz et al., 2010, Duncan et al., 2007).

It has been shown that in a Caucasian population, the urinary metabolic ratio of 6 β-hydroxycortisol to cortisol was significantly increased in females compared to males (Fig. 4c) (Lutz et al., 2010). This ratio is used as an endogenous marker for CYP3A activity.

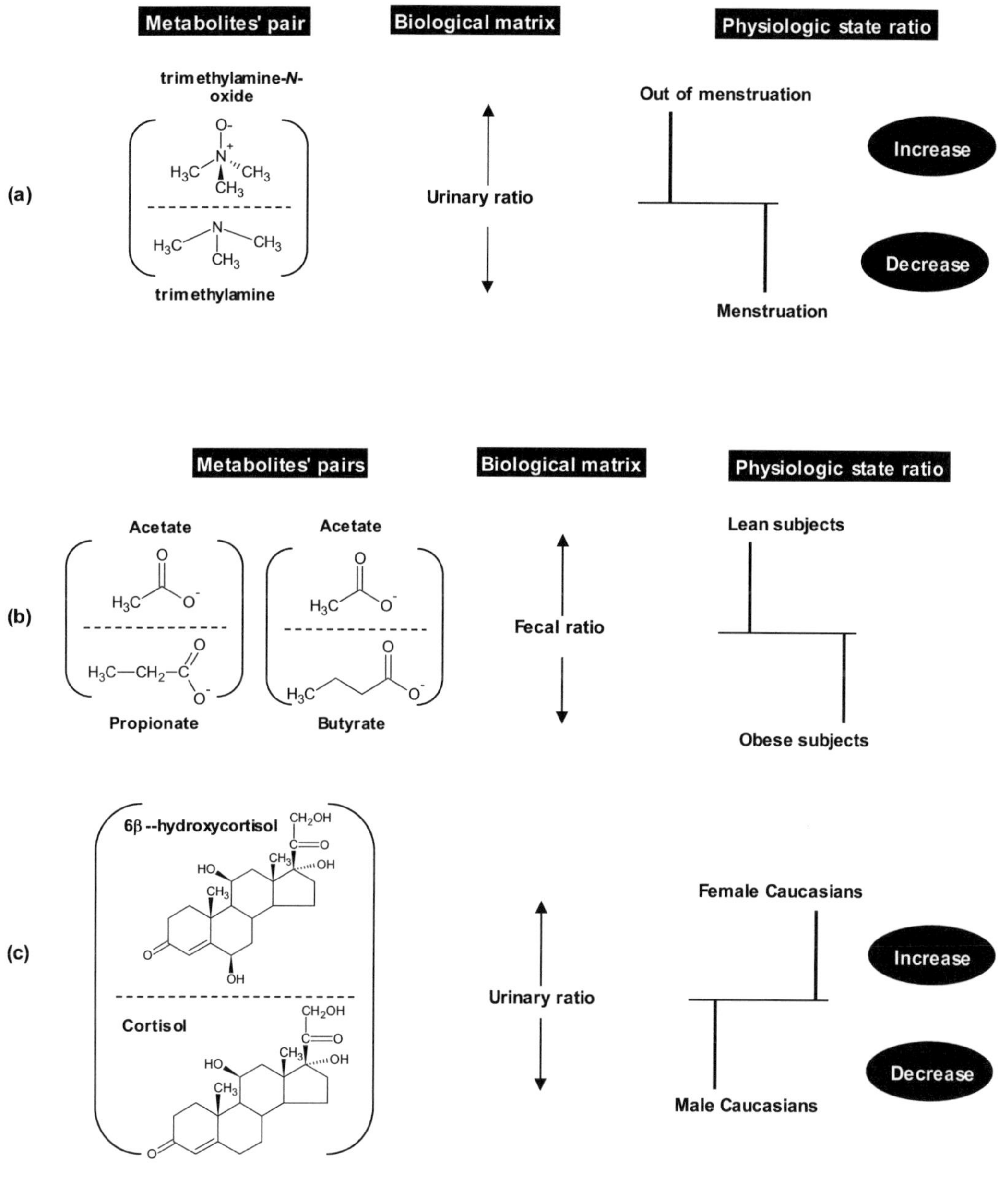

Fig. 4. Metabolic ratios between structurally close metabolites providing biochemical discriminations between different biological states in human populations.

2.4 Metabotypes based on profiling

In animals, gender, species or races have been characterized by specific profiles of organic acid derivatives (amine, tricarboxylic) which are produced at relatively high or low levels:

Metabolic profiling has been applied to characterize genders in rats: elevated levels of bile acid metabolites in urine of female rats reflected increased rate of cholesterol and bile acid synthesis compared to males (Stanley, 2002).

Metabolomic differences between rats and mice have also been highlighted by NMR: ^{1}H-NMR spectra of urines from B6C3F1 mice and Sprague-Dawley (SD) rats revealed consistently higher levels of formate, creatinine, hippurate, dimethylglycine, dimethylamine, fumarate, 2-oxoglutarate and citrate versus lower levels of taurine and betaine (i.e. *N*-trimethylglycine) in the rats compared to mice (Bollard et al., 2005).

^{1}H-NMR spectra of urine samples showed that the genetic strain Alpk:ApfCD mice had relatively higher levels of 2-oxoglutarate, citrate, trimethylamine-*N*-oxide and guanidinoacetic acid, whilst C57BL107 mice had higher levels of taurine, creatinine, dimethylamine and trimethylamine (Fig. 5) (Gavaghan et al., 1996).

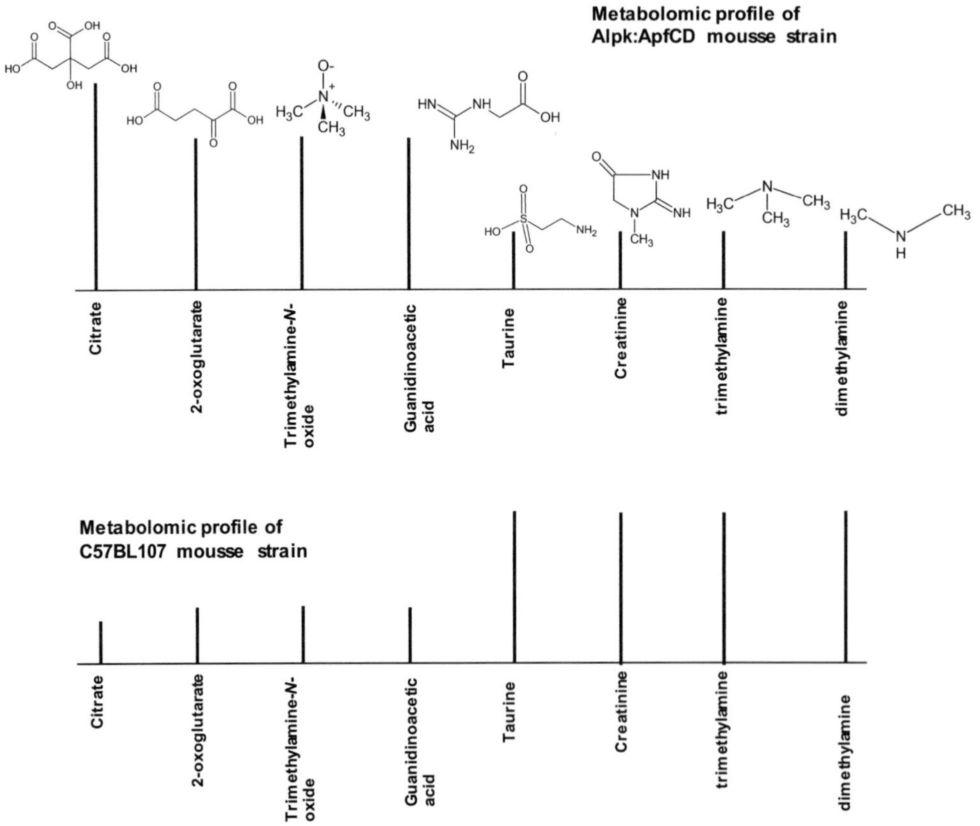

Fig. 5. Metabolomic profiles representing two genetic strains of mice on the basis of relative levels of several metabolites. Bar heights are indicative of relatively higher or lower concentrations depending on mousse strains (Gavaghan et al., 1996).

Metabolic phenotypes analysis has been applied to characterize particular laboratory animal varieties including "germfree" (GF) specimens. GF is the highest quality level of laboratory animals in which there are no any detectable microorganisms in contrast to those commonly known as "SPF", which is merely free of specific pathogens. Germfree animals are especially useful in the researches concerning genetic engineering, cancer, normal intestine flora, immunology and nutrition.

Aqueous extract profiles of gut tissues from GF mice were markedly different from those of conventional mice (Claus et al., 2008) (Fig. 6):

i. The metabolite profile of the duodenum from GF mice was mainly characterized by higher levels of tauro-conjugated bile acids (TCBAs) and alanine versus lower levels of glycerophosphocoline (GPC) (Fig. 6a) when compared with conventional mice.
ii. The jejunal tissue of GF group had higher levels of creatine and TCBAs versus lower levels of tyrosine (Fig. 6b).
iii. The ileum of GF mice was characterized by a higher level of TBCAs and lower levels of glutamate, fumarate, lacate, phophocholine and alanine when compared with the ileum from conventional mice (Fig. 6c).
iv. The metabolic profile of the colon from GF mice revealed a high level in a complex carbohydrate identified as raffinose, and lower levels of lactate, creatine, 5-aminovalerate, propionate, glutamine, myo-inositol, scyllo-inositol, GPC, phosphocholine, choline, formate, uracile and fumarate (Fig. 6d) (Monero and Arus, 1996).

Metabolic profiles of ileum and particularly colon in GF mice were markedly more affected than those of duodenum and jejunum. This reflects the higher microbial loads found in ileum and colon (Dunne, 2001). The lower levels of choline and its phosphorylated derivatives, GPC and phosphocholine (Fig. 6a, c-d) were reported to be likely due to the disturbance of the membrane of colonocytes in GF mice (Claus et al., 2008). Also, the accumulation of the trisaccharide, raffinose, can be a possible consequence of this disruption. In GF animals, raffinose seems to be able to cross the epithelial membrane and accumulates in colonocytes where it induces a rise in osmotic pressure. This phenomenon provokes a well-known signaling cascade that leads to the release of the mobile osmolytes: GPC, myo-inositol and scyllo-inositol.

Beyond static analysis, kinetic metabolic profiling is applied in chronobiology and pharmacokinetics in relation to intrinsic or extrinsic factors (e.g diurnal variations):

In SD rat, ^{1}H-NMR profiles of urinary samples collected during the day showed lower levels of hippurate, taurine, and creatinine together with elevated levels of glucose, succinate, dimethylglycine, glycine, creatine and betaine compared with urine collected during the night (Bollard et al., 2001).

Male rats secrete growth hormone in an "on-off" episodic rhythm between which there are periods when there are no detectable levels of the hormone. Growth hormone secretion in the female rat is "continuous" since hormone levels are always present (Czerniak, 2001).

In women subjects, plasmatic cortisol stimulated by synacthen (synthetic ACTH) showed obesity level-dependent kinetic profiles (Fig. 7) (Semmar et al., 2005a): the secretion and elimination of cortisol were more rapid and higher in the most obese followed by intermediate obese then non-obese subjects.

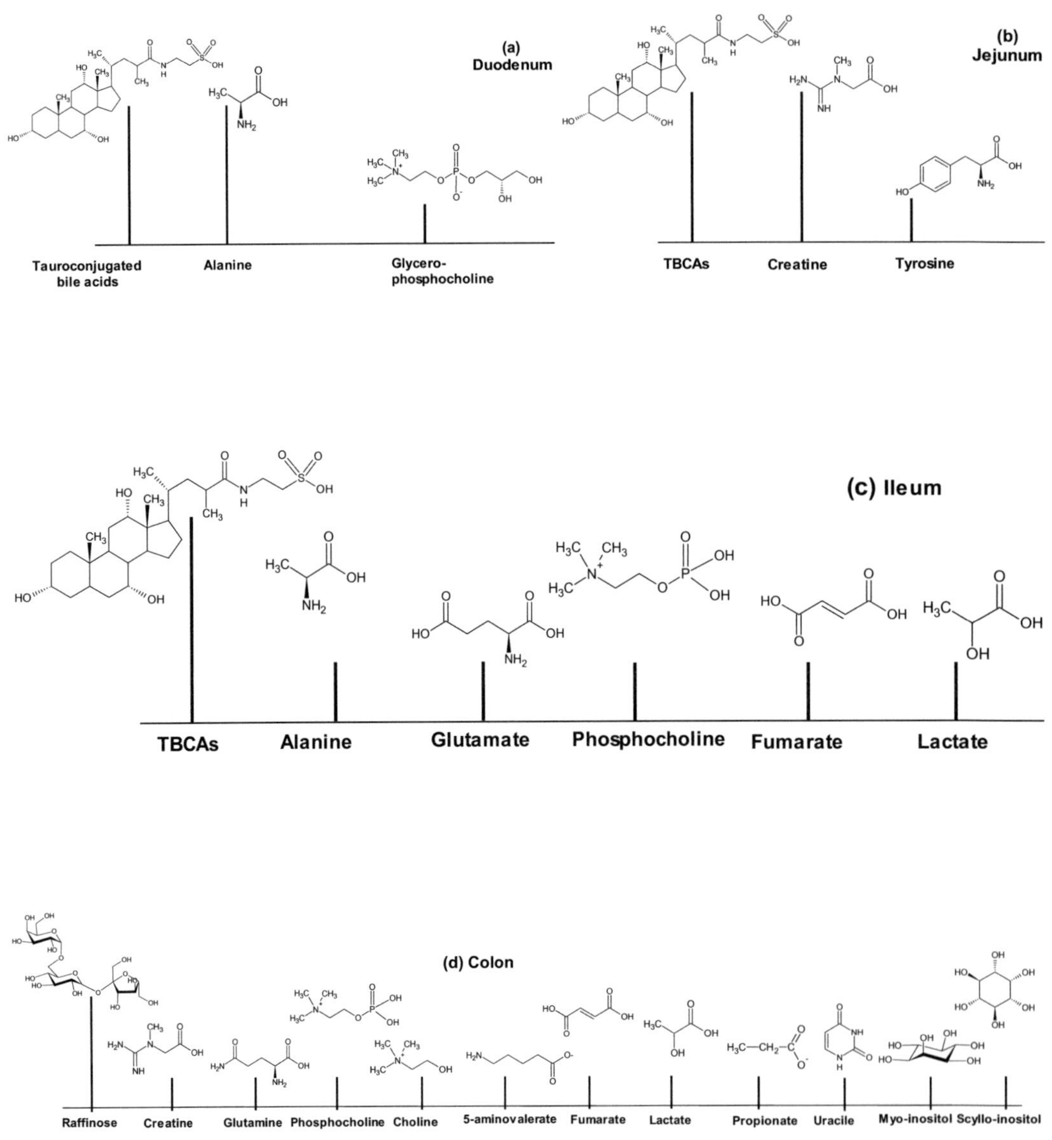

Fig. 6. Metabolomic profiling of different gut tissues of germ free (GF) mice (Monero and Arus, 1996). Bar heights are indicative of relatively higher or lower concentrations in GF compared with conventional mice (details are given in text).

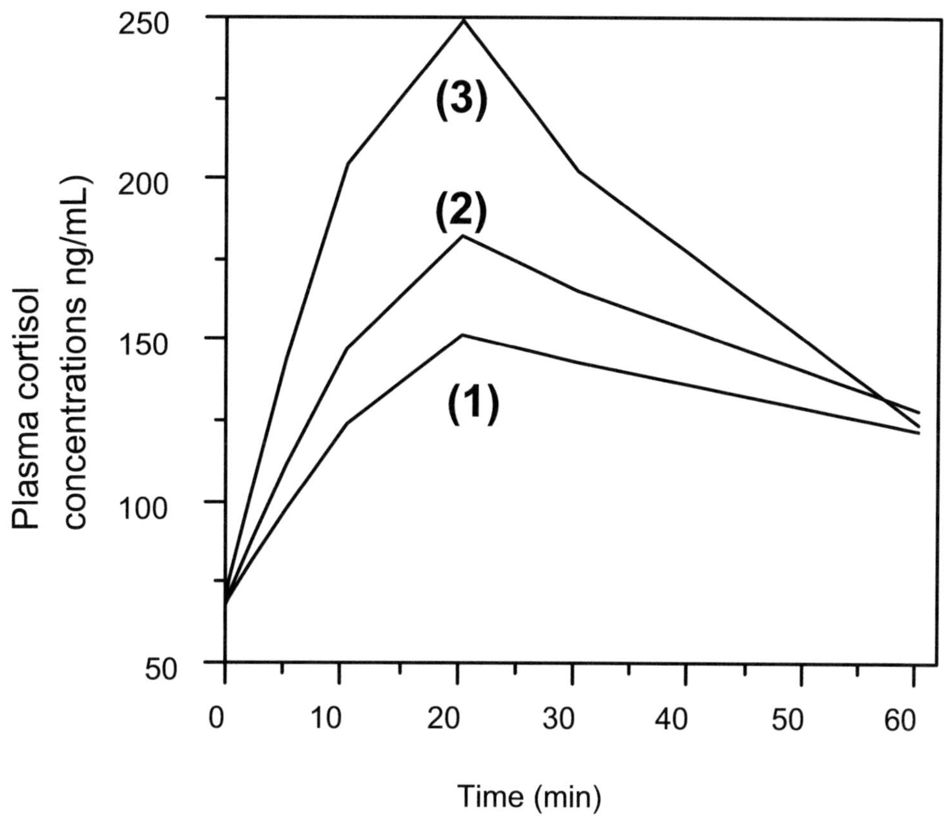

Fig. 7. Kinetic profiles of plasma cortisol concentrations of three body weight levels women populations, highlighting compensatory process between secretion and elimination in relation to obesity levels. (1), (2), (3) corresponds to non-obese, intermediate obese and extreme obese populations, respectively (Semmar et al., 2005).

3. Dietary metabotypes

3.1 Metabotypes based on occurrence of metabolites

Fruits and legumes can be generally characterized by occurrences of specific or abundant secondary metabolites belonging to flavonoids and terpenoids. For instance, in flavonoid class, flavonols, flavones, flavanones, isoflavones, flavanols and anthocyanins are widely present in onions, parsley, citrus fruits, leguminous plants, green tea and blackberry, respectively (Fig. 8) (Majewska et al., 2011; Holden et al., 2005; Kaufman et al., 1997).

Among the flavonols, quercetin is widely present in the plant world. It occurs as different glycosidic forms with quercetin-3-rhamnoglucoside (or rutin) being one of the most widespread forms (Fig. 9d). The different forms of quercetin glycosides have been found to be good markers in food quality control: in onions, quercetin is bound to one or two glucoses to give quercetin-4′-glucoside and quercetin-3,4′-glucoside (Fig. 9c); apples and berries, however, have been characterized by the occurrence of quercetin-3-galactoside and quercetin-3- arabinoside, respectively (Fig. 9b, a) (Kühnau, 1976; Zheng et al., 2003).

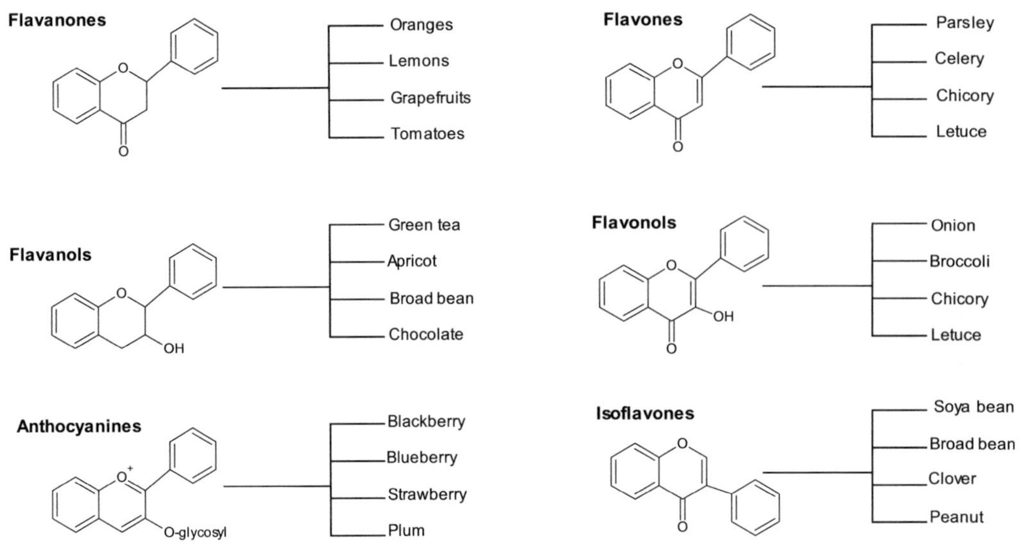

Fig. 8. Metabolic characterization of different dietary plants on the basis of high occurrences of produced flavonoid classes in their tissues.

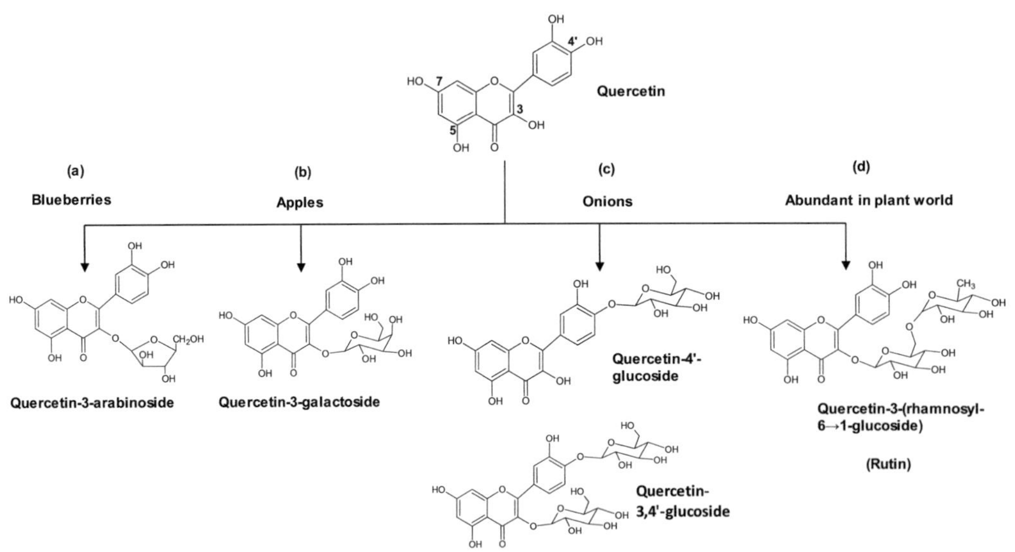

Fig. 9. Metabolic characterization of some dietary plants on the basis of abundant quercetin glycosides in their tissues.

Flavanones are flavonoids particularly abundant in citrus and vary qualitatively in relation to fruit types (Mouly et al., 1998; Kawaii et al., 1999; Gattuso et al., 2007): The lemon (*Citrus limon*) can be distinguished by production of eriocitrin and hesperidin (Fig. 10 a), whereas in grapefruits, naringin predominates in presence of narirutin (Fig. 10c). In oranges (*Citrus sinensis*) and mandarins (*Citrus reticulata*), hesperidin is the major flavanone in presence of narirutin (Fig. 10b).

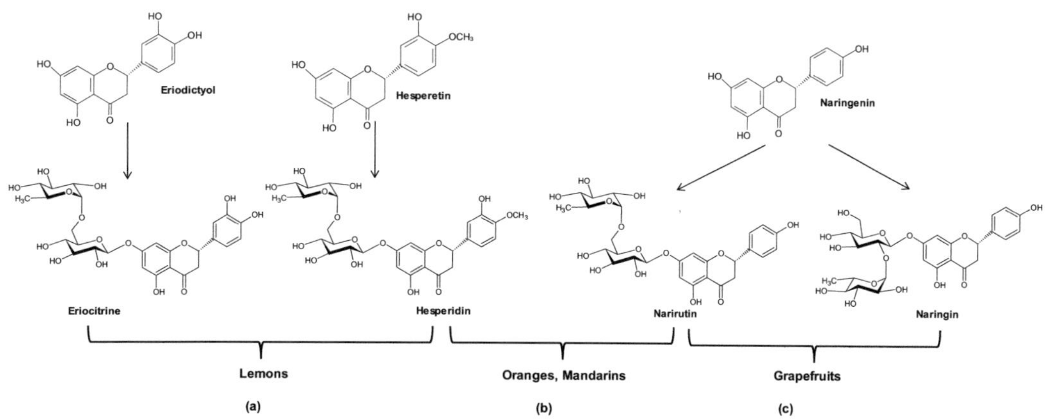

Fig. 10. Metabolomic distinction between different *Citrus* species based on major flavanone glycosides in fruit juices.

3.2 Metabotypes based on concentration levels

High amounts or concentrations of secondary metabolites have been used to control dietary plant varieties as well as to diagnose animal or human diets:

Onions are rich source of quercetin-4′-*O*-glucoside and quercetin-3,4′-*O*-diglucoside. Mullen et al. (2006) reported that 270 g of lightly fried onions contains 275 μmol of flavonol glucosides with the main constituents being 143 μmol of the 4′-*O*-glucoside and 107 μmol of the 3,4′-*O*-diglucoside (Fig. 11).

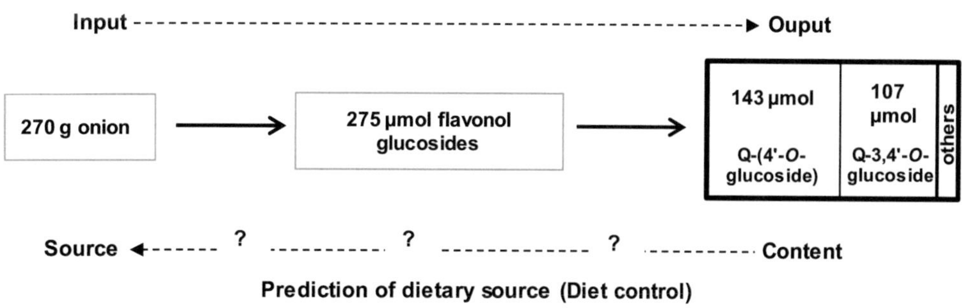

Fig. 11. Abundance of flavonol glucosides in onion particularly dominated by 4′-*O*-glucoside and 3,4′-glucoside of quercetin (Q) (Mullen et al., 2006).

In human subjects, plasma quercetin was found to be a good marker of dietary intake because its concentrations increase with increasing ingested dose (Radtke et al., 2002). In a strictly controlled dietary intervention study, 77 health human subjects consumed either 170 or 850 g of fruits, vegetable and berries daily. Quercetin intake was calculated to be 3 to 24 mg/d on the respective diets. The mean ± SD of plasma quercetin concentration was 78 ± 56 nmol/L during the habitual diet; it decreased to 70% during the low-vegetable diet and increased to 170% during the high-vegetable diet (Freese et al., 2002) (Fig. 12).

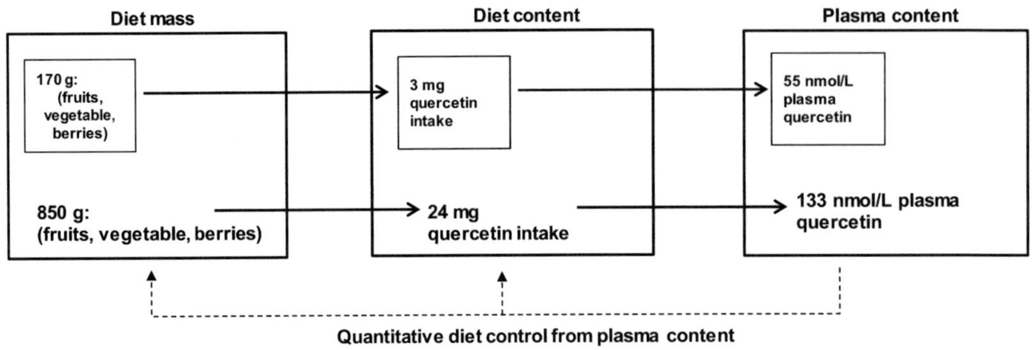

Fig. 12. Increase in plasma quercetin concentration (nmol/L) in relation to diet mass (g) and its total quercetin content (mg) in human population (Freese et al., 2002).

Ingestion of 200 mL of coffee has been reported to increase the plasma conjugated caffeic acid level in human subjects (Nardini et al., 2002).

Following consumption of both green and black tea, human urinary samples showed significant increases in level of hippuric acid and 4-hydroxyphenylacetic acid (Mulder et al., 2005).

Excretion of creatinine in the urine of rats, per unit of skeletal muscle mass, was found to be promoted by food deprivation (Rikimaru et al., 1989).

3.3 Metabotypes based on ratios between concentration levels

Legumes are known to be important dietary sources of isoflavones (Liggins et al., 2000). Women subjects having consumed isoflavones from soymilk powder were classified into two excretion levels according to the relative contents of daidzein and genistein recovered in feces and urine (Fig. 13) (Xu et al., 1995): in strong isoflavone excreters, the percentages of daidzein and genistein recovered in feces were 10 and 20 times greater than those weak excreters. Isoflavones recovered in urine for 48h revealed excretions of daidzein and genistein which were 2 and 3.5 times in high excreters than low excreters.

In analogous study, human subjects consumed a single dose of strawberries (250 g), raspberries (225 g) and walnuts (35 g), all of which contain ellagitannins (hydrolysable tannins). Intakes resulted in urinary excretion of a derivative, urolithin B-3-*O*-glucuronide, in quantities equivalent to 2.8% (strawberries), 3.4% (rasperries) and 16.6% (walnuts) regarding the ingested ellagitannins (Fig. 14) (Cerda et al., 2005).

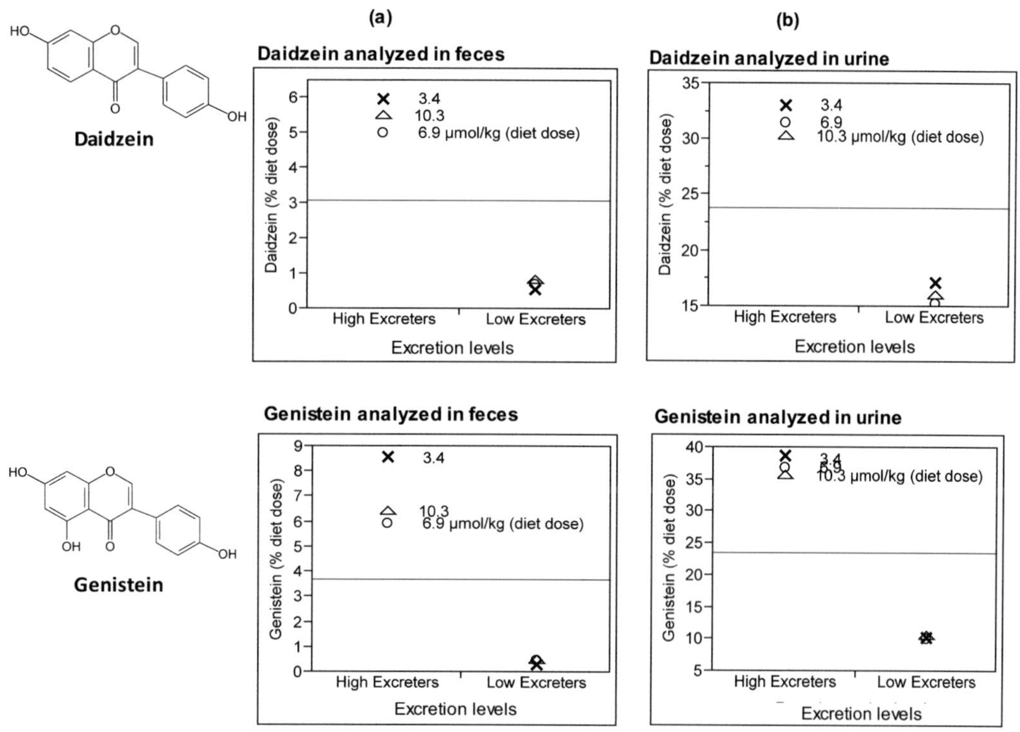

Fig. 13. Metabolomic classification of human subjects into two excretion levels of isoflavones (daizein and genistein) according to percentages of faecal and urinary isoflavones compared to diet dose. Low and high excreters had low and high excretion percentages of isoflavones, respectively.

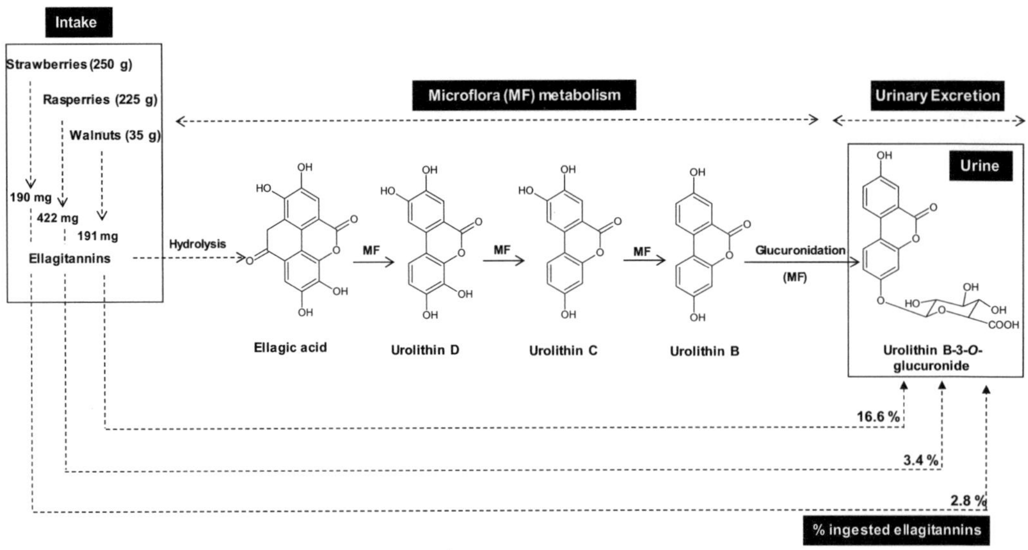

Fig. 14. Characterization of three tannin-rich diets by the excretion percentages of a urinary metabolite (urolithin B-3-*O*-glucuronide) referred to the ingested dose of ellagitannins (Cerda et al., 2005).

3.4 Metabotypes based on profiling

Flavanone enantiomers profiles have been analysed in different fruit juices and their concentrations have been found to be efficient biomarkers of the plant dietary source (Fig. 15) (Yanez et al. 2008):

- Orange juices contained the highest concentrations of (2R)- and (2S)-hesperidin;
- Conventional and organic grapefruit juices contained the highest concentrations of R(+)- and S(-)-hesperetin, and (2R)- and (2S)-naringin.
- Conventional and organic tomato juices showed the highest levels of R(+)- and S(-)-naringenin. Also, the chemical profiles of tomato juices showed the co-occurrences of the eight enantiomers. Organic juice can be distinguished by relatively higher levels of naringin enantiomers.

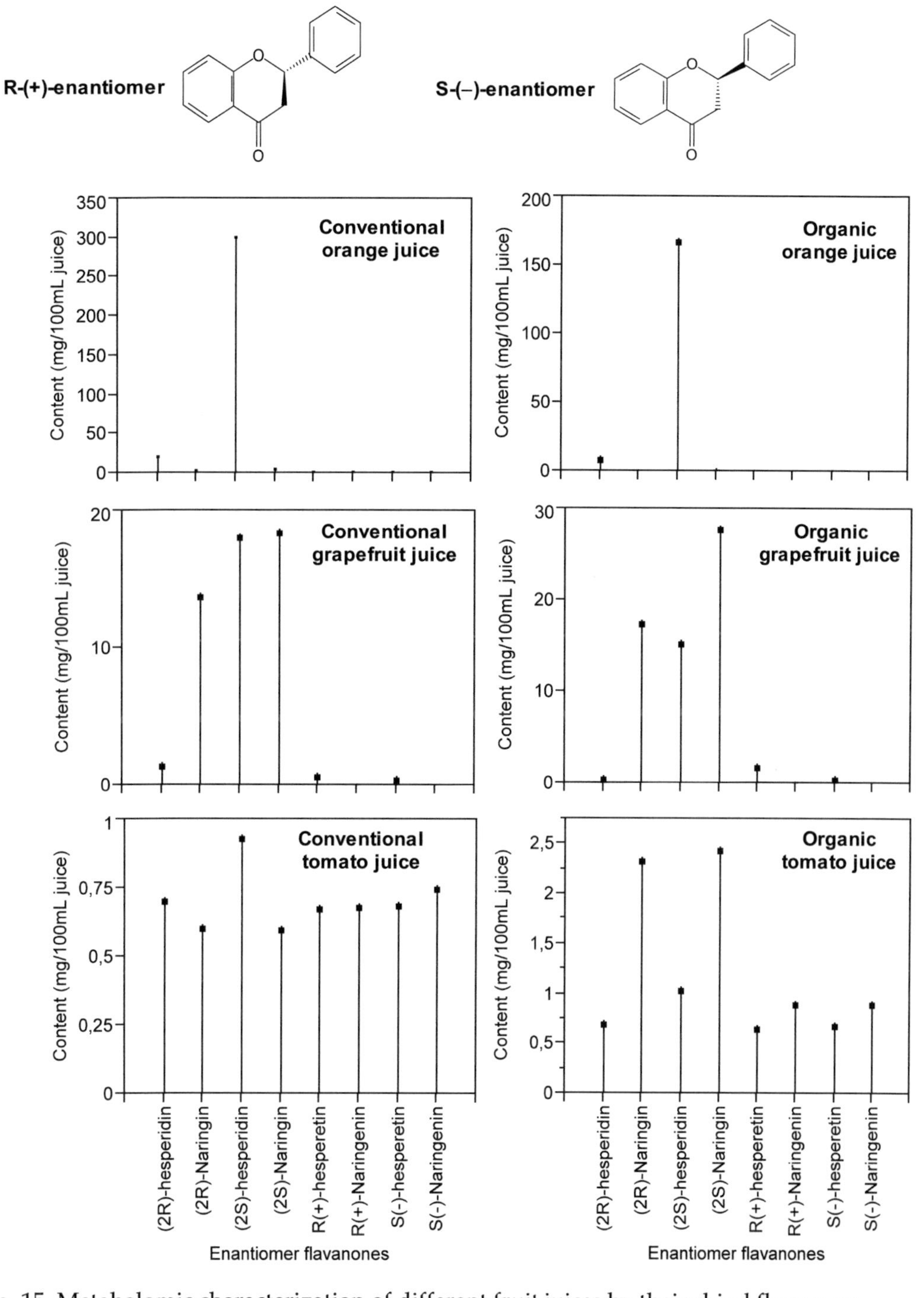

Fig. 15. Metabolomic characterization of different fruit juices by their chiral flavanone profiles.

Human subjects ingesting three times two cups of coffee at 4-h intervals had urinary profiles containing ferulic, isoferulic, dihydroferulic, 3-methoxy-4-hydroxybenzoic, hippuric and 3-hydroxyhippuric acids (Rechner et al., 2001).

After ingestion of 270 g of lightly fried onion by human subjects, plasma and urine samples collected over 24h showed very different metabolic profiles of concentrations (Fig. 16) (Mullen et al., 2006):

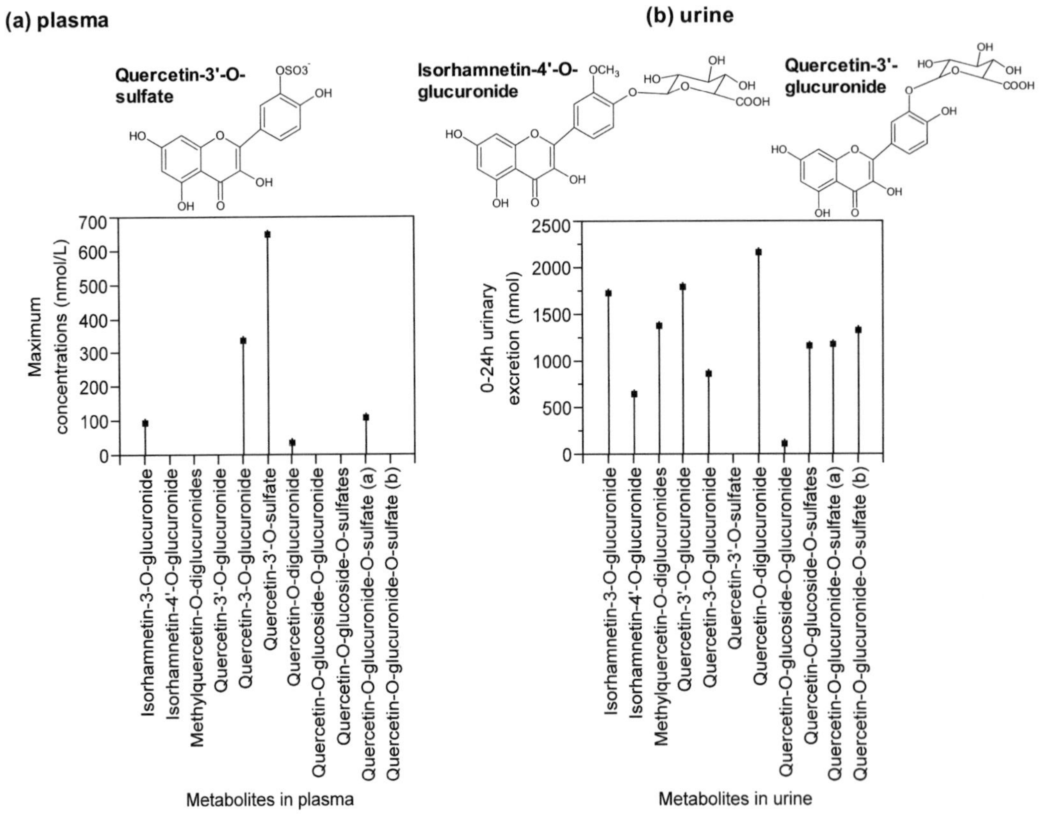

Fig. 16. Metabolic profiles of conjugated quercetin metabolites in plasma (a) and urine (b) following ingestion of lightly fried onion (270 g) by health human subjects (Mullen et al., 2006).

The main plasma metabolite, quercetin-3′-*O*-sulfate, was excreted only in trace quantities in urine while isorhamnetin-*O*-glucuronide and quercetin-*O*-diglucuronide that were minor components in plasma were major urinary metabolites (Fig. 16). Several other metabolites, including quercetin-3′-*O*-glucuronide and isorhamnetin-4′-*O*-glucuronide, which were present in trace quantities or absent from plasma were excreted in urine in substantial amounts.

In two separated human studies, and following the consumption of 200 g of strawberries (Mullen et al., 2008) or 200 g of blackberries (Felgines et al., 2005), the urinary contents were characterized by pelargonidin and cyanidin metabolites profiles, respectively: following the consumption of strawberries, urine samples were characterized by the predominance of pelargonidin-3-*O*-glucose in presence of pelargonidin-*O*-glucuronides with small quantities of pelargonidin aglycone and a pelargonidin-*O*-sulfate. Following the consumption of blackberries, the urine had a cyanidin-based profile containing unmetabolized cyanidin-3-*O*-glucoside, a cyanidin-*O*-glucuronide and a 3'-*O*-methyl-cyanidin-*O*-glucuronide.

In obese human population, low and medium carbohydrate diets reduced the total SCFA concentrations in fecal matrix, compared with a maintenance (high carbohydrate) diet (Duncan et al., 2007). This decrease concerned also acetate, propionate, and valerate concentrations analysed in saddles (Fig. 17). Low and medium carbohydrate diets can be distinguished by lower butyrate concentrations; such a reduction is more marked under low diet than intermediate one.

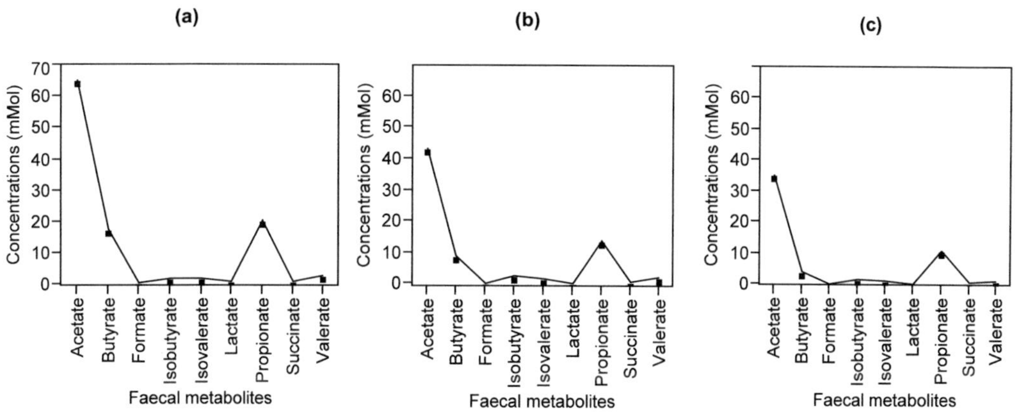

Fig. 17. Variations in concentrations of short chain fatty acids (SCFA) and lactate in stool samples of obese humans in relation to three carbohydrate levels diets (Ducan et al., 2007). (a), (b), (c) correspond to high, medium and low carbohydrate diet, respectively.

A previous study showed that SD rats deprived of water for 48h had elevated levels of creatinine and depleted levels of taurine, hippurate, 2-oxoglutarate, succinate and citrate (Clausing and Gottschalk, 1989). Water deprivation has a direct effect on osmoregulation implying variations in osmoregulators' levels as taurine.

4. Clinical metabotypes

Pathologies are known to induce changes in concentrations, regulation ratios and overall profiles of different metabolites that could be used to diagnose or characterize different diseases. Some examples will be given to illustrate the interest of different metabolomic criteria in clinical cases.

4.1 Metabotypes based on occurrences of metabolites

Occurrences of particular metabolites in biological matrices represent a strong metabolomic parameter to identify intoxication or doping sources:

Among endogenous metabolites, a particular attention has been paid to glutathione conjugates as potential markers of exposure (Van Weli et al., 1992). This is because glutathione (GS-H) detoxifies reactive chemicals (R-X) to which biological systems are exposed. The result of this conjugation (GSR) is the excretion of a variety of sulphured metabolites.

Concerning xenobiotics, chemicals that are reactive or are metabolized to intermediates reacting with DNA are of particular concern in relation to genotoxicity, and therefore, possible carcinogenicity:

In styrene industry, exposure to reactive alkylating agents can be diagnosed by analyzing DNA adducts, such as styrene oxide-O-6-guanine detected in white blood cells of exposed workers (Hemminki and Vodicka., 1995). Exposure to cyanide can be identified by rapid analysis of this toxin in blood of exposed individuals. However, free cyanide disappears rapidly from blood suggesting that a biological sample should be collected quickly, and analysis should be performed as soon as possible. If analysis of cyanide cannot be performed quickly, then cyanide exposure can be identified from the detection of three major markers of cyanide in the blood or urine: cyanide ion (CN-), thiocyanate (SCN-), and 2-aminothiazoline-4-carboxylic acid (ATCA) or its tautomer 2-iminothiazolidine-4-carboxylic acid (ITCA) (Logue et al., 2005).

Adducts such as *N*-(2-hydroxyethyl)valine have also been detected in haemoglobin from hospital workers exposed to ethylene oxide (Van Welie et al., 1992).

In clinical field, DNA adducts have also been detected in the white blood cells and urine of patients treated with anti-cancer drug such as N-methyl-*N*-nitrosourea (Prevost et al., 1996). Also, the *N*-7-guanyl aflatoxin B_1 adduct can be detected in urine, and used as a biomarker of exposure to the carcinogen aflatoxin B_1, which may be present in the diet (Groopman et al., 1994).

Oxidative damage to DNA can be detected by urinary 8-hydroxy-2′-deoxyguanosine (Van Welie et al., 1992). Also, urinary 8-hydroxy-2′-deoxyguanosine has been proposed as a biomarker of oxidative stress in humans (Bianchini et al., 1996).

In sport, doping controls are generally carried out on the basis of two complementary tests: (i) urine analysis and blood analysis provide short-term information on drug use by an individual; (ii) however, long term histories are accessible through hair analysis, because drug appears to be incorporated into the hair. For instance, in complement of testosterone determination, the identification of unique testosterone esters in hair enables an unambiguous charge for doping because the esters are certainly exogenous substances (Gaillard et al., 1999).

For drugs of abuse like cocaine and opiates, the threshold dose for detecting cocaine in hair appears to be approximately 25-35 mg cocaine, administered intravenously (Henderson, et al.; Kintz et al., 1999). Once incorporated into hair, a single dose of cocaine can be detected for 2 to 6 months. Codeine was detected in hair for 8 weeks after a single oral dose of 60 mg (Thieme et al., 2000).

Following intranasal absorption of 1.5 mg/kg BW of cocaine hydrochloride by humans, urinary excretion of unchanged cocaine was detected for only 8h (maximal excretion within

2 h) (Hamilton et al., 1977). However, benzoylecgonine (the major hepatic metabolite of cocaine) was generally detected in urine for 48 to 72 h (maximal excretion 4 to 8 h following cocaine administration).

4.2 Metabotypes based on concentration levels

The NMR spectrum of a biofluid can be conveniently thought of as a series of 'biomarker windows', which are spectral regions that contain signals from metabolites associated with specific targets for toxicity or disease (Fig. 18). For instance, metabolic response of the multimammate mousse (*Mastomys natalensis*) to 2-bromoethanamine (C_2H_6BrN) and propyleneimine (C_3H_7N) treatment was the induction of taurinuria (Fig. 18a). Taurine is amino acid known to protect renal medullary cells from osmotic stress and therefore may accumulate in the inner medulla of the *Mastomys*, protecting it from nephrotoxicity (Holmes et al. 1997).

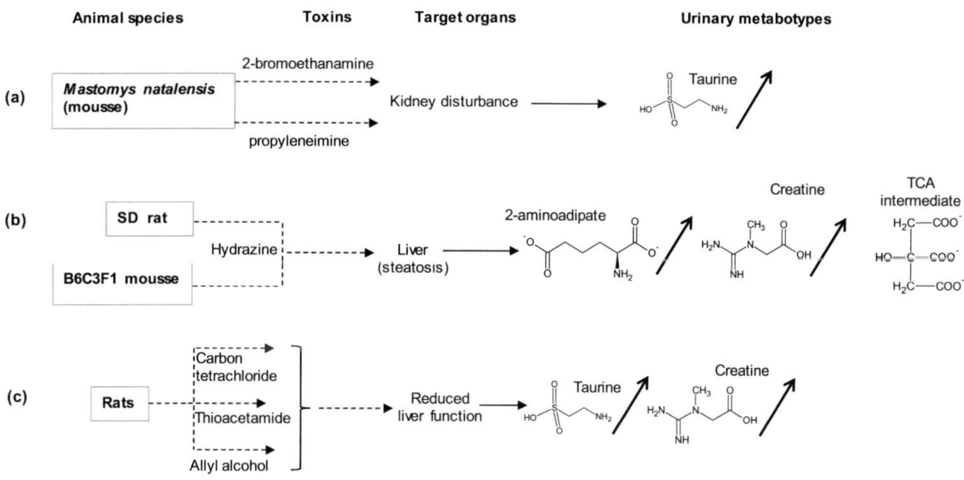

Fig. 18. Variation of urinary concentrations of metabolites, indicating kidney or liver perturbations in animal species previously exposed to some toxins.

Combining ^{1}H-NMR urine spectra with multivariate statistical analysis, metabolic responses of SD rats and B6C3F1 mice to hydrazine (NH_2NH_2) exposure were investigated (Bollard et al., 2005). Several common metabolic responses to hydrazine consisted of elevated levels of 2-aminoadipate and creatine versus depletion of the TCA cycle intermediates in the urine (Fig. 18b). Combined increases in taurine and creatine concentrations in urine have been associated with reduced liver function after exposure of rats to carbon tetrachloride (CCl_4), thioacetamide (C_2H_5NS) and allyl alcohol (C_3H_6O) (Fig. 18c) (Holmes et al., 1998).

In oncology, human prostatic epithelial cells are unique in accumulating zinc that blocks citrate degradation. In cancer cells, however, the prostate tissue contains low levels of citrate because it does not accumulate zinc and because most of citrate is used for fatty acid synthesis. 1H-NMR of prostate tissues confirmed the dramatic decrease in citrate levels in prostate gland during malignancy (Raina et al., 2009). A recent study showed a lower risk of developing high-grade prostate cancer for men with low serum cholesterol levels (Platz et al., 2009). Inversely, high levels of cholesterol in metastasic bone tissue (>70mg/g tissue) were found to be revelator of prostate cancer, compared to normal bone tissue (50-60mg cholesterol/g tissue) as well as to bone metastases from other cancers (<70mg/g) (Thysell et al., 2010).

4.3 Metabotypes based on ratios between concentration levels

Induction of cytochrome P450 isozymes may be used as biomarker of the effect of exposure of many species to a variety of chemicals, such as organochlorine compounds and polycyclic hydrocarbons. There are well-established urinary markers for cytochrome P450 induction, such as increased ratio of 6-β-hydroxycortisol/17-hydroxycorticosteroids (6β-OHF/17-OHCS) (Hugget et al., 1992). For instance, in a recent study, serum carbamazepine level was inversely associated with the urinary 6β-OHF/17-OHCS ratio (Konishi et al., 2004).

Elevated urinary ratio of creatine/creatinine has been proposed as marker of testicular damage (Timbrell et al., 1994).

Important increase of the ratio of lactate level on glucose level can be a biomarker of cancer cells in biological body: the conversion of glucose to lactate in the presence of oxygen represents a critic aerobic pathway that allows cancer cells to proliferate rapidly (Kim and Milner, 2011; Mazurek et al., 2011). Cancer cells metabolize glucose and glutamine more than normal cells to support the *de novo* biosynthesis of nucleotides and energy required for the high rate of cell proliferation.

4.4 Metabotypes based on profiling

Metabolomic profiling has been used to reliably identify different diseases including cancers and cardiovascular disturbances:

Concerning breast cancer, malignant cells (MDA-MB-435) content showed significant increase in glutathione (GSH) , *m*-inositol, creatine and phosphocholine concentrations and decrease in isoleucine, leucine, valine, and taurine concentrations, compared to normal mammary epithelial cells (MCF-10A) (Fig. 19) (Yang et al., 2007).

However, free choline and glycerophosphocholine were below the detection level in MDA-MB-435.

In patient suffering from coronary artery disease or left ventricular dysfunction, preischemia state was characterized by higher alanine levels versus lower concentrations of glucose, lactate, free fatty acids, total ketones, 3-hydroxybutyrate, pyruvate, leucine and glutamate analysed in the coronary sinus compared with arterial sample contents (Turer et al., 2009).

In patients suffering from non-ST-segment elevation acute coronary syndrome, plasma samples showed decrease in citric acid, 4-hydroxyproline, aspartic acid and fructose versus increase in lactate, urea, glucose and valine, compared to control healthy subjects (Vallejo et al., 2009).

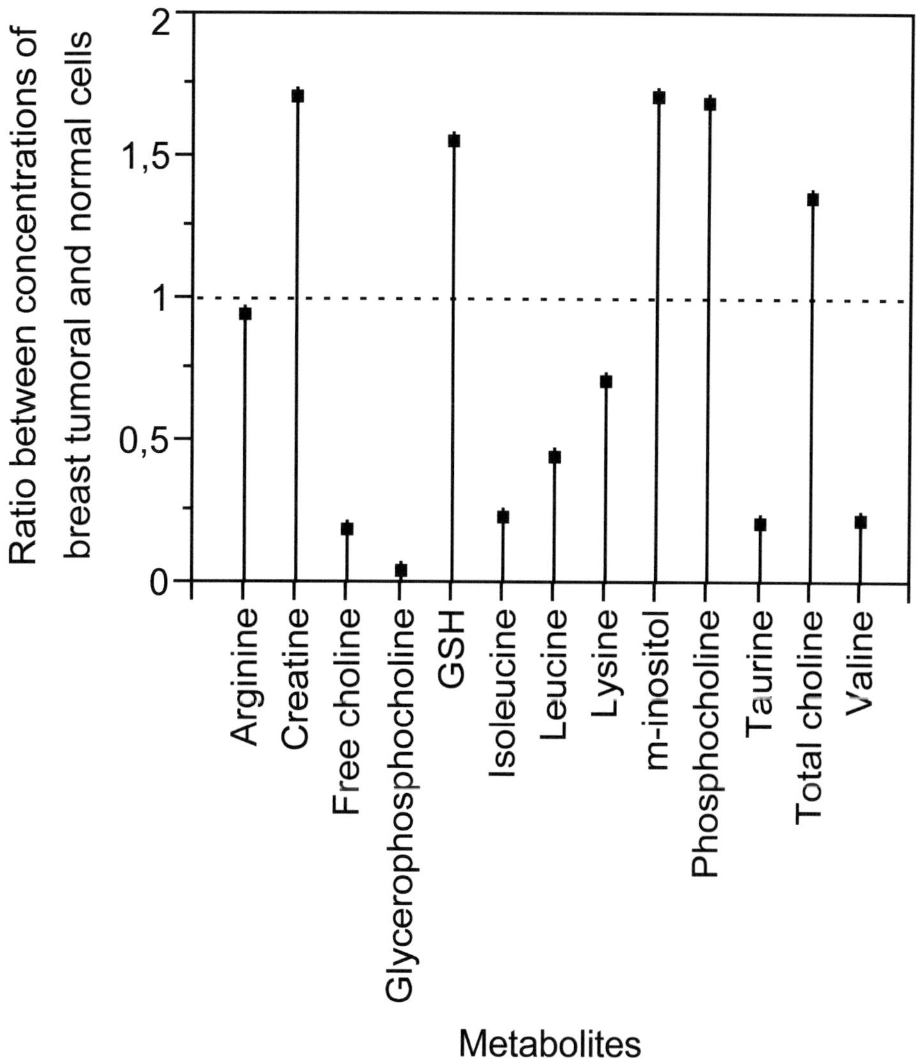

Fig. 19. Metabolomic profile characterizing breast tumor cells (MDA-MB-435) from normal cells (MCF-10A) based on ratios between metabolite concentrations MDA-MB-435/MCF-10A. High metabolic regulations are indicated by ratios >1, and inversely.

5. Metabotypes linked to biodiversity and environment conditions

5.1 Metabotypes based on occurrences of metabolites

In plant world, presence-absence of some secondary metabolites have high chemotaxonomic values. For instance, in monocotyledons, some families (Poaceae and Cyperaceae) are characterized by *C*-glycosyl flavonoids, i.e. flavonoids in which aglycone and saccharidic moiety have a *C-C* link, in addition to the *C-O* link which is abundant in the plant world (Fig.20a) (Semmar, 2010).

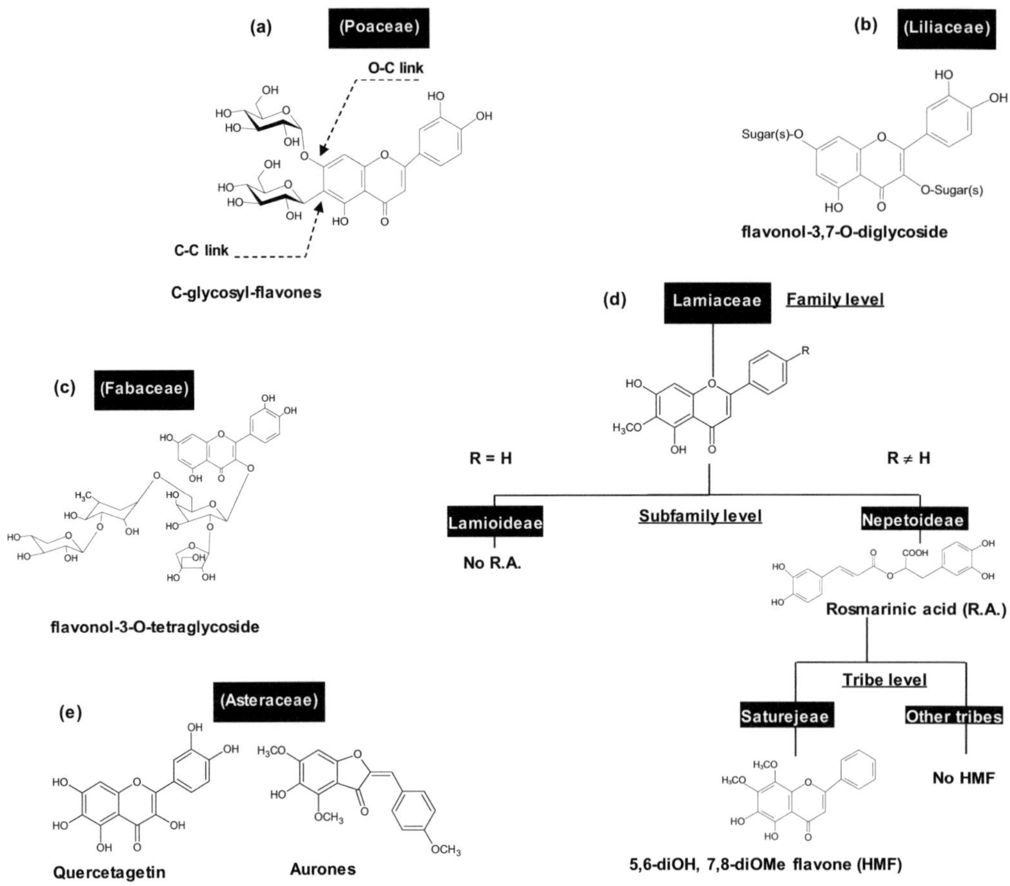

Fig. 20. Phytochemical characterizations of different plant taxons on the basis of specific or abundant phenolic compounds in their tissues.

Apart the link type between aglycone and sugar, glycosylation degree provides good chemotaxonomical criterion to a general characterization of plant families. For instance, leaf-tissues of Liliaceae species (monocotyledons) were characterized by the presence of di- and tri-*O*-glycosides of flavonoids, and a rare occurrence of monoglycosides (Williams, 1975). Flavonol 3, 7-diglycosides seem to be common constituents of the Liliaceae (Fig.20b) (Budzianowski, 1991; Williams, 1975). The family Fabaceae (dicotyledons) has been shown to be productive of multiglycosylated flavonols (Fig.20c) (Semmar, 2010).

Within the Lamiaceae family (dicotyledons), the presence-absence of rosmarinic acid (Fig. 20d) has been shown to be an excellent chemotaxonomic marker because of its presence in the subfamily Nepetoideae and absence in the subfamily Lamioideae (Janicsák et al., 1999; Harborne, 1966b). Members (tribes and genera) of these two subfamilies have been phytochemically characterized by the presence-absence of hydroxyl and methyl groups substituted on the A-ring of flavone aglycones. For instance, the presence of 5,7-dihydroxy-6-methoxyflavones with a substituted B-ring is characteristic of the subfamily Nepetoideae (Fig. 20d), particularly of *Salvia, Rosmarinus* and *Ocimum* species (Tomás-Barberán and Wollenweber, 1990); in Lamioideae, a 5,7-dihydroxy 6-methyl ether flavone has been found in the genus *Scutellaria* but with unsubstituted B-ring (Fig. 20d). Moreover, in the subfamily Nepetoideae, the genera *Thymus, Satureja, Micromeria, Acinos, Calamintha, Origanum* and *Mentha* were characterized by the production of the 5,6-dihydroxy-,7,8-dimethoxyflavone (Tomás-Barberán and Wollenberg, 1990); all these genera belong to the tribe Saturejeae (Fig. 20d).

Among the dictotyledons, the family Asteraceae has been characterized by the production of aurones and quercetagetin which is a flavonol almost entirely found in this family (Fig. 20e) (Iwashina, 2000).

In *Tulipa* (Liliaceae), the flower colors are fundamentally defined by the anthocyanidin type: orange and flesh pink colors are linked to pelargonidin, black-red and red-orange are due to cyanidin, and black-blue-violet-purple are governed by delphinidin (Fig. 21) (Shibata and Ishikura, 1960; Torskangerpoll et al., 2005).

Moreover in tulips, the shade of flower tepals was showed to be dependent on some chemical substitutions of anthocyanins: substitutions of anthocyanins by aromatic acyl groups have been reported to be responsible for bluing effect (Torskangerpoll et al., 2005). Also, combinations of cyanidin and pelargonidin with carotenoids generally induced attractive red and orange colours (van Eijk et al., 1987).

5.2 Metabotypes based on concentration levels

In plant world, accumulation of anthocyanins has been shown to be a good marker of cold stress: low temperatures have been shown to induce anthocyanin synthesis in many plant species, e.g. in leaves of *Arabidopsis thaliana* (Leyva et al., 1995), *Cotinus coggygria* (Oren-Shamir and Levi-Nissim, 1997), *Pinus banksiana* (Krol et al., 1995), etc. .

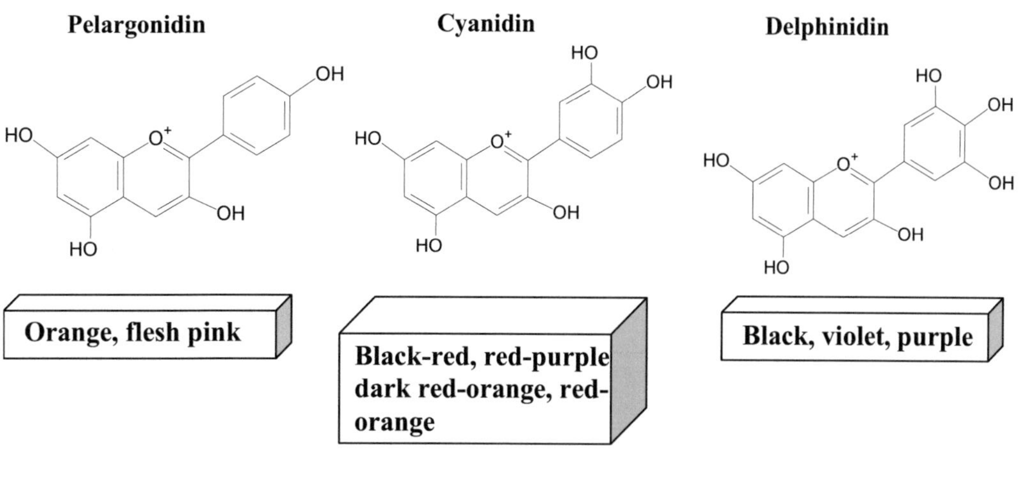

Fig. 21. Variation of tulip flower colours in relation to occurrence and abundance of anthocyanidin type.

Apart anthocyanins, sesquiterpenes (SQT) were found to be reliable indicator of thermic stress in emitting plant species: for instance, in young orange tree, emission of *β*-caryophyllene (*β*-Car) increased 5-6 fold for air temperature increase of 10°C (Hansen and Seufert, 1999). Temperature experiment performed on young corn plants showed that the proportion of *β*-Car (as a percentage of total emitted biogenic volatile organic compounds) was maximal at 37°C (Gouinguené and Turlings, 2002).

SQT levels have been also found to be biomarkers of diurnal and seasonal rhythmicity: *β*-Car emissions from various *Citrus* varieties were found to increase during the morning with generally a concentration peak around noon (Ciccioli et al., 1999). In potatos, emitted SQT increased steadily throughout the day and peaked in the afternoon (Agelopoulos et al., 2000). In Finnish scots pine, *β*-Car emissions exhibited significantly seasonal variability, with maximum emissions observed during summer months (Tarvainen et al., 2005).

Volatile terpenes (isoprenes, monoterpenes and sesquiterpenes) were found to be good markers of water stress in some plants. In *Pinus halepensis* (Alepo pine), water stress induced monoterpenes emissions by the leaves (Ormeño et al., 2007a). In young orange tree, severe drought reduced β-Car emissions to 6% of pre-drought levels, but emissions were unaffected by mild drought conditions (Hansen and Seufert, 1999).

Different works concluded relationships between metabolic variability in plants and soil composition: higher concentrations of aluminium in soil resulted in increase in exuded phenolic compounds by the roots of maize (Kidd et al., 2001). Aluminium resistant variety of maize exuded 15-fold higher level of flavonoids when pre-treated with silicon than when no such pre-treatment was applied. In scots pine (*Pinus sylvestris* L.), tree exposed to nickel had higher concentrations of condensed tannins compared with control (Roitto et al., 2005). Calcareous soils stimulated emissions of α-humulene from Aleppo pine, whereas siliceous soils favored α-humulene and β-bourbonene from Rock Rose (Ormeño et al., 2007b).

In plant world, emissions of some volatile compounds were found to be positively correlated to biotic disturbance such as parasite or herbivory: in Black Sage, SQT emissions increase significantly under infection with aphids (Arey et al., 1995). In corn seedlings, SQT emissions increase as a response to caterpillar feeding, and it has been demonstrated that such emission attracted wasps which parasitize caterpillars (Turlings et al., 1995).

5.3 Metabotypes based on ratios between concentration levels

Several plant species have been biogenically characterized by the ratios of individual sesquiterpenes (SQT) relative to the overall emitted (volatile) SQT (Duhl et al., 2008). In SQT emission profiles, β-caryophyllene (β-Car) was the most frequently reported and abundant; α- and β- farnesene as α-humulene are also prominent components to observed profiles (Fig. 22). The results concern studies where the plants were not disturbed because, disturbance is known to affect the variability of emitted SQT blends:

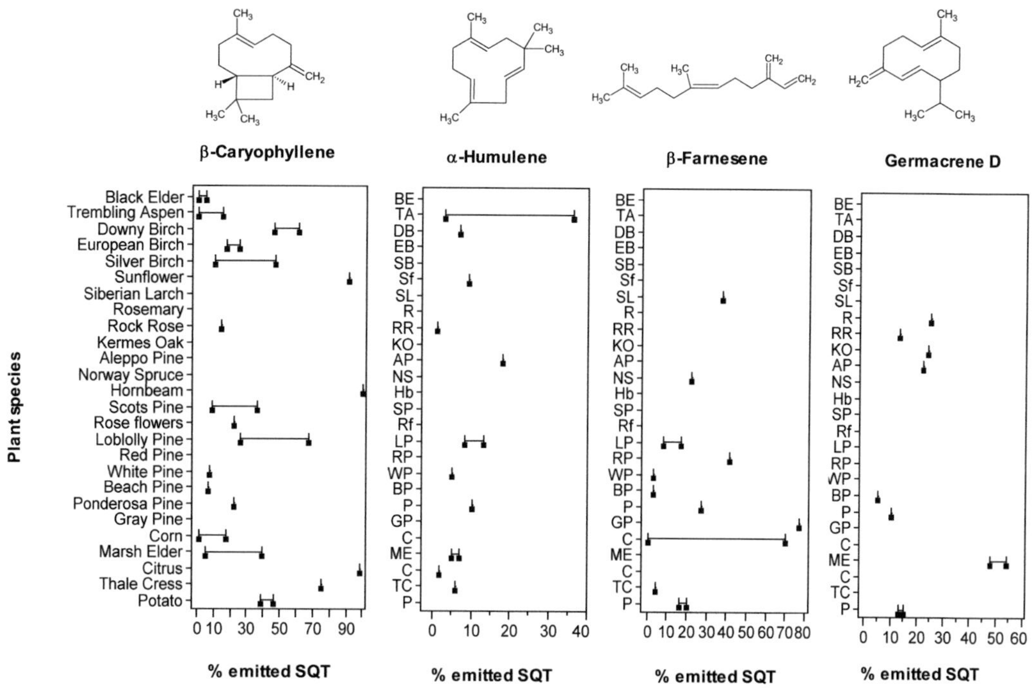

Fig. 22. Characterization of different plant species by the percentages (%) of some volatile sesquiterpenes (SQT) (β-caryophyllene, α-humulene, β-farnesene, germacrene D) referred to the overall SQT emissions.

SQT ratios showed that plant species can be characterized by dominance or relatively high levels of some SQT : sunflower, hormbeam and citrus species are highly productive of *β*-caryophyllene (*β*-car) (>90%). The SQT pool of gray pine seems to be dominated by high relative levels of β-farnesene (77%); the corn shows wide inter-individual variation range of β-farnesene going from 0 to 70%. Marsh elder can be characterized by germacrene D representing 48 to 54% of overall emitted SQT. Trembling aspen seems to emit relatively more α-humulene with high inter-individual variability (3-36%). Apart from *α*-humulene, *β*-humulene (not presented) is also frequent in nature and was reported to represent 55-57% of SQT in red and white pines (Duhl et al., 2008).

In tulips (Liliaceae), the flower colors acquire higher variability governed by the relative levels of mixed anthocyanins (Shibata and Ishikura, 1960): cultivars having "magenta nuances" showed anthocyanin content in which the relative amounts of cyanidin 3-rutinoside increased at the expense of delphinidin-3-rutinoside. Garden varieties with blue nuance (black, black-purple, fade-sky, violet and purple) have relatively high content of delphinidin type in tepals (i.e. the delphinidin content was more than 50% of the total anthocyanin content). Orange colored tepals were to a large extent correlated with high relative amounts of the pelargonidin derivatives at the expense of the two other aglycone types.

Apart from the chemotaxonomic characterization of plants, metabolic ratios were analysed in relation to different environmental conditions to characterize adaptive responses of biological species:

By opposition to monoterpene emissions, sesquiterpene emissions were found to be reduced by water deficit in *Pinus halepensis* among other typical Mediterranean species (Ormeno et al., 2007a). On the basis of negative effect on sesquiterpene emissions versus positive effect on monoterpene emissions, the drought stress can induce a shift in terpene composition leading to an increase in the ratio monoterpenes/sesquiterpenes in plant.

5.4 Metabotypes based on profiling

Metabolic profiling is used in chemotaxonomy to highlight chemical polymorphisms in plant species leading to better understand the biochemical origins of biodiversity:

For instance, analysis of flavonoid glycosides in the leaves of *Astragalus caprinus* (Fabaceae) highlighted four chemotypes (Chtp) characterized by high relative levels of different compounds among 14 in all (Fig. 23) (Semmar et al., 2005b): Chtp I was exclusively characterized by the presence of diglycosides of methylated flavonols (rhamnazine, rhamnocitrin) (11-14) which are acylated by a methyl-glutaric acid. Chtp II was characterized by high regulation of a tetraglycoside of quercetin (1). Chtp III was characterized by high regulations of a tetraglycoside of kaempferol (2) and its acylated derivatives (acylated by *p*-coumaric or ferrulic acid) (6-10). Chtp IV was characterized by a relatively high regulation of a triglycoside of kaempferol (5).

Correlation analyses between the relative levels of the 14 flavonoids showed positive trends between flavonoids based on a same aglycone and negative trends between flavonoids having different aglycones (kaempferol, quercetin and methylated aglycones) (Semmar et al., 2007). This compatibility between statistical correlations and chemical structures helped to analyse the network of metabolic links between flavonol glycosides (Fig. 23).

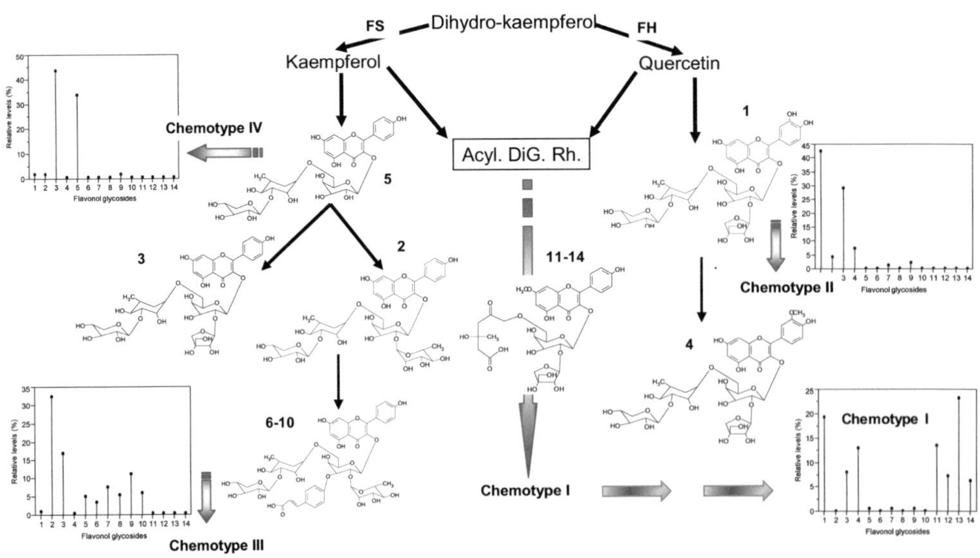

Fig. 23. Four chemotypes of *Astragalus caprinus* (Fabaceae) based on different metabolic regulations between flavonol glycosides pathways. FS: Flavonol Synthase; FH: Flavonol Hydroxylase; Acyl. Dig. Rh.: Acylated Diglycosyl of Rhamnazin or Rhamnocitrin

Sampling of 404 plants of *A. caprinus* from North to South of Tunisia highlighted a significant link between metabotypes' (chemotypes') abundances and geographical area (climatic conditions): Chemotype I rich in less hydrophilic compounds was abundant in the south (arid). The north (humid) seemed to offer favorable conditions to chemotypes III and IV (both based on kaempferol derivatives). The center of Tunisia, with intermediate climatic conditions, allowed a co-evolution of the four chemotypes with a relatively higher abundance for chemotype II (based on quercetin metabolism).

6. References

Arey, J., Crowley, D. E., Crowley, M., Resketo, M., and Lester, J. (1995). Hydrocarbon emissions from natural vegetation in California's South Coast Air Basin. *Atmos. Environ.*, Vol. 29, pp. 2977–2988, 13522310

Agelopoulos, N. G., Chamberlain, K. & Pickett, J. A. (2000). Factors affecting volatile emissions of intact Potato plants, *Solanum tuberosum*: variability of quantities and stability of ratios. *J. Chem. Ecol.*, Vol. 26, pp. 497–511, 0098-0331.

Barderas, M.G., Laborde, C.M., Posada, M., de la Cuesta, F., Zubiri, I., Vivanco, F. & Alvarez-Llamas, G. (2011). Metabolomic Profiling for Identification of Novel Potential Biomarkers in Cardiovascular Diseases. *J. Biomed. Biotechnol.* 2011, ID 790132, pp. 1-9, 11107243

Bell, J.D., Sadler, P.J., Morris, V.C. & Levander, O.A. (1991). Effect of aging and diet on proton NMR spectra of rat urine. *Magn. Reson. Med.*, Vol. 17, No. 2, pp. 414–422, 15222594

Bianchini, F., Hall, J., Donato, F. & Cadet, J. (1996). Monitoring urinary excretion of 5-hydroxymethyluracil for assessment of oxidative DNA damage and repair. *Biomarkers*, Vol. 1, pp. 178-184, 1354-750X.

Bishop, J.H. & Green, R. (1980). Effects of pregnancy on glucose handling by rat kidneys. *J. Physiol.*, Vol. 307, pp. 491–502, 00223751

Bollard, M.E., Holmes, J.C., Lindon, S.C., Mitchell, D., Branstetter, W. & Zhang JK Nicholson, (2001). Investigations into biochemical changes due to diurnal variation and estrus cycle in female rats using high resolution 1H NMR spectroscopy of urine and pattern recognition. *Anal. Biochem.*, Vol. 295, pp. 194–202, 00032697

Bollard, M.E., Keun, H., Ebbels, T., Beckonert, O., Antti, H., Lindon, J.C., Holmes, E. & Nicholson, J.K. (2005). Comparative metabonomics of differential species toxicity of hydrazine in the rat and mouse. *Toxicol. Appl. Pharmac.*, Vol. 204, No 2, pp. 135-151, 0041008X

Budzianowski, J. (1991). Six flavonol glucuronides from Tulipa gesneriana. *Phytochemistry* 30, 1679-1682, 0031-9422.

Cerda, B., Tomas-Barberan, F.A. & Espin, J.C. (2005). Metabolism of antioxidant and chemopreventive ellagitannins from strawberries, raspberries, walnuts, and oak-aged wine in humans: identification of biomarkers and individual variability. *J. Agric. Food Chem.*, Vol. 53, pp. 227–235, 00218561

Ciccioli, P., Brancaleoni, E., Frattoni, M., Di Palo, V., Valentini, R., Tirone, G., Seufert, G., Bertin, N., Hansen, U., Csiky, O., Lenz, R. & Sharma, M. (1999). Emission of reactive terpene compounds from orange orchards and their removal by within-canopy processes, *J. Geophys. Res.*, Vol. 104, pp. 8077–8094, 01480227

Claus, S.P., Tsang, T.M., Wang, Y., Cloarec, O., Skordi, E., Martin, F.P., Rezzi, S., Ross, A., Kochhar, S., Holmes, E. & Nicholson, J.K. (2008). Systemic multicompartmental effects of the gut microbiome on mouce metabolic phenotypes. *Molecular Systems Biology*, Vol. 4, No. 219, pp. 1-14, 17444292

Clausing, P. & Gottschalk, M. (1989). Effects of drinking water acidification, restriction of water supply and individual cageing on parameters of toxicological studies in rats. *Z. Versuchstierkd.*, Vol. 32, No 3, pp. 129–134, 00443697

Czerniak, R. (2001). Gender-based differences in pharmacokinetics in laboratory animal models. *Int. J. Toxicol.*, Vol. 20, No. 3, pp. 161–163, 10915818

Davison, J.M. & Hytten, F.E. (1975). The effect of pregnancy on the renal handling of glucose. *Br. J. Obstet. Gynaecol.*, Vol. 82, No 5, pp. 374–381, 03065456

Dudareva, N., Murfitt, L.M., Mann, C.J., Gorenstein, N., Kolosova, N., Kish, C.M., Bonham, C. & Wood, K. (2000). Developmental regulation of methyl benzoate biosynthesis and emission in snapdragon flowers. *Plant Cell*, Vol. 12, pp. 949-961, 10404651

Duhl, T.R., Helmig, D. & Guenther, A. (2008). Sesquiterpene emissions from vegetation: a review. *Biogeosciences*, Vol. 5, pp. 761–777, 17264170

Duncan, S.H., Belenguer, A., Holtrop, G., Johnstone, A.M., Flint, H.J. & Lobley, G.E. (2007). Reduced Dietary Intake of Carbohydrates by Obese Subjects Results in Decreased Concentrations of Butyrate and Butyrate-Producing Bacteria in Feces. *Appl. Environ. Microbiol.*,Vol. 73, No. 4, pp. 1073-1078, 00992240

Dunne, C. (2001). Adaptation of bacteria to the intestinal niche: probiotics and gut disorder. *Inflamm. Bowel. Dis.*, Vol. 7, pp. 136–145, 10780998

Felgines, C., Talavéra, S., Texier, O., Gil-Izquierdo, A., Lamaison, J.-L. & Rémésy, C. (2005). Blackberry anthocyanins are mainly recovered from urine as methylated and glucuronidated conjugates in humans. *J. Agric. Food Chem.*, Vol. 53, pp. 7721–7727, 00218561

Freese, R., Alfthan, G., Jauhiainen, M., Basu, S., Erlund, I., Salminen, I., Aro, A. & Mutanen, M. (2002). High intakes of vegetables, berries, and apples combined with a high intake of linoleic or oleic acid only slightly affect markers of lipid peroxidation and lipoprotein metabolism in healthy subjects. *Am. J. Clin. Nutr.*, Vol. 76, pp. 950-960, 00029165

Gaillard, Y., Vayssette, F., Balland, A., Pépin, G. (1999). Gas chromatographic-tandem mass spectrometric determination of anabolic steroids and their esters in hair: application in doping control and meat quality control. *J. Chromatogr. B*, Vol. 735, pp. 189-205, 15700232

Gattuso, G., Barreca, D., Gargiulli, C., Leuzzi, U. & Caristi, C. (2007). Flavonoid Composition of Citrus Juices. *Molecules*, Vol. 12, pp. 1641-1673, 14203049

Gavaghan, C.L., Holmes, E., Lenz, E., Wilson, I.D. & Nicholson, J.K. (2000). An NMR-based metabonomic approach to investigate the biochemical consequences of genetic strain differences: application to the C57BL10J and Alpk:ApfCD mouse. *FEBS Lett.*, Vol. 484, No. 3, pp. 169–174, 00145793

Gouinguené, S. P. & Turlings, T. C. J. (2002). The effects of abiotic factors on induced volatile emissions in corn plants. *Plant Phys.*, Vol. 129, pp. 1296–1307, 00320889

Groopman, J.D., Wogan, G.N., Roebuck, B.D. & Kensler, T.W. (1994). Molecular biomarkers for aflatoxins and theirapplication to human cancer prevention. *Cancer Res.* Vol. 54, pp. 1907-1911, 00085472

Hakola, H., Laurila, T., Lindfors, V., Hellén, H., Gaman, A., & Rinne, J. (2001). Variation of the VOC emission rates of birch species during the growing season. *Boreal Environ. Res.*, Vol. 6, pp. 237–249, 12396095

Hamilton, H.E., Wallace, J.E., Shimek E.L. Jr, Land, P., Harris, S.C. & Christenson J.G. (1977). Cocaine and benzoylecgonine excretion in humans. *J. Forensic Sci.*, Vol. 22, No. 4, pp. 697-707, 00221198

Hansen, U. & Seufert, G. (1999). Terpenoid emission from *Citrus sinensis* (L.) OSBECK under drought stress. *Phys. Chem. Earth (B)*, Vol. 24, pp. 681–687, 14747065

Harborne, J.B. (1966b). Caffeic acid Ester distribution in higher plants. *Z. Naturforsch.* 21b, 604-605.

Hemminki, K. & Vodicka, P. (1995). Styrene: from characterization of DNA adducts to application in styrene-exposed lamination workers. *Toxicol. Lett.*, Vol. 77, pp. 153-161, 03784274

Hodgkinson, A. (1962). Citric acid excretion in normal adults and in patients with renal calculus. *Clin. Sci.*, Vol. 23, pp. 203–205, 01435221

Holden, J.M., Bhagwat, S.A., Haytowitz, D.B., Gebhardt, S.E., Dwyer, J.T., Peterson, J., Beecher, G.R., Eldridge, A.L. & Balentine, D. (2005). Development of a database of critically evaluated flavonoids data: application of USDA's data quality evaluation system. *J. Food Comp. Anal.*, Vol. 18, pp. 829–844, 08891575

Holmes, E., Bonner, F.W. & Nicholson, J.K. (1995). Comparative studies on the nephrotoxicity of 2-bromoethanamine hydrobromide in the Fischer 344 rat and the multimammate desert mouse (*Mastomys natalensis*). *Arch. Toxicol.*, Vol. 70, No 2, pp. 89–95, 03405761

Holmes, E., Bonner, F.W. & Nicholson, J.K. (1996). Comparative biochemical effects of low doses of mercury II chloride in the F344 rat and the Multimammate mouse (*Mastomys natalensis*). *Comp. Biochem. Physiol.*, Vol. 114C, No. 1, pp. 7–15, 15320456

Holmes, E., Bonner, F.W. & Nicholson, J.K. (1997). 1H NMR spectroscopic and histopathological studies on propyleneimine-induced renal papillary necrosis in the rat and the multimammate desert mouse (*Mastomys natalensis*). *Comp. Biochem. Physiol. C. Pharmac. Toxicol. Endocrinol.*, Vol. 116, No. 2, pp. 125–134, 07428413

Holmes, E., Nicholls, A.W., Lindon, J.C., Ramos, S., Spraul, M., Neidig, P., Connor, S.C., Connelly, J., Damnent, S.J., Haselden, J. & Nicholson, J.K. (1998). Development of a model for classification of toxin-induced lesions using 1H NMR spectroscopy of urine combined with pattern recognition. *NMR Biomed.*, Vol. 11, No. 4–5, pp. 235–244, 09523480

Holmes, E., Nicholls, A.W., Lindon, J.C., Connor, S.C., Connelly, J.C., Haselden, J.N., Damment, S.J., Spraul, M., Neidig, P. & Nicholson, J.K. (2000). Chemometric models for toxicity classification based on NMR spectra of biofluids. *Chem. Res. Toxicol.*, Vol. 13, No. 6, pp. 471–478, 0893228X

Holmes, E., Wilson, I.D. & Nicholson, J.K. (2008). Metabolic Phenotyping in Health and Disease. *Cell*, Vol. 134, pp. 714-717, 00928674

Hugget, R.J., Kimerle, R.A., Mehrle, P.M. & Bergman, H.L. (1992). Biomarkers. Biochemical, Physiological and Histological Markers of Anthropogenic Stress. Lewis, Chelsea, MI. ISBN: 978-0873715058.

Iwashina, T. (2000). The structure and distribution of the flavonoids in plants. *J. Plant Res.*, Vol. 113, pp. 287-299, 09189440

Janicsák, G., Máthé, I., Miklóssy-Vári, V. & Blunden, G. (1999). Comparative studies of the rosmarinic and caffeic acid contents of Lamiaceae species. *Biochem. Syst. Ecol.*, Vol. 27, pp. 733-738, 03051978

Jones, A.R. (1982). Some observations on the urinary excretion of glycine conjugates by laboratory animals. *Xenobiotica*, Vol. 12, pp. 387–395, 00498254

Joossen, J.V. (1988). Mechanisms of hypercholesterolemia and atherosclerosis. *Acta Cardiol. Suppl.* Vol. 29, pp. 63–83

Kaufman, P.B., Duke, J.A., Brielmann, H., Boik, J. & Hoyt, J.E. (1997). A Comparative Survey of Leguminous Plants as Sources of the Isoflavones, Genistein and Daidzein: Implications for Human Nutrition and Health. *The Journal of Alternative and Complementary Medicine*, Vol. 3, No. 1, pp. 7-12, 10755535

Kawaii, S., Tomono, Y., Katase, E., Ogawa, K. & Yano, M. (1999). Quantitation of flavonoid constituents in citrus fruits. *J. Agric. Food Chem.*, Vol. 47, No. 9, pp. 3565-3571, 00218561

Kidd, P.S., Llugany, M., Poschenrieder, C., Gunsé, B & Barceló, J. (2001). The role of root exudates in aluminium resistance and silicon-induced amelioration of aluminium toxicity in three variety of maize (*Zea mays* L.). *J Exp Bot*, Vol. 52, pp. 1959-1967, 00220957

Kintz, P., Cirimele, V., Jeanneau, T. & Ludes, B. (1999). Identification of testosterone and testosterone esters in human hair. *J. Anal. Toxicol.*, Vol. 23, pp. 352-356, 01464760

Konishi, H., Tanaka, K., Minouchi, T. & Yamaji, A. (2004). Urinary 6β-hydroxycortisol/17-hydroxycorticosteroids ratio as a measure of hepatic CYP3A4 capacity after enzyme induction. *Ann Clin Biochem*, Vol. 41, pp. 335-337, 00045632

Krol, M., Gray, G.R., Hurry, V.M., Oquist, G., Malek, L. & Huner N.P.A. (1995). Low-temperature stress and photoperiod affect an increased tolerance to photoinhibition in *Pinus banksiana* seedlings. *Can. J. Bot.*, Vol. 73, pp. 1119-1127, 00084826

Kühnau, J. (1976). The flavonoids. A class of semi-essential food components: their role in human nutrition. *World Rev. Nutr. Diet.*, Vol. 24, pp. 117-191, 00842230

Leyva A., Jarillo J. A., Salinas J. & Martinez-Zapater J.M. (1995). Low temperature induces the accumulation of phenylalanine ammonia-lyase and chalcone synthase mRNAs of *Arabidopsis thaliana* in a light-dependent manner. *Plant Physiol.*, Vol. 108, pp. 39-46, 00320889

Liggins, J., Bluck, L.J., Runswick, S., Atkinson, C., Coward, W.A. & Bingham, S.A. (2000). Daidzein and genistein contents of vegetables. *Br. J. Nutr.*, Vol. 84, pp. 717- 25, 00071145

Logue, B.A., Kirschten, N.P., Petrikovics I., Moser, M.A., Rockwood, G.A., Baskin, S.I. (2005). Determination of the cyanide metabolite 2-aminothiazoline-4-carboxylic acid in urine and plasma by gas chromatography-mass spectrometry. *J Chromatogr B*, Vol. 819, No. 2, pp. 237-244, 15700232

Lutz, U., Bittner, N., Ufer, M. & Lutz, W. K. (2010). Quantification of cortisol and 6 beta-hydroxycortisol in human urine by LC MS/MS, and gender-specific evaluation of the metabolic ratio as biomarker of CYP3A activity. *J. Chromatogr. B*, Vol. 878, No. 1, pp. 97–101, 15700232

Majewska, M. Skrzycki, M., Podsiad, M. & Czeczot, H. (2011). Evaluation of antioxidant potential of flavonoids: an *in vitro* study. *Acta Poloniae Pharmaceutica. Drug Research*, Vol. 68, No 4, pp. 611-615, 00016837

Maslowski, K.M., Vieira, A.T., Ng, A., Kranich, J., Sierro, F., Yu, D., Schilter, H.C., Rolph, M.S., Mackay, F., Artis, D., Xavier, R.J., Teixeira, M.M. & Mackay, C.R. (2009). Regulation of inflammatory responses by gut microbiota and chemoattractant receptor GPR43. *Nature*, Vol. 461, pp. 1282-1286, 00280836

Mayr, M., Yusuf, S., Weir, G., Chung, Y.L., Mayr, U., Yin, X., Ladroue, C., Madhu, B., Roberts, N., De Souza, A., Fredericks, S., Stubbs, M., Griffiths, J.R., Jahangiri, M., Xu, Q. & Camm, A.J. (2008). Combined metabolomic and proteomic analysis of human atrial fibrillation. *Journal of the American College of Cardiology*, Vol. 51, No. 5, pp. 585–594, 07351097

Kim, Y.S., & Milner, J.A. (2011). Bioactive Food Components and Cancer-Specific Metabonomic Profiles. *Journal of Biomedicine and Biotechnology* ID 721213, pp. 1-9, 11107243

Mazurek, S. (2011). Pyruvate kinase type M2: a key regulator of the metabolic budget system in tumor cells. *Int J Biochem Cell Biol.*, Vol. 43, No. 7, pp. 969-980, 0020711X

Moreno, A. & Arus, C. (1996). Quantitative and qualitative characterization of 1H NMR spectra of colon tumors, normal mucosa and their perchloric acid extracts: decreased levels of myo-inositol in tumours can be detected in intact biopsies. *NMR Biomed.*, Vol. 9, pp. 33–45, 09523480

Mouly, P., Gaydou, E. M. & Auffray, A. (1998). Simultaneous separation of flavanone glycosides and polymethoxylated flavones in citrus juices using liquid chromatography. *J. Chromatogr.*, Vol. 800, No. 2, pp. 171–179, 0021-9673.

Mulder, T.P., Rietveld, A.G. & van Amelsvoort, J.M. (2005). Consumption of both black tea and green tea results in an increase in the excretion of hippuric acid into urine. *Am. J. Clin. Nutr.*, Vol. 81, pp. 256S–260S, 00029165

Mullen, W., Edwards, C.A. & Crozier, A. (2006). Absorption, excretion and metabolic profiling of methyl-, glucuronyl-, glucosyl and sulpho-conjugates of quercetin in human plasma and urine after ingestion of onions. *Br. J. Nutr.*, Vol. 96, pp. 107–116, 00071145

Mullen, W., Edwards, C.A., Serafini, M., Crozier, A. (2008). Bioavailability of pelargonidin-3-O-glucoside and its metabolites in humans following the ingestion of strawberries with and without cream. *J. Agric. Food Chem.*, Vol. 56, pp. 713–719, 00218561

Nardini, M.E., Cirillo, F., Natella, C. & Scaccini, C. (2002). Absorption of phenolic acids in humans after coffee consumption. *J. Agric. Food Chem.*, Vol. 50, pp. 5735–5741, 00218561

Oren-Shamir, M. & Levi-Nissim, A. (1997). Temperature effect on the leaf pigmentation of *Cotinus coggygria* 'Royal Purple'. *J. Hor. Sci. Biotech.*, Vol. 72, pp. 425-432. 14620316

Ormeño, E., Mévy, J.P., Vila, B., Bousquet-Mélou, A., Greff, S., Bonin, G. & Fernandez, C. (2007a). Water deficit stress induces different monoterpene and sesquiterpene emission changes in Mediterranean species. Relationship between terpene emissions and plant water potential. *Chemosphere*, Vol. 67, pp. 276-284, 00456535

Ormeño, E., Fernandez, C., Bousquet-Mélou, A., Greff, S., Morin, E., Robles, C., Vila, B. & Bonin, G. (2007b). Monoterpene and sesquiterpene emissions of three Mediterranean species through calcareous and siliceous soils in natural conditions, *Atmos. Environ.*, Vol. 41, pp. 629-639, 13522310

Phipps, A.N., Wright, B., Stewart, J. & Wilson, I.D. (1997). Use of proton NMR for determining changes in metabolite excretion profiles induced by dietary changes in the rat. *Pharmacy and Pharmacology Communications*, Vol. 3, pp. 143–146, 20427158

Platz, E.A., Till C., Goodman, P.J., Parnes, H.L., Figg, W.D., Albanes,D., Neuhouser, M.L., Klein, E.A., Thompson, I.M., Jr & Kristal, A.R. (2009). Men with low serum cholesterol have a lower risk of high-grade prostate cancer in the placebo arm of the prostate cancer prevention trial. *Cancer Epidemiol Biomarkers Prev*, Vol. 18, pp. 2807–2813, 10559965

Prevost, V., Likhachev, A.J., Loktionova, N.A., Bartsch, H., Wild, C.P., Kazanova, O.I., Arkhipov, A.I., Gershanovich, M.L. & Shuker, D.E.G. (1996). DNA base adducts in urine and white blood cells of cancer patients receiving combination chemotherapies which include N-methyl-N-nitrosourea. *Biomarkers*, Vol. 1, pp. 244-251, 1354750X

Radtke, J., Linseisen, J. & Wolfram, G. (2002). Fasting plasma concentrations of selected flavonoids as markers of their ordinary dietary intake. *Eur. J. Nutr.*, Vol. 41, pp.203-209, 14366207

Raina, K., Serkova, N. J. & Agarwal, R. (2009). Silibinin feeding alters the metabolic profile in TRAMP prostatic tumors:1hnmrs-based metabolomics study. *Cancer Research*, Vol. 69, No. 9, pp. 3731–3735, 00085472

Rechner, A.R., Spencer, J.P.E., Kuhnle, G., Hahn, U. & Rice-Evans, C.A. (2001). Novel biomarkers of the metabolism of caffeic acid derivatives in vivo. *Free Rad. Biol. Med.*, Vol. 30, pp. 1213–1222, 08915849

Rikimaru, T., Oozeki, T., Ichikawa, M., Ebisawa, H. & Fujita, Y. (1989). Comparisons of urinary creatinine, skeletal muscle mass and indices of muscle protein catabolism in rats fed ad libitum, with restricted food intake and deprived of food. *J. Nutr. Sci. Vitaminol.*, Vol. 35, No. 3, pp. 190–209, 03014800

Robertson, D.G., Reily, M.D., Lindon, J.C., Holmes, E. & Nicholson, J.K. (2002). Metabonomic technology as a tool for rapid throughput in vivo toxicity screening. In *Comprehensive Toxicology* Vol. XIV: Molecular and Cellular Toxicology, Heuvel JPV, Perdew GH, Mattes WB & Greenlee WF (Eds). Elsevier (2002). pp. 583-610, 9780080468846, Amsterdam

Roitto, M., Rautio, P., Julkunen-Tiitto, R., Kukkola, E. & Huttunen, S. (2005). Changes in the concentrations of phenolics and photosynthates in Scots pine (*Pinus sylvestris* L.) seedlings exposed to nickel and copper. *Environmental Pollution*, Vol. 137, pp. 603-609, 02697491

Schwiertz, A., Taras, D., Schäfer, K., Beijer, S., Bos, N.A., Donus, C. & Hardt, P.D. (2010). Microbiota and SCFA in lean and overweight healthy subjects. *Obesity*, Vol. 18, pp. 190-195, 19307381

Semmar, N., Bruguerolle, B., Boullu-Ciocca, S. & Simon, N (2005a). Cluster Analysis: An Alternative Method for Covariate Selection in population Pharmacokinetic Modeling. *Journal of Pharmacokinetics and Pharmacodynamics*, Vol. 32, No. 3-4, pp. 333-358, 1567567X

Semmar, N., Jay, M., Farman, M. and Chemli, R. (2005b). Chemotaxonomic analysis of Astragalus caprinus (Fabaceae) based on the flavonic patterns. *Biochem. Syst. Ecol.*, Vol. 33, pp. 187-200, 03051978

Semmar, N., Jay M. & Nouira S. (2007). A new approach to graphical and numerical analysis of links between plant chemotaxonomy and secondary metabolism from HPLC data smoothed by a simplex mixture design. *Chemoecology*, Vol. 17, No. 3, pp. 139-156, 09377409

Semmar, N., Nouira, S. & Farman, M. (2008). Variability and Ecological Significance of Secondary Metabolites in Terrestrial Plant World. In: *Handbook of Nature Conservation*, Aronoff, J.B., pp. 1-89. Nova Science Publishers, 9781606929933, NY

Semmar, N. (2010). *Chemotaxonomical Analysis of Herbaceous Plants Based on Phenolic and Terpenic patterns: flexible tools to survey biodiversity in grassland.* Nova Science Publishers, 9781616687892, NY.

Shibata, M. & Ishikura, N. (1960). Paper chromatographic survey of anthocyanin in tulip-flowers. *I. Jap. J. Bot.*, Vol. 17, pp. 230-238

Stanley, E.G. (2002). 1H NMR spectroscopic and chemometric studies on endogenous physiological variation in rats. *Ph.D. thesis, university of London*, pp. 43–67

Tarvainen, V., Hakola, H., Hellén, H., Bäck, J., Hari, P. & Kulmala, M. (2005): Temperature and light dependence of the VOC emissions of Scots Pine, *Atmos. Chem. Phys.*, Vol. 5, pp. 989–998, 16807367

Thieme, D., Grosse, J., Sachs, H. & Mueller, R.K. (2000). Analytical strategy for detecting doping agents in hair. *Forensic Sci. Int.*, Vol. 107, pp. 335-345, 03790738

Thysell, E., Surowiec, I., Hörnberg, E., Crnalic, S., Widmark, A., Johansson, A.I., Stattin, P., Bergh, A., Moritz, T., Antti., H. & Wikström, P. (2010). Metabolomic Characterization of Human Prostate Cancer Bone Metastases Reveals Increased Levels of Cholesterol. *PLoS One*, Vol. 5, No. 12, e14175, 19326203

Timbrell, J.A., Draper, R.P. & Waterfield, C.J. (1994). Biomarkers in toxicology: new uses for some old molecules? *Toxicol. Ecotoxicol. News*, Vol. 1, pp. 4-14, 13504592

Timbrell, J.A. (1998). Biomarkers in toxicology. *Toxicology*, Vol. 129, pp. 1–12, 0300483X

Tomás-Barberán, F.A.T. & Wollenweber, E. (1990). Flavonoid aglycones from the leaf surfaces of some Labiatae species. *Pl. Syst Evol.*, vol. 173, pp. 109-118, 03782697

Torskangerpoll, K., Nørbæk, R., Nodland, E., Øvstedal, D.O. & Andersen, Ø.M. (2005). Anthocyanin content of Tulipa species and cultivars and its impact on tepal colours. *Biochem. Syst. Ecol.*, Vol. 33, pp. 499-510, 03051978

Turer, A. T., Stevens, R. D., Bain, J. R., Muehlbauer, M.J., van der Westhuizen, J., Mathew, J.P., Schwinn, D.A., Glower, D.D., Newgard, C.B. & Podgoreanu M.V. (2009). Metabolomic profiling reveals distinct patterns of myocardial substrate use in humans with coronary artery disease or left ventricular dysfunction during surgical ischemia/reperfusion. *Circulation*, Vol. 119, No. 13, pp. 1736–1746, 00097322

Turlings, T. C. J., Loughrin, J. H., McCall, P. J., R¨ose, U. S. R., Lewis, W. J., & Tumlinson, J. H. (1995). How caterpillar-damaged plants protect themselves by attracting parasitic wasps, *Proc. Natl. Acad. Sci.*, Vol. 92, pp. 4169–4174, 00278424

Vallejo, M., Garcia, A., Tunon, J. , García-Martínez, D., Angulo, S., Martin-Ventura, J.L., Blanco-Colio, L.M., Almeida, P., Egido, J. & Barbas, C. (2009). Plasma fingerprinting with GC-MS in acute coronary syndrome. *Analytical and Bioanalytical Chemistry*, Vol. 394, No. 6, pp. 1517–1524, 16182642

van Eijk, J.P., Nieuwhof, M., van Keulen, H.A. & Keijzer, P. (1987). Flower colour analyses in tulip (*Tulipa* L.). The occurrence of carotenoids and flavonoids in tulip tepals. *Euphytica*, Vol. 36, pp. 855-862, 00142336

Van Welie, R.T.H., van Dijck, R.G.J.M., Vermeulen, N.P.E., van Sittert, N.J., 1992. Mercapturic acids, protein adducts and DNA adducts as biomarkers of electrophilic chemicals. *Crit. Rev. Toxicol.*, Vol. 22, No. 5-6, pp. 271-306, 10408444

Williams, C.A. (1975). Biosystematics of the Monocotyledoneae - Flavonoid Patterns in Leaves of the Liliaceae. *Biochem. Syst. Ecol.*, Vol. 3, pp. 229-244, 03051978

Wilson, I.D. (2009). Drugs, bugs, and personalized medicine: Pharmacometabonomics enters the ring. *PNAS*, Vol. 106, No. 34, pp. 14187-14188, 00278424

Xu, X., Harris, K.S., Huei-Ju, W., Murphy, P.A. & Hendrich, S. (1995). Bioavailability of soybean isoflavones depends upon gut microflora in women. *The Journal of Nutrition*, Vol. 125, No. 9, pp.2307-2315, 00223166

Yanez, J.A., Remsberg, C.M., Miranda, N.D., Vega-Villa, K.R., Andrews, P.K. & Davies, N.M. (2008). Pharmacokinetics of Selected Chiral Flavonoids: Hesperetin, Naringenin and Eriodictyol in Rats and their Content in Fruit Juices. *Biopharm. Drug Dispos.*, Vol. 29, pp. 63–82, 01422782

Yang, C., Richardson A.D., Smith, J.W. & Osterman A. (2007). Comparative metabolomics of breast cancer. *Pacific Symposium on Biocomputing* 12, 181-192, 17935091

Zhang, A.Q., Mitchell, S.C. & Smith, R.L. (1996). Exacerbation of symptoms of fish-odour syndrome during menstruation. *Lancet*, Vol. 348, No. 9043, pp. 1740–1741, 01406736

Zheng, W. & Wang, S.Y. (2003). Oxygen radical absorbing capacity of phenolics in blueberries, cranberries, chokeberries, and lingonberries. *J. Agric. Food Chem.*, Vol. 51, pp. 502-509, 00218561

Zoetendal, E.G., Rajilic-Stojanovic, M. & de Vos, W.M. (2008). High throughput diversity and functionality analysis of the gastrointestinal tract microbiota. *Gut.*, Vol. 57, pp. 1605-1615, 00175749

7

Software Techniques for Enabling High-Throughput Analysis of Metabolomic Datasets

Corey D. DeHaven, Anne M. Evans, Hongping Dai and Kay A. Lawton

Metabolon, Inc.
United States of America

1. Introduction

In recent years, the study of metabolomics and the use of metabolomics data to answer a variety of biological questions have been greatly increasing (Fan, Lane et al. 2004; Griffin 2006; Khoo and Al-Rubeai 2007; Lindon, Holmes et al. 2007; Lawton, Berger et al. 2008). While various techniques are available for analyzing this type of data (Bryan, Brennan et al. 2008; Scalbert, Brennan et al. 2009; Thielen, Heinen et al. 2009; Xia, Psychogios et al. 2009), the fundamental goal of the analysis is the same - to quickly and accurately identify detected molecules so that biological mechanisms and modes of action can be understood. Metabolomics analysis was long thought of as, and in many aspects still is, an instrumentation problem; the better and more accurate the instrumentation (LC/MS, GC/MS, NMR, CE, etc.) the better the resulting data which, in turn, facilitates data interpretation and, ultimately, the understanding of the biological relevance of the results.

While the quality of instrumentation does play a very important role, the rate-limiting step is often the processing of the data. Thus, software and computational tools play an important and direct role in the ability to process, analyze, and interpret metabolomics data. This situation is much like the early days of automated DNA sequencing where it was the evolution of the software components from highly manual to fully automated processes that brought about significant advances and a new era in the technology (Hood, Hunkapiller et al. 1987; Hunkapiller, Kaiser et al. 1991; Fields 1996). Currently, software tools exist for the automated initial processing of metabolomic data, especially chromatographic separation coupled to mass spectrometry data (Wilson, Nicholson et al. 2005; Nordstrom, O'Maille et al. 2006; Want, Nordstrom et al. 2007; Patterson, Li et al. 2008). Samples can be processed automatically; peak detection, integration and alignment, and various quality control (QC) steps on the data itself can be performed with little to no user interaction. However, the problem is that the generation of data, together with peak detection and integration, is the relatively simple part; without a properly engineered system for managing this part of the process the vast number of data files generated can quickly become overwhelming.

Two major processes in metabolomic data processing are the verification of the accuracy of the peak integration and the verification of the accuracy of the automated identification of the metabolites that those peaks represent. These two processes, while vitally important to the accuracy of the results, are very time consuming and are the most significant bottlenecks in processing metabolomic data. In fact, the peak integration verification step is often omitted due to the extremely large number of peaks whose integration must be verified.

2. Background

At the outset, running a metabolomics study is actually simple and straightforward. Samples are prepared for running on a signal detection platform, signal data is collected on samples from the instrumentation, the signals are translated into peaks, the peaks are compared to reference libraries for the identification of metabolites and those identified metabolites are then statistically analyzed with whatever metadata may exist for the samples. Alternatively, the entirety of the detected peaks resulting from the instrument signal data are statistically analyzed without metabolite identification prior to the statistical analysis.

Once statistical analysis is completed and the significant signals have been stratified and metabolites identified, biochemical pathway analysis is performed to gain insight into the original biological questions the study asked. Too often, when the metabolomic experiments do not provide meaningful biological results, the realization may come that there's so much variability in the data, it can't be used to address the original objectives of the study. Despite the methods and software provided by the various instrument vendors, it turns out that running a global, non-targeted analysis of small molecules in a complex mixture that generates high-quality data and provides answers to biological questions is challenging. Doing so in a high-throughput environment is significantly more challenging.

However, a high-throughput metabolomics platform that produces reliable, precise, reproducible, and interpretable data is possible. It simply requires the right process coupled with the right software tools. As with any high throughput process it is important to have a logical, consistent workflow that is simple, reproducible, and expandable without negatively impacting the efficiency of the process. It is important to know when human interaction is required and when it is not. Well designed and integrated software can efficiently handle the majority of the mundane workload, allowing human interaction to be focused only where required.

3. Approach

Metabolite identification is essential for chemical and biological interpretation of metabolomics experiments. Two approaches to metabolomic data analysis have been used and will be described in detail below. The main difference between the two approaches is when the metabolite is identified, either before or after statistical analysis of the data.

To date, the most commonly used method of processing metabolomic data has been to statistically analyze all of the detected ion-features ('ion-centric'). Ion features, defined here as a chromatographic peak with a given retention time and m/z value, are analyzed using a statistical package such as SAS or S-plus to determine which features vary statistically significantly and are related to a test hypothesis (Tolstikov and Fiehn 2002; Katajamaa and Oresic 2007; Werner, Heilier et al. 2008) . The significant ion feature changes are then used to prioritize metabolite identification. One issue with this type of approach is the convoluted nature of the data being analyzed. In many cases the "statistically significant ion-features" are various forms of the same chemical and are therefore redundant information. Most biochemicals detected in a traditional LC- or GC-MS analysis produce several different ions, which contributes to the massive size and complexity of metabolomics data. In addition, there are an even larger number of measurements for each experimental sample which impacts the false discovery rate (Benjamini and Hochberg 1995; Storey and Tibshirani 2003).

In the 'chemo-centric' approach to metabolomics data analysis discussed here, metabolites are identified on the front-end through the use of a reference library comprised of spectra of authentic chemical standards(Lawton, Berger et al. 2008; Evans, Dehaven et al. 2009). Then, instead of treating all detected peaks independently, as is done in the ion-centric approach, the chemo-centric method selects a single ion ('quant-ion') to represent that metabolite in all subsequent analyses. The other ions associated with the metabolite are essentially redundant information that only add to data complexity. Furthermore, the statistical analysis may be skewed since a single metabolite may be represented by multiple ion peaks, and the false discovery rate increased due to the large number of measurements relative to the number of samples in the experiment. Accordingly, by taking a chemo-centric approach any extraneous peaks can be identified and removed from the analysis based on the authentic standard library/database. Since the number of features analyzed statistically contributes to the probability of obtaining false positives, analyzing one representative ion for each metabolite reduces the number of false positives. Further, the chemo-centric data analysis method is powerful because a significant amount of computational processing time and power can be saved simply due to data reduction.

The majority of work and complexity with the chemo-centric approach are: first, the generation of the reference library of spectra from authentic chemical standards; second, the actual identification of the detected metabolites using the reference library; and third, the ability for quality control (QC) of the automated metabolite identification, peak detection and integration. Notably, the QC of the automated processes is often overlooked. However, the QC step is critical to ensure that false identifications and poor or inconsistent peak integrations do not make their way into the statistical analysis of the experimental results. The generation of a reference library entry made up of the spectral signature and chromatographic elution time of an authentic chemical standard is relatively straightforward, as is the generation of spectral-matching algorithms that use the reference library to identify the experimentally detected metabolites. In contrast, performing the QC step on the automated processes, including peak detection, integration and metabolite identification, is time and human resource intensive.

Not to be overlooked, an issue with using a reference library comprised of authentic standards is dealing with metabolites in the samples that are not contained within the reference library. The power of the technology would be significantly reduced if it was limited to identifying only compounds contained in the reference library. Through intelligent software algorithms, it is possible to analyze data of similar characteristics across multiple samples in a study to find those metabolites that are unknown by virtue of not matching a reference standard in the library, and, in the process, group all the ion-features related to that unknown together by examining ion correlations across the sample set (Dehaven, Evans et al. 2010). One such method capitalizes on the natural biological variability inherent in the experimental samples, using this variation the metabolites and their respective ion-features can reveal themselves and be entered into the chemical reference library as a novel chemical entity (Dehaven, Evans et al. 2010). The unknown chemical can then be tracked in future metabolomics studies, and, if important, can be identified using standard analytical chemistry techniques.

Without going into detail, it is important to note that the sample preparation process is critical. High quality samples that have been properly and consistently prepared for analysis on sensitive scientific instrumentation are of extreme importance. Ensuring this high quality starts with the collection and preparation of the samples. No software system is going to be able to produce high-quality data unless ample effort is focused on consistently following standardized protocols for preparing high quality samples for analysis.

The following discussion, examples and workflow solutions make use of GCMS or LCMS (or both) platforms for metabolomic analysis of samples, although the concepts in general could apply to a variety of data collection techniques. Software tools are also presented to demonstrate the application of the concepts that are discussed but the tools themselves will not be discussed in great detail. It is also important to note that achieving the greatest operation efficiency of the process relies on treating all of the experimental samples in a study as a set and not as individual files. By using tools to analyze and perform quality control on the samples as a single group or set it becomes much easier to spot patterns that can be useful to determine what is going on in the overall process.

4. Processing data files, peak detection, alignment, and metabolite identification

4.1 File processing can become a major hurdle

There is no shortage of software available on the market to read spectral data and detect the start and the stop of peaks, and the baseline, and then calculate the area inside of those peaks. Each instrument vendor provides some flavor of detection and analysis software with their instrument and several open-source and commercial efforts to read spectral data and produce integrated peak data regardless of vendor format are available (Tolstikov and Fiehn 2002; Katajamaa and Oresic 2005; Katajamaa and Oresic 2007). In almost all cases, these packages do a complete job of finding and integrating peaks and do so in a reasonable amount of time. Thus, the peak detection and integration process is not the rate-limiting step when it comes to data quality and automated processing.

As it turns out, the file processing problem is primarily a file management problem that is the result of two issues - human and machine. The first problem stems from human interaction, in that a human being can introduce more error and inconsistency than is acceptable. Optimally, a human should play no role in the naming or processing of instruments data files. Naming of instrument data files should take place within the system used to track sample information, a LIMS for instance. The LIMS or other sample tracking system should generate a sample list and run order for the samples to be run on the instrument using a consistent naming convention that can be easily associated with the sample in question. The second problem stems from both machine and human. The software performing the peak detection and integration must have the capability of automatically processing a data file when presented with it, then archiving the file when completed. And, in high-throughput mode, it is best not to have humans manage data files, either in storage locations, or, as noted above, in naming. For consistency, it is imperative that the machines control this step; running one experiment on one machine may be manageable manually but running experiments in tandem or on more than one instrument can easily result in misnaming, file version problems, location mishaps, etc. if file management is not automated.

4.2 Manual integration of peak data is inadequate for high-throughput processing

Processing metabolomics data in a high-throughput setting requires automated processing of data files. While an SDK (software development kit) is provided by many instrument vendors, and there are commercial and open source packages for creating this functionality available (Smith, Want et al. 2006), not all vendor software permits this functionality. One of the main reasons automated peak integration works well is because it allows data to be rapidly uploaded and processed. Manual integration, while perhaps more accurate, dramatically slows the peak analysis process. Further QC and refinement of the automated peak integration can be performed more optimally later in the process, where, in practice, the bar for peak detection can be slightly lower. The reasons that the bar for peak detection can be reduced will be discussed below.

4.3 Alignment based on peak similarity inadequate, retention index should be used

Many of the software packages provide capabilities to align the chromatograms to account for time drift in an instrument. In many instances internal standards and/or endogenous metabolites are used across the analyzed samples to align chromatography based on their retention times, such that there is confidence that the same peak at the same mass is consistent among the data files. This approach should be avoided because while it works fine for peak analysis and chromatographic alignment on a single, small study it will only be applicable within that one study where retention times are quite consistent. This type of alignment approach makes it much harder to do a comparison to a reference standard library where a retention profile is used as matching criteria. The better choice is to opt for retention index (RI) calculation, which can correctly align chromatograms even over long periods of time where conditions can be vastly different dependent on the condition in these systems. Using a retention index method, each RT marker is given a fixed RI value (Evans, Dehaven et al. 2009). The retention times for the retention markers can be set in the integrator method and the time at which those internal standards elute are used to calculate an adjustment RI ladder. All other detected peaks can then use their actual retention time and adjustment index to calculate a retention index. In this way, all detected peaks are aligned based on their elution relative to their flanking RT markers. An RI removes any systematic changes in retention time by assuming that the compound will always elute in the same relative position to those flanking markers. Because of this, a unique time location and window for a spectral library entry can be set in terms of RI, thereby ensuring that metabolites don't fall outside the allowed window over a much longer period of time. Retention indices have predominately been used for GC/MS methods however this approach can also have great success for LC/MS data alignment as well. LC/MS is certainly more complex as certain metabolites and classes of metabolites show more chromatographic shift in their RI markers than others, in these cases increasing the expected RI window of the library entry in conjunction with mass and fragmentation spectrum data is sufficient for accurate identification. The advantages over many of the widely available chromatographic alignment tools, eg. XCMS (Smith, Want et al. 2006), as it can be used to match against a RI locked library over long periods of time and can align data from different biological matrices without potential distortion from structural isomers.

4.4 Identifying metabolites

Metabolite identification is essential to the biochemical and biological interpretation of the results of metabolomic studies. Lists of integrated peak data are of little use unless a library of spectra is available to compare peak data with to identify the metabolites represented by those peaks. Publicly created and maintained databases do exist (Wishart, Tzur et al. 2007; Wishart 2011). However, the utility of these databases to identify metabolites of interest from metabolomics studies is currently limited for a number of reasons. First, due to the significant number of different instrument types, methods, and runtimes it is a nearly impossible task to account for every possible representative of the spectra and retention time for a given metabolite under all of these diverse conditions. Second, metabolomics experiments utilize a global non-targeted approach where the method is optimized to measure as many metabolites as possible in a wide range of biological sample types (i.e., matrices). Certain metabolites behave differently in one matrix than in another, or differently in the same matrix under different conditions, for example in response to an experimental treatment versus when non-treated. Third, there may be areas of the chromatogram with a high-degree of co-eluting metabolites. Public databases of metabolite spectra can provide useful information in many cases, especially when no existing library exists. However, the public information is limited and certainly not as informative or reproducible as generating an in-house chemical reference library using the same equipment and protocols as used to analyze the experimental samples.

While requiring a significant resource commitment, the generation of an internal library of authentic chemical standards is a worthwhile task with significant advantages for high-throughput metabolomics. An in-house library of authentic standards provides a clear representation of the spectra resulting from a metabolite on the same instrument and method used to analyze the experimental sample. A retention index for the internal library can be calculated and set, resulting in library entries that are fixed in time. Consequently consistent, reliable, standard spectra that do not change over time are ensured which, in turn, facilitates automated, high confidence metabolite identification.

Software for performing spectral library matching, much like peak integration software discussed above, is readily accessible (Scheltema, Decuypere et al. 2009). From open-source applications to commercial packages there are numerous choices. Many software packages use some type of forward or reverse (or both) fitting algorithm that use mass and time components to match peaks to metabolites of similar mass and peak shape within a time window. Due to their global, non-targeted nature, metabolomic studies are not optimized for any metabolite in particular, so a positive metabolite identification in a metabolomics analysis is almost never a binary decision. It is highly unlikely to simply have a positive yes or no for a metabolite identification, instead it is more likely to have a probability score associated with the identification. Quality control of the scoring is essential and one of the most important aspects of metabolomics analysis, especially for running studies in high throughput.

4.5 Unnamed metabolites

A chemo-centric approach, based on a reference library, to high-throughput metabolomics is a powerful method to identify metabolites within biological samples. If there is any weakness to using in-house generated reference libraries it would be in the realm of identifying the redundant ion peaks that originate from metabolites that do not exist in the library. Methods available to identify and group these redundant ion peaks are limited (Bowen and Northen; Dunn, Bailey et al. 2005; Wishart 2009).

The most common approach is to rely on the chromatographic elution similarity between these redundant ions as well as looking for user defined mass relationships between the ions that are consistent with known chemical modifications. The effectiveness of this approach is limited in highly complex samples where metabolite co-elution is common. In such situations, there can be multiple metabolites eluting simultaneously which confounds identifying their respective ions based on elution. Another shortcoming of this method is the inability to identify unique modifications or fragments that are not known to occur.

A method that has yielded very good results for analyzing spectrometry data and fits well within the framework of high-throughput metabolomics is the QUICS method (Dehaven, Evans et al. 2010) This method to identify and quantify individual components in a sample, (QUICS), enables the generation of chemical library entries from known chemical standards and, importantly, from unknown metabolites present in experimental samples but without a corresponding library entry. The fundamental concept of this method is that by looking at detected ion features across an entire set of related samples, it is possible to detect subtle spectral trends that are indicative of the presence of one or more obscure metabolites. In other words, because of the natural biological variability of the metabolite in the study samples, by performing an ion-correlation analysis across all samples within a given dataset it is possible to detect ion features that are both reproducible and related to one another. Using the cross sample correlation analysis it is then possible to add the spectral features for that metabolite to the reference library. Then the metabolite can be detected in the future using that library entry, even though the metabolite is unknown, i.e., without an exact chemical identification. Importantly, this method captures any unknown metabolite because it does not require chemical adducts and/or fragment products to be previously known or expected. Another advantage is that statistical analysis can be used to determine whether or not the metabolite is significant or of interest. In this way the important unnamed metabolites can be focused on for the work of performing an actual identification which enhances efficiency and reduces the work to identifying the most important metabolites.

5. Quality control

The ability to perform thorough quality control on identified metabolites in metabolomics studies is extremely important. The higher the quality of data entering statistical analysis, the higher the probability that the study will provide answers to the questions being asked. This section will focus on three aspects of quality control - quality control samples (i.e., blanks, technical replicates), software for assessing the quality of metabolite identification, and software for assessing the original peak detection and integration. This last point may seem out of order but for reasons to be described results in an invaluable check of the peak quality.

5.1 Blanks – Identify the artifacts of the process

A commonly overlooked issue in biological data collection is the presence of process artifacts. A process artifact is defined as any chemical whose presence can be attributed to sample handling and processing and not originating from the biological sample. In all analytical methods chemicals are inadvertently added to samples. Artifacts can include releasing agents and softeners present in plastic sample vials and tubing, solvent contaminants, etc. One of the easiest and most efficient means of identifying artifacts is to run a "water blank" sample interspersed throughout the entire process alongside the true experimental samples. In this way, the water blank will acquire all the same process-related chemicals as the experimental samples. Consequently, identification and *in silico* removal of artifacts can be accomplished by identifying those chemicals detected at significant levels in the water blank when compared to the signal intensity in the experimental samples. If not identified and removed, process artifacts can inadvertently arise as false discoveries.

5.2 Technical replicates – Find the total process variation

The intrinsic reproducibility of a method is critical since it has considerable impact on the significance and interpretation of the results. For example, if a 20% change was detected between treatment and control samples but the analytical method had a 20% coefficient of variation (CV) for that measurement, concerns regarding the accuracy of the measurement would call into question the biological relevance of that change in measurement. On the other hand, if the analytical method had a 2% CV for that same measurement it is much more likely that the same 20% change is of "real" biological significance. Clearly, smaller analytical variability of the method enables small, yet meaningful, biological changes to be detected accurately and consistently. It is therefore critical to determine the analytical reproducibility/variability of a method for every compound/measurement.

By far the most common way to assess system stability and reproducibility is by use of internal standards. Internal standards can be measured throughout a study to monitor system reproducibility and stability. The drawbacks to this approach are that the number of standards is typically small and do not represent the myriad of chemical classes typically observed in a metabolomics analysis.

Another common approach to address method variability is by the use of technical replicates. With this approach the same biological sample is run multiple times, e.g., in triplicate, to determine method reproducibility. The advantage of this method over internal standards is the ability to determine the CV of the method for each compound detected within the matrix of the samples being analyzed. However, the disadvantage is that, while the replicate approach is extremely effective, it is also very time-consuming and of limited practicality in a high-throughput setting.

An extremely practical and efficient approach is to run a technical replicate of a sample composed of a small aliquot from all the samples in a study interspersed among individual experimental samples. An aliquot of each experimental sample is pooled, then an aliquot of the pooled sample mixture is run at regular intervals—every *n* number of experimental sample injections (*n* to be set by operator). An advantage of this pooled sample is that it provides CV information for all compounds detected in the study, in the matrix under study. Another advantage is that far less instrument analysis time is required which makes it far more practical in a high-throughput laboratory.

5.3 Quality control of automated metabolite identifications

Performing quality control (QC) for a given metabolite identification can be an exhaustive and time-consuming task. The work to perform QC on every metabolite identification in every sample within a metabolomics study can seem to be a nearly-impossible task. Considering a relatively small metabolomics study of 50 samples, with an average of 800 identified metabolites per sample, there would be 40,000 spectra to review for just that one study. Yet, as time-consuming as this process is, quality control of automated library calls is vital for ensuring accuracy and high confidence in the data which, in turn, enables meaningful biological interpretation of the results. A software package that can permit this process to proceed quickly and efficiently is critical in a high-throughput setting.

Visual inspection of all the samples in a study simultaneously enables rapid metabolite identification QC. By representing the sample data within a study as a single set in a visual manner and creating tools that quickly allow an analyst to investigate and manually accept or reject an automated metabolite identification, the task of performing quality control on even extremely large datasets can be accomplished rapidly and easily. An example of a visual data display is shown in Figure 1. In this example the panel across the top (Figure 1A) contains a list of all of the metabolites identified by the software in the experimental samples being analyzed. By highlighting one chemical, the structure for that compound is displayed in an adjacent window (Figure 1B). The default visualization for viewing a highlighted metabolite is broken down into a distinct method chart for each analytical platform method that was used to identify that metabolite. The display also shows the multiple analytical platforms where the metabolite was identified. In this example, the same metabolite identified on a GC/MS platform (Figure 1C), and LC/MS negative ion platform (Figure 1D) is shown. Within each chart, the individual sample injections, each with a unique identifier, make up the y-axis (Figure 1E). The x-axis represents the retention index (RI) time scale. Navigation of the interface involves scrolling down through the data table window (Figure 1A). From the interface it is also possible to review annotation regarding the highlighted metabolite (Figure 1F), view the analytical characteristics (e.g., Mass, RI) of the metabolite as well as toggle through RI windows containing ions characteristic of that metabolite (Figure 1G).

An example plot of data from the LC/MS negative platform is illustrated in Figure 2. In this example the samples are initially sorted by the sample type, namely process blank, technical replicate, or experimental sample. The dots within each method chart represent the detected ion peaks, and each point has associated peak area, mass to charge (m/z), chromatographic start and stop data which can be accessed by clicking on the individual dots, as shown in Figure 3.

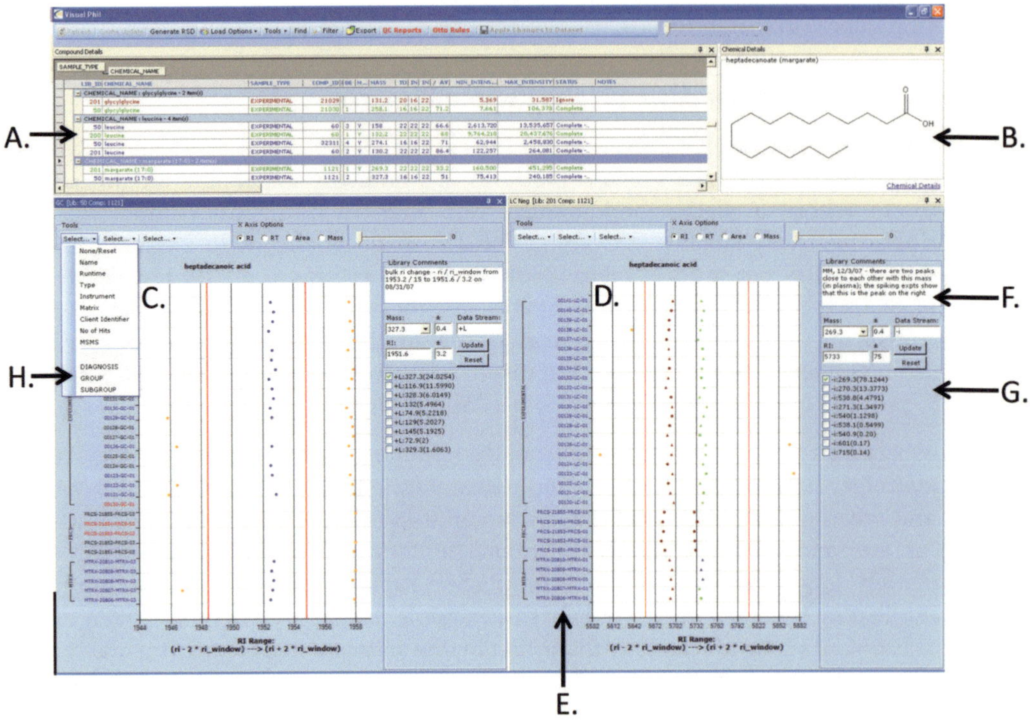

Fig. 1. Graphical user interface showing the view for the proposed identification of heptadecanoic acid. (A) Distinct list of identified metabolites for the loaded sample set. This list includes any metabolite identified at least once in any sample with the set. It also includes summary statistics such as averages for spectral scoring and chromatographic peak intensities, number of times detected, and status.(B) Chemical structure for displayed metabolite. (C) Data for the posed library identification heptadecanoic acid from the GC/MS method. (D) Data for the posed library identification heptadecanoic acid from the LC/MS negative ion method. (E) List of unique sample identifiers comprising the study. (F) Comment field for storing and displaying annotations that are relevant to the currently displayed metabolite. (G) List of other ion peaks that exist as part of the spectral library entry. (H) List of sample sorting options including associated sample metadata; diagnosis, group and subgroup.

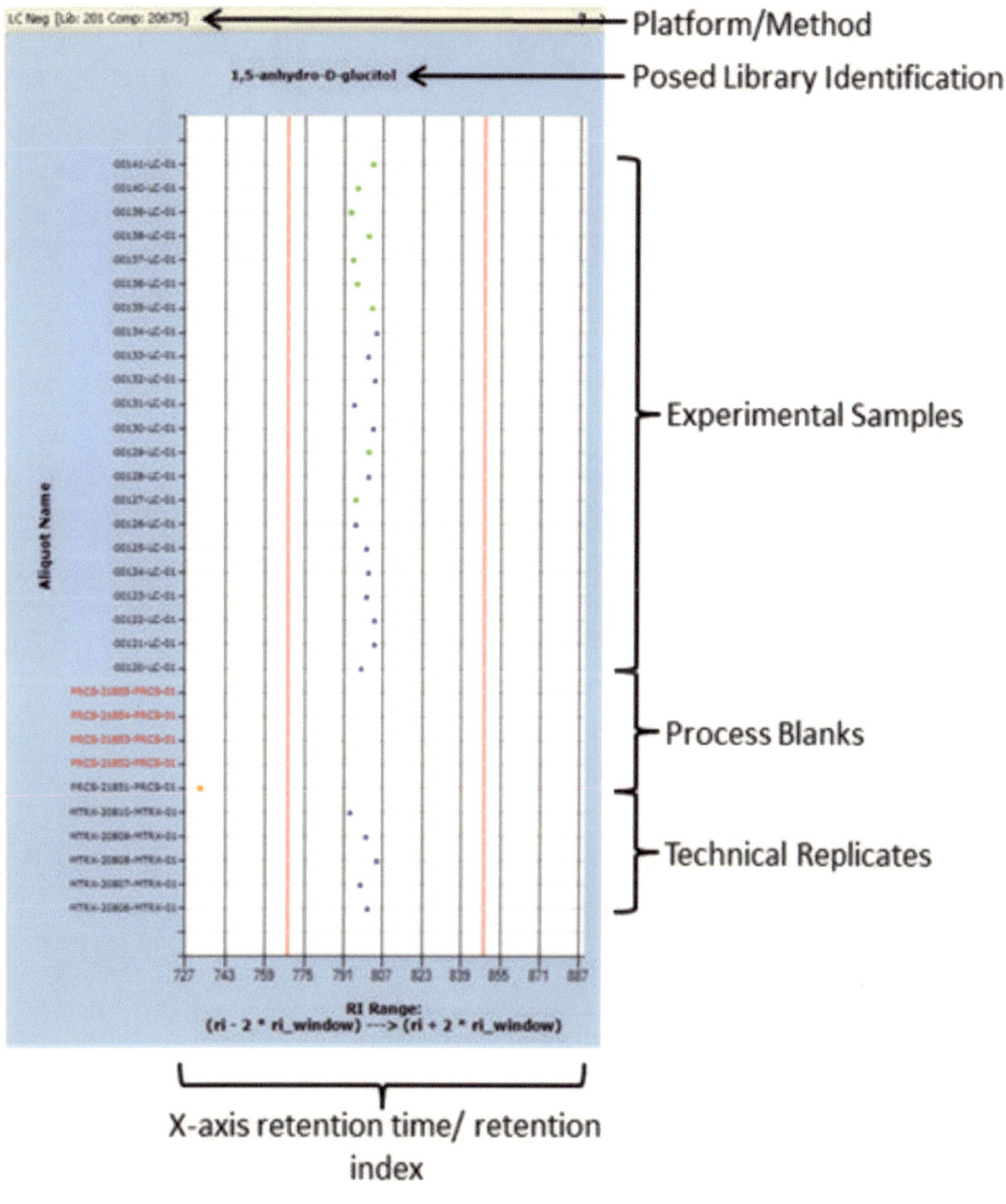

Fig. 2. Plot for LC/MS negative method. Individual samples in the sample set are displayed and sorted on the y-axis. Chromatographic retention time is presented on the x-axis.

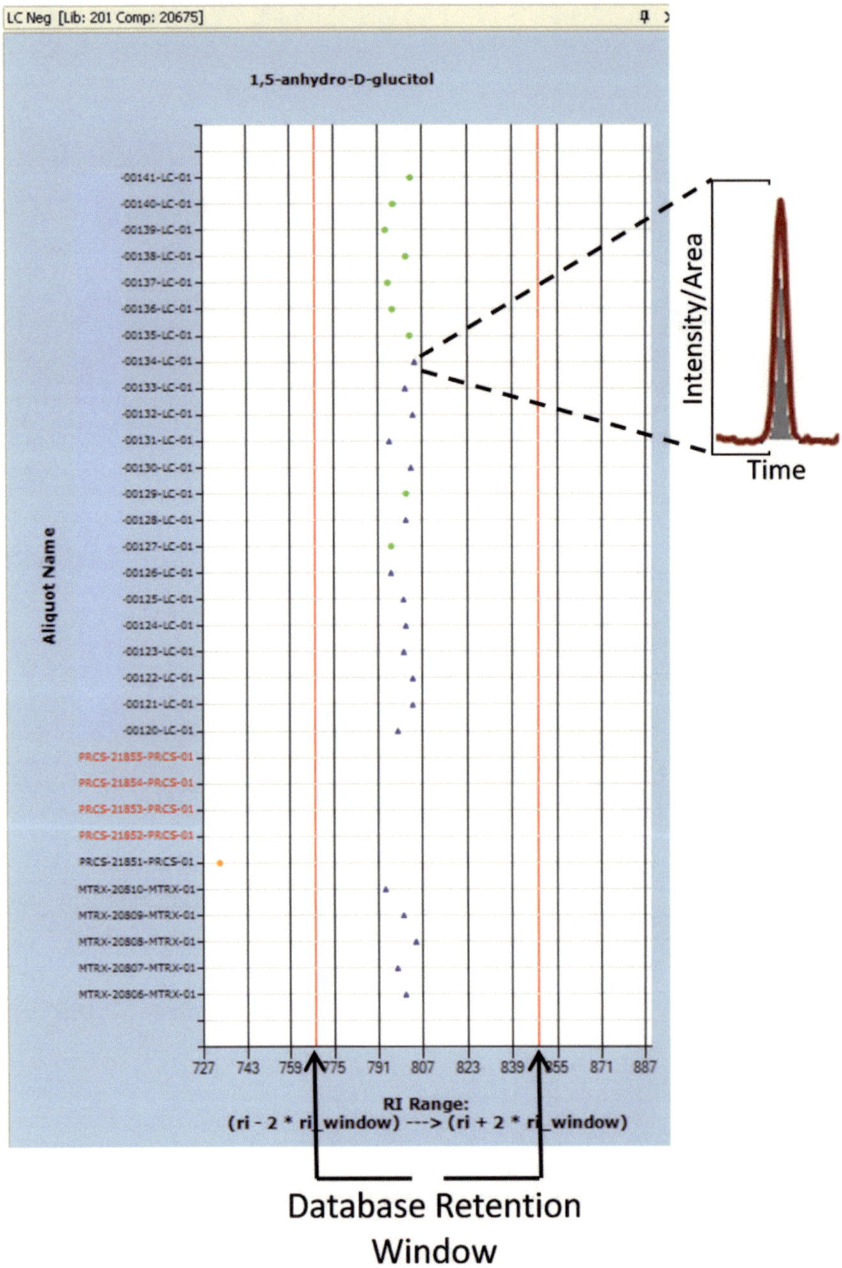

Fig. 3. Raw MS data can be accessed for each sample via graphical user interface links. Each "dot" represents the detected and integrated ion peak in the individual sample listed on the y-axis. Thus, each "dot" has an associated area, height intensity, chromatographic start, stop and apex retention time/ retention index. Color and shape of "dot" are indicative of the quality of the match to the posed library identification (see chart 1) and can be used to launch underlying data such as raw MS data (insert). Colors of the samples listed on the y-axis also hold meaning (see chart 1).

This type of visualization permits the analyst to quickly verify the quality (i.e., QC) of the automated peak detection and integration software. Each dot that is representative of an ion peak can be individually removed/rejected from the proposed library identification. In this way, extraneously detected ion peaks in the window can be visualized and individually removed, as is the case shown in Figure 4. In this particular example there are two closely eluting ion peaks with the same mass. One of the peaks is 2-stearoylglycerophosphocholine (Figure 4, panel A) and the other is 1-stearoylglycerophosphocholine (Figure 4, panel B). In both panels the correct peak must be manually approved and the incorrect peak rejected (indicated by red dots). Stray detected ions can also be individually rejected from the identifications. In addition, the interface permits the interrogation of the integration quality of individual ion peaks since each dot is linked to the raw ion data as illustrated in Figure 4, panel C and Figure 5, panel B. In this fashion any potential inaccuracy in the automated detection and integration of individual peaks can be readily determined.

In addition to being able to curate each sample individually, the automated library identification for an entire sample set can be rejected. An example of this is shown in Figure 5. The presence of multiple dots for each sample in the RI window (Figure 5 A) coupled with the ability to view the underlying ion data (Figure 5B) makes it apparent that the automated metabolite identification was based on erroneous ion peaks that resulted from the integration of noise. As a result the automated call for the entire sample set was manually rejected by the analyst. Accordingly, with this visualization tool the analyst can rapidly determine the quality of the automated detection and integration and remove from the dataset any peaks which are of questionable quality.

In addition to being able to QC the automated peak detection and integration software, an interface such as this allows an analyst to visually inspect the quality of the library identification in each individual sample. In the graphical plot the "dot" representing the detected ion peak for the proposed metabolite identification is displayed in various color and shape combinations. Each combination of color and shape within each plot is an indicator of the quality of the automated metabolite identification, which greatly aids the analyst in making the quality assessment rapidly. Listed in Table 1 are possible color and shape combinations with the meaning for each. The quality assessment is based on spectral library matching logic (Evans, Dehaven et al. 2009). This graphical display allows the analyst to look at a proposed identification for a given metabolite made by the software and immediately determine its quality and confidence based on spectral match scores. In this way, the automated metabolite identifications for large datasets can be quickly evaluated by the analyst. An example of a proposed call for a group of ions in a sample set where the MS/MS spectral match was poor is shown in Figure 6. The low data quality is readily apparent by the preponderance of the red colored dots in the plot.

5.4 Quality control of automated integration

GC/MS and LC/MS/MS measurements of a bio-sample usually produce millions of ions, which are fragments and/or adducts/aggregates of the metabolites, artifacts from the system, and potentially false ions from background noise. Ideally, the false ions from the background noise are removed, using, for example, a Gaussian smoothing algorithm to filter them out of the dataset. The remaining ions are then integrated across time within a mass window to identify ion chromatographic peaks. Thousands of such ion chromatographic

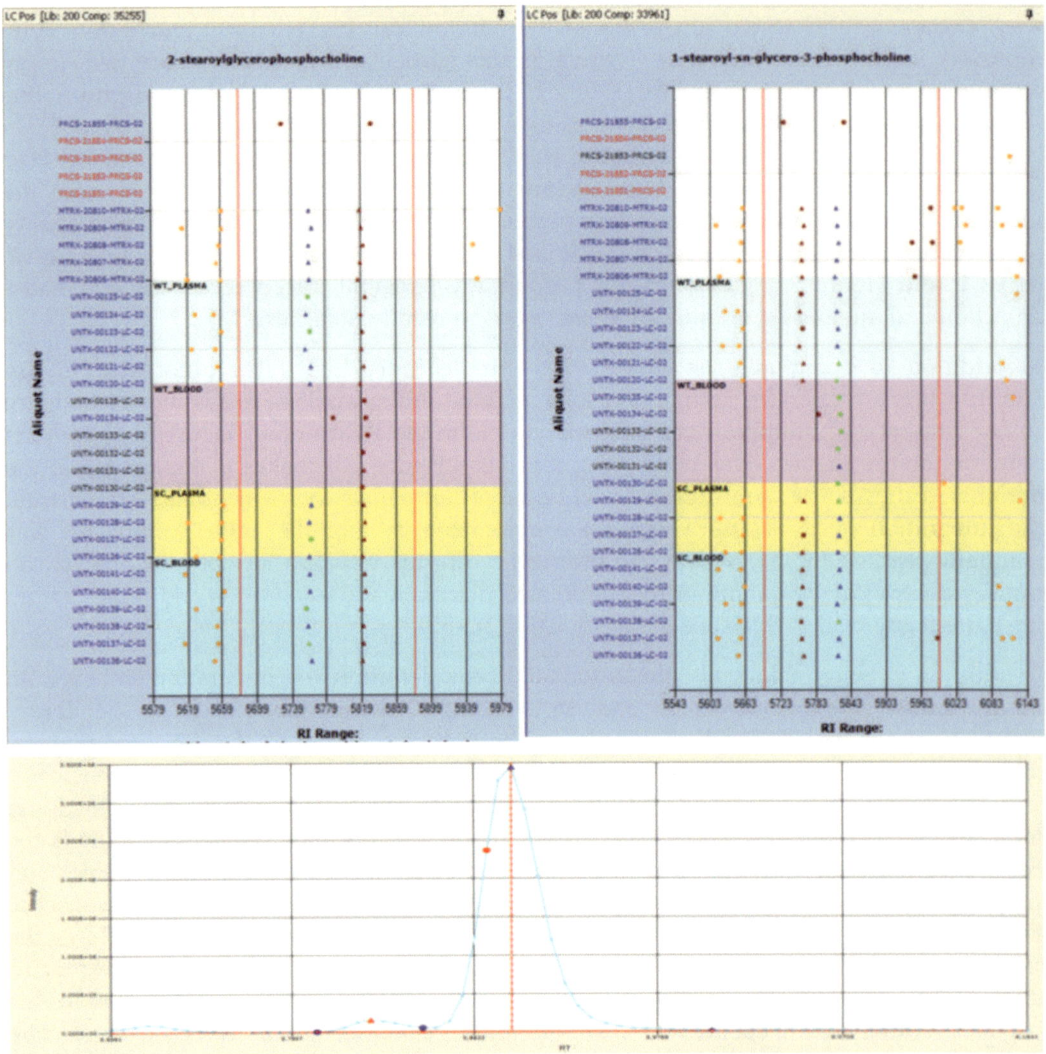

Fig. 4. Example of the visualization of closely related ion peaks. (A) Two possible ion peaks were detected in same retention window for 2-stearoylglycerophosphocholine. In this case the ion peak to the left is the correct peak (green and blue dots with arrow) and the peak on the right was rejected (red). (B) The peak on the right is actually 1-stearoylglycerophosphocholine. Therefore, the peak on the right is correct (green and blue with arrow) while the peak on the left is rejected (red). In addition ion peaks in dashed boxes are stray detected ion peaks not associated with the peak for 1-stearoylglycerophosphocholine ion peaks that were rejected (red). (C) The extracted ion chromatogram for one of the samples in the sample set for this ion shows the two peaks are well separated and accurately integrated.

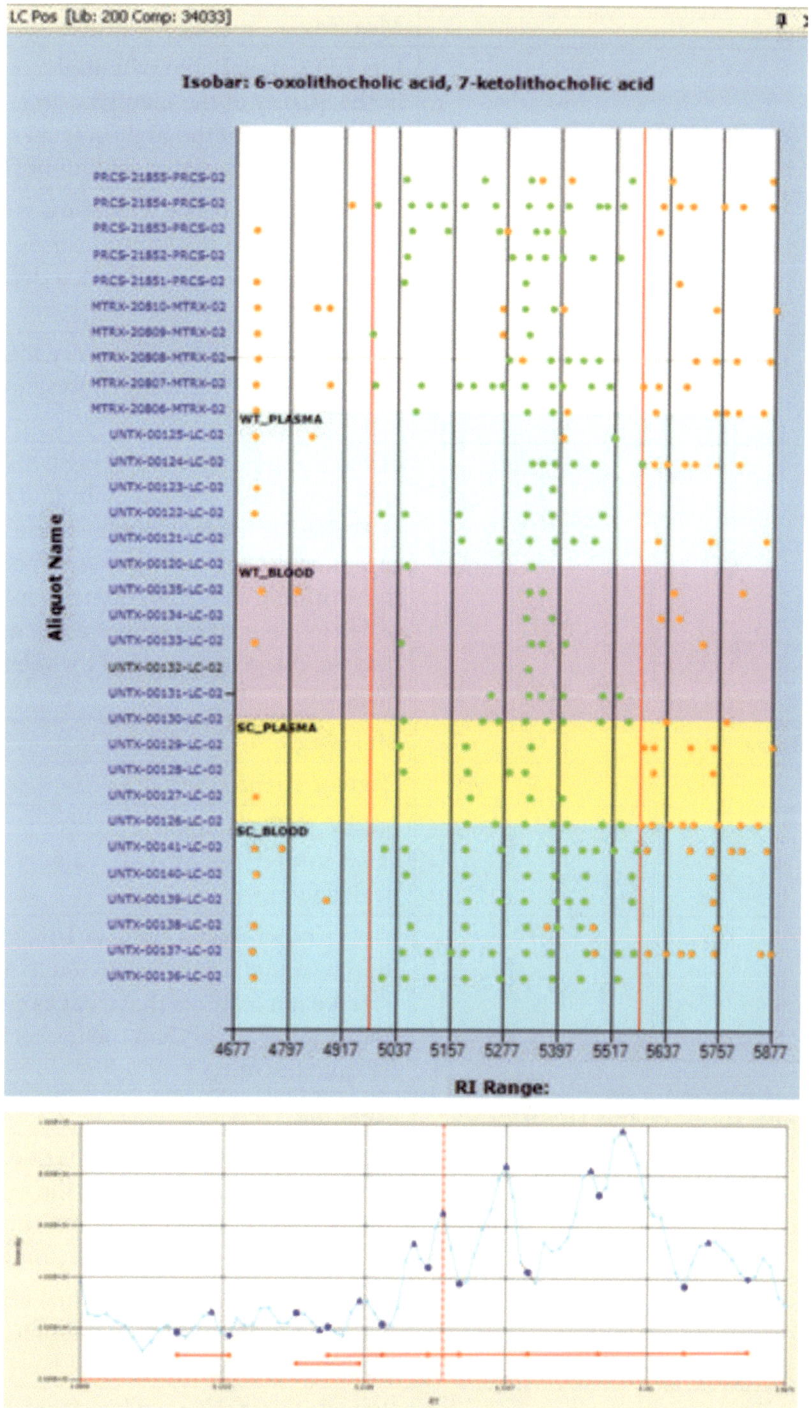

Fig. 5. Example of rejected identification for entire sample set. (A) Detected ion-peaks that result from (B) a noisy baseline as seen in one injection; the entire library identification for this LC/MS positive method is rejected.

Feature Color	Meaning
Blue	Identified metabolite with high confidence in the quality of the identification. These points represent the highest level of confidence for a metabolite identification.
Green	Identified metabolite with some confidence in the quality of the identification. These points require further research depending on the context.
Red	Identified metabolite with very low confidence in the quality of the identification.
Orange	Orange points represent a peak that was within and near the time window but was not called as the metabolite. There may be cases such as for very low-level peaks that the scoring was not adequate to make a call but taken in context with the rest of the dataset, the peak is indeed the metabolite in question.
Feature Shape	**Meaning**
Circle	Regular peak.
Triangle	For LC/MS a triangle represent that there exists underlying MS/MS data to confirm a peak's identity.
Other shape	Shapes other than a circle or triangle represent that a user has opted to also observe ion features that exist as part of the library entry other than the ion feature used for quantification.
Sample Label Color (y-axis label)	**Meaning**
Black	Black labels mean that there is one and only one ion feature representing the metabolite identification in the given time window.
Blue	Blue labels indicate that there is more than one ion-feature within the time window that may represent the metabolite identification.
Red	Red labels are shown when there are no possible hits for the given metabolite.

Table 1. Color/Shape combinations to demonstrate peak quality

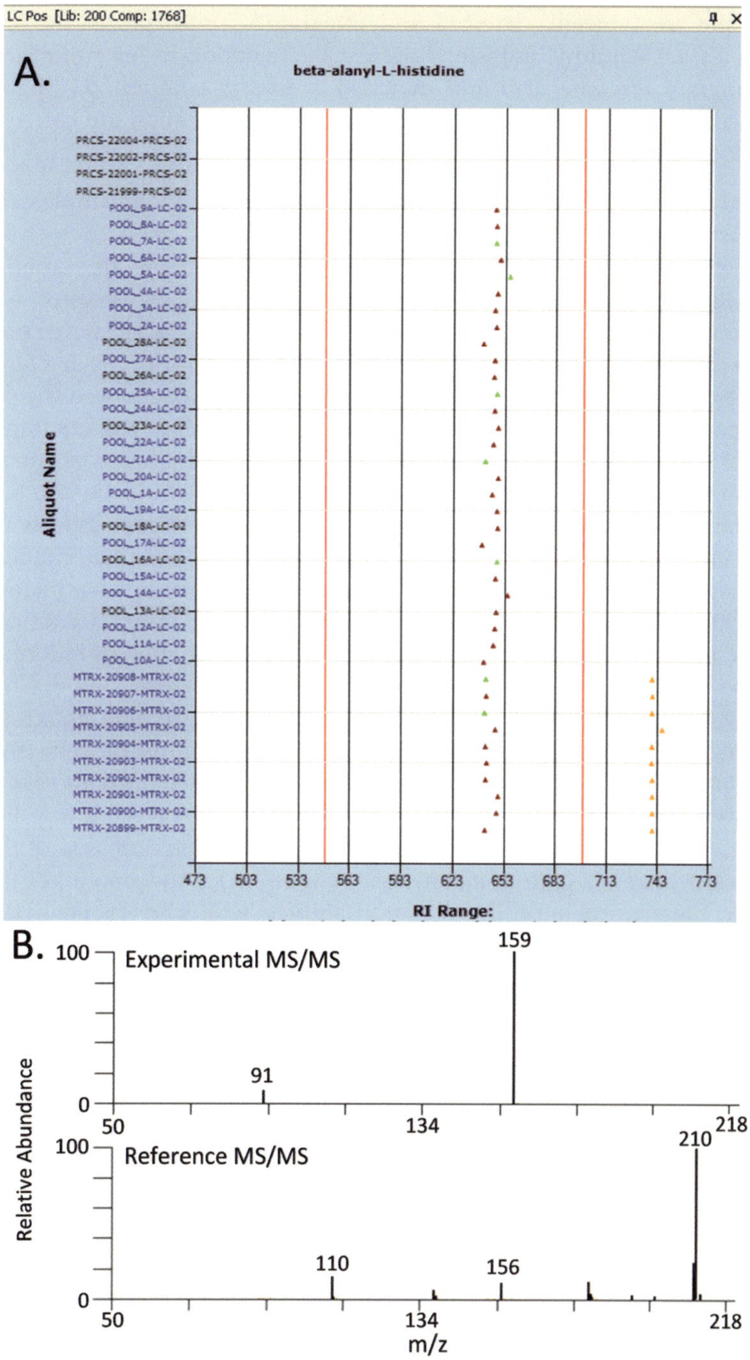

Fig. 6. An example of an entire metabolite call that was rejected because the MS/MS spectral match was poor. (A) Red color of dots indicates that the MS/MS spectral match was of low quality. (B) Experimental MS/MS spectrum from one injection compared to the reference library spectrum for beta-alanyl-L-histidine (carnosine).

peaks per sample are typically detected and integrated. Those peaks are organized into groups within a time window, adjusted/aligned by retention index from known internal standards to account for time drift and matched to library compounds (metabolites). For LC/MS/MS, secondary fragment ions of the primary quantification ion can also be used to match library compounds.

Obviously, all the profiling and analysis of metabolites in biological samples are dependent on the accuracy and consistency of ion chromatographic peak detection and peak integration. However, GC/MS and LC/MS/MS measurements are complicated by a number of factors; for example, the co-elution of metabolites because of incomplete separation, the existence of artifacts from the system, the background noise, and the potential wide concentration ranges of metabolites in the sample. Such complexities can affect the detection and determination of the peak start, the peak end and the peak baseline. Incomplete separation can lead to shoulders on peaks on either the leading edge or the trailing edge of the main peak from metabolites present at higher concentrations. When compared to the baseline, the peak start and the peak end would be characterized by a baseline peak or a drop peak. Because of the complexity and variance inherent in biological samples, the same metabolite in different samples may have been automatically detected differently in regard to peak start, peak end and peak background. For example, in some cases, the ion chromatographic peaks for metabolites present in only trace amounts may not be well shaped, especially when a noisy background is present, so integration of such peaks might be quite variable from sample to sample. In other cases, the major ions for a metabolite may appear as a small shoulder on a larger ion peak and, as a result, may not even be detected during the automatic peak detection/integration process from sample to sample. In still other instances, a metabolite present in high concentrations may overload the column and distort the chromatographic profile, leading to peak splitting. For such high concentration metabolites the automatic library match may pick only one of the two peaks for quantification which will give an erroneously low value for the amount of the metabolite in the sample. Clearly, each of the above examples will lead to peak detection and integration inconsistency and inaccuracy across the samples in a sample set, which will potentially lead to wrong conclusions and wrong decisions in later analysis.

Global metabolomics has other challenges when it comes to peak detection. Unlike targeted metabolomics, global metabolomic profiling cannot be optimized for each metabolite that is present within a biological sample. Chromatography methods must be broad enough to detect as many of the metabolites in the sample as possible, regardless of chemical characteristics. Consequently, chiral compounds cannot be resolved and structural isomers are usually not well resolved in global metabolomics profiling. In downstream analysis for identified metabolites, structural isomers might better be combined to represent the metabolites, or, if one isomer is more crucial in elucidating the metabolism or biochemical pathway, consistently picking that one form across samples would ensure analytical consistency.

A software solution to detect and correct such inconsistencies in ion peak integration and library matching across samples in a sample set could be developed. After the quality control phase of the identification of the detected metabolites is completed, a deeper examination of the consistency of peak detection and integration could be performed to ensure consistency and accuracy. The quality control phase of automatically detected metabolites involves providing a high-quality, filtered list of identified metabolites devoid of noise and artifactual metabolites to the end user (Figure 7). Sample set sizes range from a few samples to hundreds and even thousands of samples. Because hundreds of metabolites can be detected and measured in each sample, this type of quality control operates on the 'quant' ion peaks -those peaks detected in the samples that are used for quantification of those metabolites.

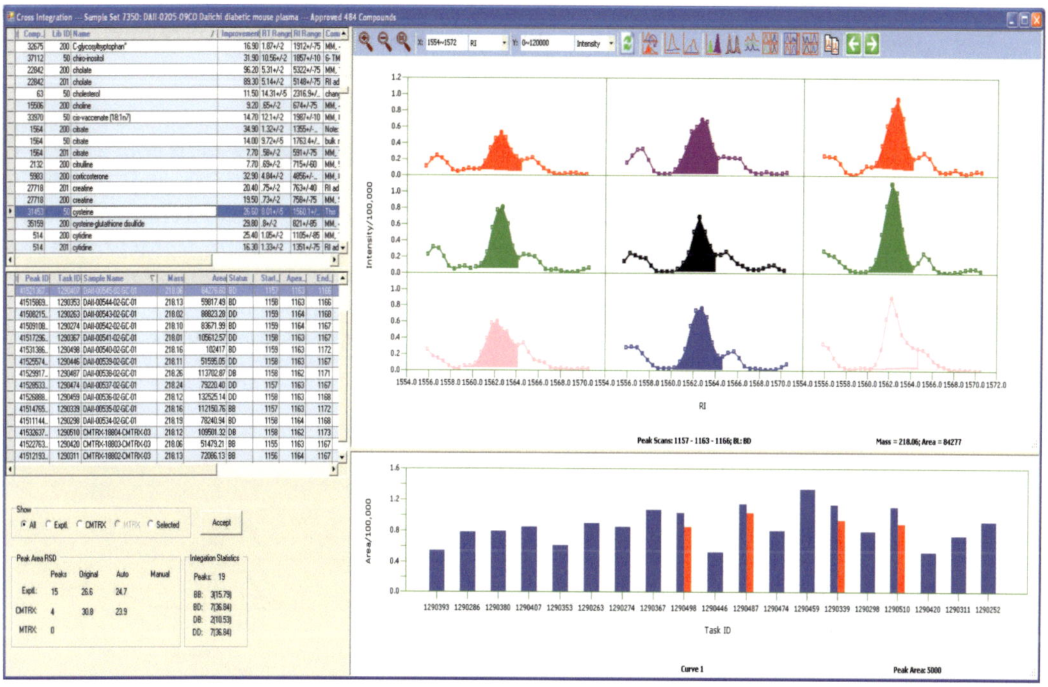

Fig. 7. Graphical View of Peak Integration for an Identified Metabolite: Identified metabolites (200~600) in the specified sample set (Upper Left); Quantitation peaks for selected metabolite in the samples in the sample set (Middle Left); Type of samples and Information about the sample peaks (Lower Left); Peak chromatograms (Upper Right); Sample peak area (blue for original integration and red for re-integrated (Lower Right).

The chromatograms of the ion peaks representing the quantitative mass from all of the samples in a set must be evaluated to determine if:

- the majority of the sample peaks are on the trailing edge of another peak,
- the majority of the sample peaks are on the leading edge of another peak,
- the majority are peaks that encompass two peaks in other samples, as a result of peak splitting.

Peak integration ranges are evaluated with alignment by retention index and the statistics of peak limits across the sample set. Accordingly, in addition to user specified manual correction, corrections in consistency and re-integration would be suggested and presented to the analyst for review and approval. Functionally, this type of software would give the end user a variety of methods to both investigate the automated integration and peak calls and to correct them as necessary. The software features must include:

- Automatic merging of approved peaks from the sample that match to the same library compound.
- Detection of shoulder peaks based on RI-aligned peak start or peak end distribution across the samples.
- Manual integration
- Manual peak splitting
- Show peak chromatograms in overlay mode or tabular mode for easy review/manual re-integration.
- Update peak integrations, peak recovery and library rematch

When an identified metabolite in a biological sample is at a sufficiently high concentration, it can overload the column and distort the chromatographic peak. Even though it may be out of the linear range, a consistent integration of the peak is still needed to characterize the group of samples. Distorted peaks tend to drive the integration software to identify a less than optimal peak to be used for quantification. In Figure 7, the peak for glucose was incorrectly split in a handful of samples by the automated peak integrator. By examining the consistency of the peak integration across the set of samples it is possible to easily identify and correct this situation. As shown in the example in Figure 8, this correction would improve the relative standard deviation from 20.1 to 7.4

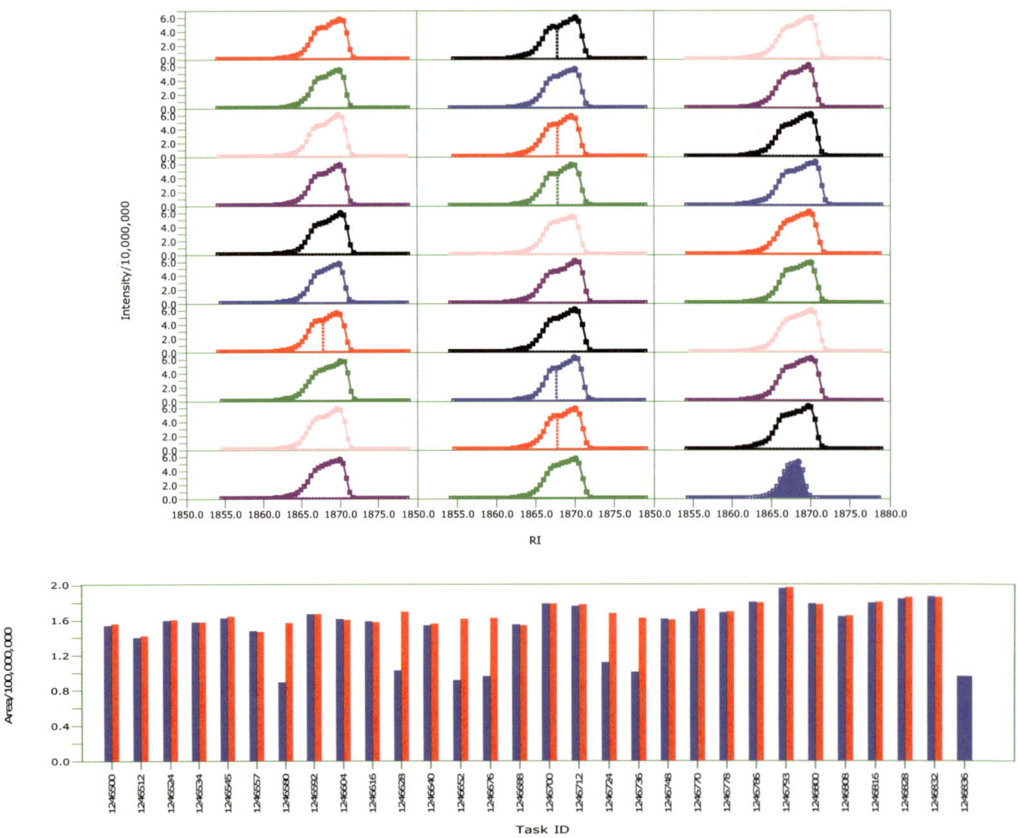

Fig. 8. Combining Peaks

As illustrated in Figures 9 and 10, small peaks on the leading or trailing side of a larger peak are often integrated inconsistently:

- Small shoulder peaks are detected
- Small shoulder peaks are not detected
- Small shoulder peaks are combined into the main peak

In Figure 8, the major peak on the left is identified as cysteine, whereas the shoulder on the right side is from threonate. In one sample, the small peak from threonate was inaccurately combined into the main peak for cysteine when it was automatically integrated, thus inadvertently increasing the response for cysteine in that sample. After re-integration the erroneous integration was corrected thereby restoring the correct integration for cysteine and permitting the independent detection of threonate in the sample as well.

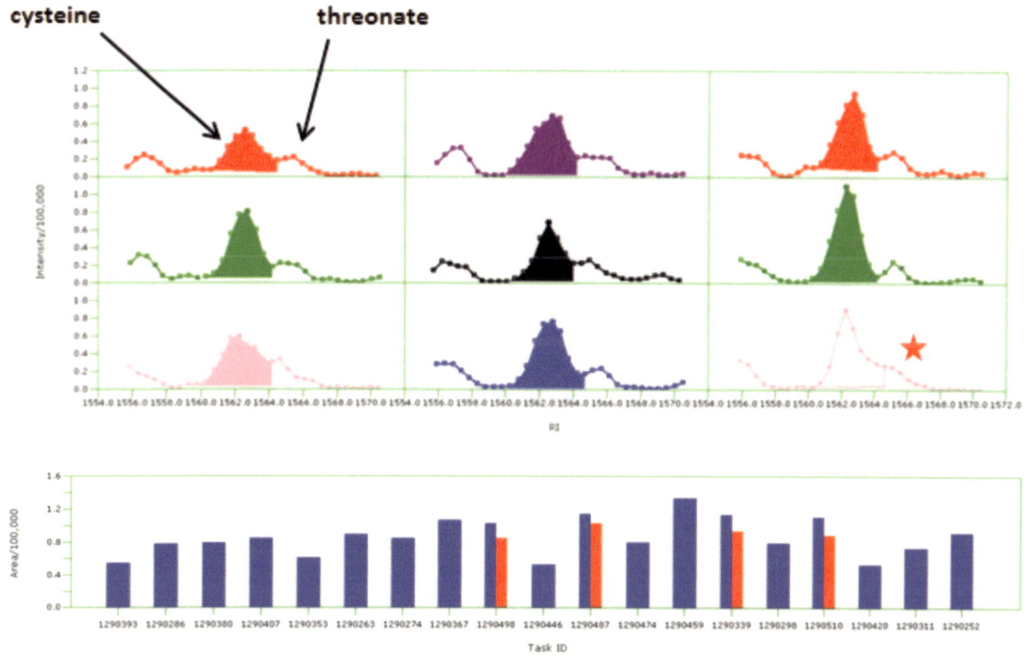

Fig. 9. Examples in inconsistent shoulder peaks. Splitting of shoulder (Upper panel); Area change after re-integration (Blue for automatic integration and red for re-evaluated integration (Lower panel).

In Figure 10, the major peak on the right is identified as 1-docosahexaenoylglycerophosphocholine (1-DHGPC), whereas the shoulder on the left side is identified as 2-docosahexaenoylglycerophosphocholine (2-DHGPC). In one sample, the peak for 2-DHGPC was inaccurately combined into the peak for 1-DHGPC when it was automatically integrated. In another sample, the baseline was not calculated consistently. The curves at the lower right show the correction. After re-integration the erroneous integration was corrected and the small peak for 2-DHGPC was recovered.

Software that can detect inconsistencies in peak detection and integration across samples in a sample set can ultimately improve the accuracy in the integration of peaks that have been identified as metabolites; this in turn leads to lower CV's and more accurate statistical analysis which can contribute significantly to the elucidation of metabolism and metabolite pathway.

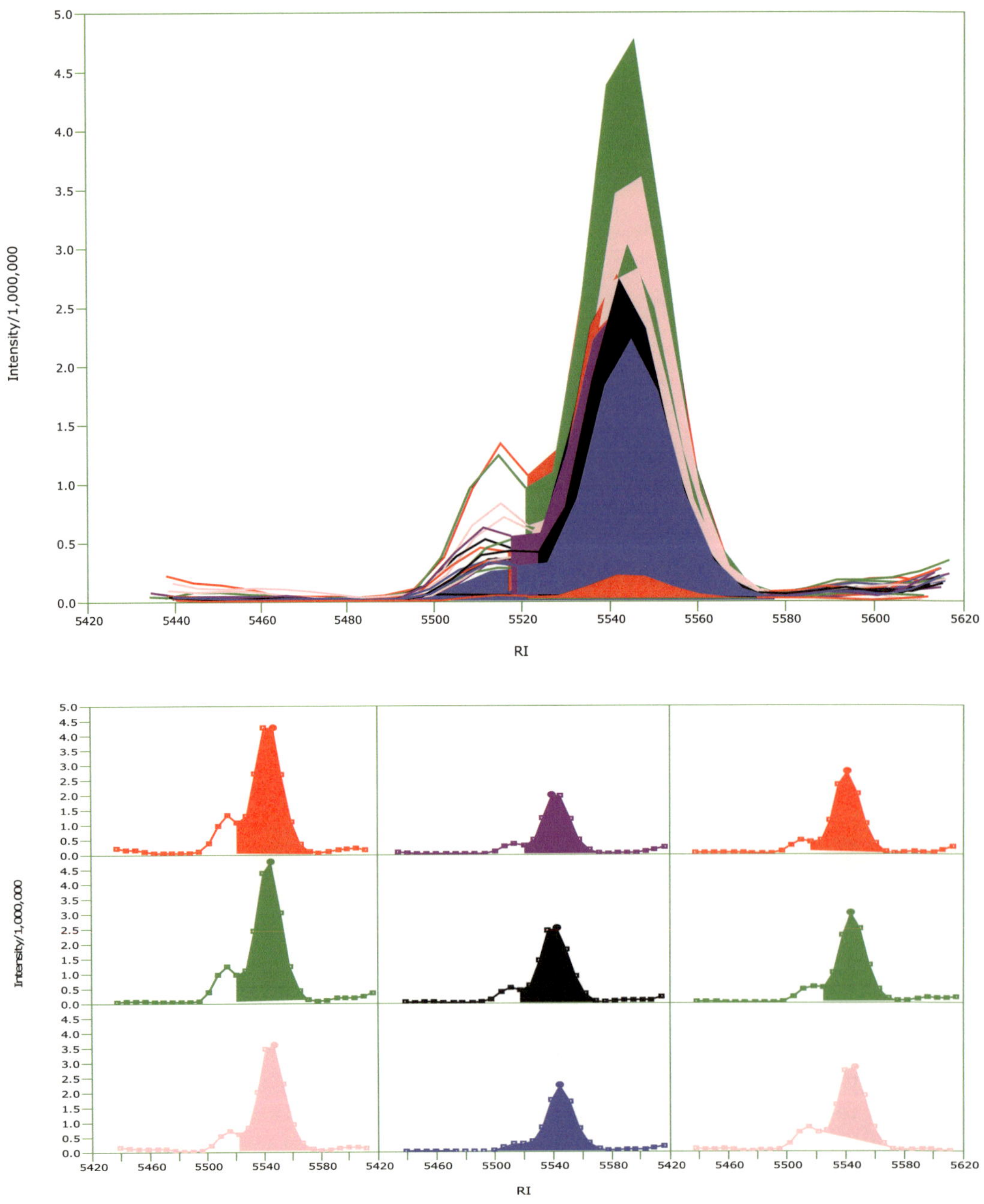

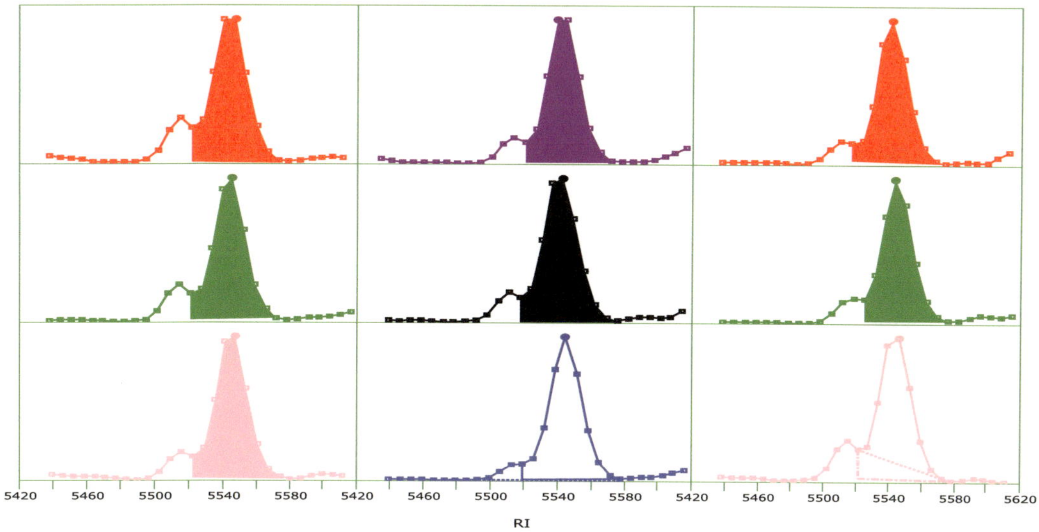

Fig. 10. Examples of inconsistent shoulder peaks

6. Conclusion

Metabolomics as a technology has demonstrated clear utility in a broad array of biological applications. The applications are not only in demonstrating simple metabolic comparisons between treated and control groups but in studies involving biomarker discovery, drug development/MOA/recovery, bio-processing, agricultural applications, consumer products, diagnostics, and so on (Sreekumar, Poisson et al. 2009; Berger, Kramer et al. 2007; Barnes, Teles et al. 2009; Boudonck, Mitchell et al. 2009; Ma, Ellet et al. 2009; Ohta, Masutomi et al. 2009; Watson, Roulston et al. 2009; Oliver, Guo et al.). The ability to run metabolomic studies in high-throughput has been a challenge thus far, not so much because of the complexity or size of the data, but because of the difficulty in generating reproducible data having low process variation that can be quantified, is devoid of artifactual components, and provides high confidence in the identification of metabolites. Without knowledge of the variability of the process on a metabolite by metabolite basis, it is not possible to determine the true biological variability and thus, cannot provide accurate answers to the questions that under investigation.

As demonstrated here, quality and the throughput of processing sample data for metabolomics studies do not need to be mutually exclusive. By taking an intelligent engineering approach to the data workflow, knowing when to automate a process and developing software solutions that are streamlined for this process, the processing of sample data for metabolomics studies can be done in significantly high volume and with high quality.

7. Acknowledgement

We gratefully acknowledge the work and contributions of all members of the Metabolon informatics, platform, project management, statistics and management teams for their dedicated work in building an enterprise metabolomics platform. CD, AE, HD, and KL are employees of Metabolon.

8. References

Barnes, V. M., R. Teles, et al. (2009). "Acceleration of purine degradation by periodontal diseases." *J Dent Res* 88(9): 851-855.

Benjamini, Y. and Y. Hochberg (1995). "Controlling the false discovery rate: a practical and powerful approach to multiple testing." *Journal of the Royal Statistical Society. Series B* 57: 289-300.

Berger, F. G., D. L. Kramer, et al. (2007). "Polyamine metabolism and tumorigenesis in the Apc(Min/+) mouse." *Biochem Soc Trans* 35(Pt 2): 336-339.

Boudonck, K. J., M. Mitchell, et al. (2009). "Characterization of the biochemical variability of bovine milk using metabolomics " *Metabolomics* 5(4): 375-386.

Bowen, B. P. and T. R. Northen "Dealing with the unknown: metabolomics and metabolite atlases." *J Am Soc Mass Spectrom* 21(9): 1471-1476.

Bryan, K., L. Brennan, et al. (2008). "MetaFIND: a feature analysis tool for metabolomics data." *BMC Bioinformatics* 9: 470.

Dehaven, C. D., A. M. Evans, et al. (2010). "Organization of GC/MS and LC/MS metabolomics data into chemical libraries." *J Cheminform* 2(1): 9.

Dunn, W. B., N. J. Bailey, et al. (2005). "Measuring the metabolome: current analytical technologies." *Analyst* 130(5): 606-625.

Evans, A. M., C. D. Dehaven, et al. (2009). "Integrated, Nontargeted Ultrahigh Performance Liquid Chromatography/Electrospray Ionization Tandem Mass Spectrometry Platform for the Identification and Relative Quantification of the Small-Molecule Complement of Biological Systems." *Anal Chem* 81(16): 6656-6667.

Fan, T. W., A. N. Lane, et al. (2004). "The promise of metabolomics in cancer molecular therapeutics." *Curr Opin Mol Ther* 6(6): 584-592.

Fields, C. (1996). "Informatics for ubiquitous sequencing." *Trends Biotechnol* 14(8): 286-289.

Griffin, J. L. (2006). "Understanding mouse models of disease through metabolomics." *Curr Opin Chem Biol* 10(4): 309-315.

Hood, L. E., M. W. Hunkapiller, et al. (1987). "Automated DNA sequencing and analysis of the human genome." *Genomics* 1(3): 201-212.

Hunkapiller, T., R. J. Kaiser, et al. (1991). "Large-scale and automated DNA sequence determination." *Science* 254(5028): 59-67.

Katajamaa, M. and M. Oresic (2005). "Processing methods for differential analysis of LC/MS profile data." *BMC Bioinformatics* 6: 179.

Katajamaa, M. and M. Oresic (2007). "Data processing for mass spectrometry-based metabolomics." *J Chromatogr A* 1158(1-2): 318-328.

Khoo, S. H. and M. Al-Rubeai (2007). "Metabolomics as a complementary tool in cell culture." *Biotechnol Appl Biochem* 47(Pt 2): 71-84.

Lawton, K. A., A. Berger, et al. (2008). "Analysis of the adult human plasma metabolome." *Pharmacogenomics* 9(4): 383-397.

Lindon, J. C., E. Holmes, et al. (2007). "Metabonomics in pharmaceutical R&D." *Febs J* 274(5): 1140-1151.

Ma, N., J. Ellet, et al. (2009). "A single nutrient feed supports both chemically defined NS0 and CHO fed-batch processes: Improved productivity and lactate metabolism." *Biotechnol Prog* 25(5): 1353-1363.

Nordstrom, A., G. O'Maille, et al. (2006). "Nonlinear Data Alignment for UPLC-MS and HPLC-MS Based Metabolomics: Quantitative Analysis of Endogenous and Exogenous Metabolites in Human Serum." *Anal Chem* 78(10): 3289-3295.

Ohta, T., N. Masutomi, et al. (2009). "Untargeted metabolomic profiling as an evaluative tool of fenofibrate-induced toxicology in Fischer 344 male rats." *Toxicol Pathol* 37(4): 521-535.

Oliver, M. J., L. Guo, et al. "A sister group contrast using untargeted global metabolomic analysis delineates the biochemical regulation underlying desiccation tolerance in Sporobolus stapfianus." *Plant Cell* 23(4): 1231-1248.

Patterson, A. D., H. Li, et al. (2008). "UPLC-ESI-TOFMS-based metabolomics and gene expression dynamics inspector self-organizing metabolomic maps as tools for understanding the cellular response to ionizing radiation." *Anal Chem* 80(3): 665-674.

Scalbert, A., L. Brennan, et al. (2009). "Mass-spectrometry-based metabolomics: limitations and recommendations for future progress with particular focus on nutrition research." *Metabolomics* 5(4): 435-458.

Scheltema, R., S. Decuypere, et al. (2009). "Simple data-reduction method for high-resolution LC-MS data in metabolomics." *Bioanalysis* 1(9): 1551-1557.

Smith, C. A., E. J. Want, et al. (2006). "XCMS: processing mass spectrometry data for metabolite profiling using nonlinear peak alignment, matching, and identification." *Anal Chem* 78(3): 779-787.

Sreekumar, A., L. M. Poisson, et al. (2009). "Metabolomic profiles delineate potential role for sarcosine in prostate cancer progression." *Nature* 457(7231): 910-914.

Storey, J. D. and R. Tibshirani (2003). "Statistical significance for genomewide studies." *Proc Natl Acad Sci U S A* 100(16): 9440-9445.

Thielen, B., S. Heinen, et al. (2009). "mSpecs: a software tool for the administration and editing of mass spectral libraries in the field of metabolomics." *BMC Bioinformatics* 10: 229.

Tolstikov, V. V. and O. Fiehn (2002). "Analysis of highly polar compounds of plant origin: combination of hydrophilic interaction chromatography and electrospray ion trap mass spectrometry." *Anal Biochem* 301(2): 298-307.

Want, E. J., A. Nordstrom, et al. (2007). "From exogenous to endogenous: the inevitable imprint of mass spectrometry in metabolomics." *J Proteome Res* 6(2): 459-468.

Watson, M., A. Roulston, et al. (2009). "The small molecule GMX1778 is a potent inhibitor of NAD+ biosynthesis: strategy for enhanced therapy in nicotinic acid phosphoribosyltransferase 1-deficient tumors." *Mol Cell Biol* 29(21): 5872-5888.

Werner, E., J. F. Heilier, et al. (2008). "Mass spectrometry for the identification of the discriminating signals from metabolomics: current status and future trends." *J Chromatogr B Analyt Technol Biomed Life Sci* 871(2): 143-163.

Wilson, I. D., J. K. Nicholson, et al. (2005). "High resolution "ultra performance" liquid chromatography coupled to oa-TOF mass spectrometry as a tool for differential metabolic pathway profiling in functional genomic studies." *J Proteome Res* 4(2): 591-598.

Wishart, D. S. (2009). "Computational strategies for metabolite identification in metabolomics." *Bioanalysis* 1(9): 1579-1596.

Wishart, D. S. (2011). "Advances in metabolite identification." *Bioanalysis* 3(15): 1769-1782.

Wishart, D. S., D. Tzur, et al. (2007). "HMDB: the Human Metabolome Database." *Nucleic Acids Res* 35(Database issue): D521-526.

Xia, J., N. Psychogios, et al. (2009). "MetaboAnalyst: a web server for metabolomic data analysis and interpretation." *Nucleic Acids Res* 37(Web Server issue): W652-660.

Part 3

Metabolomics to Identify Health Promoting Factors and New Bioactives

8

Metabolic Pathways as Targets for Drug Screening

Wai-Nang Paul Lee[1,2*], Laszlo G. Boros[1,2,3] and Vay-Liang W. Go[1,4]
[1]*UCLA Center of Excellence for Pancreatic Diseases, Los Angeles Biomedical Research Institute,*
[2]*Department of Pediatrics, Harbor-UCLA Medical Center,*
[3]*SiDMAP, LLC, Los Angeles, CA,*
[4]*Department of Medicine, David Geffen School of Medicine at UCLA, Los Angeles, CA, USA*

1. Introduction

In the drug development process, candidate compounds are first screened for desirable biological properties such as effects on gene expression, signal transduction, or enzyme activity. The genetic and metabolic pathways used in the readouts are known as targets of the drug screening process. Despite advances in molecular targeting, proteomics and metabolomics, drug screening with molecular or metabolic targets have not produced the results that meet the need of the pharmaceutical industry in the selection of small molecules leads/targets for clinical testing. The relative lack of success in applying the -omics in drug screening is partly due to the inability of the –omics to account for metabolic regulation, a property of the cellular metabolic network. More recently, tracer-based metabolomics has been developed as an experimental approach for the study of cellular metabolic networks. Interconversion of metabolites are measured in terms of "extreme pathways" of the metabolic network which can be used for drug screening purposes. In this paper, these approaches for drug screening targeting genetic pathways (transcriptomics), biochemical pathways (metabolomics and fluxomics) and 'extreme pathways" (tracer-based metabolomics) are compared. The advantages and limitations of these approaches for metabolic research and drug screening are discussed.

2. Genetic/signaling pathways as targets for drug screening

In the days of the genomic era, scientists are eager to apply the knowledge of genomics and the advances in genetic/molecular engineering in clinical and translational research. The general concept is that a genetic signal acts as an on-off switch in controlling metabolic processes. However, in order to successfully apply genetic pathways (gene switches) for drug screening, one has to establish genotype-phenotype correlation. The generally accepted dogma of genotype-phenotype correlation is that metabolism is the final expression of the genetic information, and peptide molecules act as signaling switches for

*Corresponding author

the regulation of metabolism. This popular molecular genetic approach to drug screening is based on the assumption that the effect of drugs on metabolism and metabolic regulation is determined by gene transcription and translation alone.

The rationale for choosing gene switches as targets for drug screening can be illustrated by the example of the action of the tumor suppressor gene (P53) in cancer metabolism. Cancer cells have metabolic characteristics that are distinct from normal cells in that there is an overall increased macromolecular syntheses to sustain cell growth and proliferation. These metabolic characteristics are generally grouped under the Warburg effect which consists of increased anaerobic glycolysis, decreased glucose oxidation and increased glutamine utilization (1). A representation of the model of gene switches is depicted in Figure 1. The signals that orchestrate these metabolic changes originate from the balance between oncogenes (growth promoting factors) that turn on signaling pathways regulating the utilization of substrates for growth and tumor suppressor genes such as P53 that modulate energy utilization. The loss of a cancer suppressor gene or the over-expression of an oncogene may be sufficient to generate genetic signals to switch on or off (or modulate) metabolic pathways resulting in the cancer cell metabolic phenotype. The interaction between molecular pathways and metabolic pathways in cancer has recently been reviewed (1). At the molecular level, P53 regulates transcription of genes that modulate PI3K, Akt and mTOR pathways (growth promoting pathways) to reduce cancer growth. Excessive growth induces expression of P53 in cells keeping cell growth and cell death in balance. Independently, P53 inhibits glucose uptake, ribose synthesis and glycolysis thus modulating cellular metabolism. When the action of P53 is lost due to mutation, cells take up more glucose for ribose synthesis and glycolysis, the key elements of the Warburg effect. The fact that the actions of P53 can be used to explain the cancer metabolic phenotype suggests that any signaling pathway that interacts with P53 is a potential target for anticancer drug screening.

The use of genetic pathways for the understanding of metabolism and drug screening has its limitations. The interactions among signaling pathways are often based on demonstrations using artificial overexpression or underexpression of these pathways. The real actions of these signaling pathways in normal physiology are not exactly known. The quantitative relationship connecting gene expression to metabolism has not been worked out. Therefore, the genetic switch hypothesis is only one possible explanation for the expression of the cancer metabolic phenotype. Conceptual limitations of genetic switches in the understanding of metabolisms or the metabolic effect of drugs have been noted by D. E. Koshland Jr (2) almost half a century ago. He pointed out that overproduction or underproduction of enzymes by molecular manipulation may sometimes have dramatic effects on an organism and other times with only minor effects. The overall effect of genetic manipulation on cellular metabolism cannot always be predicted. The lack of observable effect when an enzyme concentration is changed is analogous to the "silent" phenotypes (3) of the carrier states of many recessive diseases when enzyme or protein concentrations of the affected genes can be substantially reduced.

Discrepancies in genotype phenotype correlation between signaling pathways and metabolism when it occurs may be explained by our incomplete knowledge of the feedback regulation of the signaling pathways as well as metabolic regulations of cellular metabolism. However, the lack of genotype-phenotype correlation in many cases can be attributed to conceptual difficulties of using genetic switches to the understand metabolism. First, metabolic regulation is rarely an "all-or-none" type of control. According to metabolic control analysis, the regulation of metabolic pathway is distributed over many enzymes of the biochemical reaction. Transcriptional or post-translational modification of an enzyme potentially changes its Km and/or Vmax of the reaction. However, the change in Km or Vmax of one enzyme may be compensated by either a change in precursor substrate concentration or by a shift in the locus of control of the reaction to other enzymes such that net flux remains unchanged. Secondly, the model of metabolic switches does not take into account how the change in one metabolic pathway may impact on many other pathways that are connected by shared substrates or co-factors and vice versa. The lack of quantitative relationship between genotype and phenotype is the Achille's heel of the gene switching hypothesis† and the use of genetic pathways for drug screening.

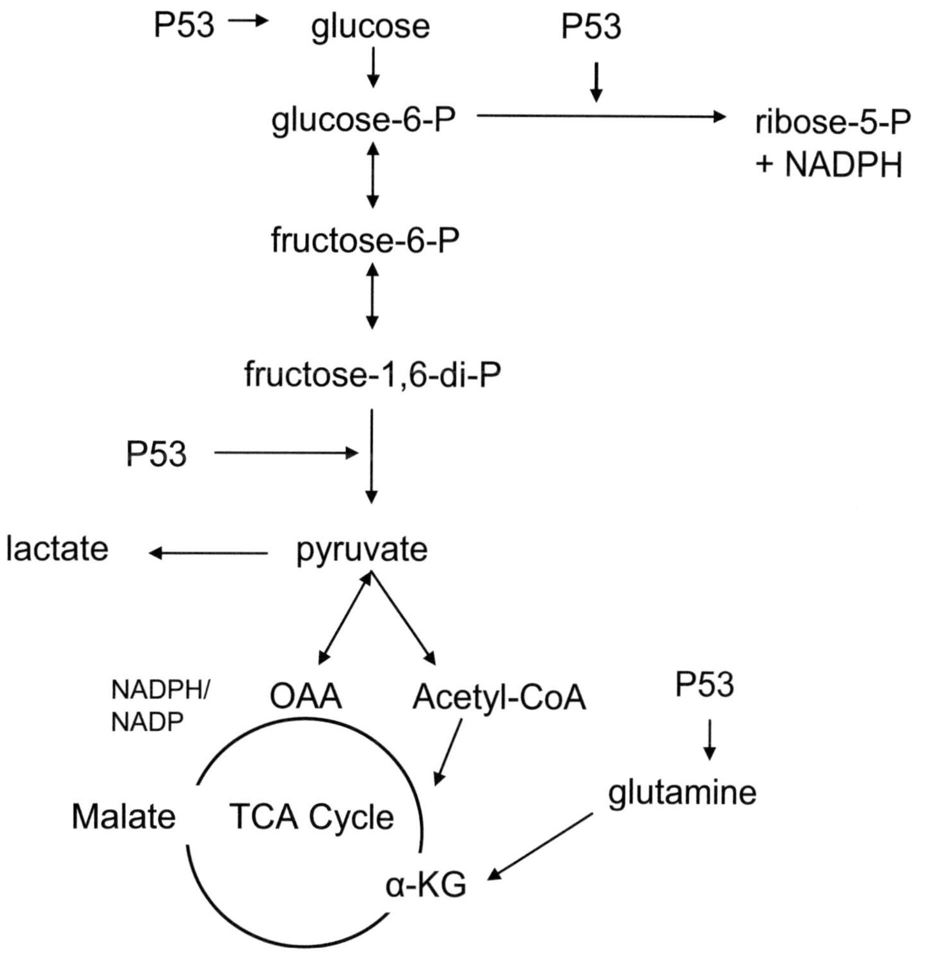

Fig. 1. A system of glucose metabolic pathways in a traditional format. The biochemical pathways potentially affected by P53 as a gene-switch are indicated.

†Gene expression can be quantitatively determined using RTPCR method. Results are reported in folds of change. Even though there may be a correlation between the fold of change and the observed metabolic effect, the correlation is not a quantitative one.

3. Biochemical pathways as targets for drug screening

In the past decades, metabolite profiling (metabolomics) and fluxomics have been developed to fill in the knowledge gaps of gene regulations of metabolism. Thus genomics, transcriptomics, metabolomics and fluxomics are the popular -omics of systems biology‡. The advances in mass spectrometry and nuclear magnetic resonance spectroscopy have enabled the new industry of metabolomics. These technologies provide quantitative and qualitative analyses of organic compounds in biological fluids and specimens. Quantitation of metabolites at different time points is the basis of fluxomics. Research in metabolite profiling and fluxomics is based on our understanding of metabolic control analysis (MCA). MCA provides quantitative measures of degree of influence of a change in enzyme kinetics or a change in substrate concentration can affect the consumption or production of a metabolite in terms of metabolic control coefficients (5, 6). Thus, measurements of substrate concentrations by metabolite profiling and flux analyses allow detail information regarding metabolic changes when the system is perturbed by drug treatment. In metabolomics and fluxomics, traditional biochemical pathways can be considered as targets for drug candidate screening. Such screening has the limitation in that the effect on the metabolic system as a whole is not evaluated in the screening process. Technically, current metabolomics technologies do not permit characterization of substrate concentrations at the subcellular level and reactions that are compartmentalized cannot be properly evaluated. The sampling processes usually do not separate the contribution from background environment such as the culture medium or neighboring cells to the metabolic processes of the cell. For these reasons, metabolomics has not been successfully used as targets for drug screening.

Measurements of flux (fluxomics) depend on the use of isotopes (7). Since ^{13}C labeled isotopes can be distributed widely among many metabolites, and not all of these metabolites can be measured in the same analytical method, there are always fewer data than needed to give precise quantification of flux. Nevertheless, this approach has its appeal in that once a mathematical model is constructed, literature values can be fitted into the model to give insight into possible changes in the system, and whether the model is robust or not can be tested (8). Such an approach was used by Selivanov et al. to model of the pathways of pentose phosphate cycle. The interconnection of these pathways is shown in Figure 2 (9). After incubating cells with [1, 2 $^{13}C_2$]-glucose, ribose was found to be labeled in many carbon positions. Using mass isotopomer distribution in ribose and known sugar phosphate concentrations and Km values of enzymes from the literature, these authors were able to simulate the fluxes of the pentose phosphate pathways. They were able to identify three reactions among other transketolase mediated reactions that were significantly inhibited when cells were treated with oxythiamine, a tranketolase inhibitor. These are xylulose-5p to glyceraldehyde-3-P, sedoheptulose-7-p to ribulose-5-P and xyluose-5-P to sedoheptulose-7-P (reactions 14, 15 and 13 in figure 2). The differences in response among tranketolase enzymes inhibited by oxythiamine are the consequence of stoichiometric constraints.

---

‡Systems biology as commonly defined is the enumeration of a collection of biologically related objects (genomics, proteomics and metabolomics) or characteristics (transcriptomics and fluxomics) within the boundary of a cell. However, in actuality the context of a cellular boundary i.e. how these objects or characteristics separate the cell from its environment is often absent in the definition of these systems (4).

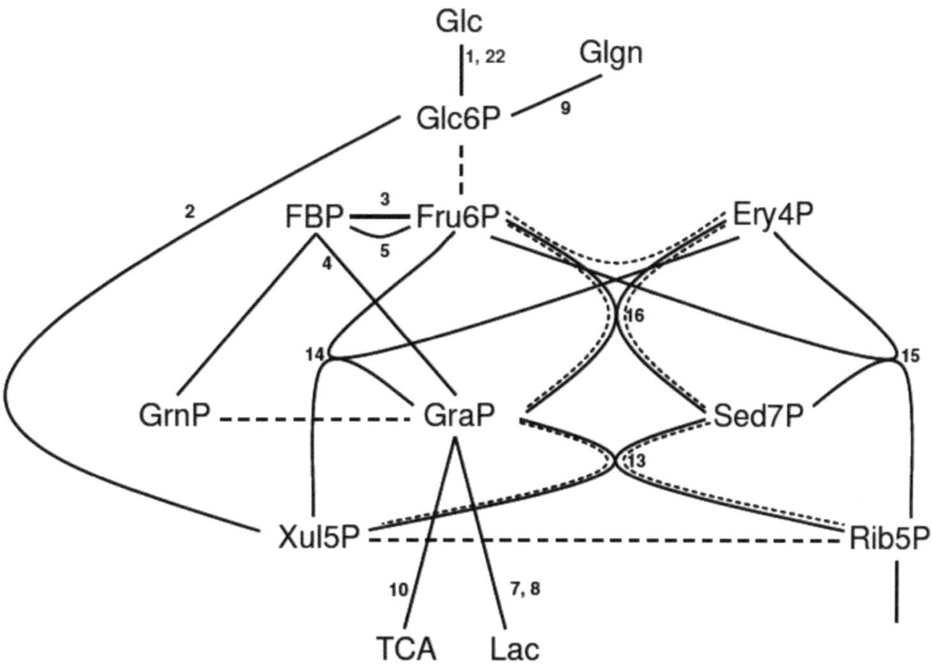

Fig. 2. A model of pentose phosphate cycle used for carbon tracing and flux analysis. Using experimentally determined isotopomer distribution in ribose and lactate, the fluxes of the numbered reactions can be calculated (from reference 9 with permission).

The fluxomic approach targeting traditional biochemical reactions provides more specific information regarding the metabolic system than metabolite profiling (metabolomics). The use of fluxomics allows the simultaneous assessment of the effect of a drug on multiple metabolic pathways and permits a better understanding of metabolism than the gene-targeting approach. However, in order to take into account futile cycling or stoichiometric constraints, stable isotope tracing (carbon tracing) is required as illustrated in the above example. Even though it is possible to construct a complex model for mammalian metabolic networks to take into account of futile cycles and stoichiometric constraints, such a model requires a very large data set and extensive programming. In the best case scenario, there is never sufficient data for solving all the parameters of the system and the results are model dependent and are difficult to verify for practical reasons (2). Nonetheless, the fluxomics approach definitely provides better correlation with phenotype than the gene switch targeting approach.

4. "Extreme pathways" and metabolic network

The "extreme pathways" of a metabolic network can also be used as targets for drug screening. "Extreme pathways" are elements of the well known constraint-based modeling (8) which has been applied to the study of cellular homeostasis. (The definition of "extreme pathway" is given in the next section.) Living organisms (cells) are metabolic systems (networks) continuously exchanging energy substrates with their environments to maintain the biological systems in a homeostatic state. The main metabolic function of a cell is to utilize substrates from its environment to produce energy and building material for the synthesis of macromolecules. Excess intermediates are returned to the surrounding environment to maintain a relatively constant internal environment. The boundaries of metabolic activities represented by "extreme pathways" within which the cell functions define the homeostatic state (10, 11). These boundaries are the result of constraints by the stoichiometry of competing reactions, synchronization of shared pathways and/or intermediates, and balance of energy production and utilization.

The role of "extreme pathways" in the maintenance of homeostasis can be illustrated by the example of glucose metabolism via the TCA cycle. Pyruvate from glycolysis is metabolized via pyruvate carboxylation leading to the conservation of 3-carbon species or pyruvate decarboxylation leading to production of 2-carbon species (via acetyl-CoA) and energy production (beta-oxidation and tricarboxylic acid (TCA) cycle) (16). These two processes are concurrent in cells and the activity of one pathway constrains the activity of the other. For a given homeostatic state, the observed utilization of pyruvate via these pathways is the optimal[§] pyruvate utilization and can be represented by a vector in the pyruvate phenotypic phase plane. The operation of the TCA cycle is an example of metabolic constraint due to synchronization of shared pathways or intermediates. A full turn of the TCA cycle oxidizes a mole of acetate into two moles of carbon dioxide with production of reducing equivalents and/or high energy phosphates. At the same time each of the TCA cycle intermediate may have its respective substrate cycle such as the malate cycle and the citrate lyase cycle. These individual substrate cycles perform separate metabolic functions in conveying reducing equivalents (malate shuttle) and acetyl-CoA (citrate lyase cycle) from the mitochondria to the cytosol. The operations of these cycles are usually synchronized for efficiency. When there is a lack of synchrony of these cycle, abnormal substrate and energy balance can result and a loss of homeostasis in the cell occurs. The imbalance of energy metabolism in the mitochondria due to imbalance of substrate cycles is a frequent cause for reactive oxygen species generation and apoptosis. Changes in these boundaries consisting of "extreme pathways" are sensitive to metabolic or therapeutic perturbations and are excellent markers of therapeutic effects.

The differnces between a metabolic network and a traditional biochemical reaction model can best be shown by representing a metabolic network as an engineering system. The working of such a system is illustrated in figure 3 in which pathways shown in Figure 2 are represented as belts and wheels connecting glycolytic/gluconeogenic substrates to those of the pentose cycle. The enzymes that drive the belts are indicated and the role of energy production and utilization are included. Figure 2 is model of pentose cycle intermediates linked by enzymatic reaction. The fluxes of these reactions can be modeled mathematically using a set of ordinary differential equations. Figure 3 shows the production and consumption of different classes of compounds connected to the production and consumption of ATP and reducing equivalents. These models are conceptually different. The input-output model of cellular homeostasis of tracer-based metabolomics can account for stoichiometric constraints and synchronization of substrate cycles thus overcoming limitations of the previous approaches in metabolic studies.

[§]Optimality is sometimes thought of as a teleological concept. The optimal metabolic function of a cell is not for its purpose to survive, but is defined by the internal organization of the metabolic network.

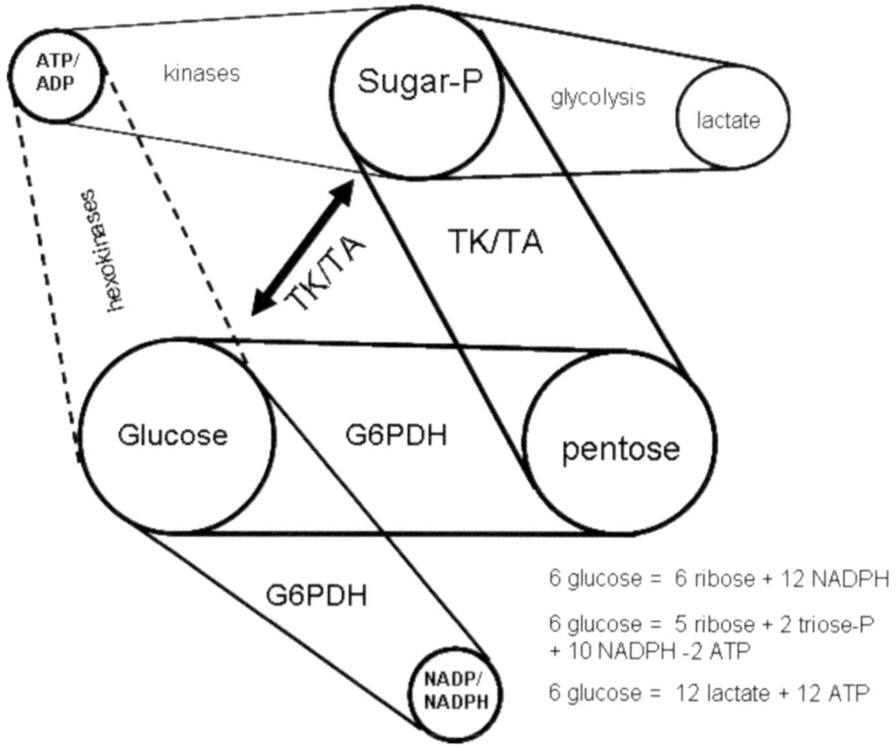

Fig. 3. An engineering model of the system of reactions depicted in Figure 1 and Figure 2. The relationship among the different substrate pools is represented by different circles. Stoichiometric relationships are provided by the mass balance equations. The productions of these metabolites from one another are indicated by respective drive belts. Energy substrate consumption and production is also included in the model. The metabolic network and its function is shown as a factory production model with sources and sinks of the raw materials and products.

The basic concept and tracer methodology of tracer-based metabolomics have been reviewed (4, 10-12). A key feature that distinguishes tracer-based metabolomics from metabolite profiling (metabolomics) and fluxomics is the inclusion of a system boundary that permits input-output analysis and a balance of flux** model in which substrate input is link to its output (products) by "extreme pathways" (12, 13, 14). "Extreme pathways" are pathways that elements (carbon, oxygen and nitrogen) from compounds (precursors) introduced into the system travel over to the final products. The basic elements of "extreme pathways" form the axes of a high dimension phenotypic space, any two of these axes forms a phenotypic phase plane and the line of optimality which is a vector within the space (or a plane) representing the metabolic phenotype. The relationship among any three "extreme

**A balance of flux analysis requires a steady state or quasi-steady state assumption. For most cellular processes involving cell growth and division, these processes are slow relative to the experimental study period and quasi-steady state of metabolic reactions may be safely assumed. However, in biological processes that are fast such as muscle contraction, or nerve conduction the balance of flux model cannot be applied and a dynamic model is required.

pathways" can be described by isoclines (15). Therefore, tracer-based metabolomics offers a graphical representation of a quantitative metabolic phenotype not available by gene-switching model or fluxomic model.

Instead of measuring substrate fluxes over specific biochemical reactions, the experimental model focuses on fluxes over "extreme-pathways" which are pathways linking the precursor substrates to the specific products. There may be many "extreme pathways" for the formation of a product depending on how many interconnecting reactions between precursor and end-product. In the synthesis of glutamic acid from glucose, there are at least two "extreme pathways" namely the pyruvate carboxylase and pyruvate dehydrogenase reactions (4). Pentose is synthesized from glucose at least through two "extreme pathways" either by oxidative (G6PDH) or non-oxidative (TK/TA) pathways (10).

5. Measuring "extreme pathways" – Carbon tracing in tracer-based metabolomics

A unique feature of tracer-based metabolomics as the name implies is the application of stable isotope labeled metabolites and mass spectrometry or magnetic resonance (NMR) spectroscopy. The ^{13}C or ^{2}H label from the labeled substrate is distributed into metabolic intermediates in specific positions according to the "extreme pathways". Tables 1a and 1b show some of the examples of labeling in amino acids, glycogen, ribose and lactate from uniformly labeled glucose [$U^{13}C_6$]-glucose (carbon tracing from glucose) (16-24). The tables show the potential mass isotopomers that can be generated, the positions that are labeled in the products, and the corresponding glucose carbon that the ^{13}C originates. For example, three mass isotopomers (M3, M2 and M1 can be found in alanine or lactate from an experiment with [$U^{13}C_6$]-glucose. M3 of alanine comes from glycolysis of glucose. The ^{13}C in carbon 1 of alanine comes from carbon 3 of glucose (G3). There are two M2 isotopomers with ^{13}C on either C3 and C2 or C2 and C1 of alanine. The sources of ^{13}C's are from glucose carbon 2 (G2) and carbon 1 (G1). X represents a ^{12}C carbon originated from exchange at the level of the TCA cycle. The mass isotopomers and positions with ^{13}C label in these glucose metabolic intermediates are indicated. These tables can be used as a guide to design tracer studies or interpret results from such studies. The mass and position isotopomers in these metabolites represent individual "extreme pathways" from glucose carbon to the respective products. It should be noted that some of the isotopomers are products of the same "extreme pathways" thus providing redundancy in the information on the "extreme pathways". In the example of labeled glucose forming labeled amino acids (Table 1a), we can gain insight into the simultaneous reactions of pyruvate carboxylation, pyruvate dehydrogenase, malate cycle, gluconeogenic cycle relative to TCA cycle flux. When the distribution of isotopomers is determined using mass spectrometry or NMR, we can use these isotopomer ratios to construct phenotypic phase planes (11, 15). Such a database of mass isotopomers can easily be managed with computational algorithm (subroutines) which can compare distances between individual phenotypes on different phenotypic phase planes. Phenotypic differences can also be quantitatively compared using isocline analysis. Thus, tracer-based metabolomics is a quantitative experimental approach to the study of metabolism and metabolic regulation.

Pathway	Isoto-pomer	Alanine C3-C2-C1	Serine C3-C2-C1	Glycine C2-C1	Asparate C4-C3-C2-C1	Glutamate C5-C4-C3-C2-C1
Glycolysis Plus TCA Cycle	M5	---	---	---	---	G2-G1-G1-G2-G3@
	M4	---	---	---	G2-G1-G1-G2@	G2-G1-G1-G2-X; G2-G1-G1-G1-X
	M3	G1-G2-G3	G1-G2-G3	---	X-G1-G2-G3	
	M2	G2-G1-X X-G2-G1	G2-G1-X	G2-G1	X-G1-G2-X G2-G1-X-X	G2-G1-X-X-X
	M1	G1-X-X X-G1-X X-X-G1	X-X-G1	---	G1-X-X-X, X-G1-X-X X-X-G1-X	X-X-G1-X-X X-X-X-G1-X X-X-X-X-G1
1-Carbon Metabolism	M3	---	G1-G2-G3	---	---	---
	M2	---	G2-G3	G2-G3	---	---
	M1	---	G3	G3	---	---

The orientations of the amino acid molecules are shown in the top row. Mass isotopomers are designated as M1 to M5 indicating the number of ^{13}C per molecule of the amino acid. The corresponding position of glucose carbon within the amino acid is designated as G1 to G3. The glucose molecule is symmetrical around C3-C4. In the table, G1-G2-G3 is the same as G6-G5-G4, if these positions are labeled equally. X represents ^{12}C from exchange within the TCA cycle.
@ When glucose enrichment is high, there is a likelihood of labeled OAA condensing with labeled acetyl-CoA resulting in M5 α–ketoglutarate and subsequently M5 glutamate and M4 aspartate.

Table 1a. Examples of Position and Mass Isotopomer Distribution in Gluconeogenic Amino Acids from [$U^{13}C_6$]-glucose (16-24)

Pathway	Isotopomer	Glycogen C6-C5-C4-C3-C2-C1	Ribose C5-C4-C3-C2-C1	Lactate C3-C2-C1
Hexose-P cycles	M6	G6-G5-G4-G3-G2-G1	---	---
Glycolysis	M3	G6-G5-G4-X-X-X X-X-X-G3-G2-G1		G1-G2-G3
Glycolysis Plus TCA Cycle	M2	G6-G5-X-X-X-X X-X-X-G3-G2-X		G2-G1-X X-G2-G1
	M1	G1-X-X X-X-X X-G1-X-X-X-X X-X-G1-X-X-X		G1-X-X X-G1-X X-X-G1
Oxidative	M5	Glycogen glucose are labeled as in glycolysis/ gluconeogenesis shown above	G6-G5-G4-G3-G2	
Non-oxidative	M4		G3-G2-X-G3-G2	
Oxidative plus Non-oxidative	M3		G3-G2-X-X-G3	
	M2		X-X-X-G3-G2 G3-G2-X-X-X	
	M1		G3	

The orientations of the glycogen, ribose and lactate molecules are shown in the top row. Mass isotopomers are designated as M1 to M6 indicating the number of ^{13}C per molecule of the amino acid. The corresponding position of glucose carbon within the glycogen, ribose and lactate is designated as G1 to G6. The glucose molecule is symmetrical around C3-C4. In the table, G1-G2-G3 is the same as G6-G5-G4, if these positions are labeled equally. X represents ^{12}C from exchange within the TCA cycle.

Table 1b. Examples of Position and Mass Isotopomer Distribution in Glycogen, Ribose and Lactate from [$U^{13}C_6$]-glucose (16-24)

6. "Extreme pathways" as targets for drug screening

The application of phenotypic phase plane (PPP) analysis of balance of flux data from tracer-based metabolomics is a graphical way of presenting experimental data that is unique to tracer-based metabolomics. Since metabolic phenotype of a cell is characterized by the pattern of its utilization of substrates, the phenotype of a cell is represented by the input/output characteristics which can be measured as fluxes through the "extreme pathways". Any two of these "extreme pathways" can form a phenotypic phase plane. The metabolic phenotype of a cell is given by a vector in the plane and the vector divides the plane into two regions representing regions of relative excess and relative deficiency of substrate utilization (4). The use of phenotypic phase plane analysis together with isocline analysis allows quantitative comparison of different treatment effects. An example of the use of isocline analysis is illustrated by the study of effects of fructose and glutamine on the glycolytic/gluconeogenic pathways (11).

Other applications of phenotypic phase plane analysis to investigate metabolic mechanisms of a therapeutic intervention are illustrated in Figure 4. Panel (i) of the figure shows the metabolic response of a cell to two therapeutic interventions (e.g. drugs, receptor inhibitor,

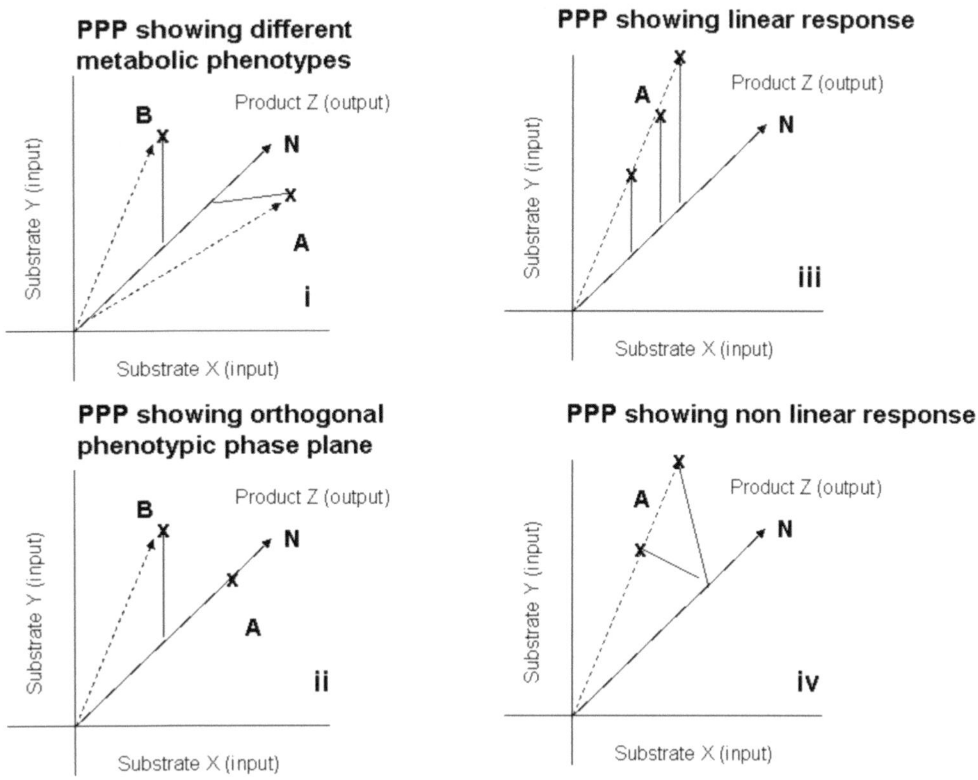

Fig. 4. Examples of phenotypic phase plane analysis showing the quantitative relationship among phenotypes by isocline analysis. Panel (i) shows effects of two drugs (A and B) with different mechanisms on metabolic pathways X and Y. N represents the normal phenotype. Panel (ii) shows the effect of A is orthogonal to the X-Y plane. Panel (iii) shows dose dependent effect of A. Panel (iv) shows non-linear response to two different doses of A.

or siRNA) A and B. These treatments result in changes in phenotypes (decrease in production of Z) accompanied by different metabolic compensations in substrate utilization. Treatment B results in a decreased utilization of substrate X (or its metabolic pathways) which is compensated by a slight increase in the utilization of substrate Y. Whereas treatment A results in the increase utilization of X and decreased utilization of Y as compared to control (N). If the mechanism of action of treatment B is known (such as inhibition of a specific kinase), one can conclude that treatment A must act on a different set of metabolic and/or signaling pathways. Such an approach will allow an iterative approach to the discovery of new treatment or new pathways. Additional examples of using phenotypic phase plane to understand phenotype changes of fibroblasts from a patient with thiamine responsive megaloblast anemia (TRMA) and pancreatic cancer cell (MIA) and normal fibroblasts using PPP have been provided by Lee and Go in their review (15). When TRMA cells were treated with high doses of thiamine, the phenotype approached that of the normal fibroblasts. On the other hand, MIA cells had a high pentose synthesis phenotype which was corrected when they were treated with oxythiamine. (15) We have recently used phenotypic phase plane analysis to show that inhibition of histone acetylation by genetic intervention or by chemical inhibitor of the reaction had similar metabolic phenotypes (25). Panel (ii) shows result of treatment A which is orthogonal to the phenotypic phase plane of X and Y. This means that treatment A affects a different part of the metabolic system which is not linked to the utilization of substrates X and Y. Metformin and rosiglitazone, two antihyperglycemic drugs, have been shown to alter de novo lipogenesis. While inhibition of de novo lipogenesis by metformin is in the plane of ribose metabolism, meaning changes in pentose cycle metabolism is related to the decrease in lipogenesis. On the other hand, the increase in fatty acid synthesis by rosiglitazone is orthogonal to the ribose phenotypic phase plane suggesting very different mechanism of actions by these two drugs (26). The ability to detect orthogonal phenotypic phase plane is important because there are potentially many of these orthogonal phenotypic phase planes which can be discovered using tracer-based metabolomics, and each of the orthogonal pair would suggest different mechanism of action by different drugs. The finding of orthogonal planes is one of the unique capability of the metabolomics approach in generating mechanistic hypothesis. Panel (iii) shows the proportional response to inhibitor of substrate X where all of the isoclines are parallel to each other. An example of this type of response is provided by our study on the response of a methotrexate resistant colon cancer cell line (HT29) to the effect of DHEAS, oxythiamine and methotrexate treatment alone and in combination (27). Panel (iv) shows response to two inhibitors of substrate X with non-linear compensation of substrate Y. The application of PPP analysis has allowed a far better understanding of metabolic adaptation in cellular homeostasis using tracer-based metabolomics. Using PPP and isocline analysis, we can directly exploit the large dataset accumulated from tracer-based metabolomics studies for target discovery and lead identification in pharmaceutical industry.

7. Concluding comments

Study of metabolism in the post-genomic era differs from the traditional biochemistry in that the study is focused on the function of the system of biochemical reactions in a cell (or the cellular metabolic network) and its regulation. Metabolic function of a living organism (cell) is what mediates the the genetic potential of a cell and its interaction with its environment to maintain homeostasis. The maintenance of homeostasis by the cellular metabolic network in a living organism is the basis of normal physiology and histology (28). When the metabolic environment of a living organism or a cell is altered such as in diabetes or metabolic diseases, maladaptation or the lack of homeostasis in the living organism is the underlying cause for pathophysiology and histopathology (29).

Metabolic phenotype of a cell is the result of genetic and environmental interaction. Understanding metabolic phenotyping changes is important to our understanding of how cells maintain homeostasis by metabolic regulation. We have reviewed three approaches that are used in such investigations based on three different models. Of these different approaches, the gene-switch approach is the most extensively used in the pharmaceutical industry. In the gene-switch model, metabolic regulation begins with the interaction between genes and signaling pathways which eventually impact on biochemical reactions known as down-stream effects (30). This model ignores the fact that many of these signaling pathways or transcriptional factors are altered through post-translational events such as phosphorylation, acetylation, glycosylation and methylation. Since all these post-translational modifications are basic biochemical reactions, they are all subject to the stoichiometric and energy substrate constraints.

Once the downstream events are initiated, the interconversion of metabolic intermediates is subject to all the constraints as described in preceding paragraphs. We have previously shown that altered metabolic pathways can be the initiating events in gene transcription and post-translational modification of signaling pathways and enzymes (31, 32). Therefore, gene-switch model is an incomplete model to understand metabolic regulation. Because of these conceptual problems, application of gene-switch approach has had disappointing results in identifying drug candidates or targets and appalling failures in clinical trial due to unexpected toxicity or lack of efficacy. Of the two remaining approaches, tracer-based metabolomics is a practical experimental approach that does not require complicated mathematical modeling. Furthermore, the results can be graphically presented and the quantitative difference of metabolic phenotypes can be compared. Such features make the tracer-based metabolomics a powerful approach for drug screening in pharmaceutical research. Since the model does not assume any signaling pathways, it is most suitable for studies of nutriceuticals such as phytochemicals (33) or for screening of compounds in a chemical library that have no known molecular targets (34). It is also applicable to investigate the metabolic effect of drug combinations, in which the interaction of drugs can be studied. Most important of all, tracer-based metabolomics approach provides the understanding of cellular homeostasis and its changes under the influence of nutrient conditions or pharmaceuticals.

Since our first publication on metabolic profiling (35), progress in tracer-based metabolomics has been slow because there are few investigators who are trained in tracer technology. The current tracer model mainly addresses the area of glucose metabolic pathways. Methods for the investigation of other metabolic systems that may be distantly connected to glucose metabolism (orthogonal systems) are not represented. These systems include the systems of glutamine metabolism which connects glucose metabolism to nucleic acid synthesis; arginine metabolism which is part of the urea cycle and nitric oxide synthesis system; and the methyl donor pathways which are important in nucleic acid synthesis and choline synthesis. The complete development of tracer-based metabolomics is probably a decade away provided the development has the attention and adequate funding to complete the tasks to cover the metabolic pathways of the whole cellular metabolic network.

8. Acknowledgement

This work is supported by the Biomedical Mass Spectrometry Laboratory of the GCRC at the Los Angeles Biomedical Research Institute at Harbor-UCLA Medical Center (PHS M01-RR00425; UCLA; CTSI 1UL1-RR033176) and the Metabolomics Core Laboratory of the UCLA Center of Excellence in Pancreatic Diseases (P01 AT003960).

9. References

[1] Levine AJ, Puzio-Kuter AM. The Control of the Metabolic Switch in Cancers by Oncogenes and Tumor Suppressor Genes. Science. 2010 Dec 3;330(6009):1340-4.

[2] Koshland Jr. DE Control of enzyme activity and metabolic pathways. pp1-8, in "Metabolic Regulation" (eds. R.S. Ochs, R.W Hanson and J Hall), Elsevier Science Publisher, Amsterdam (1985).

[3] Raamsdonk LM, Teusink B, Broadhurst D, Zhang N, Hayes A, Walsh MC, Berden JA, Brindle KM, Kell DB, Rowland JJ, Westerhoff HV, van Dam K, Oliver SG. A Functional Genomics Strategy That Uses Metabolome Data to Reveal The Phenotype of Silent Mutations. Nat. Biotechnol 2001; 19:45–50.

[4] Cascante, M., Comin, B., Boren, J., Raïs, B., Centelles, J.J., Puigjaner, J., Lee, W.-N. P., Boros, L.G. "Application of Metabolic Control Analysis to the Design of a New Strategy for Cancer Therapy", pp. 173-180 in "Technological and Medical Implications of Metabolic Control Analysis" (ed. A. Cornish-Bowden and M. L. Cárdenas), Kluwer Academic Publishers, Dordrecht (2000)

[5] Cascante M, Boros LG, Comin-Anduix B, de Atauri P, Centelles JJ, Lee PW.Metabolic control analysis in drug discovery and disease. Nat Biotechnol. 2002 Mar;20(3):243-9.

[6] Lee WN. Characterizing Metabolic Phenotype Using Tracer-Based Metabolomics. Metabolomics 2006, 2:31-39.

[7] Zamboni N, Fendt SM, Rühl M, Sauer U. (13)C-based metabolic flux analysis. Nat Protoc. 2009;4(6):878-92. Epub 2009 May 21.

[8] Selivanov VA, Puigjaner J, Sillero A, Centelles JJ, Ramos-Montoya A, Lee PW, Cascante M. An Optimized Algorithm for Flux Estimation From Isotopomer Distribution in Glucose Metabolites. Bioinformatics 2004; 20:3387-3397.

[9] Selivanov VA, Meshalkina LE, Solovjeva ON, Kuchel PW, Ramos-Montoya A, Kochetov GA, Lee PW, Cascante M. Rapid Simulation and Analysis of Isotopomer Distributions Using Constraints Based on Enzyme Mechanisms: An Example From HT29 Cancer Cells. Bioinformatics 2005; 21:3558-3564.

[10] Lee WN, Go VL. Nutrient-Gene Interaction: Tracer-Based Metabolomics. J Nutr 2005; 135:3027S-3032S.

[11] Lee, WN, Wahjudi PN, Xu, J, Go VLW. Tracer-based Metabolomics: Concepts and Practices. Clin Biochem. 2010 Nov;43(16-17):1269-77.

[12] Maguire G, Boros LG and Lee P. Development of Tracer-Based Metabolomics and its Implications for the Pharmaceutical Industry. Int J Pharm Med 2007; 21 (3): 217-224.

[13] Schilling CH, Palsson BO. The Underlying Pathway Structure of Biochemical Reaction Networks. Proc Natl Acad Sci 1998; 95:4193-4198.

[14] Schilling CH, Letscher D, Palsson BO. Theory for the Systemic Definition of Metabolic Pathways and Their Use in Interpreting Metabolic Function From a Pathway-Oriented Perspective. J Theor Biol 2000; 203:229-248.

[15] Edwards JS, Ramakrishna R, Palsson BO. Characterizing the Metabolic Phenotype: a Phenotype Phase Plane Analysis. Biotechnol Bioeng. 2002; 77:27-36.

[16] Katz J., Lee WP. The Application of Mass Isotopomer Analysis in The Determination of Pathways of Glycogen Synthesis. Am. J. Physiol 1991; 259:E757-E762.

[17] Lee WNP, Boros LG, Puigjaner J, Bassilian S, Lim S, Cascante M. Investigation of The Pentose Cycle Using [1, 2-13C2]-Glucose and Mass Isotopomer Analysis: Estimation of Transketolase and Transaldolase Activities. Am. J. Physiol 1998; 274: E843-E851.

[18] Katz J, Lee WNP, Wals PA, Bergner EA. Studies of Glycogen Synthesis and the Krebs Cycle by Mass Isotopomer Analysis with U-13C-Glucose in Rats. J. Biol. Chem 1989; 264:12994-13001.

[19] Katz J, Wals PA, Lee WP. Determination of Pathways of Glycogen Synthesis and Dilution of the 3-Carbon Pool with [U-13C6]-Glucose. Proc. Natl. Acad. Sci 1991; 88:2103-2107.

[20] Katz J, Wals PA, Lee WP. Isotopomer Studies of Gluconeogenesis and the Krebs Cycle with 13C Labeled Lactate. J. Biol. Chem 1993; 268:25511-25521.

[21] Lee WNP. Appendix: Analysis of Tricarboxylic Acid Cycle Using Mass Isotopomer Ratios. J. Biol. Chem 1993; 268:25522-25526.

[22] Lee WP, Edmond J, Bassilian S, Morrow J. Mass Isotopomer Study of Glutamine Oxidation and Synthesis in Primary Culture of Astrocytes. Develop. Neurosci 1996; 18:469-477.

[23] Fu TF, Rife JP, Schirch V. The Role of Serine Hydroxymethyltransferase Isozymes in One-Carbon Metabolism in MCF-7 Cells as Determined by (13)C NMR. Arch Biochem Biophys 2001; 393:42-50.

[24] Solà A, Maaheimo H, Ylönen K, Ferrer P, Szyperski T. Amino Acid Biosynthesis and Metabolic Flux Profiling of Pichia Pastoris. Eur J Biochem 2004; 271:2462-2470.

[25] Alcarraz-Vizán G, Boren J, Lee WN, Cascante M. Histone deacetylase inhibition results in a common metabolic profile associated with HT29 differentiation. Metabolomics. 2010:229-237.

[26] Personal unpublished observation.

[27] Ramos-Montoya A, Lee WN, Bassilian S, Lim S, Trebukhina RV, Kazhyna MV, Ciudad CJ, Noe V, Centelles JJ, Cascante M. Pentose phosphate cycle oxidative and nonoxidative balance: A new vulnerable target for overcoming drug resistance in cancer. Int J Cancer. 2006:119(12):2733-41.

[28] Boros, LG, Lee WNP and VLW. Go. A Metabolic Hypothesis of Cell Growth and Death in Pancreatic Cancer. Pancreas 2002: 24:26-33.

[29] Boros LG, Steinkamp MP, Fleming JC, Lee WNP, EJ Neufeld. Defective RNA Ribose Synthesis in Thiamine-Responsive Megaloblastic Anemia (TRMA): Mechanism for the Syndrome. Blood 2003; 102:3556-3561.

[30] Zanuy M, Ramos-Montoya A,• Oscar Villacanas O, Canela N, Miranda A, Aguilar E, Agell N, Bachs O, Rubio-Martinez J, Pujol MD, Lee WNP, Marin S, Cascante M. Cyclin-dependent kinases 4 and 6 control tumor progression and direct glucose oxidation in the pentose cycle. Metabolomics 2011 DOI 10.1007/s11306-011-0328

[31] Zhang H, Cao R, Lee WN, Deng C, Zhao Y, Lappe J, Recker R, Yen Y, Wang Q, Tsai MY, Go VL, Xiao GG. Inhibition of protein phosphorylation in MIA pancreatic cancer cells: confluence of metabolic and signaling pathways. J Proteome Res. 2010 Feb 5;9(2):980-9. PMID: 20035555; PMC2836017

[32] Ma D, Wang J, Zhao Y, Lee WNP, Xiao J, Go VL, Wang Q, Recker R, Xiao GG Inhibition of glycogen phosphorylation induces changes in cellular proteome and signaling pathways in MIA pancreatic cancer cells Pancreas 2011 D01:10.1097/mpaOBO13E318236F022

[33] Harris DM, Li L, Chen M, Lagunero FT, VLW Go, Boros LG. Diverse mechanisms of growth inhibition by luteolin, resveratrol, and quercetin in MIA PaCa-2 cells: a comparative glucose tracer study with the fatty acid synthase inhibitor C75 Metabolomics 2011 DOI 10.1007/s11306-011-0300-9

[34] Harrigan GG, Brackett DJ, Boros LG. Medicinal chemistry, metabolic profiling and drug target discovery: a role for metabolic profiling in reverse pharmacology and chemical genetics. Mini Rev Med Chem. 2005 Jan;5(1):13-20.

[35] Boros LG, Cascante M, Go VLW, Heber D, Hidvégi M, Lee WNP. Metabolic Profiling of Cell Growth and Death in Cancer: Applications In Drug Discovery. Drug Discovery Today 2002; 7:366-374.

Part 4

Metabolomics in Plant Research

9

New Opportunities in Metabolomics and Biochemical Phenotyping for Plant Systems Biology

Gibon Yves[1,2], Rolin Dominique[2,3], Deborde Catherine[1,2], Bernillon Stéphane[1,2] and Moing Annick[1,2]

[1]*INRA, UMR1332 Fruit Biology and Pathology, Centre INRA de Bordeaux, Villenave d'Ornon*

[2]*Metabolome Facility of Bordeaux Functional Genomics Centre, Centre INRA de Bordeaux, Villenave d'Ornon*

[3]*Université de Bordeaux, UMR1332 Fruit Biology and Pathology, Centre INRA de Bordeaux, Villenave d'Ornon*

France

1. Introduction

Today's unsustainable use of fossil fuel reserves or green fuel is predicted to destabilize the global climate and lead to reduced food security. The key challenge for the coming decades are to meet local needs for food, in terms of both quantity and quality, while conserving natural resources and biodiversity (Ruane & Sonnino, 2011) and to develop a supply industry based on renewable plant-derived products. Indeed agricultural crops can be viewed as a source of or starting point for a plant based economy, potential input to a bio refinery in which all parts of the plant are processed and used to yield (i) food, both traditional and with enhanced nutritional safety, stability and processability; (ii) industrial products, including polymers, fibbers, industrial oils and packaging materials as well as basic chemical building blocks (green chemistry); (iii) fuels such as ethanol and biodiesel; (iv) molecules with pharmaceutical properties and health benefits. To reach these new agricultural perspectives, new varieties with the appropriate properties need to be selected (Tester & Langridge, 2010) through plant breeding, be it conventional, marker assisted, QTL mapping assisted, or genetically modified (GM) (Mittler & Blumwald, 2010). There are also growing demands for germplasm adapted to deal with changing climates and effective under a range of cultural practices and for foods with higher nutritional value. To decipher agronomical traits, functional genomics approaches can be of good use to understand physiological, molecular and genetic processes underlying complex traits. Appropriate functional genomics technologies such as transcriptomics, proteomics and metabolomics must be used together with detailed physiological and environmental information as a combined platform for 'candidate' gene identification or translational genomics approaches that aims to improve complex traits in plants (Sanchez et al., 2011). Without a comprehensive understanding of the plant physiology, molecular processes and genetics of the components of complex traits, the development of new varieties will remain an empirical yet uncertain procedure. This integration of functional genomics data can be viewed as the first step to systems and predictive biology serving agricultural perspectives.

Among the 'omics' technologies, metabolomics is one of the more recently introduced. The term 'metabolome' coined in 1998 (Oliver et al., 1998) refers to the richly diverse population of small molecules present in biofluids, living cells or organisms. Overall, there are two approaches to analyse small molecules, and they differ in the number of compounds analysed, the level of structural information obtained, and their sensitivity. The most common approach, metabolite profiling, is the analysis of small numbers of known metabolites in specific compound classes (e.g. sugars, amino acids or phenolics). At the other extreme, metabolic fingerprinting detects many compounds but their structures are rarely identified. Today metabolomics methods typically allow measuring hundreds of compounds, with a small number being definitively identified, a larger number being identified as belonging to particular compound classes, and many remaining unidentified.

Over the past decade, metabolomics has gone from being just a simple concept to becoming a rapidly growing discipline with valuable outputs in plant biology (Hall, 2006; Saito & Matsuda, 2010; Hall, 2011a; Shepherd et al., 2011). Metabolomics has played a key role in basic plant biology and started having a potentially broad field of applications. Plants produce an astonishing wealth of metabolites estimated to figures ranging from 200,000 to 1,000,000 metabolites (Dixon & Strack, 2003; Saito & Matsuda, 2010). The first significant advances have been made in the area of analytical technology for metabolite identification in order to increase our capacity to simultaneously analyse a chemically diverse range of metabolites in complex mixtures. The metabolomics community has set up analytical platforms with complementary analytical technologies (Moing et al., 2011) after having realized that no single technology currently available (or likely in the close future) will be able to detect all compounds found in living cells. Today these analytical platforms provide a combination of multiple analytical techniques such as gas chromatography (GC), liquid chromatography (LC) or capillary electrophoresis (CE) coupled to mass spectrometry (MS), or nuclear magnetic resonance spectroscopy (NMR) and much more (Kim et al., 2011; Lei et al., 2011).

Considering metabolomics as a combination of knowledge and know-how in biochemistry, signal processing, data and metadata handling, and data mining, the challenge remains to perform in a cohesive and coordinated manner these multidisciplinary approaches to solve biological questions (Ferry-Dumazet et al., 2011; Hall, 2011b). Recently, plant biologists have used metabolomics approaches to understand fundamental plant processes (Leiss et al., 2010; Sulpice et al., 2010), to make a link between genotype and biochemical phenotype and to study plant responses to biotic or abiotic stresses by combining genomics and biochemical phenotyping capabilities (Redestig & Costa, 2011; Villiers et al., 2011). While full genome sequence annotations of the major crops have been published, many post-genomic studies using metabolomics approaches have tried to bridge the phenotype-genotype gap in order to link gene to function (Smith & Bluhm, 2011). Such integrated approaches have been helpful in assigning functions to a large class of function-unknown genes and their interactions with other pathways and also useful in applications such as metabolic engineering (Liu et al., 2009) and assessment of GM plants (Kusano et al., 2011b).

As part of a more recent emerging area, robust data generated from metabolomics can be combined with computationally-intensive approaches based on modelling of pathways to steer this field towards systems biology, which promises to provide an integrated view of cellular processes (Joyce & Palsson, 2006; Wang et al., 2006). Bringing metabolomics data into the forefront of system biology is a challenging opportunity that implies using quantitative metabolomics data in the context of models to improve our understanding of metabolism and drive the biological discovery process. So far, computational studies on metabolomics data have often been restricted to multivariate statistical analyses such as principal component analysis or PLS discriminant analysis to look at trends among different data sets. Such work has proven useful in discovering potential biomarkers of stress and identifying key metabolic difference in GM plants, but provides minimal insight into the underlying biology or the means to modulate it for agronomic or industrial purposes. Now researchers are rising to the challenge by using omics data integration and specially high-throughput metabolomics data within a constraint-based framework to address fundamental questions that would increase our understanding of systems as a whole.

This article provides an overview of the technological trends in plant metabolomics to optimize the characterization of a large number of metabolites with accurate and absolute quantification in a few samples (concept of vertical high-throughput metabolomics) and present the needed technologies to increase the analysis capacity of samples for large-scale studies (concept of horizontal high-throughput metabolomics). This article also outlines how these technological developments in plant metabolomics can be used for systems biology, quantitative genetics and the emerging field of meta-phenomics to answer the key challenges of plant biology and agriculture in the future, and which technological and computational developments are necessary to meet these challenges.

2. Technological trends in plant metabolomics

For plant metabolomics, the analytical strategies reviewed a few years ago (Weckwerth, 2007) are still widely used. Major improvements over the past five years have targeted spectra resolution and processing (http://www.metabolomicssociety.org/software.html), and the emergence of databases (http://www.metabolomicssociety.org/database.html). Thanks to technological and methodological progress, numbers of analytes and compound families that can be determined in a given sample are still increasing, but usually at the expense of the number of samples that can be analysed due to increasing costs and/or labour (Fig. 1). Conversely, novel experimental strategies produce increasing numbers of samples. Thus, not only the best compromise between analyte number (vertical high-throughput approach) and sample throughput (horizontal high-throughput approach) has to be found, but also synergisms between such approaches.

2.1 Vertical approaches

Vertical high-throughput approaches, also called high-density approaches, are defined as strategies that promote sample variables over sample numbers. They are especially interesting for studies in plants given their enormous metabolic diversity. In the plant kingdom, the species number is estimated between 270,000 (observed) and 400,000 and the number of metabolites produced between 200,000 and 1,000,000 (Dixon & Strack, 2003;

Saito & Matsuda, 2010). Even the number of primary metabolites, defined as the type of compounds synthesized by all or most plant species, may exceed the number of compounds found in other eukaryotes since plants are true autotrophs (Pichersky & Lewinsohn, 2011). In addition, different plant lineages synthesize distinct sets of "specialized metabolites", often mis-named "secondary metabolites" (Pichersky & Lewinsohn, 2011), with *Arabidopsis thaliana* estimated to make up to 3,500 of such specialized metabolites. Capturing such diversity is one of the challenges for plant metabolomics compared to animal metabolomics, which has to deal with 'only' 5,000 to 25,000 different metabolites (Trethewey, 2004). However, the consumption of plant-derived food is known to lead to a strong increase in metabolite diversity in animal or human derived samples, e.g. blood or urine. This implies that plant and nutrition scientists face a similar challenge. Indeed, specific plant metabolites are attracting attention due to their role/impact on health and nutrition. Vertical metabolomics mainly relies on sophisticated instrumentation such as NMR and MS, with or without hyphenation of chromatography or capillary electrophoresis (LC-NMR, LC-SPE-NMR, LC-MS, GC- MS, GC- SPE-MS, CE-MS, Fourier Transform-MS (FT-MS), Table 1).

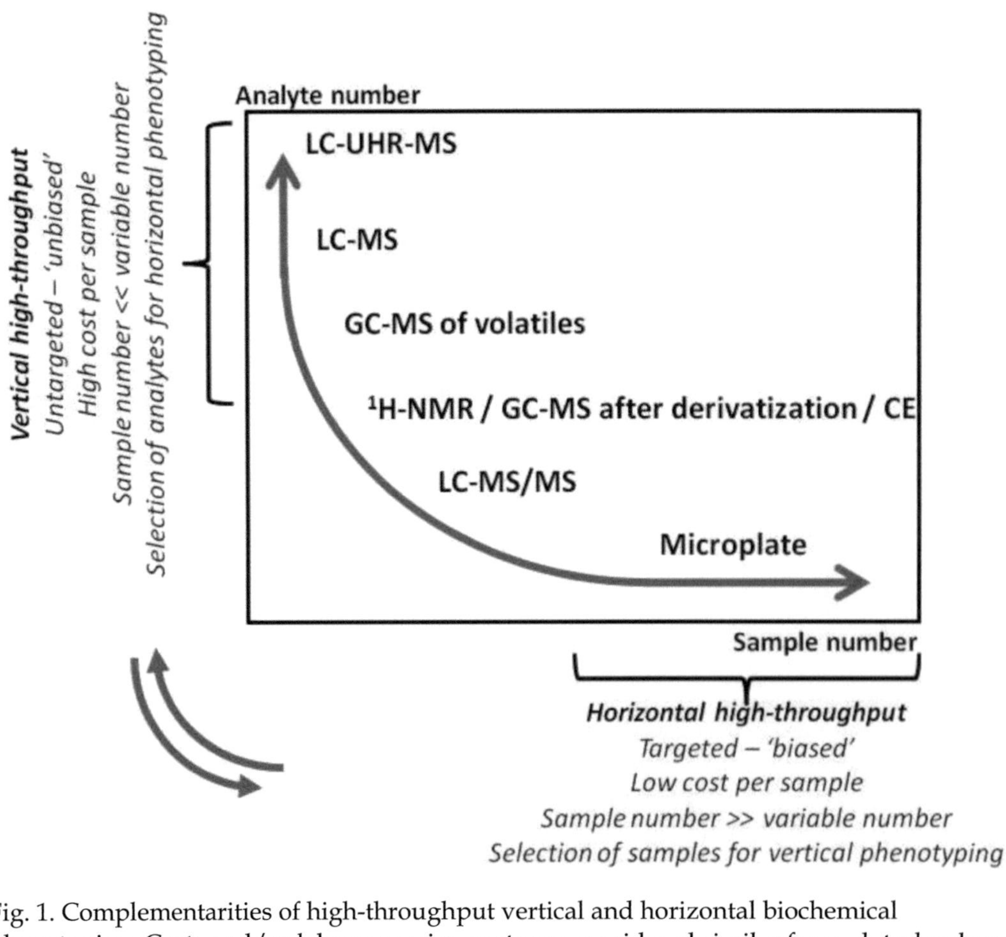

Fig. 1. Complementarities of high-throughput vertical and horizontal biochemical phenotyping. Costs and/or labour requirements are considered similar for each technology.

Metabolite classes	Typical metabolites	Instruments
Amino acids and their derivatives	Amino acids, beta-alanine, GABA, oxoproline	CE-MS, GC-MS (after derivatization), LC-MS, NMR
Amines	Polyamines (putrescine, spermine, spermidine) Betaines, choline	CE-MS, GC-MS, NMR NMR, LC-MS
Organic acids in central metabolism	TCA cycle intermediates	CE-MS, GC-MS (after derivatization), NMR, LC-MS (partially)
Other organic acids	Quinic acid, shikimic acid	NMR, LC-MS, GC-MS (after derivatization), CE-MS
Sugars and their derivatives	Mono-, di- and trisaccharides, sugar alcohols, sugar mono- and diphosphates Phytic acid	GC-MS (after derivatization), CE-MS (sugar phosphates), CE-PDA (partially), NMR NMR, LC-MS
Alkaloids	Polar alkaloids (e.g. pyrrolizidine alkaloids)	LC-MS, NMR
Fatty acids and their derivatives	Saturated and unsaturated aliphatic monocarboxylic acids and their derivatives	GC-MS (after derivatization)
Polar lipids	Phospholipids, mono-, di-, and triacylglycerols	LC-MS
Isoprenoids	Terpenoids and their derivatives	GC-MS (non-polar), LC-MS (polar)
Nucleic acids and their derivatives	Purines, pyrimidines, mono-, di-, and triphosphate nucleosides	CE-MS, GC-MS (partially), NMR
Pigments	Carotenoids, chlorophylls, anthocyanins	LC-PDA, LC-MS
Volatiles	Phenylpropanoid volatiles, aliphatic alcohols, aldehydes, ketones	GC-MS, GCxGC-MS
Other specialized metabolites	Polar phenylpropanoids (e.g. chlorogenic acids), flavonols Phytohormones (e.g. auxins)	LC-MS, LC-(SPE)-NMR, NMR LC-MS

Table 1. A selection of examples of plant primary and specialized metabolites detected by a variety of analytical techniques. Adapted from (Kusano et al., 2011a). PDA, photodiode array detection; GABA, gamma-aminobutyrate; TCA, tricarboxylic acid cycle; SPE: solid-phase extraction. See (Saito & Matsuda, 2010) to have an overview of plant metabolomics pipelines.

Vertical approaches have to deal with a wide variety of chemical structures, which implies wide ranges of solubility, polarity and stability, as well as a high dynamic range of metabolite concentrations (>10^{12}, (Sumner, 2010); 10^6, (Saito & Matsuda, 2010)). In addition, plant metabolites are usually extracted with sometimes sophisticated protocols including steps like heating or fractionation that may lead to losing or modifying metabolites, but also promote the synthesis or import of chemical artefacts. This is why the term analyte, which might be a metabolite or an artefact, is preferred. For example, during the derivatization process, which is required for non-volatile compounds when performing GC-MS, a single metabolite may produce multiple derivatives leading to different peaks. Similarly, adducts and product ions are formed during the desolvation step following the ionization process in LC-MS analyses (Werner et al., 2008). To cover the wide range of chemical diversity and concentrations of plant metabolites, careful experimental design is definitely required, including special care for harvest (Ernst, 1995), several extraction protocols and multi-analytical platforms (see (Ryan & Robards, 2006; Allwood et al., 2011) and Tables 1-2).

Plant Species	**Analytical instruments**	**References**
Arabidopsis	NMR, GC-MS, CE-MS, LC-MS, DI-FT-MS	(Beale & Sussman, 2011) review
Aspen	GC-MS	(Bylesjo et al., 2009)
Broccoli	NMR	(Ward et al., 2010)
Grape	GC-MS NMR	(Deluc et al., 2007) (Mulas et al., 2011)
Tomato	NMR, GC-FID, LC-MS NMR, GC-MS, LC-MS, LC-FT-MS HRMAS-NMR GC-MS, LC-MS, CE-MS	(Mounet et al., 2009) (de Vos et al., 2011) review (Sanchez-Perez et al., 2010) (Kusano et al., 2011b)
Maize	NMR GC-MS	(Cossegal et al., 2008; Broyart et al., 2010) (Skogerson et al., 2010)
Medicago	LC-MS	(Farag et al., 2008)
Melon	NMR, GC-MS, LC-MS	(Moing et al., 2011)
Palm trees	NMR, GC-FID	(Bourgis et al., 2011)
Potato	GC-MS	(Urbanczyk-Wochniak et al., 2005)
Strawberry	LC-MS, DI-MS GC-MS, LC-MS	(McDougall et al., 2008) (Fait et al., 2008)
Rice	GC-MS, CE-MS, CE-DAD, FT-MS	(Oikawa et al., 2008)
Vanilla	NMR	(Palama et al., 2009)
Medicinal species	NMR, LC-MS, GC-MS, HPLC	(Okada et al., 2010)

Table 2. Representative examples of model-, crop- and medicinal-plant metabolomics studies using different analytical platforms. DI : Direct Infusion. FID, Flame Ionization Detection. HRMAS-NMR: High resolution Magic-Angle Spinning NMR. ICR: Ion Cyclotron Resonance.

Currently the number of quantified analytes in a given sample and in one shot is approximately 50 with proton NMR, 100-200 with GC-MS, >1000 with LC-High-Resolution-MS (LC-HR-MS). This expansion of scale has been made possible through improved analytical capabilities, dissemination of routine procedures between laboratories, but also implementation of dedicated statistical and data mining strategies. However, a large proportion of the analytes detected in plant extracts cannot be annotated and identified based on chemical shift and multiplicity for NMR analysis, or on elemental formula (based on m/z ratio and isotopic ratio) and chromatographic retention time for GC- or LC-MS analysis, alone. Hence metabolite identification, which uses a variety of analytical techniques along with analyte/metabolite databases, remains difficult (Moco et al., 2007). Achieving standardization for naming compounds at the plant metabolomics community level is also an important issue, as it will enable researchers to share knowledge and speed up metabolite identification (Saito & Matsuda, 2010; Kim et al., 2011). Another challenge is the development of chimiotheques, where trusted reference compounds would be available for the community to validate analyte identifications, for example via spiking experiments.

2.1.1 Optimization and combination of the different current techniques

As already mentioned (see Tables 1-2), a combination of different current techniques is needed to cover the wide diversity of metabolites found in plants. Thus, a combination of different MS technologies is helpful for identification purpose. Among HR-MS technologies, LC- Time-Of-Flight (TOF) (resolution of 8,000-20,000, accuracy 1-5 ppm), Orbitrap® (resolution of 100,000, accuracy< 1 ppm) and FT-ICR-MS (resolution> 100,000, accuracy< 1 ppm) are currently the most powerful ultra-high resolution (UHR) mass spectrometers. They provide molecular formula information, thus offering great possibilities in terms of metabolite identification (see (Werner et al., 2008) for the strategy, pitfalls and bottleneck of metabolite identification). Nevertheless, the poor reproducibility and fragmentation variability between instruments from the same brand require a home-made metabolite database for each instrument. In addition it should be kept in mind that plant extracts contain many isomers, i.e. with identical elemental compositions and accurate masses. UHR-MS analysis of a selection of extracts may help to identify marker-metabolites revealed using HR-MS on a larger range of samples. Besides, multidimensional separation techniques have emerged in order to enhance metabolite coverage in the GC-MS (Gaquerel et al., 2009; Allwood et al., 2011) and LC-MS (Lei et al., 2011) fields. Further methodologies such as Ion Mobility MS (Dwivedi et al., 2008), which have not been tested in plants so far, might also prove useful. Anyway, processing and integrating data still remain the major bottlenecks and thus the most labour intensive steps for all these different analytical platforms. Developments are nevertheless underway to automate them (see (Redestig et al., 2010) for MS and hyphenated technologies).

2.1.2 From relative to absolute quantification of biological variability

Whereas the "convenient" relative quantification is often used in MS studies, absolute quantification will be increasingly required. For example, various modelling approaches require precise concentrations of metabolites. Furthermore, the sharing and integration of data obtained on different analytical platforms will be greatly facilitated if expressed as absolute quantities. To face the challenge of quantification, in GC-MS and LC-ESI-MS impaired by ion suppression or enhancement and matrix effects, a solution is to use stable-isotopomers of target metabolites or to run whole ^{13}C metabolome isotope labelling (Feldberg et al., 2009; Giavalisco et al., 2009). However, even with a stable isotope the matrix effects may impair the quantification (Jemal et al., 2003) and few isotopically-labelled metabolites are currently commercially available (Lei et al., 2011). In contrast to MS-based technologies, NMR, although less sensitive, provides ease of quantitation since the resonance intensity is only determined by the molar concentration, and high reproducibility (Ward et al., 2010; Kim et al., 2011).

Surprinsingly, a unique extraction protocol (sometimes one-step protocol) is typically used for a given analytical technique, regardless of the vast variety of plant matrices (plant species, organs and tissues). Very few metabolomic publications are prolix on extraction recovery and stability. Running blanks (solvent blank and extraction blank) in the same conditions as the biological samples is also important, as it is needed to identify impurities originating from solvents (Kaiser et al., 2009) or consumables (i.e., phthalates from plastic ware) (Allwood et al., 2011; Weckwerth, 2011). Although metabolomics is by definition an untargeted approach, highly selective extraction protocols along with targeted analysis should not be forgotten, especially to reach high and reproducible extraction recovery as well as quantification accuracy (Sawada et al., 2009). Then, replication is required to achieve statistical reliability. Biological replicates should be preferred to technological replicates assuming biological variance almost always exceeds analytical variance (Shintu et al., 2009). Five biological replicates of five pooled-tissue samples or of five individuals and two to three technological replicates are recommended in plant metabolomics to get statistically reliable information (Tikunov et al., 2007). Quality control samples should also be run (Fiehn et al., 2008; Allwood et al., 2011).

2.2 Horizontal approaches

Horizontal approaches are defined as strategies that promote sample number over number of variables being measured. Mutant screens and quantitative genetics are typical examples requiring horizontal high-throughput, as they typically involve experiments with hundreds to thousands of samples. Targeted assays are usually preferred due to their low needs in terms of labour and/or costs, although several untargeted strategies such as bucketing and fingerprinting are also amenable to very high numbers of samples. While the processing of raw data still represents the slowest step in vertical strategies, the toughest bottleneck in horizontal high-throughput approaches is probably sample logistics.

2.2.1 Sample logistics

In large scale experiments, harvesting, grinding and weighing become extremely work intensive (>75% of the time), especially when samples need to be kept at very low temperatures to avoid alteration of their biochemical composition. Due to the highly dynamic nature of the metabolome, harvesting and quenching of samples into liquid nitrogen should also be achieved as quickly as possible (Ap Rees et al., 1977). Unfortunately, fast solutions are very limited (e.g. leaf punchers), and thus recruiting as many as possible helpers probably remains the best way to achieve a reliable large-scale harvest. Sample storage may also become problematic when sample turnover dramatically increases. A good way to avoid losses of samples and overfilling of -80°C freezers is to use software enabling sample management. Although costly, available automated storage solutions may also dramatically improve the handling of samples.

Most analytical technologies require sample grinding prior to extraction and analysis. Mills enabling the parallel grinding of large numbers of samples (e.g., 192 samples) are now available at affordable prices. However, they usually do not allow multiparallel grinding of samples of large size, suggesting that further developments are needed to enable large-scale studies with organs such as fruits or ears and with most crops. Last but not least, the weighing of aliquots is a tedious task, especially when the material needs to be kept at very low temperature. A robot combining grinding and weighing of up to 96 samples has been developed recently (http://www.labman.co.uk), opening the way for unprecedented horizontal high-throughput.

2.2.2 Microplate technology

The first microplate was fabricated in 1951 by the Hungarian Gyula Takátsky (Takatsy, 1955). It was made of 72 wells machined in a polymethyl methacrylate block and was used to speed up serial dilutions. This invention was driven by the need for a fast and reliable diagnostic for influenza, as Hungary was facing a major epidemic at that time. Sixty years later, the microplate format has driven the development of a huge diversity of labware and equipment, and hundreds of millions of microplates are sold every year. Sample storage, extractions and dilutions can be achieved in microplates, which remain the fastest and cheapest solution to process large numbers of samples in parallel. The quantification of various metabolites can be achieved in microplates via chemical or enzymatic reactions yielding products that can be quantified in a wide range of dedicated readers. The most common and cheapest readers are filter-based UV-visible spectrophotometers. They enable the quantification of a wide range of metabolites, including major sugars, organic acids and amino acids using endpoint methods (Bergmeyer, 1983, 1985, 1987), and metabolic intermediates that are present at much lower concentrations using kinetic assays (Gibon et al., 2002). Fluorimetry (Hausler et al., 2000) and luminometry (Roda et al., 2004) also provide high sensitivity and benefit from many commercially available fluorigenic substrates. Their use is nevertheless restricted in plants, due to the quenching of the emitted light that occurs in the presence of e.g. polyphenols that are usually present in plant extracts.

Throughput on microplates can be dramatically increased by using pipetting robots, which can handle up to 1536 samples in parallel and down to the nanoliter scale, depending on the brand. Thus, using one 96-head robot equipped with microplate handling and a series of microplate readers, a single person can run the determination of a given metabolite in thousands of samples per week. Increasing the number of analytes would nevertheless result in a decrease in sample throughput, roughly by a factor 2 at each supplemental analyte. It is estimated that at equal costs, such an approach might be of advantage for 10 to 20 analytes over other targeted technologies such as LC-MS/MS, which have already proven efficient for the capture of relatively large numbers of metabolites from the same class at high-throughput (Rashed et al., 1997) and 2.2.3 Section below. Microplates of increasing density formats (up to 9600 wells per plate) have been released to increase the overall throughput of analyses and decrease the costs per assay. Such miniaturisation nevertheless faces physical constraints of delivering very small volumes to wells and of detecting responses in a manner that is both sensitive and rapid (Battersby & Trau, 2002). The use of volumes in the nanoliter range is also limited by quick evaporation of the solvent used for analysis. A further drawback is high costs in terms of equipment (e.g., pipetting robots and readers able to handle high density plates), which implies that very high numbers of samples will have to be processed before decreasing costs per assay. These limitations probably explain why the use of high density microplates has not been adopted by a wide research community so far.

2.2.3 Targeted MS technologies: Quantification of selected biochemical markers using LC-MS

Targeted analysis for small molecules using MS may use different technologies: GC-MS (Koek et al., 2011) , CE-MS (Ramautar et al., 2011), LC-MS and more recently MALDI-MS (Shroff et al., 2009). Here only LC-MS will be dealt with. LC-MS technology has been used for quantification long before the ages of the "omics". Despite its high-skilled technical need and its expensive cost, it has gained popularity in the metabolomics field. Triple quadrupoles analyzers (TQMS) are the workhorse of LC-MS quantification. They are mostly operated in multiple reaction monitoring (MRM) mode to achieve high selectivity and sensitivity. A new promising approach is the use of high resolution extracted ion chromatograms from full scans of high resolution instruments (Lu et al., 2008). Main advantages over MRM are the virtually unlimited number of monitored compounds and the possibility to reanalyze data after acquisition by extracting ion chromatograms corresponding to new compounds of interest.

Calibration of these methods involves most of the time internal calibration, with or without use of stable isotope analogs (Ciccimaro & Blair, 2010). For instance, quantification of amino acids by LC-MS (MRM) in barley was calibrated using d_2-Phe as an internal standard. The interday precision of the method ranged from 3.7 to 9.4 % RSD, depending on the amino-acid (Thiele et al., 2008). However an isotopic dilution calibration is not always possible due to the lack of the corresponding labelled metabolite or its cost. These targeted LC-MS methods must undergo a complete method validation. They need fast separation, high selectivity, linearity range and limits of quantification in agreement with the metabolite level. For methods involving atmospheric pressure ionization, a careful evaluation of matrix effect on quantification and its minimization should be addressed (Trufelli et al., 2011). Moreover, to be relevant, these methods must obviously be applied after an exhaustive extraction evaluated by recovery procedures.

Targeted approaches have been applied in functional genomics. A "widely targeted" metabolomics approach based on LC-MS (MRM) has been proposed (Sawada et al., 2009). It consisted in repeated UPLC-TQMS analyses performed on a same sample. Each 3 min analytical run allowed simultaneous detection of 5 compounds. Expected throughput was estimated 1,000 biological samples per week for quantification of about 500 metabolites. This methodology was later applied on mature seeds of 2656 mutants and 225 *Arabidopsis* accessions for 17 amino-acids, 18 glucosinolate derivatives and one flavonoid, leading to characterization of amino-acids hyper-accumulating genotypes (Hirai et al., 2010). They have also been applied in phytochemistry and phytomedicine. For instance, Chinese medicinal herbs were tested for two secondary metabolites inducing nephrotoxicity. The UPLC-MS (MRM) run was 5 min and was amenable for high-throughput analyses (Jacob et al., 2007) This method was however impaired by a strong matrix effect that could not be prevented and another approach was preferred. At last, these techniques have been also shown to be a must for some classes of compounds such as hormones (Kojima et al., 2009), intermediates of central metabolism (Arrivault et al., 2009) and pesticides (Kmellar et al., 2010). In fact, they provide together the appropriate selectivity, sensitivity and throughput.

2.2.4 Further technologies

Other technologies involve miniaturization of the separation step used prior to detection. A key step in miniaturization and automation of chromatography is the development of microfluidic systems, which process or manipulate very small volumes (down to 10^{-18} L) using channels of micrometre dimensions (Whitesides, 2006). The fact that factors such as surface tension and viscosity are getting very different in such systems brings many new possibilities to control concentrations and behaviours of molecules, particles or even cells (Nagrath et al., 2007) in space and time. Thus, the performance of soft lithography on e.g. poly(dimethylsiloxane) (McDonald et al., 2000) or polypropylene (Vengasandra et al., 2010) enables the design of reservoirs, channels, valves, and reaction chambers that can be used to separate and transform a wide range of molecules. Combined to detection systems such as laser induced fluorescence (Jiang et al., 2000), infrared spectroscopy (Shaw et al., 2009) or electrochemical electrodes (Eklund et al., 2006), they are well suited for massively parallel assays and provide the advantage of using very small amounts of reagents and samples. Further advantages are high resolution and sensitivity as well as fast analysis.

The use of microfluidic systems for metabolite analysis has just begun. Whereas applications targeting one molecule, e.g. glucose (Atalay et al., 2009), have been developed, the possibility to separate molecules has already enabled the profiling of classes of metabolites such as glucosinolates (Fouad et al., 2008) or flavonoids (Hompesch et al., 2005). Furthermore, the ease of creating systems able to distribute fluids into multiple channels enables the performance of several assays in parallel (Moser et al., 2002), or even n-multidimensional separations (Tomas et al., 2008) that would eventually be coupled to various detection devices, opening unprecedented possibilities for targeted and untargeted metabolomics.

Unfortunately, microfluidics have not yet benefited from standardisation, which hampers their adoption by a wide research community. Besides involving complex designs and fabrication techniques prohibiting widespread use due to cost and/or time for production, microfluidic systems may require unfamiliar laboratory habits. Therefore, one logical next step is the integration with the standardized microplate layout, thus taking advantage of the extensive work of the lab automation community (Choi & Cunningham, 2007; Halpin & Spence, 2010). Strikingly, such integration might ultimately result in methodologies enabling high density analyses on very large numbers of samples, thus breaking the relationship depicted in Figure 1. Finally, microfluidics and more generally nanotechnologies have almost certainly much more to offer, as future developments could for example lead to portable systems that would allow metabolite profiling directly in the field, thus shortcutting sample handling, or even to chips embarked on growing plants that would be able to monitor fluxes *in situ*.

2.3 Complementarities of vertical and horizontal approaches

The combination of horizontal and vertical high-throughput approaches (Fig. 1) is of particular interest, as it has the potential to dramatically speed up the process of discovery. Thus, depending on the objectives of the study, an untargeted approach can be used first on a selection of samples to identify the most discriminating biomarkers that would then be analyzed on a much greater number of samples using a targeted approach (Tarpley et al., 2005). For example, such strategy has been successfully used in maize where a number of enzymes were first profiled in a small panel of eight highly diverse maize inbred lines, revealing a highly heritable variation in NAD-dependent isocitrate dehydrogenase activity. The use of a panel of about hundred lines then allowed the identification of a novel amino-acid substitution in a phylogenetically conserved site, which is assoaciated with isocitrate activity variation (Zhang et al., 2010). On the contrary, a horizontal approach can be used to screen large numbers of samples, thus revealing the most extremes or representative ones, on which a vertical approach can then be used to search for unexpected modifications, to study the system as a whole in the best possible matrix of samples, or simply to find novel biomarkers. As an example, the easy to measure glucose-6-phosphate, which is a good temporal marker of carbon depletion (Stitt et al., 2007), has been used to define a precise time frame to study transcriptomic and metabolomic responses to carbon starvation in Arabidopsis leaves (Usadel et al., 2008), thus avoiding unnecessary and costly analyses.

3. Key challenges for plant metabolomics

An increasing number of approaches benefit from plant metabolomics. Among them, systems biology, quantitative genetics and meta-phenomics offer particularly exciting yet challenging perspectives.

3.1 Systems biology

In the context of plant functional genomics, the combination of metabolomics, proteomics, and transcriptomics has permitted to decipher and understand dynamic interactions in metabolic networks and to discover new correlations with biochemically characterized pathways as well as pathways hitherto unknown (Zhang et al., 2009; Williams et al., 2010). The main lesson from the latter or similar studies is that metabolic pathways are highly interactive rather than operating as separate units. In each biological system (cell, tissue, organism) there are metabolic networks in place, which are highly flexible and present a huge capacity to provide compensatory mechanisms through regulatory process. These observations actually explain why many dedicated GMO strategies ended up with silent phenotype (Weckwerth et al., 2004) and also convinced a large set of researchers to study metabolic networks as a whole, and not as a sum of parts, moving from reductionist to holistic approaches.

Although systems biology may mean different things to different people, there is a common understanding that this discipline is a comprehensive quantitative analysis of the manner in which all the components of a biological system (cell, tissue, organisms, communities) interact functionally over time. Systems biology aims at combining omics data resulting from complex networks into computational models. Besides integration with upstream levels (genome, transcriptome, proteome), metabolite data also have to be integrated with downstream levels (e.g. growth, performance) data. The quantitative data are the initial point for the formulation of mathematical models, which are refined by hypothesis-driven, iterative systems perturbations and data integrations. Cycles of iteration result in a more accurate model and ultimately the model explains emergent properties of the biological system of interest. Once the model is sufficiently accurate and detailed, it allows biologists to accomplish two tasks (1) predict the behavior of the system given any perturbation such as a modification of the environment, and (2) redesign or perturb the gene regulatory network to create completely new emergent systems properties (Vidal, 2009; Westerhoff et al., 2009; Arkin & Schaffer, 2011).

Exciting examples of integrated system biology to solve biological questions in plant science have been published such as identification of key players in the branched amino acid metabolism in *A. thaliana* (Curien et al., 2009), analysis of carbohydrate dynamics during acclimation to low temperature in *A. thaliana* (Nagele et al., 2011), or understanding the metabolism of tobacco grown on media containing different cytokins (Lexa et al., 2003). Systems biology will benefit from close collaborations between different teams covering complementary sectors of metabolism, e.g. central metabolism and different sectors of secondary metabolism. The challenges in establishing such systems approaches rely on collecting reliable, quantitative and systemic "omics" data, including metabolomics data, for developing modelling able to predict de novo biological outcomes given the list of the components involved. Advances in plant genome sequencing, transcriptomics and proteomics have paved the way for a systematic analysis of cellular processes at gene and protein levels. For metabolomics, some limitations remain for real system biology approaches, in terms of analytical sensitivity, throughput and access to specific tissue or subcellular compartments. Moreover, the high turn-over rate of many metabolic intermediates has to be taken into consideration. In addition, the absolute quantification of metabolites under physiological, *in vivo* and dynamic conditions remains a major challenge. The combination of existing multiparallel analytical platforms with special attention to metabolite quantification (see Sections 2.1.2 and 2.2.3) in a cohesive manner may not be sufficient and emerging microtechnologies such as microfluidics will certainly help (see Section 2.2.4 and (Wurm et al., 2010)).

Recently, plant systems biology has been redefined from cell to ecosystem (Keurentjes et al., 2011). For these authors, in a holistic systems-biology approach, plants have to be studied at six levels of biological organization (from subcellular level to ecosystem) in an orchestrated way, with special attention to the interdependence between the various levels of biological organization. The corresponding challenge will be to generate accurate experimental data for communities, populations, single whole plants, down to cell types and their organelles that can be used to feed new modelling concepts. For example, at the subcellular level molecular signaling pathways are crucial to understand cell development, defense against pathogens and many more intermediate processes in plants. The highly sensitive and high-throughput method developed for the simultaneous analysis of 43 molecular species of cytokinins, auxins, ABA and gibberellins (Kojima et al., 2009) has opened a big opportunity to routinely describe basic molecular signaling pathways in plant cells. Others challenges need to be considered in terms of dry labs. Because systems biology heavily relies on information stored in public databases for the different levels of biological organization, which is often incomplete, not standardized or improperly annotated, it is essential that collective efforts are developed for the validation of large data sets. Plant network biology is in its infancy and other current needs range from the development of new theoretical methods to characterize network topology, to insights into dynamics of motif clusters and biological function.

3.2 Quantitative genetics

Quantitative genetics, which aim at associating quantitative traits with genomic regions called quantitative trait loci (QTL), represent a great opportunity to understand the diversity of plant metabolism and its relationship to nutritional value or biomass production. Studies combining metabolomics and quantitative genetics performed in Arabidopsis seedlings (Keurentjes et al., 2006) and tomato fruits (Schauer et al., 2006) have shown that variations in metabolite levels are for a large part heritable, and have identified large numbers of metabolite QTL, implying that levels of metabolites of interest could be controlled by manipulating small genome regions (Saito & Matsuda, 2010). Conversely, genetic diversity has been used to study the behaviour of metabolic networks and the way they integrate with whole plant traits, eventually revealing links between metabolic composition and growth (Meyer et al., 2007). Such findings appeal for multivariate QTL mapping (Calinski et al., 2000), thus opening exciting perspectives for the manipulation of plant performance.

The identification of the molecular bases underlying QTL has usually been a major challenge, and several years of hard work were typically necessary to unravel just one of them. However, thanks to the development of increasingly powerful methodologies exploiting genetic diversity that combine linkage and/or association mapping and high density genotyping, the elucidation of such molecular bases can now be achieved much quicker (Myles et al., 2009). The other side of the coin is that these methodologies require experiments of increasing sizes. Thus, the nested association mapping (NAM) approach recently developed in maize (Yu et al., 2008) already involves 5,000 genotypes (25 mapping populations of 200 genotypes each), which would represent at least 5,000 samples to process. Unfortunately, due to technical and financial limitations, the processing of so many samples remains very unusual in plant metabolomics. Furthermore, taking into account different growth scenarios, temporal aspects or different organs or tissues would result in factorial increases in numbers of samples. As mentioned above, combinations of horizontal and vertical metabolomics might nevertheless be very useful to decrease costs and labour. For instance, small sub-panels with high genetic diversity can be used first to assess heritability for a large number of metabolic traits, selected ones being then evaluated in full panels using inexpensive and fast methods.

Finally and importantly, genetic divergence and phenotypic divergence are too different things (Kozak et al., 2011). Accordingly, one single gene can be responsible for huge phenotypic variations and one single trait can be controlled by many QTL. One consequence is that molecular marker-assisted breeding might not always be the best and/or cheapest solution to select genotypes yielding phenotypes of interest. Therefore, it is pertinent to explore the possibility to use alternative biomarkers, including metabolites that can be measured at reasonable costs in very large populations.

3.3 Meta-phenomics

Comparing different species is a powerful way to extend knowledge about biological processes. Thus, comparative genomics facilitate the assignation of gene function in non sequenced organisms, enable the quick annotation of newly sequenced genomes and greatly contribute to studies of gene function and evolution. For example, extensive synteny between genomes of Graminae species has been shown (Salse, 2004) and QTL controlling similar traits have been found in orthologous regions of e.g., maize and sorghum (Figueiredo et al., 2010). Conversely, the fact that orthologous genes do not necessary have the same functions in different species (Buckler et al., 2009) opens fascinating perspectives regarding evolution of gene function (Wang et al., 2009).

Finding common and divergent phenotypes among large numbers of species is also a promising way to better understand biological functions in the context of evolution. Meta-phenomics, which has recently been proposed by Poorter and colleagues (Poorter et al., 2009; Poorter et al., 2010), defines as the study of plant responses to environmental factors by performing meta-analyses. This novel ecophysiological approach aims at generalising plant responses by integrating phenotypic and environmental data gathered for large numbers of species. Thus, by using accurate normalisation procedures generic response curves were found for surface leaf area as related to major abiotic factors. Noteworthy, data for >300 species had to be collected and curated manually throughout 60 years of literature. One exciting finding is divergences between groups of species could be pinpointed, for example C3 and C4 species. There is no doubt that meta-phenomics is amenable to the cellular level, and in particular to metabolic pathways, and C3 and C4 metabotypes are indeed easy to distinguish when comparing their respective metabolomes. However, this might be considerably complicated given the heterogeneity of available metabolic data (in terms of e.g., annotation and normalisation). Furthermore, descriptions of environmental conditions found in literature are almost always text-based, and thus very difficult to compute. Fortunately, the adoption and use of standardised conceptualisations with explicit specifications to report data and metadata (i.e. minimum checklists) is progressing in the field of metabolomics (Fiehn et al., 2007a; Fiehn et al., 2007b). It will nevertheless be of central importance to prefer absolute quantification and to enable quantitative descriptions of environmental factors, which will probably be facilitated via collaborations with ecophysiologists.

4. Conclusion

As metabolomics in general (Hall et al., 2011), plant metabolomics is moving towards biology with a growing variety of applications from 'simple' diagnostic of culture practices to translational studies towards systems biology. However, for some of the emerging applications, the optimization of analytical and computational technologies for the acquisition, handling and mining of metabolomics data remains necessary. Some of the crucial bottlenecks that still have to be adressed concern quantification for modelling, time and spatial resolved experiments, multi-experiments and data sharing.

The promotion of multi-experiments and multi-labs combined analyses (Allwood et al., 2009; Ward et al., 2010) for high sample numbers, indispensable for some ecology or quantitative genetics studies for instance, requires shared plant biological standards (labeled or non-labeled) and standardization of their use. The absolute quantification data, needed for metabolism modelling in systems bioloy approaches, also requires isotopically labelled plant standards or at least labelled reference compounds for MS approaches. The generalisation of time-resolved experiments for instance for the study of fine metabolism regulation or short-term responses to stresses will need further increases in horizontal high-throughput using microplate, microfluidics or other technologies. Besides increased throughput, increased sensitivity for all the analytical technologies listed in this review may open new insights into the use of metabolomics for plant development studies. Spatial-resolved experiments with analysis of laser-microdissected samples by NMR or MS (Moco et al., 2009; Kim et al., 2011) will be particularly useful for the study of plant-pathogen interactions. The generalization of metabolite compartmentation studies in plant tissues at the cellular and subcellular levels, possibly with non-aqueous fractionation (Krueger et al., 2011), will also request increases in both horizontal high-throughput and sensitivity.

Moreover, the systematic sharing, combining, and re-exploring of the data produced using targeted metabolic phenotyping or untargeted metabolomics will produce new knowledge. Cataloging the metabolome itself by experimental data and literature data, stored in curated databases can complement genomic reconstructions of metabolism (Fiehn et al., 2011). Access to the regulation of the plasticity and flexibility of metabolic networks implies that the metadata of each experiment, including environment metadata (Hannemann et al., 2009) have to be carefully documented and uploaded into a central or distributed network repository dedicated to plants. This suggests that the MSI initiative (Fiehn et al., 2007a) has to continue to propose and promote standardization criteria that will be integrated by the bioinformatics developments of open repositories and used by the community. In addition, sophisticated but easy-to-use tools for metabolomics data combining, integration with other phenotyping or omics data, and integrated statistical analyses and modelling are needed. The plant metabolome community may benefit from more interaction with the human metabolome community for the use and development of such tools, and both may address combined analyses of food quality determinants (Hall et al., 2008) and food human consumption monitoring (Wishart, 2008).

5. Acknowledgment

Financial supports of ERA-NET ERASysBio+ (FRIM) and FP7 KBBE (DROPS, grant agreement number FP7-244374) are acknowledged. All authors acknowledge support from the Metabolome Facility of Bordeaux Functional Genomics Centre.

6. References

Allwood, J.W.; de Vos, C.H.R.; Moing, A.; Deborde, C.; Erban, A.; Kopka, J.; Goodacre, R. & Hall, R. (2011). Plant metabolomics and its potential for systems biology research: background concepts, technology and methodology, In: *Methods In Systems Biology*, Westerhoff, H., Hayes, N., (Eds) in press, Elsevier Inc., isbn 978-0-12-385118-5, Amsterdam, Netherlands

Allwood, J.W.; Erban, A.; de Koning, S.; Dunn, W.B.; Luedemann, A.; Lommen, A.; Kay, L.; Loscher, R.; Kopka, J. & Goodacre, R. (2009). Inter-laboratory reproducibility of fast gas chromatography-electron impact-time of flight mass spectrometry (GC-EI-TOF/MS) based plant metabolomics. *Metabolomics,* Vol.5, No. 4, (Dec 2009), pp. 479-496, issn 1573-3882

Ap Rees, T.A.; Fuller, W.A. & Wright, B.W. (1977). Measurements of glycolytic intermediates during onset of thermogenesis in spadix of *Arum maculatum. Biochimica Biophysica Acta,* Vol.461, No. 2, (Aug 1977), pp. 274-282, issn 0006-3002

Arkin, A.P. & Schaffer, D.V. (2011). Network News: Innovations in 21st Century Systems Biology. *Cell,* Vol.144, No. 6, (Mar 2011), pp. 844-849, issn 0092-8674

Arrivault, S.; Guenther, M.; Ivakov, A.; Feil, R.; Vosloh, D.; van Dongen, J.T.; Sulpice, R. & Stitt, M. (2009). Use of reverse-phase liquid chromatography, linked to tandem mass spectrometry, to profile the Calvin cycle and other metabolic intermediates in Arabidopsis rosettes at different carbon dioxide concentrations. *Plant Journal*, Vol.59, No. 5, (Sep 2009), pp. 824-839, issn 0960-7412

Atalay, Y.T.; Witters, D.; Vermeir, S.; Vergauwe, N.; Verboven, P.; Nicolai, B. & Lammertyn, J. (2009). Design and optimization of a double-enzyme glucose assay in microfluidic lab-on-a-chip. *Biomicrofluidics*, Vol.3, No.4, (Oct-Dec 2009), issn 1932-1058

Battersby, B.J. & Trau, M. (2002). Novel miniaturized systems in high-throughput screening. *Trends in Biotechnology*, Vol.20, No. 4, (Apr 2002), pp. 167-173, issn 0167-7799

Beale, M.H. & Sussman, M.R. (2011). Metabolomics of *Arabidopsis Thaliana*, In: *Biology of Plant Metabolomics*, Hall, R.D., (Ed.) 157-180, Wiley-Blackwell, isbn 978-1-4051-9954-4, Chichester, UK

Bergmeyer, H.U. (1983). Metabolites 1: Carbohydrates, In: *Methods of enzymatic analysis*, Bergmeyer, J., GraBl, M., (Eds) 701, VCH Verlagsgesellschaft mbH, isbn 3-527-26046-3, Weinheim, Germany

Bergmeyer, H.U. (1985). Metabolites 2: Tri and dicarboxylic acids, purines, pyrimidines and derivatives, coenzymes, inorganic compounds, In: *Methods of enzymatic analysis*, Bergmeyer, J., GraBl, M., (Eds) 656, VCH Verlagsgesellschaft mbH, isbn 3-527-26047-1, Weinheim, Germany

Bergmeyer, H.U. (1987). Metabolites 3 - Lipids, Amino Acids and Related Compounds In: *Methods of enzymatic analysis*, Bergmeyer, J., GraBl, M., (Eds) VCH Verlagsgesellschaft mbH, isbn 3-527-26048-X, Weinheim, Germany

Bourgis, F.; Kilaru, A.; Cao, X.; Ngando-Ebongue, G.F.; Drira, N.; Ohlrogge, J.B. & Arondel, V. (2011). Comparative transcriptome and metabolite analysis of oil palm and date palm mesocarp that differ dramatically in carbon partitioning. *Proceedings of the National Academy of Sciences of the United States of America*, Vol.108, No. 30, (Jul 2011), pp. 12527-12532, issn 0027-8424

Broyart, C.; Fontaine, J.X.; Molinie, R.; Cailleu, D.; Terce-Laforgue, T.; Dubois, F.; Hirel, B. & Mesnard, F. (2010). Metabolic profiling of maize mutants deficient for two glutamine synthetase isoenzymes using (1)H-NMR-based metabolomics. *Phytochemical Analysis*, Vol.21, No. 1, (Jan-Feb 2010), pp. 102-109, issn 0958-0344

Buckler, E.S.; Holland, J.B.; Bradbury, P.J.; Acharya, C.B.; Brown, P.J.; Browne, C.; Ersoz, E.; Flint-Garcia, S.; Garcia, A.; Glaubitz, J.C.; Goodman, M.M.; Harjes, C.; Guill, K.; Kroon, D.E.; Larsson, S.; Lepak, N.K.; Li, H.H.; Mitchell, S.E.; Pressoir, G.; Peiffer, J.A.; Rosas, M.O.; Rocheford, T.R.; Romay, M.C.; Romero, S.; Salvo, S.; Villeda, H.S.; da Silva, H.S.; Sun, Q.; Tian, F.; Upadyayula, N.; Ware, D.; Yates, H.; Yu, J.M.; Zhang, Z.W.; Kresovich, S. & McMullen, M.D. (2009). The genetic architecture of maize flowering time. *Science*, Vol.325, No. 5941, (Aug 2009), pp. 714-718, issn 0036-8075

Bylesjo, M.; Nilsson, R.; Srivastava, V.; Gronlund, A.; Johansson, A.I.; Jansson, S.; Karlsson, J.; Moritz, T.; Wingsle, G. & Trygg, J. (2009). Integrated analysis of transcript, protein and metabolite data to study lignin biosynthesis in hybrid aspen. *Journal of Proteome Research,* Vol.8, No. 1, (Jan 2009), pp. 199-210, issn 1535-3893

Calinski, T.; Kaczmarek, Z.; Krajewski, P.; Frova, C. & Sari-Gorla, M. (2000). A multivariate approach to the problem of QTL localization. *Heredity,* Vol.84, No. 3, (Mar 2000), pp. 303-310, issn 0018-067X

Choi, C.J. & Cunningham, B.T. (2007). A 96-well microplate incorporating a replica molded microfluidic network integrated with photonic crystal biosensors for high throughput kinetic biomolecular interaction analysis. *Lab on a Chip,* Vol.7, No. 5, (Mar 2007), pp. 550-556, issn 1473-0197

Ciccimaro, E. & Blair, I.A. (2010). Stable-isotope dilution LC-MS for quantitative biomarker analysis. *Bioanalysis,* Vol.2, No. 2, (Feb 2010), pp. 311-341, issn 1757-6180

Cossegal, M.; Chambrier, P.; Mbelo, S.; Balzergue, S.; Martin-Magniette, M.L.; Moing, A.; Deborde, C.; Guyon, V.; Perez, P. & Rogowsky, P. (2008). Transcriptional and metabolic adjustments in ADP-glucose pyrophosphorylase-deficient bt2 maize kernels. *Plant Physiology,* Vol.146, No. 4, (Apr 2008), pp. 1553-1570, issn 0032-0889

Curien, G.; Bastlen, O.; Robert-Genthon, M.; Cornish-Bowden, A.; Cardenas, M.L. & Dumas, R. (2009). Understanding the regulation of aspartate metabolism using a model based on measured kinetic parameters. *Molecular Systems Biology,* Vol.5, (May 2009), issn 1744-4292

de Vos, R.C.H.; Hall, R. & Moing, A. (2011). Metabolomics of a model fruit: tomato In: *Biology of Plant Metabolomics,* Hall, R., (Ed.) 109-155, Wiley-Blackwell Ltd, isbn 978-1-4051-9954-4, Oxford

Deluc, L.G.; Grimplet, J.; Wheatley, M.D.; Tillett, R.L.; Quilici, D.R.; Osborne, C.; Schooley, D.A.; Schlauch, K.A.; Cushman, J.C. & Cramer, G.R. (2007). Transcriptomic and metabolite analyses of Cabernet Sauvignon grape berry development. *BMC Genomics,* Vol.8 (Nov 2007), issn 1471-2164

Dixon, R.A. & Strack, D. (2003). Phytochemistry meets genome analysis, and beyond. *Phytochemistry,* Vol.62, No. 6, (Mar 2003), pp. 815-816, issn 0031-9422

Dwivedi, P.; Wu, P.; Klopsch, S.J.; Puzon, G.J.; Xun, L. & Hill, H.H. (2008). Metabolic profiling by ion mobility mass spectrometry (IMMS). *Metabolomics,* Vol.4, No. 1, (Mar 2008), pp. 63-80, issn 1573-3882

Eklund, S.E.; Snider, R.M.; Wikswo, J.; Baudenbacher, F.; Prokop, A. & Cliffel, D.E. (2006). Multianalyte microphysiometry as a tool in metabolomics and systems biology. *Journal of Electroanalytical Chemistry,* Vol.587, No. 2, (Feb 2006), pp. 333-339, issn 0022-0728

Ernst, W.H.O. (1995). Sampling of plant material for chemical analysis. *Science of the Total Environment,* Vol.176, No. 1-3, (Dec 1995), pp. 15-24, issn 0048-9697

Fait, A.; Hanhineva, K.; Beleggia, R.; Dai, N.; Rogachev, I.; Nikiforova, V.J.; Fernie, A.R. & Aharoni, A. (2008). Reconfiguration of the achene and receptacle metabolic networks during strawberry fruit development. *Plant Physiology,* Vol.148, No. 2, (Oct 2008), pp. 730-750, issn 0032-0889

Farag, M.A.; Huhman, D.V.; Dixon, R.A. & Sumner, L.W. (2008). Metabolomics reveals novel pathways and differential mechanistic and elicitor-specific responses in phenylpropanoid and isoflavonoid biosynthesis in *Medicago truncatula* cell cultures. *Plant Physiology,* Vol.146, No. 2, (Feb 2008), pp. 387-402, issn 0032-0889

Feldberg, L.; Venger, I.; Malitsky, S.; Rogachev, I. & Aharoni, A. (2009). Dual labeling of metabolites for metabolome analysis (DLEMMA): A new approach for the identification and relative quantification of metabolites by means of dual isotope labeling and liquid chromatography-mass spectrometry. *Analytical Chemistry,* Vol.81, No. 22, (Nov 2009), pp. 9257-9266, issn 0003-2700

Ferry-Dumazet, H.; Gil, L.; Deborde, C.; Moing, A.; Bernillon, S.; Rolin, D.; Nikolski, M.; de Daruvar, A. & Jacob, D. (2011). MeRy-B: a web knowledgebase for the storage, visualization, analysis and annotation of plant NMR metabolomic profiles. *BMC Plant Biology,* Vol.11 (Jun 2011), pp. 12, issn 1471-2229

Fiehn, O.; Barupal, D.K. & Kind, T. (2011). Extending biochemical databases by metabolomic surveys. *Journal of Biological Chemistry,* Vol.286, No. 27, (Jul 2011), pp. 23637-23643, issn 0021-9258

Fiehn, O.; Robertson, D.; Griffin, J.; van der Werf, M.; Nikolau, B.; Morrison, N.; Sumner, L.W.; Goodacre, R.; Hardy, N.W.; Taylor, C.; Fostel, J.; Kristal, B.; Kaddurah-Daouk, R.; Mendes, P.; van Ommen, B.; Lindon, J.C. & Sansone, S.A. (2007a). The metabolomics standards initiative (MSI). *Metabolomics,* Vol.3, No. 3, (Sep 2007), pp. 175-178, issn 1573-3882

Fiehn, O.; Sumner, L.W.; Rhee, S.Y.; Ward, J.; Dickerson, J.; Lange, B.M.; Lane, G.; Roessner, U.; Last, R. & Nikolau, B. (2007b). Minimum reporting standards for plant biology context information in metabolomic studies. *Metabolomics,* Vol.3, No. 3, (Sep 2007), pp. 195-201, issn 1573-3882

Fiehn, O.; Wohlgemuth, G.; Scholz, M.; Kind, T.; Lee, D.Y.; Lu, Y.; Moon, S. & Nikolau, B. (2008). Quality control for plant metabolomics: reporting MSI-compliant studies. *Plant Journal,* Vol.53, No. 4, (Feb 2008), pp. 691-704, issn 0960-7412

Figueiredo, L.F.D.; Sine, B.; Chantereau, J.; Mestres, C.; Fliedel, G.; Rami, J.F.; Glaszmann, J.C.; Deu, M. & Courtois, B. (2010). Variability of grain quality in sorghum: association with polymorphism in Sh2, Bt2, SssI, Ae1, Wx and O2. *Theoretical and Applied Genetics,* Vol.121, No. 6, (Oct 2010), pp. 1171-1185, issn 0040-5752

Fouad, M.; Jabasini, M.; Kaji, N.; Terasaka, K.; Tokeshi, M.; Mizukami, H. & Baba, Y. (2008). Microchip analysis of plant glucosinolates. *Electrophoresis,* Vol.29, No. 11, (Jun 2008), pp. 2280-2287, issn 0173-0835

Gaquerel, E.; Weinhold, A. & Baldwin, I.T. (2009). Molecular interactions between the specialist herbivore *Manduca sexta* (Lepidoptera, Sphigidae) and its natural host *Nicotiana attenuata.* VIII. An unbiased GCxGC-ToFMS analysis of the plant's elicited volatile emissions. *Plant Physiology,* Vol.149, No. 3, (Mar 2009), pp. 1408-1423, issn 0032-0889

Giavalisco, P.; Kohl, K.; Hummel, J.; Seiwert, B. & Willmitzer, L. (2009). C-13 isotope-labeled metabolomes allowing for improved compound annotation and relative quantification in liquid chromatography-mass spectrometry-based metabolomic research. *Analytical Chemistry,* Vol.81, No. 15, (Aug 2009), pp. 6546-6551, issn 0003-2700

Gibon, Y.; Vigeolas, H.; Tiessen, A.; Geigenberger, P. & Stitt, M. (2002). Sensitive and high throughput metabolite assays for inorganic pyrophosphate, ADPGlc, nucleotide phosphates, and glycolytic intermediates based on a novel enzymic cycling system. *Plant Journal*, Vol.30, No. 2, (Apr 2002), pp. 221-235, issn 0960-7412

Hall, R.D. (2006). Plant metabolomics: from holistic hope, to hype, to hot topic. *New Phytologist*, Vol.169, No. 3, (Jan 2006), pp. 453-468, issn 1469-8137

Hall, R.D. (2011a). Biology of Plant Metabolomics, In: *Annual Plant Reviews*, 420, Wiley-Blackwell, isbn 978-1-4051-9954-4, Chichester, UK

Hall, R.D. (2011b). Plant Metabolomics in a Nutshell: Potential and Future Challenges, In: *Biology of Plant Metabolomics*, Hall, R.D., (Ed.) 1-24, Wiley-Blackwell, isbn 978-1-4051-9954-4, Chichester, UK

Hall, R.D.; Brouwer, I.D. & Fitzgerald, M.A. (2008). Plant metabolomics and its potential application for human nutrition. *Physiologia Plantarum*, Vol.132, No. 2, (Feb 2008), pp. 162-175, issn 0031-9317

Hall, R.D.; Wishart, D. & Roessner, U. (2011). Metabolomics and the move towards biology. *Metabolomics*, Vol.7, No. 3, (Sep 2011), pp. 454-456, issn 1573-3882

Halpin, S.T. & Spence, D.M. (2010). Direct plate-reader measurement of nitric oxide released from hypoxic erythrocytes flowing through a microfluidic device. *Analytical Chemistry*, Vol.82, No. 17, (Sep 2010), pp. 7492-7497, issn 0003-2700

Hannemann, J.; Poorter, H.; Usadel, B.; Blasing, O.E.; Finck, A.; Tardieu, F.; Atkin, O.K.; Pons, T.; Stitt, M. & Gibon, Y. (2009). Xeml Lab: a tool that supports the design of experiments at a graphical interface and generates computer-readable metadata files, which capture information about genotypes, growth conditions, environmental perturbations and sampling strategy. *Plant Cell and Environment*, Vol.32, No. 9, (Sep 2009), pp. 1185-1200, issn 0140-7791

Hausler, R.E.; Fischer, K.L. & Flugge, U.I. (2000). Determination of low-abundant metabolites in plant extracts by NAD(P)H fluorescence with a microtiter plate reader. *Analytical Biochemistry*, Vol.281, No. 1, (May 2000), pp. 1-8, issn 0003-2697

Hirai, M.Y.; Sawada, Y.; Kanaya, S.; Kuromori, T.; Kobayashi, M.; Klausnitzer, R.; Hanada, K.; Akiyama, K.; Sakurai, T.; Saito, K. & Shinozaki, K. (2010). Toward genome-wide metabolotyping and elucidation of metabolic system: metabolic profiling of large-scale bioresources. *Journal of Plant Research*, Vol.123, No. 3, (May 2010), pp. 291-298, issn 0918-9440

Hompesch, R.W.; Garcia, C.D.; Weiss, D.J.; Vivanco, J.M. & Henry, C.S. (2005). Analysis of natural flavonoids by microchip-micellar electrokinetic chromatography with pulsed amperometric detection. *Analyst*, Vol.130, No. 5, (May 2005), pp. 694-700, issn 0003-2654

Jacob, S.S.; Smith, N.W. & Legido-Quigley, C. (2007). Assessment of Chinese medicinal herb metabolite profiles by UPLC-MS-based methodology for detection of aristolochic acids. *Journal of Separation Science*, Vol.30, No. 8, (May 2007), pp. 1200-1206, issn 1615-9306

Jemal, M.; Schuster, A. & Whigan, D.B. (2003). Liquid chromatography/tandem mass spectrometry methods for quantitation of mevalonic acid in human plasma and urine: method validation, demonstration of using a surrogate analyte, and demonstration of unacceptable matrix effect in spite of use of a stable isotope analog internal standard. *Rapid Communications in Mass Spectrometry*, Vol.17, No. 15, (Jun 2003), pp. 1723-1734, issn 0951-4198

Jiang, G.F.; Attiya, S.; Ocvirk, G.; Lee, W.E. & Harrison, D.J. (2000). Red diode laser induced fluorescence detection with a confocal microscope on a microchip for capillary electrophoresis. *Biosensors & Bioelectronics*, Vol.14, No. 10-11, (Jan 2000), pp. 861-869, issn 0956-5663

Joyce, A.R. & Palsson, B.O. (2006). The model organism as a system: integrating 'omics' data sets. *Nature Reviews Molecular Cell Biology*, Vol.7, No. 3, (Mar 2006), pp. 198-210, issn 1471-0072

Kaiser, K.A.; Barding, G.A. & Larive, C.K. (2009). A comparison of metabolite extraction strategies for 1H-NMR-based metabolic profiling using mature leaf tissue from the model plant *Arabidopsis thaliana*. *Magnetic Resonance in Chemistry*, Vol.47, No. S1, (Dec 2009), pp. S147-S156, issn 0749-1581

Keurentjes, J.J.B.; Angenent, G.C.; Dicke, M.; Dos Santos, V.; Molenaar, J.; van der Putten, W.H.; de Ruiter, P.C.; Struik, P.C. & Thomma, B. (2011). Redefining plant systems biology: from cell to ecosystem. *Trends in Plant Science*, Vol.16, No. 4, (Apr 2011), pp. 183-190, issn 1360-1385

Keurentjes, J.J.B.; Fu, J.Y.; de Vos, C.H.R.; Lommen, A.; Hall, R.D.; Bino, R.J.; van der Plas, L.H.W.; Jansen, R.C.; Vreugdenhil, D. & Koornneef, M. (2006). The genetics of plant metabolism. *Nature Genetics*, Vol.38, No. 7, (Jul 2006), pp. 842-849, issn 1061-4036

Kim, H.K.; Choi, Y.H. & Verpoorte, R. (2011). NMR-based plant metabolomics: where do we stand, where do we go? *Trends in Biotechnology*, Vol.29, No. 6, (Jun 2011), pp. 267-275, issn 0167-7799

Kmellar, B.; Abranko, L.; Fodor, P. & Lehotay, S.J. (2010). Routine approach to qualitatively screening 300 pesticides and quantification of those frequently detected in fruit and vegetables using liquid chromatography tandem mass spectrometry (LC-MS/MS). *Food Additives and Contaminants Part a-Chemistry Analysis Control Exposure & Risk Assessment*, Vol.27, No. 10, (Oct 2010), pp. 1415-1430, issn 1944-0049

Koek, M.; Jellema, R.; van der Greef, J.; Tas, A. & Hankemeier, T. (2011). Quantitative metabolomics based on gas chromatography mass spectrometry: status and perspectives. *Metabolomics*, Vol.7, No. 3, (Sep 2011), pp. 307-328, issn 1573-3882

Kojima, M.; Kamada-Nobusada, T.; Komatsu, H.; Takei, K.; Kuroha, T.; Mizutani, M.; Ashikari, M.; Ueguchi-Tanaka, M.; Matsuoka, M.; Suzuki, K. & Sakakibara, H. (2009). Highly sensitive and high-throughput analysis of plant hormones using MS-probe modification and liquid chromatography tandem mass spectrometry: An application for hormone profiling in *Oryza sativa*. *Plant and Cell Physiology*, Vol.50, No. 7, (Jul 2009), pp. 1201-1214, issn 0032-0781

Kozak, M.; Bocianowski, J.; Liersch, A.; Tartanus, M.g.; Bartkowiak-Broda, I.; Piotto, F. & Azevedo, R. (2011). Genetic divergence is not the same as phenotypic divergence. *Molecular Breeding*, Vol.28, No. 2, (May 2011), pp. 277-280, issn 1380-3743

Krueger, S.; Giavalisco, P.; Krall, L.; Steinhauser, M.C.; Bussis, D.; Usadel, B.; Flugge, U.I.; Fernie, A.R.; Willmitzer, L. & Steinhauser, D. (2011). A topological map of the compartmentalized *Arabidopsis thaliana* leaf metabolome. *Plos One,* Vol.6, No. 3, (Mar 2011), pp. 16, issn 1932-6203

Kusano, M.; Fukushima, A.; Redestig, H. & Saito, K. (2011a). Metabolomic approaches toward understanding nitrogen metabolism in plants. *Journal of Experimental Botany,* Vol.62, No. 4, (Feb 2011a), pp. 1439-1453, issn 0022-0957

Kusano, M.; Redestig, H.; Hirai, T.; Oikawa, A.; Matsuda, F.; Fukushima, A.; Arita, M.; Watanabe, S.; Yano, M.; Hiwasa-Tanase, K.; Ezura, H. & Saito, K. (2011b). Covering chemical diversity of genetically-modified tomatoes using metabolomics for objective substantial equivalence assessment. *Plos One,* Vol.6, No. 2, (Feb 2011b), issn 1932-6203

Lei, Z.; Huhman, D. & Sumner, L.W. (2011). Mass spectrometry strategies in metabolomics. *Journal of Biological Chemistry,* (Jun 2011), issn 0021-9258

Leiss, K.A.; Choi, Y.H.; Verpoorte, R. & Klinkhamer, P.G.L. (2010). An overview of NMR-based metabolomics to identify secondary plant compounds involved in host plant resistance. *Phytochemistry Reviews,* Vol.10, No. 2, (Jun 2010), pp. 205-216, issn 1568-7767

Lexa, M.; Genkov, T.; Malbeck, J.; Machackova, I. & Brzobohaty, B. (2003). Dynamics of endogenous cytokinin pools in tobacco seedlings: A modelling approach. *Annals of Botany,* Vol.91, No. 5, (Apr 2003), pp. 585-597, issn 0305-7364

Liu, G.N.; Zhu, Y.H. & Jiang, J.G. (2009). The metabolomics of carotenoids in engineered cell factory. *Applied Microbiology and Biotechnology,* Vol.83, No. 6, (Jul 2009), pp. 989-999, issn 0175-7598

Lu, W.; Bennett, B.D. & Rabinowitz, J.D. (2008). Analytical strategies for LC-MS-based targeted metabolomics. *Journal of Chromatography B-Analytical Technologies in the Biomedical and Life Sciences,* Vol.871, No. 2, (Aug 2008), pp. 236-242, issn 1570-0232

McDonald, J.C.; Duffy, D.C.; Anderson, J.R.; Chiu, D.T.; Wu, H.K.; Schueller, O.J.A. & Whitesides, G.M. (2000). Fabrication of microfluidic systems in poly(dimethylsiloxane). *Electrophoresis,* Vol.21, No. 1, (Jan 2000), pp. 27-40, issn 0173-0835

McDougall, G.; Martinussen, I. & Stewart, D. (2008). Towards fruitful metabolomics: High throughput analyses of polyphenol composition in berries using direct infusion mass spectrometry. *Journal of Chromatography B-Analytical Technologies in the Biomedical and Life Sciences,* Vol.871, No. 2, (Aug 2008), pp. 362-369, issn 1570-0232

Meyer, R.C.; Steinfath, M.; Lisec, J.; Becher, M.; Witucka-Wall, H.; Torjek, O.; Fiehn, O.; Eckardt, A.; Willmitzer, L.; Selbig, J. & Altmann, T. (2007). The metabolic signature related to high plant growth rate in *Arabidopsis thaliana. Proceedings of the National Academy of Sciences of the United States of America,* Vol.104, No. 11, (Mar 2007), pp. 4759-4764, issn 0027-8424

Mittler, R. & Blumwald, E. (2010). Genetic engineering for modern agriculture: challenges and perspectives, In: *Annual Review of Plant Biology,* Merchant, S., Briggs, W.R., Ort, D., (Eds) 443-462, isbn 978-0-8243-0661-8, Palo Alto, California

Moco, S.; Bino, R.J.; De Vos, R.C.H. & Vervoort, J. (2007). Metabolomics technologies and metabolite identification. *Trends in Analytical Chemistry,* Vol.26, No. 9, (Oct 2007), issn 01659936

Moco, S.; Schneider, B. & Vervoort, J. (2009). Plant micrometabolomics: The analysis of endogenous metabolites present in a plant cell or tissue. *Journal of Proteome Research,* Vol.8, No. 4, (Apr 2009), pp. 1694-1703, issn 1535-3893

Moing, A.; Aharoni, A.; Biais, B.; Rogachev, I.; Meir, S.; Brodsky, L.; Allwood, J.W.; Erban, A.; Dunn, W.B.; Kay, L.; de Koning, S.; de Vos, R.C.H.; Jonker, H.; Mumm, R.; Deborde, C.; Maucourt, M.; Bernillon, S.; Gibon, Y.; Hansen, T.H.; Husted, S.; Goodacre, R.; Kopka, J.; Schjoerring, J.K.; Rolin, D. & Hall, R.D. (2011). Extensive metabolic cross-talk in melon fruit revealed by spatial and developmental combinatorial metabolomics. *New Phytologist,* Vol.190, No. 3, (May 2011), pp. 683-696, issn 1469-8137

Moser, I.; Jobst, G. & Urban, G.A. (2002). Biosensor arrays for simultaneous measurement of glucose, lactate, glutamate, and glutamine. *Biosensors & Bioelectronics,* Vol.17, No. 4, (Apr 2002), pp. 297-302, issn 0956-5663

Mounet, F.; Moing, A.; Garcia, V.; Petit, J.; Maucourt, M.; Deborde, C.; Bernillon, S.; Le Gall, G.; Colquhoun, I.; Defernez, M.; Giraudel, J.-L.; Rolin, D.; Rothan, C. & Lemaire-Chamley, M. (2009). Gene and metabolite regulatory network analysis of early developing fruit tissues highlights new candidate genes for the control of tomato fruit composition and development. *Plant Physiology,* Vol.149, No. 3, (March 2009), pp. 1505-1528, issn 0032-0889

Mulas, G.; Galaffu, M.G.; Pretti, L.; Nieddu, G.; Mercenaro, L.; Tonelli, R. & Anedda, R. (2011). NMR analysis of seven selections of Vermentino grape berry: metabolites composition and development. *Journal of Agricultural and Food Chemistry,* Vol.59, No. 3, (Feb 2011), pp. 793-802, issn 0021-8561

Myles, S.; Peiffer, J.; Brown, P.J.; Ersoz, E.S.; Zhang, Z.W.; Costich, D.E. & Buckler, E.S. (2009). Association mapping: Critical considerations shift from genotyping to experimental design. *Plant Cell,* Vol.21, No. 8, (Aug 2009), pp. 2194-2202, issn 1040-4651

Nagele, T.; Kandel, B.A.; Frana, S.; Meissner, M. & Heyer, A.G. (2011). A systems biology approach for the analysis of carbohydrate dynamics during acclimation to low temperature in *Arabidopsis thaliana. Febs Journal,* Vol.278, No. 3, (Feb 2011), pp. 506-518, issn 1742-464X

Nagrath, S.; Sequist, L.V.; Maheswaran, S.; Bell, D.W.; Irimia, D.; Ulkus, L.; Smith, M.R.; Kwak, E.L.; Digumarthy, S.; Muzikansky, A.; Ryan, P.; Balis, U.J.; Tompkins, R.G.; Haber, D.A. & Toner, M. (2007). Isolation of rare circulating tumour cells in cancer patients by microchip technology. *Nature,* Vol.450, No. 7173, (Dec 2007), pp. 1235-1239, issn 0028-0836

Oikawa, A.; Matsuda, F.; Kusano, M.; Okazaki, Y. & Saito, K. (2008). Rice metabolomics. *Rice,* Vol.1, No. 1, (Sep 2008), pp. 63-71, issn 1939-8425

Okada, T.; Afendi, F.M.; Altaf-Ul-Amin, M.; Takahashi, H.; Nakamura, K. & Kanaya, S. (2010). Metabolomics of medicinal plants: The importance of multivariate analysis of analytical chemistry data. *Current Computer-Aided Drug Design,* Vol.6, No. 3, (Sep 2010), pp. 179-196, issn 1573-4099

Oliver, S.G.; Winson, M.K.; Kell, D.B. & Baganz, F. (1998). Systematic functional analysis of the yeast genome. *Trends in Biotechnology,* Vol.16, No. 9, (Sep 1998), pp. 373-378, issn 0167-7799

Palama, T.L.; Khatib, A.; Choi, Y.H.; Payet, B.; Fock, I.; Verpoorte, R. & Kodja, H. (2009). Metabolic changes in different developmental stages of *Vanilla planifolia* pods. *Journal of Agricultural and Food Chemistry,* Vol.57, No. 17, (Sep 2009), pp. 7651-7658, issn 0021-8561

Pichersky, E. & Lewinsohn, E. (2011). Convergent evolution in plant specialized metabolism, In: *Annual Review of Plant Biology,* 549-566, Annual Reviews, isbn 978-0-8243-0662-5, Palo Alto, California

Poorter, H.; Niinemets, U.; Walter, A.; Fiorani, F. & Schurr, U. (2010). A method to construct dose-response curves for a wide range of environmental factors and plant traits by means of a meta-analysis of phenotypic data. *Journal of Experimental Botany,* Vol.61, No. 8, (May 2010), pp. 2043-2055, issn 0022-0957

Poorter, H.; Walter, A.; Fiorani, F.; Schurr, U. & Niinemets, U. (2009). Meta-phenomics: Building a unified framework for interpreting plant growth responses to diverse environmental variables. *Comparative Biochemistry and Physiology a-Molecular & Integrative Physiology,* Vol.153A, No. 2, (Jun 2009), pp. S224-S224, issn 1095-6433

Ramautar, R.; Mayboroda, O.A.; Somsen, G.W. & de Jong, G.J. (2011). CE-MS for metabolomics: Developments and applications in the period 2008-2010. *Electrophoresis,* Vol.32, No. 1, (Jan 2011), pp. 52-65, issn 0173-0835

Rashed, M.S.; Bucknall, M.P.; Little, D.; Awad, A.; Jacob, M.; Alamoudi, M.; Alwattar, M. & Ozand, P.T. (1997). Screening blood spots for inborn errors of metabolism by electrospray tandem mass spectrometry with a microplate batch process and a computer algorithm for automated flagging of abnormal profiles. *Clinical Chemistry,* Vol.43, No. 7, (Jul 1997), pp. 1129-1141, issn 0009-9147

Redestig, H. & Costa, I.G. (2011). Detection and interpretation of metabolite-transcript coresponses using combined profiling data. *Bioinformatics,* Vol.27, No. 13, (Jul 2011), pp. I357-I365, issn 1367-4803

Redestig, H.; Kusano, M.; Fukushima, A.; Matsuda, F.; Saito, K. & Arita, M. (2010). Consolidating metabolite identifiers to enable contextual and multi-platform metabolomics data analysis. *BMC Bioinformatics,* Vol.11, (Apr 2010), issn 1471-2105

Roda, A.; Pasini, P.; Mirasoli, M.; Michelini, E. & Guardigli, M. (2004). Biotechnological applications of bioluminescence and chemiluminescence. *Trends in Biotechnology,* Vol.22, No. 6, (Jun 2004), pp. 295-303, issn 0167-7799

Ruane, J. & Sonnino, A. (2011). Agricultural biotechnologies in developing countries and their possible contribution to food security. *Journal of Biotechnology,* Vol.In Press, Corrected Proof, (2011), issn 0168-1656

Ryan, D. & Robards, K. (2006). Analytical chemistry considerations in plant metabolomics. *Separation and Purification Reviews,* Vol.35, No. 4, (Nov 2006), pp. 319-356, issn 1542-2119

Saito, K. & Matsuda, F. (2010). Metabolomics for Functional Genomics, Systems Biology, and Biotechnology, In: *Annual Review of Plant Biology, Vol 61,* 463-489, Annual Reviews, isbn 1543-5008, Palo Alto, California

Salse, J. (2004). New in silico insight into the synteny between rice (*Oryza sativa* L.) and maize (*Zea mays* L.) highlights reshuffling and identifies new duplications in the rice genome (Vol.38, (May 2004), pp. 396-409). *Plant Journal,* Vol.38, No. 5, (Jun 2004), pp. 873-873, issn 0960-7412

Sanchez-Perez, E.M.; Iglesias, M.J.; Lopez-Ortiz, F.; Sanchez-Perez, I. & Martinez-Galera, M. (2010). Study of the suitability of HRMAS NMR for metabolic profiling of tomatoes: Application to tissue differentiation and fruit ripening. *Food Chemistry,* Vol.122, No. 3, (Oct 2010), pp. 877-887, issn 0308-8146

Sanchez, D.H.; Pieckenstain, F.L.; Szymanski, J.; Erban, A.; Bromke, M.; Hannah, M.A.; Kraemer, U.; Kopka, J. & Udvardi, M.K. (2011). Comparative functional genomics of salt stress in related model and cultivated plants identifies and overcomes limitations to translational genomics. *Plos One,* Vol.6, No. 2, (Feb 2011), issn 1932-6203

Sawada, Y.; Akiyama, K.; Sakata, A.; Kuwahara, A.; Otsuki, H.; Sakurai, T.; Saito, K. & Hirai, M.Y. (2009). Widely targeted metabolomics based on large-scale ms/ms data for elucidating metabolite accumulation patterns in plants. *Plant and Cell Physiology,* Vol.50, No. 1, (Jan 2009), pp. 37-47, issn 0032-0781

Schauer, N.; Semel, Y.; Roessner, U.; Gur, A.; Balbo, I.; Carrari, F.; Pleban, T.; Perez-Melis, A.; Bruedigam, C.; Kopka, J.; Willmitzer, L.; Zamir, D. & Fernie, A.R. (2006). Comprehensive metabolic profiling and phenotyping of interspecific introgression lines for tomato improvement. *Nature Biotechnology,* Vol.24, No. 4, (Apr 2006), pp. 447-454, issn 1087-0156

Shaw, R.A.; Rigatto, C.; Reslerova, M.; Ying, S.L.; Man, A.; Schattka, B.; Battrell, C.F.; Matthewson, J. & Mansfield, C. (2009). Toward point-of-care diagnostic metabolic fingerprinting: quantification of plasma creatinine by infrared spectroscopy of microfluidic-preprocessed samples. *Analyst,* Vol.134, No. 6, (Jun 2009), pp. 1224-1231, issn 0003-2654

Shepherd, L.V.T.; Fraser, P. & Stewart, D. (2011). Metabolomics: a second-generation platform for crop and food analysis. *Bioanalysis,* Vol.3, No. 10, (May 2011), pp. 1143-1159, issn 1757-6180

Shintu, L.; Le Gall, G. & Colquhoun, I.J. (2009). Metabolomics and the detection of unintended effects in genetically modified crops, In: *Plant-derived Natural Products,* Osbourn, A.E., Lanzotti, V., (Eds) 505-531, Springer New York, isbn 978-0-387-85497-7, New-York

Shroff, R.; Rulisek, L.; Doubsky, J. & Svatos, A. (2009). Acid-base-driven matrix-assisted mass spectrometry for targeted metabolomics. *Proceedings of the National Academy of Sciences of the United States of America,* Vol.106, No. 25, (Jun 2009), pp. 10092-10096, issn 0027-8424

Skogerson, K.; Harrigan, G.G.; Reynolds, T.L.; Halls, S.C.; Ruebelt, M.; Iandolino, A.; Pandravada, A.; Glenn, K.C. & Fiehn, O. (2010). Impact of genetics and environment on the metabolite composition of maize grain. *Journal of Agricultural and Food Chemistry,* Vol.58, No. 6, (Mar 2010), pp. 3600-3610, issn 0021-8561

Smith, J.E. & Bluhm, B.H. (2011). Metabolic fingerprinting in *Fusarium verticillioides* to determine gene function, In: *Fungal Genomics : Methods and Protocols*, Xu, J.R., Bluhm, B.H.H., (Eds) 237-247, Humana Press, isbn 978-1-61779-039-3, Heidelberg

Stitt, M.; Gibon, Y.; Lunn, J.E. & Piques, M. (2007). Multilevel genomics analysis of carbon signalling during low carbon availability: coordinating the supply and utilisation of carbon in a fluctuating environment. *Functional Plant Biology*, Vol.34, No. 6, (Jun 2007), pp. 526-549, issn 1445-4408

Sulpice, R.; Trenkamp, S.; Steinfath, M.; Usadel, B.; Gibon, Y.; Witucka-Wall, H.; Pyl, E.T.; Tschoep, H.; Steinhauser, M.C.; Guenther, M.; Hoehne, M.; Rohwer, J.M.; Altmann, T.; Fernie, A.R. & Stitt, M. (2010). Network analysis of enzyme activities and metabolite levels and their relationship to biomass in a large panel of Arabidopsis accessions. *Plant Cell*, Vol.22, No. 8, (Aug 2010), pp. 2872-2893, issn 1040-4651

Sumner, L. (2010). Recent advances in plant metabolomics and greener pastures. *F1000 Biology Reports*, Vol.2 (Jan 2010), pp. 7, issn 1757-594X

Takatsy, G. (1955). The use of spiral loops in serological and virological micro-methods. *Acta Microbiologica Academiae Scientiarum Hungaricae*, Vol. 3, No. 1-2, (Jun 1955), pp. 191-202, issn 0001-6187

Tarpley, L.; Duran, A.L.; Kebrom, T.H. & Sumner, L.W. (2005). Biomarker metabolites capturing the metabolite variance present in a rice plant developmental period. *BMC Plant Biology*, Vol.5, (May 2005), issn 1471-2229

Tester, M. & Langridge, P. (2010). Breeding technologies to increase crop production in a changing world. *Science*, Vol.327, No. 5967, (Feb 2010), pp. 818-822, issn 0036-8075

Thiele, B.; Fullner, K.; Stein, N.; Oldiges, M.; Kuhn, A.J. & Hofmann, D. (2008). Analysis of amino acids without derivatization in barley extracts by LC-MS-MS. *Analytical and Bioanalytical Chemistry*, Vol.391, No. 7, (Aug 2008), pp. 2663-2672, issn 1618-2642

Tikunov, Y.M.; Verstappen, F.W.A. & Hall, R.D. (2007). Metabolomic profiling of natural volatiles headspace trapping: GC-MS, In: *Metabolomics*, Weckwerth, W., (Ed.) 39-53, Humana Press, isbn 1064-3745, Totowa, New Jersey

Tomas, R.; Kleparnik, K. & Foret, F. (2008). Multidimensional liquid phase separations for mass spectrometry. *Journal of Separation Science*, Vol.31, No. 11, (Jun 2008), pp. 1964-1979, issn 1615-9306

Trethewey, R.N. (2004). Metabolite profiling as an aid to metabolic engineering in plants. *Current Opinion in Plant Biology*, Vol.7, No. 2, (Apr 2004), pp. 196-201, issn 1369-5266

Trufelli, H.; Palma, P.; Famiglini, G. & Cappiello, A. (2011). An overview of matrix effects in liquid chromatography-mass spectrometry. *Mass Spectrometry Reviews*, Vol.30, No. 3, (May-Jun 2011), pp. 491-509, issn 0277-7037

Urbanczyk-Wochniak, E.; Baxter, C.; Kolbe, A.; Kopka, J.; Sweetlove, L.J. & Fernie, A.R. (2005). Profiling of diurnal patterns of metabolite and transcript abundance in potato (*Solanum tuberosum*) leaves. *Planta*, Vol.221, No. 6, (Aug 2005), pp. 891-903, issn 0032-0935

Usadel, B.; Blasing, O.E.; Gibon, Y.; Retzlaff, K.; Hoehne, M.; Gunther, M. & Stitt, M. (2008). Global transcript levels respond to small changes of the carbon status during progressive exhaustion of carbohydrates in Arabidopsis rosettes. *Plant Physiology*, Vol.146, No. 4, (Apr 2008), pp. 1834-1861, issn 0032-0889

Vengasandra, S.; Cai, Y.K.; Grewell, D.; Shinar, J. & Shinar, R. (2010). Polypropylene CD-organic light-emitting diode biosensing platform. *Lab on a Chip,* Vol.10, No. 8, 2010), pp. 1051-1056, issn 1473-0197

Vidal, M. (2009). A unifying view of 21st century systems biology. *Febs Letters,* Vol.583, No. 24, (Dec 2009), pp. 3891-3894, issn 0014-5793

Villiers, F.; Ducruix, C.; Hugouvieux, V.; Jarno, N.; Ezan, E.; Garin, J.; Junot, C. & Bourguignon, J. (2011). Investigating the plant response to cadmium exposure by proteomic and metabolomic approaches. *Proteomics,* Vol.11, No. 9, (May 2011), pp. 1650-1663, issn 1615-9861

Wang, Q.Z.; Wu, C.Y.; Chen, T.; Chen, X. & Zhao, X.M. (2006). Integrating metabolomics into a systems biology framework to exploit metabolic complexity: strategies and applications in microorganisms. *Applied Microbiology and Biotechnology,* Vol.70, No. 2, (Mar 2006), pp. 151-161, issn 0175-7598

Wang, X.Y.; Gowik, U.; Tang, H.B.; Bowers, J.E.; Westhoff, P. & Paterson, A.H. (2009). Comparative genomic analysis of C4 photosynthetic pathway evolution in grasses. *Genome Biology,* Vol.10, No. 6, (Jun 2009), issn 1474-760X

Ward, J.; Baker, J.M.; Miller, S.; Deborde, C.; Maucourt, M.; Biais, B.; Rolin, D.; Moing, A.; Moco, S.; Vervoort, J.; Lommen, A.; Schäfer, H.; Humpfer, E. & Beale, M.H. (2010). An inter-laboratory comparison demonstrates that [1H]-NMR metabolite fingerprinting is a robust technique for collaborative plant metabolomic data collection. *Metabolomics,* Vol.6, No. 2, (Jun 2010), pp. 263-273, issn 1573-3882

Weckwerth, W. (2007). *Metabolomics. Methods and protocols.* Humana Press, isbn 1-59745-244-0, Totowa, USA

Weckwerth, W. (2011). Unpredictability of metabolism-the key role of metabolomics science in combination with next-generation genome sequencing. *Analytical and Bioanalytical Chemistry,* Vol.400, No. 7, (Jun 2011), pp. 1967-1978, issn 1618-2642

Weckwerth, W.; Loureiro, M.E.; Wenzel, K. & Fiehn, O. (2004). Differential metabolic networks unravel the effects of silent plant phenotypes. *Proceedings of the National Academy of Sciences of the United States of America,* Vol.101, No. 20, (May 2004), pp. 7809-7814, issn 0027-8424

Werner, E.; Heilier, J.F.; Ducruix, C.; Ezan, E.; Junot, C. & Tabet, J.C. (2008). Mass spectrometry for the identification of the discriminating signals from metabolomics: Current status and future trends. *Journal of Chromatography B-Analytical Technologies in the Biomedical and Life Sciences,* Vol.871, No. 2, (Aug 2008), pp. 143-163, issn 1570-0232

Westerhoff, H.V.; Winder, C.; Messiha, H.; Simeonidis, E.; Adamczyk, M.; Verma, M.; Bruggeman, F.J. & Dunn, W. (2009). Systems Biology: The elements and principles of Life. *Febs Letters,* Vol.583, No. 24, (Dec 2009), pp. 3882-3890, issn 0014-5793

Whitesides, G.M. (2006). The origins and the future of microfluidics. *Nature,* Vol.442, No. 7101, (Jul 2006), pp. 368-373, issn 0028-0836

Williams, T.C.R.; Poolman, M.G.; Howden, A.J.M.; Schwarzlander, M.; Fell, D.A.; Ratcliffe, R.G. & Sweetlove, L.J. (2010). A genome-scale metabolic model accurately predicts fluxes in central carbon metabolism under stress conditions. *Plant Physiology,* Vol.154, No. 1, (Sep 2010), pp. 311-323, issn 0032-0889

Wishart, D.S. (2008). Metabolomics: applications to food science and nutrition research. *Trends in Food Science & Technology,* Vol.19, No. 9, 2008), pp. 482-493, issn 0924-2244

Wurm, M.; Schopke, B.; Lutz, D.; Muller, J. & Zeng, A.P. (2010). Microtechnology meets systems biology: The small molecules of metabolome as next big targets. *Journal of Biotechnology,* Vol.149, No. 1-2, (Aug 2010), pp. 33-51, issn 0168-1656

Yu, J.M.; Holland, J.B.; McMullen, M.D. & Buckler, E.S. (2008). Genetic design and statistical power of nested association mapping in maize. *Genetics,* Vol.178, No. 1, (Jan 2008), pp. 539-551, issn 0016-6731

Zhang, N.Y.; Gur, A.; Gibon, Y.; Sulpice, R.; Flint-Garcia, S.; McMullen, M.D.; Stitt, M. & Buckler, E.S. (2010). Genetic analysis of central carbon metabolism unveils an amino acid substitution that alters maize NAD-dependent isocitrate dehydrogenase activity. *Plos One,* Vol.5, No. 3, (Apr 2010), issn 1932-6203

Zhang, Y.; Thiele, I.; Weekes, D.; Li, Z.W.; Jaroszewski, L.; Ginalski, K.; Deacon, A.M.; Wooley, J.; Lesley, S.A.; Wilson, I.A.; Palsson, B.; Osterman, A. & Godzik, A. (2009). Three-dimensional structural view of the central metabolic network of *Thermotoga maritima. Science,* Vol.325, No. 5947, (Sep 2009), pp. 1544-1549, issn 0036-8075

10

Metabolomics of Endophytic Fungi Producing Associated Plant Secondary Metabolites: Progress, Challenges and Opportunities

Souvik Kusari and Michael Spiteller
Institute of Environmental Research (INFU) of the Faculty of Chemistry, Chair of Environmental Chemistry and Analytical Chemistry, TU Dortmund, Dortmund, Germany

1. Introduction

Microorganisms are indispensable for every aspect of human life, in fact all life on earth, although they cannot be seen by the naked eye. Since time immemorial, every process in the biosphere has been affected by the apparently unending ability of microbes to renovate the world around them. More recently, many discoveries have been made in isolating a special class of microorganisms, mainly fungi but also bacteria, commonly called endophytes, which have been shown to have the natural potential for accumulation of various bioactive metabolites which may directly or indirectly be used as therapeutic agents against a plethora of maladies (Kusari & Spiteller, 2010, 2011). Bioprospecting endophytes have led to exciting possibilities to explore and utilize their potential. Several bioprospecting strategies might be employed in order to discover potent endophytes with desirable traits (Figure 1). These include randomly sampling different plants from any population to isolate the associated endophytes, or first performing a detailed study of an ecosystem in order to determine its features with regard to its natural population of plant species, their relationship with the environment, soil composition, and biogeochemical cycles. Another approach is to evaluate the evolutionary relatedness among groups of plants at a particular sampling site, correlating to species, genus, and populations, through morphological data matrices and molecular sequencing, followed by isolation of endophytes from the desired plants. Traditional medicinal plants are also bioprospected for endophytes, especially for the ones capable of producing one or more of the bioactive secondary metabolites present in the host plants. Finally, the valuable information obtained using the different bioprospecting schemes can be pooled together, comparatively evaluated, and stored for further use applying suitable data mining approaches.

2. What is an endophytic fungus?

Endophytes are microorganisms that internally infect living plant tissues without causing any visible manifestation of disease, and live in mutualistic association with plants for at least a part of their life cycle (Bacon & White, 2000). The term 'endophyte' (Gr. *endon*, within; *phyton*, plant) was first contrived by de Bary (1866). All types of microorganisms (fungi,

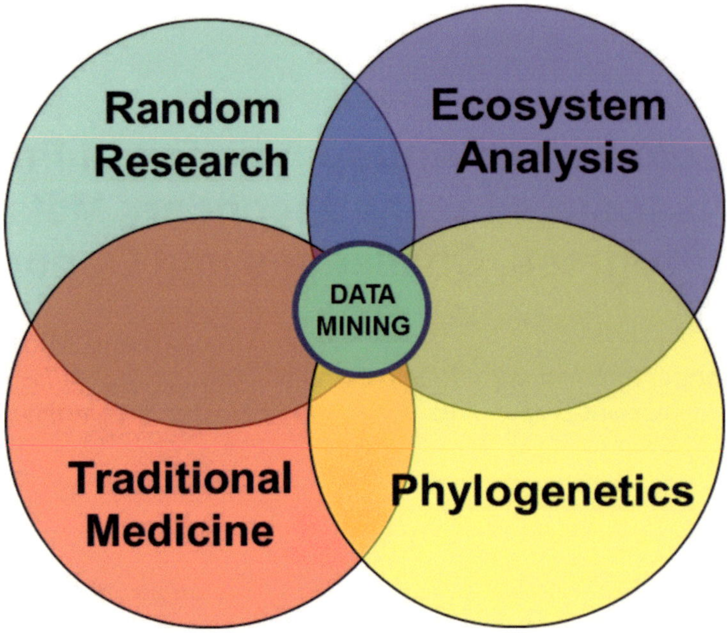

Fig. 1. Different bioprospecting strategies that might be utilized in order to discover novel or competent endophytes with desirable features.

bacteria, and actinomycetes) have been discovered as endophytes. The most frequently encountered endophytes are fungi (Staniek et al., 2008). Fungal endophytes constitute an inexplicably diverse group of polyphyletic fungi ubiquitous in plants, and maintain an indiscernible dynamic relationship with their hosts for at least a part of their life cycle (Figure 2a,b). The existence of fungi inside the tissues of asymptomatic plants has been known since the end of the nineteenth century (Guerin, 1898). Evidence of plant-associated microorganisms found in the fossilized tissues of stems and leaves has revealed that endophyte-plant associations may have evolved from the time higher plants first appeared on the earth (Redecker et al., 2000). However, except for some infrequent studies, it was not until the end of the twentieth century that fungal endophytes began to receive more attention from scientists. Since endophytes were first described in the Darnel (Freeman, 1904), various investigators have isolated endophytes from different plant species. These discoveries led to a worldwide search for novel endophytes for the better understanding and applicability of such a promising group of microorganisms. On the one hand, the ecological aspects of endophytic fungi such as host range, evolutionary relatedness, infection, colonization, transmission patterns, tissue specificity, and mutualistic fitness benefits have been investigated relating to a plethora of plants (Arnold et al., 2003, 2007, Arnold, 2005, 2007; Stone et al., 2004; Schulz &Boyle, 2005; Rodriguez et al., 2009) (Figure 2c). On the other hand, many discoveries have been made in isolating endophytic fungi, which have been shown to have the potential for *de novo* synthesis of various bioactive metabolites that may directly or indirectly be used as therapeutic agents against numerous ailments (Strobel and Daisy, 2003; Strobel et al., 2004; Zhang et al., 2006; Gunatilaka, 2006; Staniek et al., 2008; Suryanarayanana et al., 2009; Aly et al., 2010; Kharwar et al., 2011; Kusari & Spiteller, 2010, 2011).

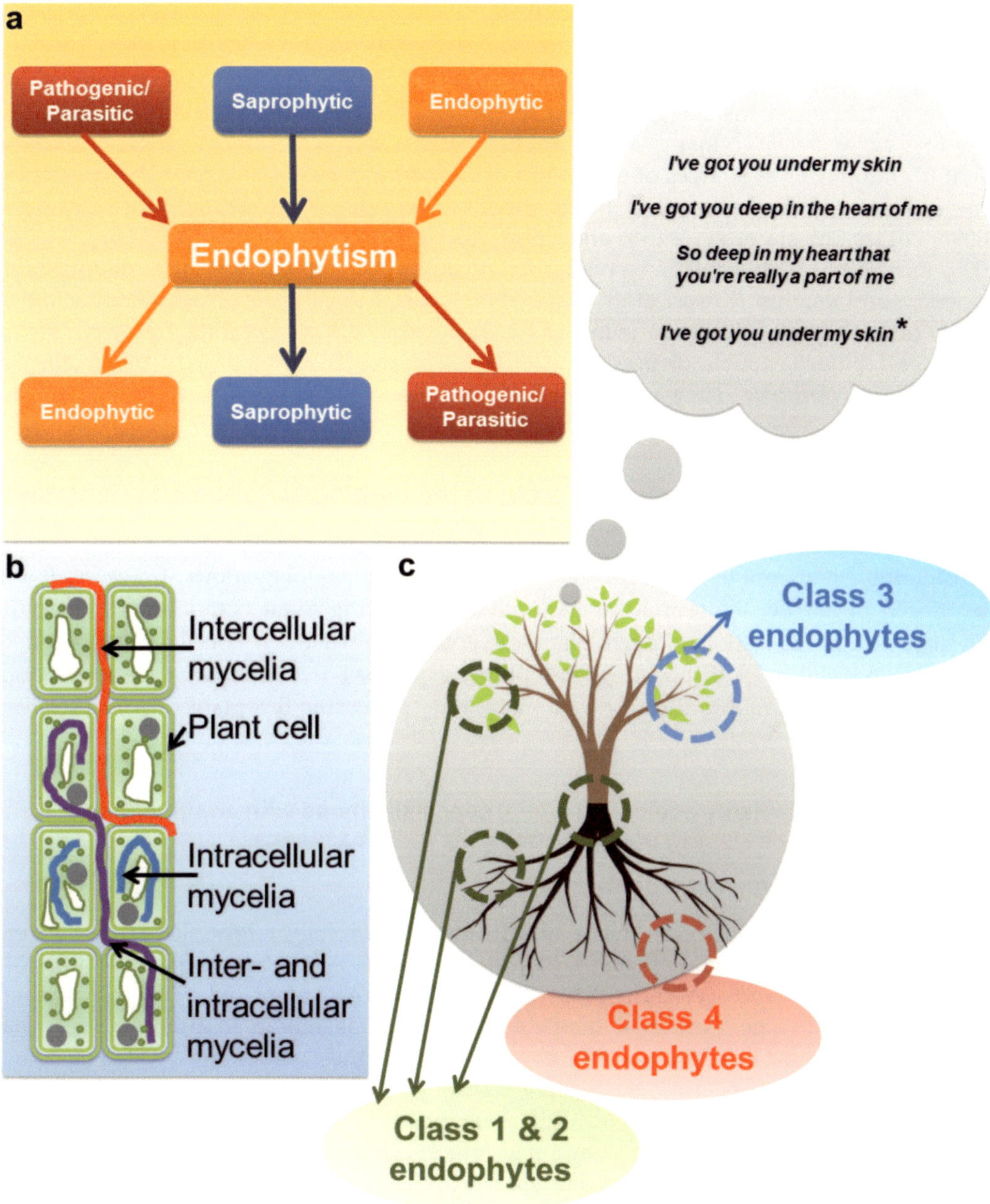

Fig. 2. A diagrammatic representation of endophytic fungal association with their host plants. (a) Possible life histories of endophytic fungi pre- and post-'endophytism'. (b) Different localization patterns of fungal endophytes within plant tissues. (c) Location of the different classes of endophytes (according to Rodriguez et al., 2009). *An artistic rendition of the possible thoughts of a plant about its endophytic counterparts, presented in memory of Cole Albert Porter (1891-1964) and Frank Sinatra (1915-1998).

3. Endophytic fungi producing host plant secondary metabolites

The possibility that endophytes biosynthesize associated plant compounds was first comprehended and published by Stierle et al. (1993), following the highly heralded discovery of endophytic *Taxomyces andreanae* that produces the multi-billion dollar anticancer compound Taxol® (generic name: paclitaxel), which was isolated from the Pacific yew tree *Taxus brevifolia*. Inspired by this discovery, numerous efforts have been made to identify endophytes as sources of associated plant natural products. Endophytes producing antineoplastic camptothecin (CPT) and its structural analogs (Puri et al., 2005; Kusari et al., 2009b, 2011b; Shweta et al., 2010), anticancer pro-drugs podophyllotoxin (Eyberger et al., 2006; Puri et al., 2006) and deoxypodophyllotoxin (Kusari et al., 2009a), antidepressants hypericin and emodin (Kusari et al., 2008, 2009c), and natural insecticides azadirachtin A and B (Kusari et al., 2011c) are some of the interesting discoveries that followed. Novel fungal endophytes capable of producing other associated plant secondary metabolites of therapeutic importance, such as artemisinin, morphine, cannabinoids, and many more, remain to be discovered and characterized.

4. Rationale for plant selection to provide the best opportunities for isolating endophytic fungi producing associated plant natural products

Considering the enormous numbers and the diversity of plants, ingenious strategies should be utilized to narrow the search for endophytes producing plant compounds. A specific rationale for the collection of each plant for endophyte isolation could be proposed to maximize possibility of discovering endophytes equipped with the capacity to produce associated plant natural products. Several hypotheses governing this plant selection strategy might be exploited.

4.1 Plants from inimitable ecological niche, especially those with an uncommon morphology and possessing unusual strategies for subsistence

4.1.1 Case study: *Hypericum perforatum*

Plants from a distinct ecological niche or with unusual biology might also harbor potent endophytes. A fine example of such a plant is *Hypericum perforatum*, which is commonly called St. John's wort (Wichtl, 1986) (Figure 3a). This plant is a pseudogamous, facultatively apomictic, perennial medicinal plant that is native to Europe, West and South Asia, North Africa, North America, and Australia (Hickey and King, 1981; Wichtl, 1986). In general, *Hypericum* has always been a very important medicinal plant occupying a significant place in ancient history. Pedanius Dioscorides, the foremost ancient Greek herbalist, mentioned four species of *Hypericum - uperikon, askuron, androsaimon*, and *koris*, which he recommended for sciatica, "when drunk with 2 heim of hydromel (honey water)" (Gunther, 1959). *H. perforatum* has also been in use, at least from the time of ancient Greece (Tammaro and Xepapadakis, 1986), as an antidepressant, in healing of wounds and menstrual disorders, due to the presence of the-then unknown bioactive compounds in the plant. This plant has also found historical use in India, China, Egypt and many countries of Europe, where the tribal peoples have been burning this plant to represent sun, light, vitality and strength (Hickey & King, 1981). We know now that this plant produces the widely used antidepressant compound hypericin (Brockmann et al., 1950). Hypericin is a photodynamic compound (Kubin et al., 2005) which is localized and probably also synthesized in the 'dark glands' (Onelli et al., 2002), which are small specialized glandular structures dispersed over all above-ground parts of the plant (flowers, capsules, leaves, stems) but not in the roots (Hölzl et al., 2003) (Figure 3b,c). Therefore, using the rationale that a plant with such an uncommon biology (dark glands) for protecting itself from the photodynamic effects of its own metabolite might also contain endophytes that have been evolutionarily co-adapted to accumulate the same or similar molecules, we undertook bioprospecting endophytic fungi from *H. perforatum* sampled from various populations across Europe and the Himalayan region. This led to the discovery of an endophytic fungus associated with this plant capable of producing hypericin and emodin in axenic cultures (Kusari et al., 2008, 2009c).

Fig. 3. *Hypericum perforatum* as a suitable example of bioprospecting plants with an unusual morphology, possessing unusual strategies for existence, and from unique environments. (a) Wild *H. perforatum* growing at the Himalayan environments (Harwan, Jammu and Kashmir, India) from where the hypericin- and emodin-producing endophyte was isolated. (Photograph courtesy of M. Spiteller). (b) A representative leaf of *H. perforatum* where the dark glands (arrows) can be seen. (c) A brightfield microscopic image of the leaf adaxial surface where the dark glands (arrow) can be seen as black spots (using Leica S8 APO Greenough stereo microscope, Leica Microsystems GmbH, Wetzlar, Germany; scale = 1 mm. The image was captured using Leica EC3 digital camera and processed using the Leica Application Suite LAS EZ ver. 1.6.0).

4.2 Plants that have an ethno-botanical history that is associated with the specific practices or applications of interest

4.2.1 Case study: *Juniperus* species

Juniperus plants (Figure 4) serve an excellent example to describe this rationale, which contain the therapeutically important anticancer lignans podophyllotoxin and deoxypodophyllotoxin (Hartwell et al., 1953). This species was in use as early as in the first century A.D., when Gaius Plinius Secundus mentioned that the smaller species of *Juniperus* could be used, among other things, to stop tumors or swelling (Imbert, 1998). The use of the oil of *Juniperus* species (*J. sabina, J. phoenicea* and *J. communis*) for the treatment of ulcers, carbuncles and leprosy was also mentioned by Dioscorides (Gunther, 1959). Generally, the dried needles, called *savin*, or the derived oil was used. In 47 A.D., Scribonius Largus wrote that *savin* oil was used to soften "hard female genital parts" (Sconocchia, 1983). *Savin* was later also used to treat uterine carcinoma, venereal warts and polyps. Based on such historical use by indigenous people, we recently isolated and characterized endophytic fungi harbored in *Juniperus* plants sampled from the natural populations in Dortmund and Haltern, Germany, and Jammu and Kashmir, India. This resulted in the discovery of a deoxypodophyllotoxin-producing endophytic fungus harbored in *J. communis* (Kusari et al., 2009a).

Fig. 4. *Juniperus* as an example of bioprospecting plants having an ethnobotanical history associated with the specific practices or applications of interest. Some *Juniperus* species growing at Rombergpark, Dortmund, Germany are shown. (Photographs courtesy of S. Kusari and M. Spiteller).

4.3 Endemic or endangered plants

4.3.1 Case study I: *Camptotheca acuminata*

Endemic plants are frequently consorted with ecological peculiarities or typical locations that are geographically distinct. Many of these plants are getting vulnerable owing to their therapeutic, agricultural, environmental, and commercial value. These distinct plants might also harbor a plethora of unique endophytes. One out of many examples is the plant *Camptotheca acuminata* (Figure 5a). This plant grows in mainland China, and is commonly called the 'happy tree', which is a direct translation of the Chinese word '*Xi Shu*'. *Camptotheca* was first recorded in 1848 and scientifically described and named by Decaisne (1873). The genus name *Camptotheca* is from the Greek *Campto* (meaning, bent or curved) and *theca* (meaning, a case) referring to the anthers, which are bent inward. The species name *acuminata* is derived from *acuminate*, which refers to the tips of leaves. This plant has been in use as traditional medicine in China for treatment of psoriasis, liver and stomach ailments and the common cold (Sung et al., 1998). The present application of this plant is on account of the fact that it contains substantial quantities of an important antineoplastic drug, namely camptothecin (CPT). This plant is uprooted and harvested by various sectors, including medical groups, pharmaceutical companies and scientists from around the world, to isolate CPT for numerous purposes (Lorence & Nessler, 2004; Sankar-Thomas, 2010). In addition to the difficulties of the practical total synthesis of this natural compound, the unpredictable problems of nature such as erratic weather and pests (Kusari et al., 2011d) have rendered this plant species vulnerable to extinction. As such, in 2000 and again in 2006, *C. acuminata* was proposed for protection in the CITES (Convention for International Trade in Endangered Species), World Conservation Monitoring Centre, appendix II (Anonymous, 2000, 2006). This appendix lists species that are not necessarily now threatened with extinction but that may become so unless trade is closely controlled. There are, of course, some nurseries growing *C. acuminata* for commercial purposes. These few nurseries, however, cannot meet the demand for CPT production (Sankar-Thomas, 2010). Furthermore, the yields of CPT from field trees vary widely and depend on factors that are difficult to control. For instance, plant diseases such as leaf spot and root rot are some of the major fungal diseases that can limit the cultivation of *Camptotheca* plants (Li et al., 2005) and diminish the production of CPT. Cultivation of *Camptotheca* plants is limited to subtropical climates and it takes about ten years for plants to produce a stable fruit yield (Li et al., 2005; Sankar-Thomas, 2010). The combination of a high demand for CPT and its scarcity from natural plant sources has, therefore, led to a different strategy of bioprospecting the endophytic fungi associated with the *C. acuminata* as alternate sources of CPT and related metabolites (Kusari et al., 2009b, 2011b).

Fig. 5. *Camptotheca acuminata* and *Nothapodytes nimmoniana* as fitting examples of bioprospecting endemic or endangered plants. (a) *C. acuminata* growing at the Southwest Forestry University campus, Kunming, Yunnan Province, China. (b) *N. nimmoniana* growing at the Western Ghats, India. (Photographs courtesy of M. Spiteller).

4.3.2 Case study II: *Nothapodytes nimmoniana*

Another plant containing CPT, *Nothapodytes nimmoniana* of Indian origin (Figure 5b) has also been subjected to extensive harvesting resulting in vulnerability of the plant. Unlike *C. acuminata*, there are no major commercial plantations of *N. nimmoniana* (Shaanker et al., 2008). This has led to harvesting of the plants sampled directly from the wild populations in India to meet the present and projected demands of CPT. In fact, it is estimated that in the last decade alone, there has been at least 20% decline in the population of *N. nimmoniana*, leading to the red-listing of this species (Kumar & Ved, 2000; Hombe Gowda et al., 2002). Therefore, a need for the preservation of this endangered plant species as well as to ensure a continuous supply of CPT has been felt. Endophyte research on this plant has yielded notable results (Puri et al., 2005).

4.4 Plants growing in areas of abundant biodiversity

4.4.1 Case study: *Azadirachta indica*

Plants growing in areas where a multitude of biotic and abiotic factors play essential roles and generating bioactive natural products might harbor diverse endophytic population. *Azadirachta indica* A. Juss. (synonym *Melia azadirachta*), commonly known as the Indian neem or Indian lilac (Butterworth, 1968) is well known in India and its neighboring countries for more than 2000 years as one of the most versatile medicinal plants growing abundantly in regions having high biodiversity of plants (Biswas et al., 2002). Traditionally, neem-based formulations have been used to cure a variety of ailments such as fever, pain, leprosy, and malaria in *Ayurvedic* and *Unani* medical treatments, but the most striking property of neem tree reported to date is its insect-repellent property (Veitch et al., 2008). It is now established that neem plants contain the natural product azadirachtin (i.e., azadirachtin A and structural analogues) that distresses insects as an antifeedent and insect growth regulator (Lay et al., 1993). Numerous natural bioactive compounds have been isolated and characterized from different parts of the neem tree, and new natural products are being discovered every year. Thus, it would seem that *A. indica* is a potential natural resource sheltering unique and competent endophytes having a multitude of desirable traits. As such, neem plants have not only been studied for their endophytic microflora concerning the composition, diversity, and distribution of endophytic microbes within the plants (Rajagopal & Suryanarayanan, 2000; Mahesh et al., 2005; Verma et al., 2007, 2009), but also for endophytes producing bioactive natural products (Li et al., 2007; Kharwar et al., 2009; Wu et al., 2008, 2009). Recently, we isolated an endophytic fungus from this plant that is capable of producing azadirachtin A and B under *in vitro* axenic conditions (Kusari et al., 2011c).

5. Metabolomics of endophytic fungi producing associated plant natural products

5.1 Anticancer compounds

5.1.1 Case study I: Camptothecin (CPT) and structural analogues

Camptothecin (CPT), a pentacyclic quinoline alkaloid, is a potent antineoplastic agent, which was first isolated from the wood of *Camptotheca acuminata* Decaisne (Nyssaceae), a plant native to mainland China (Wall et al., 1966). CPT and its structural analogues (Figure

6) have emerged as one of the most promising agents for cancer treatment owing to the typical action mechanism involving DNA-Topoisomerase I, i.e., they cause DNA damage by stabilizing a normally transient covalent complex between Topoisomerase I (Topo 1) and DNA (Hsiang et al., 1985; Kusari et al., 2011a,d). CPT interacts with the Topo 1-DNA complex, thereby forming a ternary complex that stabilizes the trans-esterification intermediate (Hertzberg et al., 1990; Pommier et al., 1995). Thus, by stabilizing the cleavable complex, CPT transforms the normally useful enzyme Topo 1 into an intracellular, cytotoxic poison, and hence, CPT and structural analogues are called 'topoisomerase poisons' or 'topoisomerase inhibitors' (Lorence & Nessler, 2004).

Endophytic *Entrophospora infrequens* (Puri et al., 2005; Amna et al., 2006) and *Neurospora crassa* (Rehman et al., 2008) isolated from *N. nimmoniana* were initially reported to produce CPT. However, in both cases, there have been no further studies on how the fungi are able to produce CPT and prevent self-toxicity from the intracellular accumulated CPT. Further, no follow-up work on up-scaling the production of CPT has been performed, and there is no published breakthrough in the commercial exploitation of these endophytic fungi as a source of CPT.

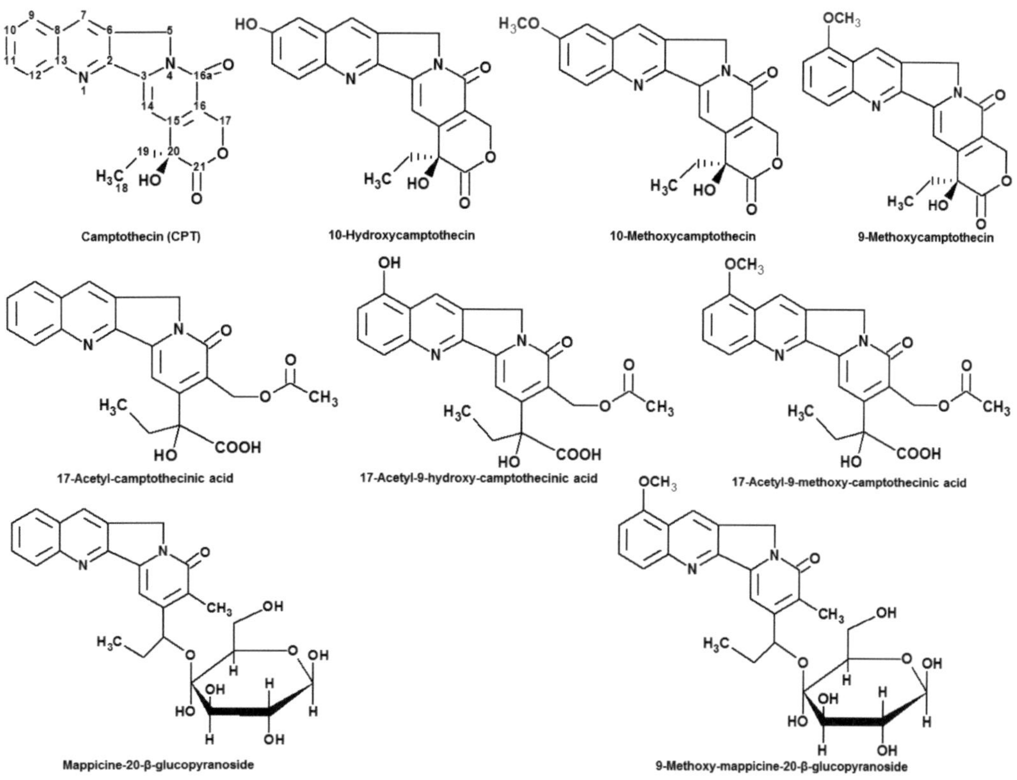

Fig. 6. Camptothecin (CPT) and some important natural analogues found in plants.

Recently, we isolated an endophytic fungus, *Fusarium solani*, from the inner bark of *Camptotheca acuminata* Decaisne, obtained from the Southwest Forestry University (SWFU) campus, Kunming (Yunnan Province), People's Republic of China (Figure 7a). This endophyte is capable of indigenously producing CPT, 9-methoxycamptothecin (9-MeO-CPT), and 10-hydroxycamptothecin (10-OH-CPT) in submerged *in vitro* axenic culture (Kusari et al., 2009b). We further investigated how this endophytic fungus, capable of producing CPT, ensures self-resistance before being incapacitated by its own and the associated plant's CPT biosyntheses. We discovered its survival strategy by examining the fungal *Top1* (Topo 1) structure with emphasis on the CPT-binding and catalytic domains (Kusari et al., 2011a). The typical amino acid residues Asn352, Glu356, Arg488, Gly503, and Gly717 (numbered according to human *Top1*) were identified that ensure fungal resistance towards intracellular CPT. The substitution Met370Thr is identical to that found in CPT-resistant human cancer cells, but different from the host *C. acuminata*. This work denoted the significance of resistance mechanisms employed by endophytes to bear toxic host metabolites, and provided a deeper understanding of plant-microbe coevolution.

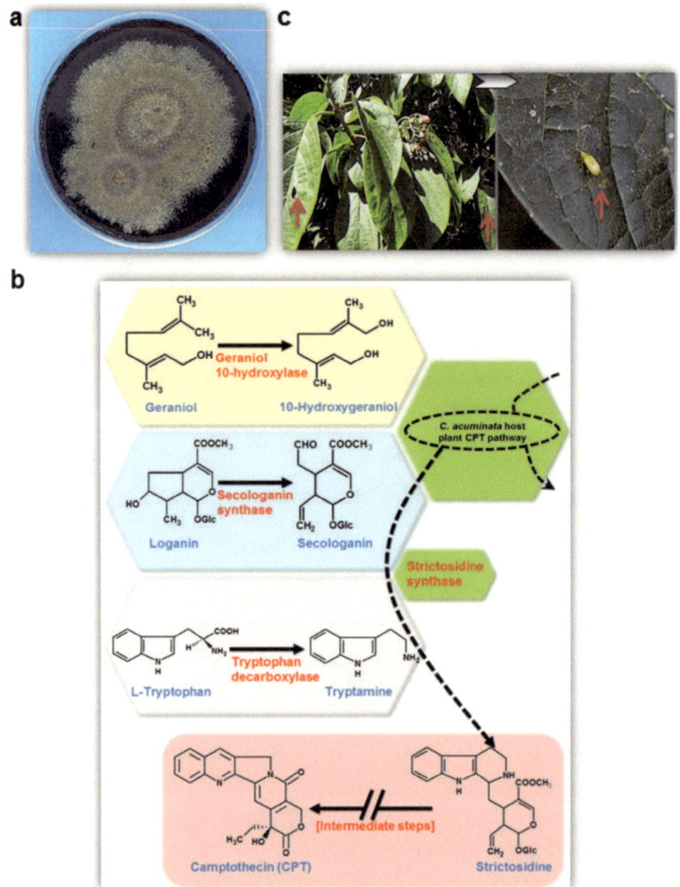

Fig. 7. (a) Endophytic *Fusarium solani* that produces CPT, 9-MeO-CPT and 10-OH-CPT. (b) A cross-species CPT biosynthetic pathway where the endophyte utilizes indigenous enzymes to biosynthesize CPT precursors (10-hydroxygeraniol, secologanin, and tryptamine), but requires the host strictosidine synthase to complete the pathway. (c) Perforations on the leaves *N. nimmoniana* plant (Western Ghats, India) (red arrows) are caused by feeding of Chrysomelid leaf beetles. (Photographs courtesy of M. Spiteller).

The discovery of this endophytic fungus that is capable of producing CPT led us to envisage the possibility of using this organism to produce CPT under controlled fermentation conditions in an economical, environment-friendly, and reproducible manner amenable to industrial scale-up. Unfortunately, it was observed that a substantial decrease occurred in the production of CPT and 9-MeO-CPT by this *in vitro*-cultured endophyte following repeated subculturing (i.e., in successive subculture generations) (Kusari et al., 2009b). Optimized fermentation conditions and the addition of precursors as well as various host plant tissue extracts did not restore the production. We then deciphered the chemical ecology of the endophyte-host interaction, where the fungal endophyte utilizes indigenous *G10H* (geraniol 10-hydroxylase), *SLS* (secologanin synthase), and *TDC* (tryptophan decarboxylase) to biosynthesize CPT precursors. However, to complete the cross-species CPT biosynthetic pathway, the endophyte requires the host *STR* (strictosidine synthase) (Kusari et al., 2011b) (Figure 7b). The fungal CPT biosynthetic genes destabilized *ex planta* over successive subculture generations. The seventh subculture predicted proteins exhibited reduced homologies to the original enzymes proving that such genomic instability leads to dysfunction at the amino acid level. The endophyte with an impaired CPT biosynthetic capability was artificially inoculated into the living host plants and then recovered after colonization. CPT biosynthesis could still not be restored. This demonstrated that the observed phenomenon of genomic instability was possibly irreversible (Kusari et al., 2011b). Following our discovery of the endophytic fungus *F. solani*, another endophytic fungus has been isolated from *Apodytes dimidiata* capable of producing the same compounds (Shweta et al., 2010). Furthermore, an endophytic *Xylaria sp.* has recently been isolated from *C. acuminata* capable of producing only 10-OH-CPT, and remarkably, not the parent compound CPT (Liu et al., 2010). In both these cases, no further follow-up studies have been reported so far. Recently, it was reported that chrysomelid beetles (*Kanarella unicolor* Jacobby) feeds on the leaves of *N. nimmoniana* without any apparent adverse effect (Ramesha et al., 2011) (Figure 7c). Interestingly, most of the CPT in the insect body was found in the parental form without any major metabolized products.

5.1.2 Metabolomics resources used for CPT and structural analogues

For analyzing both the host plants and the isolated endophytes, we employed a number of state-of-the-art analytical tools and methodologies (Kusari et al., 2009b, 2011b,d). CPT, 9-MeO-CPT, and 10-OH-CPT were identified by high-performance (pressure) liquid chromatography (HPLC) coupled to multicomponent high-resolution tandem mass spectrometry (LC-HRMS and LC-HRMS^n fragment spectra) using a LTQ-Orbitrap spectrometer, Thermo Scientific. The compounds were quantified using TSQ Quantum Ultra AM mass spectrometer (Thermo Finnigan, U.S.A.) equipped with an ESI ion source (Ion Max). The mass spectrometer was equipped with a Dionex HPLC system Ultimate 3000 consisting of pump, flow manager, and autosampler (injection volume 0.6 µL). Nitrogen was used as sheath gas (6 arbitrary units), and helium served as the collision gas. The separations were performed by using a Phenomenex Gemini C_{18} column (3 µm, 0.3 × 150 mm) (Torrance, CA) with a H_2O (+ 0.1% HCOOH) (A)/acetonitrile (+ 0.1% HCOOH) (B) gradient (flow rate 4 µL min^{-1}). Samples were analyzed by using a gradient program as follows: 95% A isocratic for 5 min, linear gradient to 60% A within 12 min, and to 100% B in 29 min. After 100% B isocratic for 5 min, the system returned to its initial condition (95% A) within 1 min and was equilibrated for 7 min. The spectrometer was operated in positive mode (1 spectrum s^{-1}; mass range: 200-800) with nominal mass resolving power of 60000 at *m/z* 400 with a scan rate of 1 Hz, with automatic gain control to provide high-accuracy mass measurements within 2 ppm deviation using one internal lock mass; *m/z* 391.284286; bis-(2-ethylhexyl)-phthalate. MS^2 led to the corresponding CO_2 loss of the precursor (CID of 45). The final MS^3 measurement was performed under CID of 45 and resulted in characteristic fragments of the compounds. The compounds were additionally confirmed using ^1H NMR, performed at 298 K with a Bruker DRX-400 spectrometer using 5 mm tubes with $CDCl_3$ (Merck, Darmstadt, Germany) as solvent.

For the host *C. acuminata* plants, the LC-MS/MS data were subjected to a number of different chemometric evaluations for metabolite profiling and correlating the phytochemical loads among the various *Camptotheca* plants (infraspecific), among the organic and aqueous phases, and among the different aerial tissues (dry and fresh in parallel) to reflect the metabolome profiles of the studied plants (Kusari et al., 2011d). The analyses included multivariate analysis (MVA), Kruskal's multidimensional scaling (MDS), principal component analysis (PCA), linear discriminant analysis (LDA), and hierarchical agglomerative cluster analysis (HACA). All analyses were performed using the statistical software XLSTAT-Pro (Addinsoft, NY, U.S.A.), except for MVA which was performed using the statistical software QI Macros (KnowWare International Inc., CO, U.S.A.). Both the statistical software packages were used in combination with Microsoft Excel (part of Microsoft Office Professional, Microsoft Corporation, U.S.A.).

Furthermore, we used the high-precision isotope-ratio mass spectrometry (HP-IRMS) by compound-specific carbon isotope (CSCI) and compound-specific nitrogen isotope (CSNI) modules, to confirm that the endophytic fungus actually utilizes host strictosidine synthase, as detailed above (Kusari et al., 2011b). The CPT produced by the cultured endophyte (first generation) outside the host plant in a nitrogen-free media was compared to CPT from the tissue (not containing the same *F. solani*) of original *C. acuminata* host (from SWFU) to check both the $\delta^{13}C/^{12}C$ (by CSCI) and the $\delta^{15}N/^{14}N$ (by CSNI). It was possible to trace the exact pattern of the accumulation of both 'carbons' and 'nitrogens' with the source of the enzyme(s) (fungal or plant) concerned up to and including the formation of CPT in the endophytic fungus and in the host plant. Briefly, the samples were readied for HP-IRMS in each case by placing 0.5 mg CPT in 3.5 × 5 mm tin capsules (HEKAtech GmbH, Germany), lyophilizing completely and finally rolling the capsules into small spheres. The HP-IRMS measurements were performed in compound-specific carbon isotope (CSCI) and compound-specific nitrogen isotope (CSNI) modules, using a FlashEA 1112 elemental analyzer (Thermo Fisher, Italy) coupled to a DELTA V Plus isotope-ratio mass spectrometer (Thermo Fisher, Bremen, Germany) interfaced through a ConFlo IV universal continuous flow interface (Thermo Fisher, Bremen, Germany) (Kusari et al., 2011b). The combustion furnace (oxidation reactor) was maintained at 1020°C, and flash combustion was initiated by injecting a pulse of O_2 at the time of sample drop. Helium was used as the carrier with a flow rate of 120 mL min^{-1}. NO_x species were reduced to N_2 in a reduction furnace at 680°C. Water was removed by phosphorus pentoxide in a water trap and CO_2 was separated from N_2 using a Porapak-packed N_2/CO_2-separation column (3 m × 6.5 mm, Thermo Electron S. p. A.) operated isothermally at 85°C. Each sample was analyzed in quadruplet. Acetanilide (Fisons Instruments) was used as the reference standard.

5.1.3 Case study II: Podophyllotoxin and deoxypodophyllotoxin

The first literature report on the extraction of 'Podophyllum' was that of King, who called the resin he obtained from alcohol extraction as 'podophyllin' (King, 1857). 'Podophyllum' is the dried roots and rhizomes of species of *Podophyllum* (Figure 8), which was described and its first modern botanical name given by Linnaeus (1753). The first successful chemical investigation was later carried out by Podwyssotzki (1881, 1882, 1884). The correct empirical formula for podophyllotoxin was first advanced by Borsche and Niemann (1932), and later confirmed (Gensler et al., 1954; Gensler and Wang, 1954; Petcher et al., 1973).

Strikingly, the first documented proof of the discovery of deoxypodophyllotoxin was not from *Podophyllum*. The Leech book of Bald, 900-950 A.D., an early English medicinal book, has reported on the use of root of *Anthriscus sylvestris* (Imbert, 1998). These roots were reported contain lignans such as deoxypodophyllotoxin and were used in ointments prepared from a large number of plants and plant extracts like *savin* to cure cancer (Cockayne, 1961). Podophyllotoxin and deoxypodophyllotoxin share the same action mechanism based on the core structure of deoxypodophyllotoxin as evidenced by the SAR studies. They inhibit the formation of the microtubules, i.e., inhibit the formation of the mitotic spindle, resulting in an arrest of the cell division process in metaphase and clumping of the chromosomes (Imbert, 1998; Canel et al., 2000; Liu et al., 2007). Under *in vitro* conditions, they bind to tubulin dimers giving lignan-tubulin complexes. This stops further formation of the microtubules at one end but does not stop the disassembly at the other end leading to the degradation of the microtubules. This mode of action is comparable to the alkaloid colchicin, and for their mode of action these compounds are called 'spindle poisons' (Liu et al., 2007).

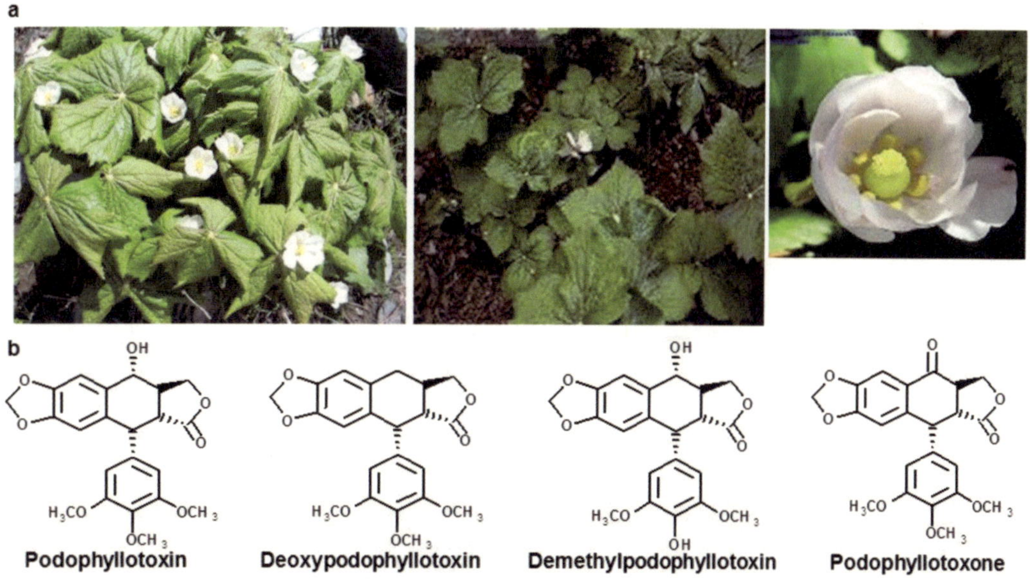

Fig. 8. (a) *Podophyllum hexandrum* Royale. (Photographs courtesy of M. Spiteller). (b) Podophyllotoxin and some important structural analogues found in plants.

Discoveries of podophyllotoxin-producing endophytic fungi include *Phialocephala fortinii* isolated from *P. peltatum* (Eyberger et al., 2006), *Trametes hirsuta* isolated from *P. hexandrum* (Puri et al., 2006), and *Fusarium oxysporum* isolated from *Juniperus recurva* (Kour et al., 2008). Unfortunately, in all the above cases, there has been no follow-up work on scale-up, and there is no published breakthrough in the commercial exploitation of these endophytic fungi as a source of podophyllotoxin. We recently isolated an endophytic fungus, *Aspergillus fumigatus* Fresenius, from *Juniperus communis* sampled from the Rombergpark botanical gardens, Dortmund, Germany, which produced deoxypodophyllotoxin under *in vitro* axenic conditions (Kusari et al., 2009a). The growth and production kinetics showed the potential of the endophyte in indigenous production of deoxypodophyllotoxin, but *in vitro* subculturing showed no production from the third subculture generation.

5.1.4 Metabolomics resources used for podophyllotoxin and structural analogues

The *Juniperus* and *Podophyllum* plants sampled by us from diverse populations were extracted and subjected to metabolomics analyses (Kusari et al., 2011e). HPLC analysis of the extracts was performed using a Surveyor HPLC system. Compounds were separated on a Hydro-RP column (150 × 2 mm, 4 µm particle size) from Phenomenex (Torrance, CA). The mobile phase consisted of 10 mM ammonium acetate in distilled water (A) and acetonitrile with 0.1% formic acid (B). Gradient elution was performed using the following solvent gradient: from 85A/15B (held for 3 min) in 16 min to 17A/83B, then in 1 min to 0A/100B and after 7 min, back to the initial conditions (85A/15B); each run was followed by an equilibration period of 8 min. The flow rate was 0.22 mL min^{-1} and the injection volume was 5 µL. All separations were performed at 30°C. Mass spectra were obtained using a TSQ Quantum Ultra AM mass spectrometer equipped with an ESI ion source (Ion Max) operating in positive mode (Kusari et al., 2011e). Nitrogen was employed as both the drying and nebulizer gas (40 AU). Capillary temperature was 200°C and capillary voltage was 3.5 kV. The calibration curves of the available reference compounds podophyllotoxin and demethylpodophyllotoxin were constructed by dilution of external standards with methanol to give the desired concentrations. The concentrations of standard solutions were 0.1, 0.5, 1, 5, 10, 50, 80, 120, and 160 µg mL^{-1}. Correlation coefficient for the linear calibration curve was >0.99 for both podophyllotoxin and demethylpodophyllotoxin. All procedures were carried out under light protection. Concentrations of the commercially unavailable compounds deoxypodophyllotoxin and podophyllotoxone were calculated with the assumption of similar precursor ion response like that of podophyllotoxin. The LOQs were 0.05 µg mL^{-1} (demethylpodophyllotoxin) and 0.2 µg mL^{-1} (podophyllotoxin, deoxypodophyllotoxin, and podophyllotoxone), respectively. The LOD (3 times noise intensities) and LOQ (10 times noise intensities) were calculated/estimated from signal to noise ratio using signal intensities of the analytes and the noise near the retention time of the analytes. Estimation was necessary for the derivatives (deoxypodophyllotoxin and podophyllotoxone) due to unavailability of reference standards.

All the above secondary metabolites were re-verified using the highly selective and sensitive LC-ESI-$HRMS^n$. HPLC analysis of the extracts was performed using an Agilent (Santa Clara, U.S.A.) 1200 HPLC system consisting of LC-pump, PDA detector (λ = 254 nm), autosampler (injection volume 10 µL) and column oven (30°C). Compounds were separated using a Synergi Fusion RP80 column (150 x 3 mm, 4 µm particle size) from Phenomenex (Torrance,

CA) with a H_2O (+ 0.1% HCOOH, + 10 mM ammonium acetate) (A)/acetonitrile (+ 0.1% HCOOH) (B) gradient (flow rate 400 µL min^{-1}). Samples were analyzed by using gradient program: 95% A isocratic for 3 min, linear gradient to 100% B over 20 min, after 100% B isocratic for 10 min, the system returned to its initial condition (95% A) within 1 min, and was equilibrated for 5 min. The FT-full scan and MS/MS spectra were obtained with an LTQ-Orbitrap XL spectrometer (Thermo Fisher, U.S.A.) equipped with H-ESI-II source. The spectrometer was operated in positive mode (1 spectrum s^{-1}; mass range: 250-1000) with nominal mass resolving power of 60000 at *m/z* 400 with a scan rate of 1 Hz with automatic gain control to provide high-accuracy mass measurements within 2 ppm deviation using an internal standard; bis(2-ethylhexyl)phthalate: *m/z* 391.284286. MS/MS experiments were performed in HCD (higher-energy C-trap dissociation, 35 eV) mode. The following parameters were used for experiments: spray voltage 5 kV, capillary temperature 260°C, and tube lens 70 V. Nitrogen was used both as sheath gas (45 AU) and auxiliary gas (10 AU). Helium served as the collision gas.

The LC-MS/MS data were subjected to a number of different chemometric evaluations for metabolite profiling and correlating the phytochemical loads among the various plants of the studied *Juniperus* and *Podophyllum* species (infraspecific), between the organic and aqueous extracts, among populations of the same species from different locations, among populations of different species from the same location, among populations of different species from different locations, as well as among populations of different genera (infrageneric) from the same and different locations (Kusari et al., 2011e). The chemometric algorithms and methodologies were used similar to those used for the *C. acuminata* plants (*vide supra*).

The extracts of the isolated endophytic fungi, both from *Juniperus* and *Podophyllum* plants, were evaluated with an LTQ-Orbitrap spectrometer (Kusari et al., 2009a). The spectrometer was operated in positive mode (1 spectrum s^{-1}; mass range: 50–1000) with nominal mass resolving power of 60000 at *m/z* 400 with a scan rate of 1 Hz with automatic gain control to provide high accuracy mass within 1 ppm deviation using one internal lock mass, polydimethylcyclosiloxane – $[(CH_3)_2SiO]_6$: *m/z* 445.120025. The spectrometer was equipped with a Dionex HPLC system Ultimate 3000 consisting of pump, flow manager and autosampler (injection volume 0.5 µL). Nitrogen was used as sheath gas (5 AU) and helium served as the collision gas. The separations were performed by using a Phenomenex Gemini C_{18} column (3 µm, 0.3 × 150 mm) (Torrance, CA, U.S.A.) with a H_2O (+0.1% HCOOH) (A) / acetonitrile (+0.1% HCOOH) (B) gradient (flow rate 4 µL min^{-1}). Samples were analyzed by using a gradient program as follows: 90% A isocratic for 2 min, linear gradient to 100% B over 8 min, after 100% B isocratic for 10 min, the system was returned to its initial condition (90% A) within 1 min, and was equilibrated for 9 min. Furthermore, for the deoxypodophyllotoxin-producing endophyte, quantitation of the compound was achieved by accurate mass (maximum deviation 1 ppm) single ion monitoring (SIM) of the $[M+H]^+$ ion of deoxypodophyllotoxin (Kusari et al., 2009a). Since deoxypodophyllotoxin is unavailable from commercial sources, the calibration was performed using podophyllotoxin as standard and detector response was assumed to be in the same range. The calibration graph was linear from 50 ng mL^{-1} up to 10000 ng mL^{-1}. Furthermore, a high-resolution full scan run was performed in order to check for the accumulation of structural analogues of podophyllotoxin and deoxypodophyllotoxin by that particular endophyte in axenic culture.

5.2 Antidepressants and photodynamic compounds

5.2.1 Case study: Hypericin and emodin

Hypericin (2,2'-dimethyl-4,4',5,5',7,7'-hexahydroxy-mesonaphtodianthrone) (Figure 9), a naphthodianthrone derivative, is a plant derived compound of high medicinal value. It is one of the main constituents of *Hypericum* species. The first detailed report of the isolation of hypericin was from the medicinal herb *Hypericum perforatum* L., published by Brockmann et al. (1939). The molecular formula of hypericin was first reported in 1942 by the same author as $C_{30}H_{16}O_8$ (Brockmann et al., 1942) and eight years later the correct structure was published (Brockmann et al., 1950). Various species of the genus *Hypericum* have long been used as medicinal plants in various parts of the world due to their therapeutic efficacy (Yazaki and Okada, 1994). Their main constituents are napthodianthrones, primarily represented by hypericin, pseudohypericin, protohypericin, protopseudophypericin (Brockmann et al., 1939, 1942, 1957), the anthraquinone emodin, and derivatives occurring in very low concentrations such as isohypericin, demethyl-pseudohypericin, hyperico-dehydro-dianthrone, pseudo-hyperico-dehydro-dianthrone (Brockmann et al., 1957), and cyclopseudohypericin (Häberlein et al., 1992). Hypericin has long been in use, at least from the time of ancient Greece (Tammaro and Xepapadakis, 1986), as an antidepressant due to its monoamine oxidase (MAO) inhibiting capacity, having effects similar to bupropion (Nahrstedt and Butterweck, 1997) and imipramine (Raffa, 1998). Potential uses of hypericin extend to improved wound healing, anti-inflammatory effects (Zaichikova et al., 1985), antimicrobial and antioxidant activity (Radulovic et al., 2007), sinusitis relief (Razinkov et al., 1989), and seasonal affective disorder (SAD) relief (Martinez et al., 1993). Hypericin also has remarkable antiviral activity against a number of viruses (Kusari et al., 2008). Several recent *in vitro* studies have revealed the multifaceted cytotoxic activity of hypericin as a result of its photodynamic activity (Kubin et al., 2005).

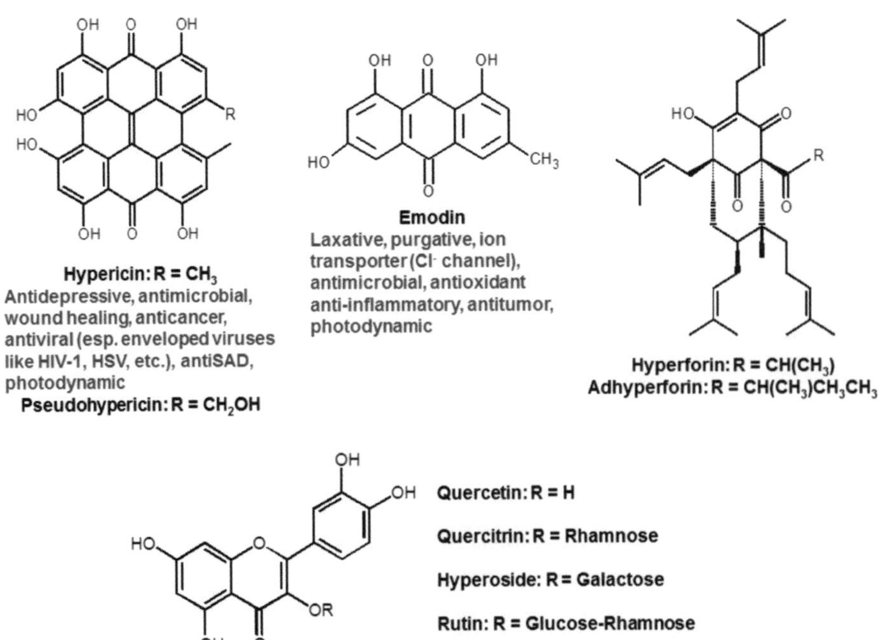

Fig. 9. Hypericin and related compounds found in plants.

Using the rationale that the plants containing hypericin may also contain endophytic fungi that are able to accumulate the same or similar molecules, a selective search for fungal endophytes was pursued. A number of endophytic fungi were isolated from various organs of the *Hypericum* plants, which were morphologically different from the strains isolated from unsterilized explants (surface-contaminating fungi). Only one endophytic fungus was able to produce hypericin and emodin under axenic submerged shake-flask fermentation (Kusari et al., 2008). The fungus was identified as *Thielavia subthermophila* by its morphology and authenticated by 28S rDNA and ITS-5.8S rDNA analyses. The growth of the endophyte and production of hypericin remained independent of the illumination conditions and media spiking with emodin. Protohypericin could not be detected, irrespective of either spiking or illumination conditions. The *hyp-1* gene, suggested to encode for the Hyp-1 phenolic coupling protein in plant cell cultures, was absent in the genome of the endophyte. Thus, it was proposed that emodin anthrone is the common precursor of both hypericin and emodin in the fungal endophyte, which is governed by a different molecular mechanism than the host plant or host cell suspension cultures (Kusari et al., 2009c).

5.2.2 Metabolomics resources used for hypericin and related compounds

The *Hypericum* plants sampled from diverse populations in Europe and the Himalayan region were extracted and subjected to a metabolic profiling (Kusari et al., 2009d). The compounds were separated on a Luna C_{18} 100 Å column (3 µm, 250 mm; Phenomenex, Torrance, CA) at 30°C. Chromatographic conditions were optimized for the separation of hypericin, pseudohypericin, hyperforin and emodin (gradient 1), and for the separation of flavonoids (gradient 2) at the Surveyor HPLC system (Thermo Finnigan, U.S.A.). The mobile phase consisted of 10 mM ammonium acetate buffer adjusted to pH 5.0 with glacial acetic acid (A) and a 9:1 mixture of acetonitrile and methanol (B). First gradient elution was performed using the following solvent gradient: from 55A/45B held for 2 min to 0A/100B in 8 min, thereafter holding for 13 min; each run was followed by an equilibration period of 6 min. The flow rate was 0.3 mL min^{-1} and injection volume was 3 µL. Second gradient elution was performed using the following solvent gradient: start for 2 min at 95A/5B, in 6 min to 75A/25B, then in 2 min to 50A/50B and in another 2 min to 100B. After holding for 13 min returned to initial conditions (95A/5B) within 1 min and held for 8 min. The eluent flow rate was 0.25 mL min^{-1} and the injection volume was 5 µL. Highly selective and sensitive selected reaction monitoring (SRM) was performed using a TSQ Quantum Ultra AM mass spectrometer (Thermo Finnigan, U.S.A.) equipped with an ESI ion source (Ion Max) operating in negative mode (Kusari et al., 2009d). Nitrogen was employed as both the drying and nebulizer gas. The capillary voltage was 5 kV and capillary temperature was set at 200°C. Sheath gas (nitrogen) was set at 45 arbitrary units and collision gas pressure was 1.5 mTorr. Each mass transition was monitored at a peak width of 0.5 and dwell time of 0.3 s. All the secondary metabolites were re-verified using the highly selective and sensitive LC-ESI-$HRMS^n$ (LTQ-Orbitrap spectrometer). External calibration was performed in the range 0.01-10 µg mL^{-1} for emodin, 0.05-50 µg mL^{-1} for hyperforin, pseudohypericin, and hypericin, as well as 0.5-100 µg mL^{-1} for hyperoside, rutin, quercetin, and quercitrin. Correlation coefficient for the calibration curves were >0.99 for all analytes. The relative standard deviation (RSD) of the analytical method was determined by eight injections of an extract and was below 6% for all compounds. The LOD (limit of detection) and LOQ were determined by minimum signal to noise ratio of 3 and 9, respectively. Instrumental LOQ of the compounds varied between 0.003 μg mL^{-1} (emodin) and 2 μg mL^{-1} (quercitrin).

Similar to the *C. acuminata* plants, the LC–MS/MS data for these plants were also subjected to a number of different statistical evaluations for metabolite profiling and correlating the phytochemical loads among the various parts of the plants of the studied *Hypericum* samples (same and different species), between the organic and aqueous extracts, among the different species, among populations of the same species from different locations, among populations of different species from the same locations, as well as among populations of different species from different locations (Kusari et al., 2009d).

For the isolated endophytic fungi, quantitation of the hypericin and emodin was performed by using a Thermo Finnigan Surveyor HPLC system consisting of Surveyor MS-pump and Surveyor Autosampler-Plus (injection volume 5 μL) (Kusari et al., 2008, 2009c). The compounds were separated on a Luna C_{18} (50 × 3 mm, 3 μm particle size) column from Phenomenex (Torrance, CA). The mobile phase consisted of water containing 10 mM ammonium acetate (pH 5.0) (A) and acetonitrile-methanol, 9:1 (B). Samples were separated using a gradient program as follows: (flow rate of 250 μL min^{-1}) 55% A isocratic for 2 min, linear gradient to 100% B over 6 min (flow rate 300 μL min^{-1}). After 100% B isocratic for 7 min, the system was returned to its initial conditions (55% A) within 1 min and was equilibrated for 4 min before the next run was started. MS detection (multiple reaction monitoring mode) was performed by using a TSQ Quantum Ultra AM spectrometer equipped with an ESI ion source (Ion Max) operating in negative mode. Nitrogen was employed as both the sheath (50 arbitrary units) and auxiliary (8 arbitrary units) gas, and argon served as the collision gas with a pressure of 1.5 mTorr. The capillary temperature was set at 250°C. External calibration was performed in the range 0.01-10.0 μg mL^{-1} for hypericin and 0.005-10.0 μg mL^{-1} for emodin. Correlation coefficients for the linear calibration curves were >0.995 for both compounds.

Hypericin and emodin were identified by HRMS fragment spectra (LTQ-Orbitrap spectrometer), which were consistent with authentic standards (Kusari et al., 2008, 2009c). The spectrometer was equipped with a Dionex HPLC system Ultimate 3000 consisting of pump, flow manager, and autosampler (injection volume 1 μL). Nitrogen was used as sheath gas (6 arbitrary units), and helium served as the collision gas. The separations were performed by using a Phenomenex Gemini C_{18} column (3 μm, 0.3 × 150 mm) (Torrance, CA) with a H_2O (+0.1% HCOOH, +1 mM ammonium acetate) (A)/acetonitrile (+0.1% HCOOH) (B) gradient (flow rate 4 μL min^{-1}). Samples were analyzed by using a gradient program as follows: 30% A isocratic for 1 min, linear gradient to 100% B over 10 min; after 100% B isocratic for 60 min, the system was returned to its initial condition (30% A) within 1 min and was equilibrated for 9 min. The spectrometer was operated in negative mode (1 spectrum s^{-1}; mass range 50–1000) with nominal mass resolving power of 60000 at *m/z* 400 with a scan rate of 1 Hz with automatic gain control to provide high-accuracy mass measurements within 2 ppm deviation using one internal lock mass (*m/z* 386.7149314; CsI_2^-).

Additional screening for emodin anthrone and protohypericin was also performed in full scan negative mode (Kusari et al., 2009c). For that, the spectrometer was equipped with a Thermo Surveyor system consisting of a LC-pump and autosampler (injection volume 5 μL). N_2 was used as sheath gas (5 arbitrary units), and He served as the collision gas. The separations were performed by using a Phenomenex Synergi Fusion RP column (4 µm, 2 × 150 mm) with a H_2O (+0.1% HCOOH, +10 mM NH_4OAc) (A)/MeCN (+0.1% HCOOH) (B) gradient (flow rate 0.25 mL min^{-1}). Samples were analyzed by using a gradient program as follows: 50% A isocratic for 2 min, linear gradient to 100% B over 8 min; after 100% B isocratic for 48 min, the system was returned to its initial condition (50% A) within 1 min and was equilibrated for 6 min. The spectrometer was operated in negative mode (1 spectrum s^{-1}; mass range 200-1000) with mass resolving power of 60000 at *m/z* 400 with a scan rate of 1 Hz with automatic gain control to provide high-accuracy mass measurements within 2 ppm deviation.

6. Current progress and future challenges

The production of bioactive compounds by endophytes, especially those exclusive to their host plants, is significant both from the molecular and biochemical perspective, and the ecological viewpoint. The production of beneficial secondary metabolites (including those produced by plants) by endophytes nurtures expectations of utilizing them as alternative and sustainable sources of these compounds. However, the commercial implication of production of desirable compounds by endophytic fungi still remains a future goal (Kusari & Spiteller, 2011). A major obstacle preventing the biotechnological application of endophytes is the perplexing problem of reduction of secondary metabolite production on repeated subculturing under axenic monoculture conditions. In addition to a constant pursuit of discovering competent endophytes with potential for pharmaceutical use, it is essential to follow-up these discoveries with advanced research to establish, restore and sustain the *in vitro* biosynthetic capability of endophytes. This can be achieved by a multifaceted approach involving complete elucidation of the dynamic endophyte-endophyte interactions pertaining to their biological, biochemical and genetic frameworks. Considering the fact that endophytes reside within plants and are constantly communicating with their hosts, it is compelling that plants would have a substantial influence on the *in planta* metabolic processes of the endophytes. Moreover, recent whole-genome sequencing strategies have shown that the known secondary metabolites of various bacteria and fungi are largely outnumbered by the number of genes encoding the biosynthetic enzymes in these microorganisms (Winter et al., 2011). This is accentuated by the fact that endophytic fungi always remain in versatile interactions with the host plant and other endophytes, and even slight variation in the *in vitro* cultivation conditions can impact the kind and range of secondary metabolites they produce (Scherlach & Hertweck, 2009). Further research to systematically understand the endophyte–endophyte and endophyte-host interspecies crosstalk is desirable for sustainable production of compounds using endophytes (Kusari & Spiteller, 2011).

7. Overcoming the obstacles

The potential of novel fungal endophytes capable of biosynthesizing plant metabolites has undoubtedly been recognized. However, there is still no known breakthrough in the biotechnological production of these bioactive secondary metabolites using endophytes. It is important to elucidate the metabolome in endophytes correlating to their associated plants on a case-by-case basis to understand how the biogenetic gene clusters are regulated and their expression is affected *in planta* and *ex planta* (by environmental changes and axenic culture conditions). Only a deeper understanding of the host-endophyte relationship at the molecular and genetic levels might help to induce and optimize secondary metabolite production under laboratory conditions to yield plant metabolites in a sustained manner using endophytes. The biosynthesis of plant metabolites in endophytes could further be manipulated to yield new lead structures which could act as pro-drugs. In addition to identifying new natural products, genome mining, metabolic engineering and metagenomics would certainly have an impact on the understanding and manipulation of secondary metabolite production by endophytic fungi. Likewise, it is well-known that an unidentified proportion of endophytic fungi are uncultivable *in vitro* under axenic conditions. For such unculturable species, environmental PCR strategies might be employed for recovery, suitably coupled with a culture-independent metagenomic approach or compound structure-based gene targeting to study their desirable biosynthetic gene clusters (Kusari & Spiteller, 2011). Further research along these directions is highly desirable in order to elucidate comprehensively the endophytic fungal biosyntheses since this knowledge can then be utilized for heterologous expression of the preferred final products in large quantities using suitable model organisms like *Saccharomyces cerevisiae* or *Escherichia coli*. Another advantage of endophytic fungi over plants is that the biosynthetic gene modules of a natural product produced by a cascade of biosynthetic steps might be arranged as an operon in highly contiguous clusters in the fungal genome. This could allow swift *in silico* detection of signature genes or gene domains that are pathway-specific followed by the possibility of expressing them in heterologous organisms. We stand at the cross-roads of time when the world's biodiversity is declining at an alarming rate. Many endemic, endangered and medicinally valuable plants are on the verge of extinction. Along with these plants, the endophytes harbored in them are also threatened. Further fundamental research must be addressed to ensure a continuous and sustained supply of bioactive pro-drugs against the present and emerging diseases.

8. Acknowledgements

We thank the International Bureau (IB) of the German Federal Ministry of Education and Research (BMBF/DLR), Germany for supporting our various research projects. We also thank the Ministry of Innovation, Science, Research and Technology of the State of North Rhine-Westphalia, Germany, and the German Research Foundation (DFG) for granting us the necessary high-resolution instruments.

9. References

Aly AH, Debbab A, Kjer J, Proksch P. (2010). Fungal endophytes from higher plants: a prolific source of phytochemicals and other bioactive natural products. *Fungal Divers.*, 41: 1-16.

Amna T, Puri SC, Verma V, Sharma JP, Khajuria RK, Musarrat J, Spiteller M, Qazi GN. (2006). Bioreactor studies on the endophytic fungus *Entrophospora infrequens* for the production of an anticancer alkaloid camptothecin. *Can. J. Microbiol.*, 52: 189–196.

Anonymous. (2000, 2006). *Consideration of proposals for amendment of appendices II: inclusion of happy tree (Camptotheca acuminata Decaisne) in CITES appendix II of convention in accordance with the provisions of article II, paragraph 2(a). Prop. 11.58,* World

Conservation Monitoring Centre, CITES Secretariat/World Conservation Monitoring Centre, Chatelâine-Genève, Switzerland.

Arnold AE. (2007). Understanding the diversity of foliar endophytic fungi: progress, challenges, and frontiers. *Fungal Biol. Rev.*, 21: 51-66.

Arnold AE, Henk DA, Eells RA, Lutzoni F, Vilgalys R. (2007). Diversity and phylogenetic affinities of foliar fungal endophytes in loblolly pine inferred by culturing and environmental PCR. *Mycologia*, 99: 185–206.

Arnold AE. (2005). Diversity and ecology of fungal endophytes in tropical forests. In: *Current trends in mycological research*, Deshmukh D (ed.), pp. 49-68, Oxford & IBH Publishing Co. Pvt. Ltd., New Delhi, India.

Arnold AE, Mejia LC, Kyllo D, Rojas EI, Maynard Z, Robbins N, Herre EA. (2003). Fungal endophytes limit pathogen damage in a tropical tree. *Proc. Natl. Acad. Sci. U. S. A.*, 100: 15649–15654.

Bacon CW, White JF. (2000). *Microbial endophytes*, Marcel Deker Inc., New York.

Biswas K, Chattopadhyay I, Banerjee RK, Bandyopadhyay U. (2002). Biological activity and medicinal properties of neem (*Azadirachta indica*). *Curr. Sci.*, 82: 1336-1345.

Borsche W, Niemann J. (1932). Über Podophyllin. *Justus Liebig's Ann. Chem.*, 494: 126–142.

Brockmann H, Falkenhausen EH, Dorlares A. (1950). Die Konstitution des Hypericins. *Naturwissenschaften*, 37: 540–540.

Brockmann H, Haschad MN, Maier K, Pohl F. (1939). Über das Hypericin, den photodynamisch wirksamen Farbstoff aus *Hypericum perforatum*. *Naturwissenschaften*, 27: 550–550.

Brockmann H, Kluge F, Muxfeldt H. (1957). Totalsynthese des Hypericins. *Chem. Ber.*, 90: 2302–2318.

Brockmann H, Pohl F, Maier K, Haschad MN. (1942). Über das Hypericin, den photodynamischen Farbstoff des Johanniskrautes (*Hypericum perforatum*). *Ann. Chem.*, 553: 1–52.

Butterworth JH. (1968). Isolation of a substance that suppresses feeding in locusts. *Chem. Commun.*, -:23-24.

Canel C, Moraes RM, Dayan FE, Ferreira D. (2000). Podophyllotoxin. *Phytochemistry*, 54: 115–120.

Cockayne TO. (1961). *Leechdoms, wortcunning, and starcraft of early England: being a collection of documents, for the most part never before printed, illustrating the history of science in this country before the Norman conquest*, vol. 2, The Holland Press, London.

de Bary A. (1866). *Morphologie und Physiologie der Pilze, Flechten, und Myxomyceten. Hofmeister's handbook of physiological botany*, vol. II, Leipzig, Germany.

Decaisne J. (1873). Caracteres et descriptions de trios genres nouveaus de plantes recueilles en chine par l'abbe a. *David Bull. Soc. Bot. France*, 20: 155–160.

Eyberger AL, Dondapati R, Porter JR. (2006). Endophyte fungal isolates from *Podophyllum peltatum* produce podophyllotoxin. *J. Nat. Prod.*, 69: 1121–1124.

Freeman EM. (1904). The seed-fungus of *Lolium temulentum*, L., the Darnel. *Phil. Trans. R. Soc. B*, 196: 1–27.

Gensler WJ, Wang SY. (1954). Synthesis of picropodophyllin. *J. Am. Chem. Soc.*, 76: 5890–5891.

Gensler WJ, Samour CM, Wang SY. (1954). Sythesis of a DL-stereoisomer of podophyllic acid. *J. Am. Chem. Soc.*, 76: 315–316.

Guerin P. (1898). Sur la presence d'un champignon dans l'ivraie. *J. Botanique*, 12: 230–238.

Gunatilaka AAL. (2006). Natural products from plant-associated microorganisms: distribution, structural diversity, bioactivity, and implications of their occurrence. *J. Nat. Prod.*, 69: 509–526.

Gunther RT. (1959). *The Greek herbal of Dioscorides*. Hafner Publishing Co., New York.

Häberlein H, Tschiersch KP, Stock S, Hölzl J. (1992). Johanniskraut (*Hypericum perforatum* L.): Nachweis eines weiteren Naphthodianthrons. *Pharm. Ztg. Wiss.*, 5/137: 169–174.

Hartwell JL, Johnson JM, Fitzgerald DB, Belkin, M. (1953). Podophyllotoxin from *Juniperus* species; Savinin. *J. Am. Chem. Soc.*, 75: 235–236.

Hertzberg RP, Busby RW, Caranfa MJ, Holden KG, Johnson RK, Hecht SM, Kingsbury WD. (1990). Irreversible trapping of the DNA-topoisomerase I covalent complex. Affinity labeling of the camptothecin binding site. *J. Biol. Chem.*, 265: 19287–19295.

Hickey M, King C. (1981). *100 Families of flowering plants* (2nd edition, Walters SM ed.), Cambridge University Press, Cambridge.

Hölzl J, Petersen M. (2003). Chemical constituents of *Hypericum* ssp. In: *Hypericum: the genus Hypericum (Series: Medicinal and Aromatic Plants - Industrial Profiles)*, vol. 31, Ernst E. (ed.), pp. 77-93, Taylor and Francis, London, UK.

Hombe Gowda HC, Vasudeva R, Mathachen GP, Shaanker RU, Ganeshaiah KN. (2002). Breeding types in *Nothapodytes nimmoniana* Graham.: An important medicinal tree. *Curr. Sci.*, 83: 1077–1078.

Hsiang YH, Hertzberg R, Hecht S, Liu LF. (1985). Camptothecin induces protein-linked DNA breaks via mammalian DNA topoisomerase I. *J. Biol. Chem.*, 260: 14873–14878.

Imbert TF. (1998). Discovery of podophyllotoxins. *Biochimie*, 80: 207–222.

Kharwar RN, Verma VC, Kumar A, Gond SK, Harper JK, Hess WM, Lobkovosky E, Ma C, Ren Y, Strobel GA. (2009). Javanicin, an antibacterial naphthaquinone from an endophytic fungus of neem, *Chloridium* sp. *Curr. Microbiol.*, 58: 233-238.

Kharwar RN, Mishra A, Gond SK, Stierle A, Stierle D. (2011). Anticancer compounds derived from fungal endophytes: their importance and future challenges. *Nat. Prod. Rep.*, 28: 1208-1228.

King J. (1857). Discovery of podophyllin. *Coll. J. M. Sci.*, 2: 557–559.

Kour A, Shawl AS, Rehman S, Sultan P, Qazi PH, Suden P, Khajuria RK, Verma V. (2008). Isolation and identification of an endophytic strain of *Fusarium oxysporum* producing podophyllotoxin from *Juniperus recurva*. *World J. Microbiol. Biotechnol.*, 24: 1115–1121.

Kubin A, Wierrani F, Burner U, Alth G, Grunberger W. (2005). Hypericin - the facts about a controversial agent. *Curr. Pharm. Des.*, 11: 233–253.

Kumar KR, Ved DK. (2000). *100 Red listed medicinal plants of conservation concern in southern India*, Foundation for Revitalisation of Local Health Traditions (FRLHT), Bangalore, India.

Kusari S, Kosuth J, Cellarova E, Spiteller M. (2011a). Survival-strategies of endophytic *Fusarium solani* against indigenous camptothecin biosynthesis. *Fungal Ecol.*, 4: 219-223.

Kusari S, Lamshöft M, Spiteller M. (2009a). *Aspergillus fumigatus* Fresenius, an endophytic fungus from *Juniperus communis* L. Horstmann as a novel source of the anticancer pro-drug deoxypodophyllotoxin. *J. Appl. Microbiol.*, 107: 1019-1030.

Kusari S, Lamshöft M, Zühlke S, Spiteller M. (2008). An endophytic fungus from *Hypericum perforatum* that produces hypericin. *J. Nat. Prod.*, 71: 159-162.

Kusari S, Spiteller M. (2010). Lessons from endophytes: peering under the skin of plants, In: *Biotechnology – Its Growing Dimensions*, Patro, LR (ed.), pp. 1-27, Sonali Publications, New Delhi, India.

Kusari S, Spiteller M. (2011). Are we ready for industrial production of bioactive plant secondary metabolites utilizing endophytes? *Nat. Prod. Rep.*, 28: 1203-1207.

Kusari S, Verma VC, Lamshöft M, Spiteller M. (2011c). An endophytic fungus from *Azadirachta indica* A. Juss. that produces azadirachtin. *World J. Microbiol. Biotechnol.*, in press, doi: 10.1007/s11274-011-0876-2.

Kusari S, Zühlke S, Borsch T, Spiteller M. (2009d). Positive correlations between hypericin and putative precursors detected in the quantitative secondary metabolite spectrum of *Hypericum*. *Phytochemistry*, 70: 1222-1232.

Kusari S, Zühlke S, Kosuth J, Cellarova E, Spiteller M. (2009c). Light-independent metabolomics of endophytic *Thielavia subthermophila* provides insight into microbial hypericin biosynthesis. *J. Nat. Prod.*, 72: 1825-1835.

Kusari S, Zühlke S, Spiteller M. (2009b). An endophytic fungus from *Camptotheca acuminata* that produces camptothecin and analogues. *J. Nat. Prod.*, 72: 2-7.

Kusari S, Zühlke S, Spiteller M. (2011b). Effect of artificial reconstitution of the interaction between the plant *Camptotheca acuminata* and the fungal endophyte *Fusarium solani* on camptothecin biosynthesis. *J. Nat. Prod.*, 74: 764-775.

Kusari S, Zühlke S, Spiteller M. (2011d). Correlations between camptothecin and related metabolites in *Camptotheca acuminata* reveal similar biosynthetic principles and *in planta* synergistic effects. *Fitoterapia*, 82: 497-507.

Kusari S, Zühlke S, Spiteller M. (2011e). Chemometric evaluation of the anti-cancer pro-drug podophyllotoxin and potential therapeutic analogues in *Juniperus* and *Podophyllum* species. *Phytochem. Anal.*, 22: 128-143.

Lay SV, Denholm AA, Wood A. (1993). The chemistry of azadirachtin. *Nat. Prod. Rep.*, 10: 109-157.

Li GH, Yu ZF, Li X, Wang XB, Zheng LJ, Zhang KQ. (2007). Nematicidal metabolites produced by the endophytic fungus *Geotrichum* sp. AL4. *Chem. Biodivers.*, 4: 1520-1524.

Li S, Zhang Z, Cain A, Wang B, Long M, Taylor J. (2005). Antifungal activity of camptothecin, trifolin, and hyperoside isolated from *Camptotheca acuminata*. *J. Agric. Food Chem.*, 53: 32–37.

Linnaeus C. (1753). *Species Plantarum: exhibentes plantas rite cognitas, ad genera relatas, cum differentiis specificis, nominibus trivialibus, synonymis selectis, locis natalibus, secundum systema sexuale digestas*, vol. 1, Laurentius Salvius, Sweden.

Liu K, Ding X, Deng B, Chen W. (2010). 10-Hydroxycamptothecin produced by a new endophytic *Xylaria sp.*, M20, from *Camptotheca acuminata*. *Biotechnol. Lett.*, 32: 689–693.

Liu YQ, Yang L, Tian X. (2007). Podophyllotoxin: current perspectives. *Curr. Bioact. Compd.*, 3: 37–66.

Lorence A, Nessler CL. (2004). Camptothecin, over four decades of surprising findings. *Phytochemistry*, 65: 2735–2749.

Mahesh B, Tejesvi MV, Nalini MS, Prakash HS, Kini KR, Subbiah V, Hunthrike SS. (2005). Endophytic mycoflora of inner bark of *Azadirachta indica* A. Juss. *Curr. Sci.*, 88: 218-219.

Martinez B, Kasper S, Ruhrmann S, Moller HJ. (1993). *Hypericum* in the treatment of seasonal affective disorders. *Nervenheilkunde*, 36: 103–108.

Nahrstedt A, Butterweck V. (1997). Biologically active and other chemical constituents of the herb of *Hypericum perforatum* L. *Pharmacopsychiatry*, 30: 129–134.

Onelli E, Rivetta A, Giorgi A, Bignami M, Cocucci M, Patrignani G. (2002). Ultrastructural studies on the developing secretory nodules of *Hypericum perforatum*. *Flora*, 197: 92–102.

Petcher TJ, Weber HP, Kuhn M, von Wartburg A. (1973). Crystal structure and absolute configuration of 2'-bromopodophyllotoxin-0.5 ethyl acetate. *J. Chem. Soc.*, Perkin Trans. 2: 288–292.

Podwyssotzki V. (1881). The active constituent of podophyllin. *Pharm. J. Trans.*, 12: 217–218.

Podwyssotzki V. (1882). On the active constituents of podophyllin. *Am. J. Pharm.*, 12: 102–115.

Podwyssotzki V. (1884). Pharmakologische Studien über *Podophyllum peltatum*. *Naunyn. Schmied Arch. Exp. Path. Phar.*, 13: 29–52.

Pommier Y, Kohlhagen G, Kohn KW, Leteurtre F, Wani MC, Wall ME. (1995). Interaction of an alkylating camptothecin derivative with a DNA base at topoisomerase I-DNA cleavage sites. *Proc. Natl. Acad. Sci. U. S. A.*, 92: 8861–8865.

Puri SC, Nazir A, Chawla R, Arora R, Riyaz-ul Hasan S, Amna T, Ahmed B, Verma V, Singh S, Sagar R, Sharma A, Kumar R, Sharma RK, Qazi GN. (2006). The endophytic fungus *Trametes hirsuta* as a novel alternative source of podophyllotoxin and related aryl tetralin lignans. *J. Biotechnol.*, 122: 494–510.

Puri SC, Verma V, Amna T, Qazi GN, Spiteller M. (2005). An endophytic fungus from *Nothapodytes foetida* that produces camptothecin. *J. Nat. Prod.*, 68: 1717–1719.

Radulovic N, Stankov-Jovanovic V, Stojanovic G, Smelcerovic A, Spiteller M, Asakawa Y. (2007). Screening of *in vitro* antimicrobial and antioxidant activity of nine *Hypericum* species from the Balkans. *Food Chem.*, 103: 15–21.

Raffa RB. (1998). Screen of receptor and uptake-site activity of hypericin component of St. John's wort reveals sigma receptor binding. *Life Sci.*, 62: 265–270.

Rajagopal R, Suryanarayanan TS. (2000). Isolation of endophytic fungi from leaves of neem (*Azadirachta indica*). *Curr. Sci.*, 78: 1375–1378.

Ramesha BT, Zuehlke S, Vijaya R, Priti V, Ravikanth G, Ganeshaiah K, Spiteller M, Shaanker RU. (2011). Sequestration of camptothecin, an anticancer alkaloid, by chrysomelid beetles. *J. Chem. Ecol.*, 37: 533-536.

Razinkov SP, Yerofeyeva LN, Khovrina MP, Lazarev AI. (1989). Validation of the use of *Hypericum perforatum* medicamentous form with a prolonged action to treat patients with maxillary sinusitis. *Zh. Ushn. Nos. Gorl. Bolezn.*, 49: 43–46.

Redecker D, Kodner R, Graham LE. Glomalean fungi from the Ordovician. (2000). *Science*, 289: 1920-1921.

Rehman S, Shawl AS, Kour A, Andrabi R, Sudan P, Sultan P, Verma V, Qazi GN. (2008). An endophytic *Neurospora sp.* from *Nothapodytes foetida* producing camptothecin. *Appl. Biochem. Microbiol.*, 44: 203–209.

Rodriguez RJ, White JFJ, Arnold AE, Redman RS. (2009). Fungal endophytes: diversity and functional roles. *New Phytol.*, 182: 314–330.

Sankar-Thomas YD. (2010). *In vitro culture of Camptotheca acuminata (Decaisne) in Temporary Immersion System (TIS): growth, development and production of secondary metabolites*, PhD thesis, Universität Hamburg, Germany.

Scherlach K, Hertweck C. (2009). Triggering cryptic natural product biosynthesis in microorganisms. *Org. Biomol. Chem.*, 7: 1753-1760.

Schulz BJE, Boyle CJC. (2005). The endophytic continuum. *Mycol. Res.*, 109: 661–687.

Sconocchia S. (1983). Scribonius largus compositions. In: *Bibliotheca Scriptorum Graecorum et Romanorum Teubneriana (B.G. Teubner)*, Hansen GC (ed.), pp. 76-77, Verlagsgesellschaft, Leipzig, Germany.

Shaanker RU, Ramesha BT, Ravikanth G, Gunaga RP, Vasudeva R, Ganeshaiah, KN. (2008). Chemical profiling of *Nothapodytes nimmoniana* for camptothecin, an important anticancer alkaloid: towards the development of a sustainable production system. In: *Bioactive molecules and medicinal plants*, Ramawat KG, Merillon JM (eds.), pp. 197-213, Springer-Verlag, Berlin and Heidelberg.

Shweta S, Zühlke S, Ramesha BT, Priti V, Kumar PM, Ravikanth G, Spiteller M, Vasudeva R, Shaanker RU. (2010). Endophytic fungal strains of *Fusarium solani*, from *Apodytes dimidiata* E. Mey. ex Arn (Icacinaceae) produce camptothecin, 10-hydroxycamptothecin and 9-methoxycamptothecin. *Phytochemistry*, 71: 117–122.

Staniek A, Woerdenbag HJ, Kayser O. (2008). Endophytes: exploiting biodiversity for the improvement of natural product-based drug discovery. *J. Plant Interact.*, 3: 75–93.

Stierle A, Strobel GA, Stierle D. (1993). Taxol and taxane production by *Taxomyces andreanae*, an endophytic fungus of Pacific yew. *Science*, 260: 214–216.

Stone JK, Polishook JD, White JF Jr. (2004). Endophytic fungi. In: *Biodiversity of fungi: inventory and monitoring methods*, Mueller G, Bills GF, Foster MS (eds.), pp. 241-270, Elsevier, Burlington, MA, USA.

Strobel GA, Daisy B. (2003). Bioprospecting for microbial endophytes and their natural products. *Microbiol. Mol. Biol. Rev.*, 67: 491–502.

Strobel GA, Daisy B, Castillo U, Harper J. (2004). Natural products from endophytic microorganisms. *J. Nat. Prod.*, 67: 257–268.

Sung CK, Kimura T, But PPH, Guo JX. (1998). *International collation of traditional and folk medicine: Northeast Asia, Part III. A project of UNESCO*, vol. 3, World Scientific Publishing Co. Pte. Ltd., Singapore.

Suryanarayananа TS, Thirunavukkarasub N, Govindarajulub MB, Sassec F, Jansend R, Murali TS. (2009). Fungal endophytes and bioprospecting. *Fungal Biol. Rev.*, 23: 9–19.

Tammaro F, Xepapadakis G. (1986). Plants used in phytotherapy, cosmetics and dyeing in the Pramanda district (Epirus, north-west Greece). *J. Ethnopharmacol.*, 16: 167–174.

Veitch GE, Boyer A, Ley SV. (2008). The azadirachtin story. *Angew. Chem. Int. Ed.*, 47: 9402-9429.

Verma VC, Gond SK, Kumar A, Kharwar RN, Strobel GA. (2007). Endophytic mycoflora of bark, leaf, and stem tissues of *Azadirachta indica* A. Juss. (neem) from Varanasi (India). *Microb. Ecol.*, 54: 119-125.

Verma VC, Gond SK, Mishra A, Kumar A, Kharwar RN, Gange AC. (2009). Endophytic actinomycetes from *Azadirachta indica* A. Juss.: isolation, diversity and anti-microbial activity. *Microb. Ecol.*, 57: 749–756.

Wall ME, Wani MC, Cook CE, Palmer KH, Mcphail AT, Sim GA. (1966). Plant antitumor agents. I. The isolation and structure of camptothecin, a novel alkaloidal leukemia and tumor inhibitor from *Camptotheca acuminata*. *J. Am. Chem. Soc.*, 88: 3888–3890.

Wichtl M. (1986). *Hypericum perforatum* L. Das Johanniskraut. *Zeitschrift Phytother.*, 3: 87–90.

Winter JM, Behnken S, Hertweck C. (2011). Genomics-inspired discovery of natural products. *Curr. Opin. Chem. Biol.*, 15: 22-31.

Wu SH, Chen YW, Shao SC, Wang LD, Li ZY, Yang LY, Li SL, Huang R. (2008). Ten-membered lactones from *Phomopsis* sp., an endophytic fungus of *Azadirachta indica*. *J. Nat. Prod.*, 71: 731–734.

Wu SH, Chen YW, Shao SC, Wang LD, Yu Y, Li ZY, Yang LY, Li SL, Huang R. (2009). Two new Solanapyrone analogues from the endophytic fungus *Nigrospora* sp. YB-141 of *Azadirachta indica*. *Chem. Biodivers.*, 6: 79-85.

Yazaki K, Okada T. (1994). *Hypericum erectum* Thunb. (St. John's wort): *in vitro* culture and the production of procyanidins. In: *Biotechnology in Agriculture and Forestry. Medicinal and Aromatic Plants VI*, Bajaj YPS (ed.), vol. 26, pp. 167-178, Springer-Verlag, Berlin.

Zaichikova SG, Grinkevich NI, Barabanov EI. (1985). Healing properties and determination of the upper parameters of toxicity of *Hypericum* herb. *Farmatsiya*, 34: 62–64.

Zhang HW, Song YC, Tan RX. (2006). Biology and chemistry of endophytes. *Nat. Prod. Rep.*, 23: 753–771.

Part 5

Metabolomics in Human Disease Research

11

Metabolomics in the Analysis of Inflammatory Diseases

Sabrina Kapoor, Martin Fitzpatrick, Elizabeth Clay, Rachel Bayley, Graham R. Wallace and Stephen P. Young
Rheumatology Research Group, School of Immunity & Infection, College of Medical and Dental Sciences, University of Birmingham United Kingdom

1. Introduction

Inflammation is a normal and extraordinarily important component of responses to infection and injury. The cardinal features of swelling, redness, stiffness and increasing temperature are strong indicators of the significant changes in tissue metabolism and the ingress of immune cells into the tissues. The increase in blood flow which underlies many of these changes may result in changes to the supply of nutrients and in particular the level of oxygen in the tissues. Inward migration of immune cells, which is also enabled by the increased blood flow, will put further stress on the metabolic environment of the tissues. The activity of macrophages and neutrophils in clearing infection and repairing tissue damage also have significant metabolic consequences particularly because of the production of cytokines and cytotoxic molecules such as reactive oxygen species and reactive nitrogen species, which are required to kill invading organisms. Production of these molecules will consume considerable quantities of oxygen, ATP and NADPH. These antimicrobial agents put considerable stress on host cells in the surrounding and distal tissues and can lead to significant loss of protective metabolites such as glutathione.

Most infections and traumatic injuries are cleared or repaired relatively rapidly and metabolic homoeostasis is soon restored. However, there is a broad range of inflammatory diseases which involve chronic activation of the immune system and, as a result, chronic persistent inflammation. We have been studying the metabolic consequences of chronic inflammatory diseases with the aim of identifying metabolic fingerprints which may provide clues about why the localised tissue disease persists. For example, why in rheumatoid arthritis does persistent inflammation lead to widespread cartilage and joint destruction? However, the metabolic consequences of chronic inflammation are much more widespread than the localised disease and can lead on to important comorbidities such as accelerated atherosclerosis and cardiovascular disease. Metabolomic analysis may be able to distinguish between localised and systemic metabolic consequences of inflammation and provide novel targets for therapeutic intervention in these important human diseases.

2. Introduction to inflammatory disease

An indication of the strong link between inflammation and metabolic processes is seen in cachexia, the loss of cellular mass associated with disease. The discovery of the involvement of tumour necrosis factor- alpha (TNFα) in this process earned it the name 'cachexin'. While TNFα is now known more generally as a mediator of inflammatory responses, the ability of inflammatory cytokines to have such profound effects on cellular and metabolic processes is informative. Systemic inflammation such as that seen in RA causes changes in metabolism and rheumatoid cachexia is a result of chronic inflammation. This is characterised by the loss of muscle mass and preservation of fat mass (Evans et al., 2008). Classically cachexia is characterised by a low BMI. Muscle wasting is a common feature of RA but low BMI is uncommon as the fat mass is preserved or even increased (Summers et al., 2008). Hence, RA patients may present with either the classic low BMI cachexia (1-13% of RA population) (Munro & Capell, 1997) or more frequently, the rheumatoid cachexia (10-20% of RA with controlled disease and 38% of patients with active RA) (Engvall et al., 2008, Metsios et al., 2009).

The muscle loss that occurs in rheumatoid cachexia is thought to be due to proinflammatory cytokines such as TNFα, IL1 and IL6. TNF promotes proteolysis through the ubiquitin-proteasome pathway. There is also some evidence that cytokines may prevent an increase in muscle protein synthesis in response to feeding (anabolic resistance) (Summers et al., 2010). In rheumatoid cachexia the degree of muscle wasting is associated with the disease activity of RA (Summers et al., 2010).

2.1 The inflammatory process

An acute inflammatory reaction is characterised by the classic cardinal signs of inflammation: heat, redness, swelling and pain. In experimental settings the temporal relationships oedema, accumulation of leukocytes and accumulation of monocytes and macrophages are well established. These events in self-limited inflammatory reactions are coupled with the release of local factors which prevent further release of leukocytes, which allows resolution (Serhan, 2009). The transition from acute inflammation to chronic inflammation is widely viewed as a result of an excess of pro-inflammatory mediators.

2.2 Inflammatory mediators

Cytokines are important regulators of inflammation. Some cytokines such as TNFα and interleukin (IL) 1 promote inflammatory responses by inducing cartilage degradation and promoting a cell-mediated immune response. Other cytokines such as IL-4, IL-10 and IL-13 function mainly as anti-inflammatory molecules (Isomaki & Punnonen, 1997). Key biological targets that have been identified as being involved in a destructive inflammatory reaction are COX-2, pro-inflammatory interleukins, TNFα, migration inhibition factor, interferon gamma and matrix metalloproteinases (Ivanenkov et al., 2008).

Several inflammatory mediators have been identified which are common to several inflammatory diseases. It has been shown that C-reactive protein (CRP) is secreted by several cell types and is capable of directly activating immune cells. This supports a role for CRP as an active inflammatory mediator which has systemic and local effects (Montecucco & Mach, 2009).

White adipose tissue has been shown to secrete several inflammatory mediators called adipokines or adipocytokines. These induce their activities by binding to selective transmembrane receptors. Leptin is the most studied adipocytokine and is thought to have an important role in the inflammatory process (Montecucco & Mach, 2009).

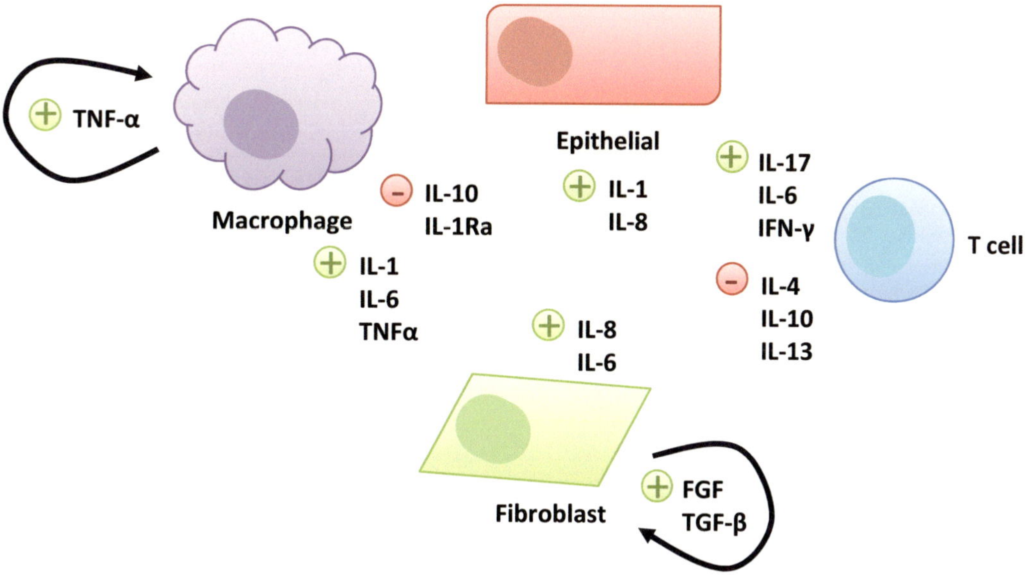

Fig. 1. Key inflammatory cytokines and the inflammatory network. Responses are a balance of pro-inflammatory tumour necrosis factor alpha (TNFα) and interleukin (IL) 1, IL-6, IL-17 and anti-inflammatory IL-1R, IL-4, IL-10 and IL-13. Expression of cytokines is dependent on activation and local signalling driving progression and eventual resolution.

2.3 Metabolic inflammation

Many factors contribute to the complex course of inflammatory reactions. Microbiological, immunological and toxic agents can initiate the inflammatory response by activating a variety of humoral and cellular mediators. In the early phase of inflammation, excessive amounts of interleukins and lipid-mediators are released and play an important role in the pathogenesis of organ dysfunction. Arachidonic acid (AA) is released from membrane phospholipids during inflammatory activation and is metabolised to prostaglandins and leukotrienes. Various strategies have been evaluated to regulate the excessive production of lipid mediators on different levels of biochemical pathways, such as inhibition of phospholipase A2, the trigger enzyme for release of AA, blockade of cyclooxygenase and lipoxygenase pathways and the development of receptor antagonists against platelet activating factor and leukotrienes. Some of these agents exert protective effects in different inflammatory disorders such as septic organ failure, rheumatoid arthritis or asthma, whereas others fail to do so. Encouraging results have been obtained by dietary supplementation with long chain omega-3 fatty acids like eicosapentaenoic acid (EPA). In states of inflammation, EPA is released to compete with AA for enzymatic metabolism inducing the production of less inflammatory and chemotactic derivatives (Heller et al., 1998).

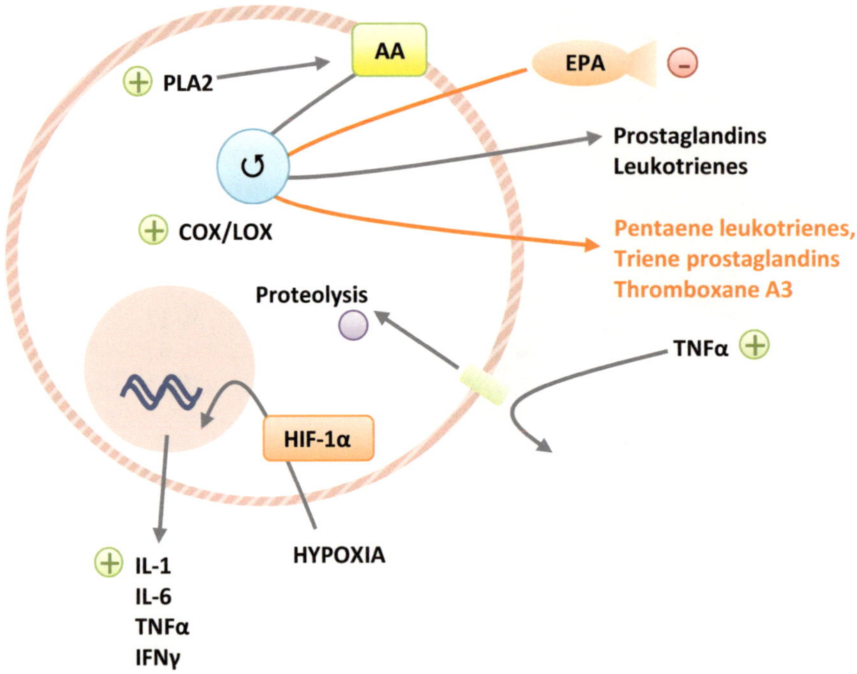

Fig. 2. Some common metabolic responses to inflammation and hypoxia. Arachidonic acid (AA) from cellular membranes is metabolised to inflammatory prostaglandins and leukotrienes. Omega-3 fatty acids (EPA) compete for the same pathway producing less inflammatory derivatives. Hypoxic conditions in the inflammatory site stabilises HIF transcription factor driving production of IL-1, IL-6, TNFα and IFNγ. TNFα in turn drives cellular proteolysis and tissue remodelling.

When investigating inflammation it is important to take into account the many facets of the inflammatory environment that have the potential to play a role in pathology. Hypoxia is known to be prevalent in the inflammatory environments such as those associated with wounds, malignant tumours, bacterial infections and autoimmunity (Eltzschig & Carmeliet, 2011, Murdoch et al., 2005). Increasing hypoxia in the inflammatory site is associated with poorer disease outcome such as increased macroscopic synovitis in rheumatoid arthritis (Ng et al., 2010).

Normal physiological oxygen levels are thought to range between 5-12% oxygen (compared to 21% atmospheric oxygen). However, hypoxic tissue oxygen levels in pathological environments can range from as little as 0.5% oxygen to around 2.5% oxygen. Local hypoxia develops as the result of either blood vessel occlusion by inflamed tissues, or when existing supply is insufficient for increased cellular density caused by infiltrating or proliferating inflammatory cells. Additionally, circulating phagocytes can block blood vessels reducing blood flow into the inflammatory site (Sitkovsky & Lukashev, 2005). Normal tissue structures can lend themselves to hypoxia where they are poorly perfused, such as the synovium or eye. Tissue alteration associated with inflammation can contribute to hypoxia by altering pressure within the blood vessels causing vessel occlusion and increasing distances between blood vessels (Jawed et al., 1997, Mapp et al., 1995).

There is increasing evidence that the inflammatory environment is hypoxic. The tumour environment is known to be hypoxic and extensive angiogenesis reveals the requirement of the tissue for a better oxygen supply. In rheumatoid arthritis, oxygen levels of synovial fluid have been directly measured revealing lower oxygen tensions compared with osteoarthritic patients and patients with traumatic joint injuries (Lund-Olesen, 1970). In systemic sclerosis, direct measurements with sensitive probes revealed lower dermal oxygen levels in fibrotic areas compared to non-fibrotic areas in both patients and healthy controls (Beyer et al., 2009). Metabolomic analysis of eye fluids from uveitis patients has shown increased levels of oxaloacetate and urea, likely derived from anaerobic respiration by locally activated macrophages (Young et al., 2009, Young & Wallace, 2009).

An elegant cellular oxygen detection system is used by cells to respond to changes in environmental oxygen. Reductions in environmental oxygen lead to the stabilisation of the transcription factor hypoxia-inducible factor (HIF), which is otherwise targeted for depletion in oxygen-rich environments. HIF expression is therefore suggestive of hypoxic exposure, and has been detected in autoimmune diseases such as rheumatoid arthritis and multiple sclerosis (Gaber et al., 2009, Hollander et al., 2001, Lassmann, 2003). HIF is known to be important in inflammatory development, for example loss of HIF-1a in macrophages is associated with impaired aggregation, motility, invasiveness and killing of bacteria (Cramer et al., 2003).

Hypoxia and HIF stabilisation has a large effect on cellular metabolism. HIF causes a preference for glycolytic metabolism over oxidative phosphorylation by inducing the expression of glycolytic enzymes. This allows ATP generation to continue in the absence of sufficient oxygen albeit at a much reduced efficiency per molecule of glucose. It also induces the upregulation of lactate dehydrogenase A, therefore promoting the conversion of pyruvate (produced during glycolysis) to lactate (Wheaton & Chandel, 2011). Lactate has been detected in many chronic inflammatory conditions such as in inflamed joints (Chang & Wei, 2011, Treuhaft & McCarty, 1971), multiple sclerosis , pulmonary inflammation (Serkova et al., 2008) and is thought to play a role in wound healing (Trabold et al., 2003). Conversely, the acidosis associated with increasing lactate concentrations is thought to play a pathogenic role in cell transformation and autoantigen development in some inflammatory environments (Chang & Wei, 2011). Recently, lactate measurements have been suggested to be useful in the diagnosis of bacterial infections in diabetic foot ulcers compared to non-infected ulcers. Both infected and non-infected ulcers revealed high lactate concentrations, but infected ulcers had significantly higher levels probably due to additional immune and bacterial cell involvement (Loffler et al., 2011). The detection of lactate in metabolomic studies of disease suggests that there may be an inflammatory component, understanding of which may help to direct future treatment.

Immune cells are thought to be highly influenced by hypoxia and HIF stabilisation especially due to the environments they normally act within. In a study performed recently by Gaber et al., peripheral blood CD4+ T cells placed under hypoxia were found to have a large induction of genes involved in metabolism and homeostasis (Gaber et al., 2009). Innate immune cells such as neutrophils and macrophages are thought to be adapted to function best at lower oxygen tensions as they preferentially use glycolysis to provide ATP even at higher oxygen levels (Cramer et al., 2003). Macrophages are known to accumulate in the hypoxic sites of chronic inflammation (Vergadi et al., 2011), and hypoxia is associated with activation of tissue-resident macrophages. Exposure of macrophages to hypoxic conditions is associated with upregulation of a whole gamut of proinflammatory cytokines such as IL-1 (Scannell, 1996), IL-6 (Albina et al., 1995), IFN-γ (Murata et al., 2002) and TNF-α (White et al., 2004). It is thought that both low oxygen levels and their downstream effects, such as lactate production, may give rise to this macrophage phenotype. That such phenotypic changes are observed in response to the hypoxic conditions of the inflammatory site is strongly suggestive of a role for metabolism in regulation of immune cells. While normal wound resolution is a tightly regulated process, the presence of long-term inflammatory diseases such as rheumatoid arthritis is indicative of the potential for this regulation to go awry. Therefore hypoxia and the resulting change in metabolism may have a profound effect on immune cell behaviour and thus influence disease onset and progression.

Adenosine is another molecule produced in response to hypoxia partly by the hypoxic inhibition of adenosine kinase (Sitkovsky & Lukashev, 2005). It is difficult to detect due to its local action, but expression of CD39 and CD73, two molecules involved in the extracellular generation of adenosine, provide a marker of its presence in the inflammatory environment. Adenosine can have profound effects on immune cells and is generally perceived to be anti-inflammatory. It is a ligand for specific receptors found on many immune and stromal cells. These receptors are upregulated by hypoxia suggesting hypoxia perpetuates both the production and action of this molecule (Hasko et al., 2008, Sitkovsky & Lukashev, 2005). These receptors have varying downstream effects, with the expression of the A2A associated with the anti-inflammatory disease but the A2B receptor expression being implicated in pro-inflammatory conditions such as colitis. Adenosine is known to cause bronchoconstriction when inhaled by asthma and COPD sufferers, but not in healthy controls (Hasko et al., 2008). Higher levels of adenosine A2 receptor are seen in asthma sufferers and these receptors are associated with a pathological role for the molecule in disease (Brown et al., 2008, Hasko et al., 2008).

2.4 Use of metabolomics in inflammatory diseases

Systemic inflammation causes changes in metabolism and many studies have investigated individual metabolites in human disease and animal models of inflammation. From these results it is apparent that the levels of many metabolites are altered by the inflammatory process and this has provided insights into the mechanisms of disease and uncovered several potential biomarkers for disease assessment.

Given these profound systemic and localised changes in metabolism provoked by inflammation and inflammatory cytokines, it is not surprising that metabolomics has been used to investigate several inflammatory diseases. Metabolomics is able to assess the changes in several hundred metabolites simultaneously to build disease metabolites profiles. NMR spectroscopy and mass spectrometry have both been used to derive these multiplexed metabolite profiles.

These metabolic "fingerprints" have proven useful in discriminating between different patient groups or identifying responses to therapy, even if the individual metabolites have not been identified. However, identification of sets of specific metabolites can be derived from these fingerprints and this has led to the identification of novel biomarkers and novel pathways in a number of inflammatory diseases. The use of metabolomic analysis of inflammatory diseases will now be discussed in further detail.

3. The inflammatory diseases

3.1 Aetiology

Chronic inflammatory diseases exist in many forms, and have the ability to affect many systems of the body. These range from localised areas of inflammation such as the gut in Crohn's disease, to more widespread systemic inflammation as in rheumatoid arthritis (RA). Although the mediators and events leading to chronic inflammation are well characterized, the precise conditions under which acute inflammation becomes chronic are poorly understood. Recent developments have highlighted the importance of genetic factors, environmental influences and the interactions between them in the development of chronic inflammatory disease (Renz et al., 2011).

Research into the genetics of inflammatory disease has been accelerated by genome wide association studies (GWAS), which has allowed identification of genetic mutations associated with an increased risk of developing specific conditions. For example, many immunologically relevant genes have been associated with an increased risk of developing RA. These include human leukocyte antigen (HLA) alleles involved in antigen recognition, and the peptidyl-arginine deiminase type IV (PADI4) gene controlling production of cyclic citrullinated proteins (CCP's) commonly seen in RA (Nishimoto et al., 2010). However, in a complex disease like RA, genetics are not the whole story, as illustrated by the fact that twin studies only report a concordance rate of around 60% (MacGregor et al., 2000). Thus the importance of external environmental factors in the development of inflammatory diseases should be considered.

Chronic inflammatory diseases have become more prevalent in recent years, and as major genetic changes are unlikely to have occurred over such a short time period, this is likely to be a result of alterations in environmental exposures and lifestyle factors. To date, several factors have been identified as significant contributors including ageing, infection, poor nutrition and smoking.

Smoking raises an individual's risk of developing inflammatory disease considerably. It has numerous effects on the body including activation of the acute inflammatory response and introduction of large amounts of reactive oxygen species (ROS) (Borgerding & Klus, 2005). It is unclear as to which particular constituent of smoke induces the inflammatory response; however studies have revealed that smoke contains large amounts of lipopolysaccharide (LPS) (Hasday et al., 1999), which could potentially trigger unwanted immune responses seen in chronic inflammatory disease. An increase in ROS is also evident, as indicated by decreased circulating antioxidants found in smokers (Alberg, 2002). This creates a pro-oxidant environment and increases the likelihood of oxidative damage to important cellular components.

It is not surprising given the complex and varied nature of chronic inflammatory diseases that the observed phenotype is a result of gene-gene and gene-environment interactions. For example, it has been shown in mice with a mutation in the Crohn's disease (CD) susceptibility gene Atg16L1 who become infected with murine norovirus develop a Crohn's-like disease (Stappenbeck et al., 2010). There was no evidence of pathology in the wild type mouse, suggesting the presence of two risk factors is required to induce disease. Another example of gene-environment interactions in disease development was found when looking at the interaction between RA susceptibility genes HLA-DRB1 and PTPN22 and their interaction with smoking (Kallberg et al., 2007). It was observed that the odds ratio (OR) of developing RA with two genetic risk factors was 13.2, which rose to 23.4 if two genetic factors were present and there was a history of smoking. These studies provides sound evidence that gene-gene and gene-environment interactions occur, and risk of inflammatory disease greatly increases with the presence of more than one additional risk factor.

3.2 The gut

Crohn's disease is a chronic debilitating inflammatory disease of the bowel. The exact aetiology is unknown but is thought to be related to the dysregulation of the immune response towards gut microflora (Strober et al., 2007). Urinary metabolite profiling was carried out on a mouse model of Crohn's disease. These samples were analysed using gas chromatography-mass spectrometry and five key metabolic differences were identified between the Crohn's disease model and controls. This suggested that there are alterations of tryptophan metabolism, fucosylation and fatty acid metabolism in Crohn's disease mice and the authors concluded that fucose and xanthurenic acid could be useful markers of gut inflammation (Lin et al., 2009).

Using a mouse model of inflammatory bowel disease (IBD) to investigate urinary metabolites using NMR, it was found that there was an increase in trimethylamine (TMA) and fucose compared to controls. The increase in TMA was parallel to the progression of IBD (Murdoch et al., 2008). A mouse model of Ulcerative Colitis (UC) was used to looked at serum and urinary metabolites (Schicho et al., 2010). These authors found that both serum and urine were equally powerful for detecting colitis but the metabolites responsible for the differences were different for serum and urine.

Metabolomics of faecal extracts have also been used to study inflammatory bowel disease (Bezabeh et al., 2009). It is sometimes difficult to distinguish Crohn's disease (CD) from UC and some cases are labelled as indeterminate. Over time these cases are usually identified by a combination of endoscopic, radiological and histological techniques. Earlier identification could aid treatment and prognostication. Metabolomic analysis of faecal extracts of patients with both inflammatory diseases showed reduced levels of butyrate, acetate, methylamine and TMA compared to control (Marchesi et al., 2007). Comparing the UC and CD samples glycerol, alanine, isoleucine, leucine, lysine and valine were present in higher quantities in CD compared to UC. Acetate was lower in CD compared to UC (Marchesi et al., 2007). Metabolic differences were more marked in CD indicating that inflammation is more extensive in CD compared to UC.

Urinary metabolites have also been used to distinguish CD and UC in humans (Williams et al., 2009). They found that specific urinary metabolites related to gut metabolism differed between CD, UC and controls. Hippurate was lowest in CD and differed significantly between CD, UC and controls. Formate levels were higher in CD than in UC or controls and 4-cresol sulphate was lower in CD than in UC or controls (Williams et al., 2009). Hippurate has been shown to be modulated according to gut microbes and this difference is likely to reflect changes in intestinal microbes.

In summary several studies have looked at IBD. The studies have shown that both in mice and in humans TMA is an important marker of IBD (Marchesi et al., 2007, Murdoch et al.,

2008, Schicho et al., 2010). This has been shown using both urine samples or faecal extracts. Hence, TMA may be a useful biomarker for IBD.

3.3 The eye

As a closed and immuno-privileged site, the eye provides an ideal system for metabolic analysis. Metabolic products of inflammatory infiltrate accumulate in the vitreous fluid of the eye and may be extracted during other corrective surgery.

Metabolomics has been used to look at vitreous humour in order to differentiate ocular inflammatory diseases (Young et al., 2009). Vitreous fluid samples were taken from patients undergoing retinal surgery and analysed using NMR. Patients had various retinal disorders including chronic non-infectious uveitis (CU), lens-induced uveitis (LIU), proliferative diabetic retinopathy, proliferative vitreoretinopathy (PVR), rhegmatogenous retinal detachment, candida endopthalmitis and varicella zoster virus acute retinal necrosis. The different disease groups showed clear separation using principle component analysis (PCA) and partial least squared discriminate analysis (PLSDA). The majority of the patients had LIU and CU. When looking at LIU and CU specifically there was clear separation and individual metabolites from the spectra showed significant differences with urea, oxaloacetate and glucose all being raised in LIU compared to CU. As urea and oxaloacetate are both involved in the urea cycle it suggests that there is more active inflammation in the LIU patients (Young et al., 2009).

NMR has also been used to look at ocular metabolism in pig eyes (Greiner et al., 1985). They used phosphorous NMR and found phosphorous containing metabolites in aqueous and vitreous fluids (Greiner et al., 1985). In addition to quantifying metabolites, phosphorous NMR can be used to monitor the rate of metabolic change in a specific biochemical reaction and the rate of change in the concentration of a particular metabolite (Greiner et al., 1985). Phosphorous NMR provides a non-invasive method to analyse ocular tissues metabolically and detect subtle biochemical changes that precede manifestations of disease. Such detection may allow for early and more effective therapeutic intervention.

3.4 Neurological disease

Multiple sclerosis (MS) is a chronic inflammatory disease affecting the nervous system. Its aetiology is still not completely understood (Ibrahim & Gold, 2005). It is characterised by demyelination, axonal loss and breakdown of the blood-brain barrier (Trapp et al., 1999). It is a heterogeneous, relapsing and remitting disease. Different treatments have been shown to work at different stages of disease (Rieckmann & Smith, 2001) so it is important to identify biomarkers that enable identification of different phases.

Interleukin-1β (IL-1β) and TNF-α, have been found to be associated with a broad spectrum of neurological diseases including MS. Griffin *et al* looked at rat urines to determine whether NMR spectroscopy could detect the presence of IL-1β and TNF-α induced lesions and distinguish between the pathology caused (Griffin et al., 2004). They used an adenoviral vector to induce chronic endogenous expression of either IL-1β or TNF-α. They found significant differences between the groups, with the IL-1β treated group showing increases in leucine, isoleucine, valine, n-butyrate and glucose whilst the TNF-α treated group showed increases of citrate, 2-oxoglutarate and succinate (Griffin et al., 2004).

NMR spectroscopy has also been used to assess cerebrospinal fluid (CSF) in patients with MS. It has been shown that there are increased CSF levels of lactate, creatinine and fructose in MS compared to control patients (Nicoli et al., 1996). Two additional unidentified signals were found to be elevated in MS. The compound responsible for both these signals has now been identified as B-hydroxyisobutyrate (Lutz et al., 2007). This is a typical partial degradation product of branched-chain amino acids. Increased B-hydroxyisobutyrate in urine is thought to be due to respiratory-chain deficiency leading to impaired oxidation of NADH (Chitayat et al., 1992). However the level of B-hydroxyisobutyrate in these experiments was much higher than the level found in CSF from MS patients, and so the precise role of B-hydroxyisobutyrate in MS needs further investigation.

In a study of metabolite fingerprints in the CSF from patients with a range of neurological conditions we have been able to differentiate between some of these conditions by comparing the metabolites found (Sinclair et al., 2010). In particular we were able to identify some novel features of idiopathic intracranial hypertension (IIH) a neurological condition, the pathogenesis of which is poorly understood (Sinclair et al., 2008). Although IIH was not thought to be an inflammatory disease, the elevated levels of lactate we observed in IIH points towards an inflammatory component since lactate has been identified in inflammatory CNS disease previously (Simone et al., 1996). Rabbits with elevated intraocular pressure also show increased levels of lactate which may reflect anaerobic metabolism resulting from decreased blood supply and this may also be an explanation for the lactate in the IIH patients' CSF due to compressed vasculature from the elevated intracranial pressure. Oxaloacetate levels were also increased in IIH and this, together with reduced citrate, suggests alterations in the citric acid cycle. Overall the observations suggest a predominantly anaerobic environment deficient in carbohydrate substrate in patients with IIH, a conclusion supported by the presence of elevated ketone bodies 3-hydroxybutyrate (Sinclair et al., 2010) often observed in hypoxic tissues.

3.5 Lung disease

Pulmonary inflammation contributes to the pathogenesis of a number of lung diseases. There is a growing need for validated experimental models that can help our understanding of disease pathogenesis and therapeutic intervention. Traditionally animal models have been used but they have their own problems in representing human disease. Genetic manipulation can greatly enhance animal models. NMR has had some application in the quantification of experimental lung injury.

Serkova et al used Magnetic Resonance Imaging (MRI) and NMR to try and detect and quantify injury in mice following intratracheal administration of inflammatory cytokines (Serkova et al., 2008). Pulmonary inflammation was induced by intratracheal administration of IL-1β and TNF-α. Lung tissue was used for the NMR metabolomics. They showed that with pulmonary inflammation there was a 50% depletion of ATP and a corresponding elevation of the lactate to glucose ratio suggesting a shift to anaerobic metabolism during inflammation. These returned to control levels at 24 hours (Serkova et al., 2008). These data show that intratracheal administration of IL-1β and TNF-α leads to profound but reversible pulmonary inflammation which is detectable by NMR.

3.6 Osteoarthritis

Osteoarthritis (OA) is a complex disease and has a multifactorial pathogenesis. It has many known risk factors such as age, sex, obesity, activity level, prior joint damage and genetic susceptibility. It is not classically thought of as an inflammatory disease but it may have an inflammatory element. There are currently no disease-modifying drugs for OA and very few are in development.

Synovial fluid (SF) has been used to look at OA via NMR. SF is felt to be a good medium to study as the SF is the first place where the degradation products, enzymes and signal transduction molecules involved in OA are released from the cartilage matrix. The SF should therefore have a higher concentration of metabolites compared to blood, lymph or urine.

Damayanovich et al used SF from a canine model of OA to look at metabolic profiles using NMR (Damyanovich et al., 1999). Metabolites from experimentally induced canine knee OA SF were compared to metabolites from SF of normal canine knees. They found large increases in lactate and sharp decreases of glucose in OA SF compared to normal SF suggesting that the intra-articular environment of an OA joint is more hypoxic and acidic than a healthy joint. They also found increased levels of pyruvate, lipoprotein associated fatty acids, glycerol and ketones in OA SF suggesting that lipolysis may be an important source of energy in OA. There were also elevated levels of N-acetylglycoproteins, acetate and acetamide in OA SF especially with progressive OA (Damyanovich et al., 1999).

In order to understand further the mechanisms behind OA progression, Damayanovich *et al* looked at the effect of joint afferent nerve injury (Damyanovich et al., 1999). They again used a bilateral canine model of OA. Paired SF samples were taken from dogs that had undergone bilateral anterior cruciate ligament transaction, unilateral knee denervation and contralateral sham nerve exposure. NMR was used to look at the SF. Increases in glycerol, hydroxybutyrate, glutamine, creatinine, acetate and N-acetyl-glycoprotein were seen in the SF from denervated compared to control knees. This suggests that the metabolite differences seen in the denervated knees are due to the aggravation of OA caused by joint denervation (Damyanovich et al., 1999). Hydroxybutyrate is also found in SF of RA patients (Naughton et al., 1993) suggesting that it is more of a marker of joint destruction rather than being specific for any joint disease.

Another group used guinea pigs to study OA metabolism (Lamers et al., 2003). They used Hartley outbred strain guinea pigs as they develop spontaneous progressive knee OA with features similar to human disease. The earliest histological features appear at 3 months but progress to extensive cartilage degeneration after 12 months. Urine samples were collected from these OA guinea pigs and from healthy animals at 10 and 12 months of age. They identified a metabolic fingerprint that reflected OA changes in the pigs. Lactic acid, malic acid, hypoxanthine and alanine contributed strongly to the fingerprint suggesting their involvement in OA (Lamers et al., 2003). The metabolic profile largely resembled that found in the guinea pig model. The presence of hypoxanthine suggests that OA may be an inflammatory disease due to the increased oxygen demand and altered purine metabolism.

Mass spectroscopy has also been used to look for novel biomarkers for knee OA (Zhai et al., 2010). They looked at serum samples of unrelated white women with and without knee OA. Knee OA was defined as radiographic, medically diagnosed or total knee replacement due to primary OA. They found that the ratio of valine to histidine and the ratio of leucine to histidine to be significantly associated with knee OA in humans (Zhai et al., 2010). These ratios have potential clinical use as an OA biomarker. OA branched chain amino acids (BCAA) are raised which may drive the release of acetoacetate and 3-hydroxybutyrate. These can result from the partial oxidation of leucine. BCAA are essential amino acids and therefore cannot be synthesised within the body. An increased level of BCAA may suggest an increased rate of protein breakdown or be secondary to collagen degradation. BCAA increase production of the cytokines IL1, IL2, TNF and interferon (Bassit et al., 2000) which could drive the collagen degradation.

3.7 Rheumatoid arthritis

Rheumatoid arthritis (RA) is a debilitating systemic inflammatory joint disease. An abnormal metabolic profile in the inflamed joint in RA may be due to the impairment of the vascular supply and/or an increase in the metabolic rate of the inflamed joint.

Hyaluronic acid is a major component of the proteoglycan aggregate of articular cartilage which is required for the functional integrity of extracellular matrix. In RA, SF hyaluronate is depolymerised by the action of reactive oxygen radical species (Parkes et al., 1991). Hyaluronidase activity is absent in both normal and inflamed SF. Generation of reactive oxygen species plays a principal part in synovial hypoxic reperfusion injury (Farrell et al., 1992). This occurs as increased intra-articular pressure during exercise exceeds synovial capillary perfusion pressure leading to impaired blood flow (Mapp et al., 1995).

In 1993, The Inflammation Research Group, The London Hospital Medical College looked at the NMR profiles of RA SF and matched serum samples (Naughton et al., 1993). The NMR profiles of SF were markedly different from their matched serum samples. There were high levels of lactate in the SF compared to the serum and low levels of glucose in the SF compared to the serum. These changes are consistent with the hypoxic status of the rheumatoid joint (Naughton et al., 1993). All the SF samples (RA and control) had lower levels of chylomicron and very-low-density-lipoprotein associated triglycerides compared to their matched serum samples. The SF samples also had high levels of ketone bodies compared to their matched serum samples. These results suggest that the intra-articular environment has an increased utilisation of fats for energy even though it is hypoxic (Naughton et al., 1993, Naughton et al., 1993). They were unable to compare the control SF to the rheumatoid SF due to the low levels of SF aspirated.

Serum from mice has been used to identify a metabolite biomarker pattern associated with RA (Weljie et al., 2007). Using NMR they found that uracil, xanthine and glycine could be used to distinguish arthritic from control animals (Weljie et al., 2007). The presence of the metabolites suggests that nucleic acid metabolism may be highly affected in RA and there may be an association with oxidative stress.

More recently, a group in Denmark have looked at the plasma of patients with RA (Lauridsen et al., 2010). They found differences in the metabolites between patients with RA and healthy controls and differences between patients with active RA and controlled RA. The metabolites that they identified were cholesterol, lactate, acetylated glycoprotein and lipids. The lactate levels represented oxidative damage and thus indirectly reflected active inflammation.

3.8 Atherosclerosis

Atherosclerosis is the thickening of arteries and is the underlying pathological process that affects the coronary, cerebral, aortic and peripheral arteries. Atherosclerosis involves the accumulation of cholesterol particles, cellular by-products, deposition of the extracellular matrix and inflammatory cell infiltration within the vessel wall (Goonewardena et al., 2010). Chronic inflammation has been recognised as one of the key components of atherogenesis (Ross, 1999) but accelerated atherosclerosis is an important confounder of chronic inflammatory diseases such as rheumatoid arthritis (Bacon et al., 2005). Animal models have been widely used to investigate the biochemical basis of atherosclerosis. Using aortas from apolipoprotein-E knockout mice Mayr *et* al concluded that inefficient vascular glucose and energy metabolism coincided with increased oxidative stress in animals with hyperlipidaemia (Mayr et al., 2007). NMR-based metabolomics of mouse urine has been used to look at atherosclerosis (Leo & Darrow, 2009). Using apolipoprotein-E knockout mice they compared untreated mice with those treated with captopril. They found elevated levels of xanthine and ascorbate in untreated mice which may be possible markers of plaque formation (Leo & Darrow, 2009). The interaction between diet and inflammation in promoting atherosclerosis has also been highlighted through metabolomic studies and Kleenmann (Kleemann et al., 2007) suggested that a high cholesterol intake lead to a switch in liver metabolism towards a pro-atherosclerotic state. Another recent example of how metabolomics can provide novel insights into inflammatory disease pathology was the observation that the metabolism of dietary lecithin by gut flora leads to the increased absorption and accumulation of choline derivatives which in turn promote cardiovascular disease (Wang et al., 2011) . Only through the use of the systematic analysis of metabolites using metabolomics was it possible to uncover these complex metabolic relationships underpinning the disease process.

4. Conclusion

As summarised above there is now a growing body of literature describing metabolomic changes in inflammatory diseases, both in humans and animal models. Several distinct metabolic changes have been identified in inflammatory disorders, but there is a core theme of increasing energy requirements coupled with decreasing oxygen supply within the inflammatory environment.

Studies in MS, RA, OA and inflammatory lung disease have all shown an increase in lactate, while studies of inflammatory eye and lung diseases have shown local reductions in glucose. Immunological responses to tissue hypoxia, such as the up-regulation of IL-1, IL-6, IFN-γ and TNF-α seen in macrophages, show the link between local metabolic changes and inflammatory responses. Here transcription factor HIF-1α may play a central co-ordinating role in both normal and pathological inflammation by regulating the underlying cellular metabolism towards anaerobic respiratory pathways and lactate production. Subsequent effects of inflammatory cytokines on tissue remodelling and perfusion further provide a mechanism for feedback driving self-sustaining inflammatory microenvironments, and potentially where resolution is disrupted, a route to chronic inflammatory disease.

Therefore, as both a by-product and mediator of local tissue conditions, metabolites offer a unique opportunity to gain an insight of local and global inflammatory processes. Metabolomics likewise, provides promising opportunities for both diagnosis of inflammatory diseases, and study of the underlying processes that may offer clues as to how the inflammatory process develops.

5. Acknowledgement

The authors were supported by grants from Arthritis Research UK (grant numbers 18552 and 19325) and the Wellcome Trust (089384/Z/09/Z and 066490/Z/01/A).

6. References

Alberg, A.J. (2002). The influence of cigarette smoking on circulating concentrations of antioxidant micronutrients, *Toxicology,* vol.180, No. 2, pp.121-37, ISSN 0300-483X.

Albina, J.E., Henry, W.L., Jr., Mastrofrancesco, B., Martin, B.A., Reichner, J.S. (1995). Macrophage activation by culture in an anoxic environment, *Journal of Immunology,* vol.155, No. 9, pp.4391-6, ISSN 0022-1767.

Bacon, P.A., Church, L.D., Young, S.P. (2005). Endothelial Dysfunction - the Link Between Inflammation and Atherosclerosis in Rheumatoid Arthritis, *Journal of the Indian Rheumatology Association,* vol.13, pp.103-6, ISSN 0971-5045.

Bassit, R.A., Sawada, L.A., Bacurau, R.F.P., Navarro, F., Rosa, L.F.B.P. (2000). The effect of BCAA supplementation upon the immune response of triathletes, *Medicine and Science in Sports and Exercise,* vol.32, No. 7, pp.1214-9, ISSN 0195-9131.

Beyer, C., Schett, G., Gay, S., Distler, O., Distler, J.H. (2009). Hypoxia. Hypoxia in the pathogenesis of systemic sclerosis, *Arthritis Research and Therapy,* vol.11, No. 2, pp.220, ISSN 1478-6362.

Bezabeh, T., Somorjai, R.L., Smith, I.C.P. (2009). MR metabolomics of fecal extracts: applications in the study of bowel diseases, *Magnetic Resonance in Chemistry,* vol.47, pp.S54-S61, ISSN 0749-1581.

Borgerding, M., Klus, H. (2005). Analysis of complex mixtures - Cigarette smoke, *Experimental and Toxicologic Pathology,* vol.57, pp.43-73, ISSN 0940-2993.

Brown, R.A., Spina, D., Page, C.P. (2008). Adenosine receptors and asthma, *British Journal of Pharmacology,* vol.153 Suppl 1, pp.S446-S56, ISSN 0007-1188.

Chang, X., Wei, C. (2011). Glycolysis and rheumatoid arthritis, *International Journal of Rheumatic Diseases,* vol.14, No. 3, pp.217-22, ISSN 1756-185X.

Chitayat, D., Meaghervillemure, K., Mamer, O.A., Ogorman, A., Hoar, D.I., Silver, K., Scriver, C.R. (1992). Brain Dysgenesis and Congenital Intracerebral Calcification Associated with 3-Hydroxyisobutyric Aciduria, *Journal of Pediatrics,* vol.121, No. 1, pp.86-9, ISSN 0022-3476.

Cramer, T., Yamanishi, Y., Clausen, B.E., Forster, I., Pawlinski, R., Mackman, N., et al. (2003). HIF-1alpha is essential for myeloid cell-mediated inflammation, *Cell,* vol.112, No. 5, pp.645-57, ISSN 0092-8674.

Damyanovich, A.Z., Staples, J.R., Chan, A.D.M., Marshall, K.W. (1999). Comparative study of normal and osteoarthritic canine synovial fluid using 500 MHz H-1 magnetic resonance spectroscopy, *Journal of Orthopaedic Research,* vol.17, No. 2, pp.223-31, ISSN 0736-0266.

Damyanovich, A.Z., Staples, J.R., Marshall, K.W. (1999). H-1 NMR investigation of changes in the metabolic profile of synovial fluid in bilateral canine osteoarthritis with unilateral joint denervation, *Osteoarthritis and Cartilage,* vol.7, No. 2, pp.165-72, ISSN 1063-4584.

Eltzschig, H.K., Carmeliet, P. (2011). Hypoxia and Inflammation REPLY, *New England Journal of Medicine,* vol.364, No. 20, pp.1977-, ISSN 0028-4793.

Engvall, I.L., Elkan, A.C., Tengstrand, B., Cederholm, T., Brismar, K., Hafstrom, I. (2008). Cachexia in rheumatoid arthritis is associated with inflammatory activity, physical disability, and low bioavailable insulin-like growth factor, *Scandinavian Journal of Rheumatology,* vol.37, No. 5, pp.321-8, ISSN 0300-9742.

Evans, W.J., Morley, J.E., Argiles, J., Bales, C., Baracos, V., Guttridge, D., et al. (2008). Cachexia: A new definition, *Clinical Nutrition,* vol.27, No. 6, pp.793-9, ISSN 0261-5614.

Farrell, A.J., Williams, R.B., Stevens, C.R., Lawrie, A.S., Cox, N.L., Blake, D.R. (1992). Exercise Induced Release of Vonwillebrand-Factor - Evidence for Hypoxic Reperfusion Microvascular Injury in Rheumatoid-Arthritis, *Annals of the Rheumatic Diseases,* vol.51, No. 10, pp.1117-22, ISSN 0003-4967.

Gaber, T., Haupl, T., Sandig, G., Tykwinska, K., Fangradt, M., Tschirschmann, M., et al. (2009). Adaptation of human CD4+ T cells to pathophysiological hypoxia: a transcriptome analysis, *Journal of Rheumatology,* vol.36, No. 12, pp.2655-69, ISSN 0315-162X.

Goonewardena, S.N., Prevette, L.E., Desai, A.A. (2010). Metabolomics and atherosclerosis, *Current Atherosclerosis Reports,* vol.12, No. 4, pp.267-72, ISSN 1534-6242.

Greiner, J.V., Kopp, S.J., Glonek, T. (1985). Phosphorus Nuclear-Magnetic-Resonance and Ocular Metabolism, *Survey of Ophthalmology,* vol.30, No. 3, pp.189-202, ISSN 0039-6257.

Griffin, J.L., Anthony, D.C., Campbell, S.J., Gauldie, J., Pitossi, F., Styles, P., Sibson, N.R. (2004). Study of cytokine induced neuropathology by high resolution proton NMR spectroscopy of rat urine, *FEBS Letters,* vol.568, No. 1-3, pp.49-54, ISSN 0014-5793.

Hasday, J.D., Bascom, R., Costa, J.J., Fitzgerald, T., Dubin, W. (1999). Bacterial endotoxin is an active component of cigarette smoke, *Chest,* vol.115, No. 3, pp.829-35, ISSN 0012-3692.

Hasko, G., Linden, J., Cronstein, B., Pacher, P. (2008). Adenosine receptors: therapeutic aspects for inflammatory and immune diseases, *Nature Reviews Drug Discovery,* vol.7, No. 9, pp.759-70, ISSN 1474-1784.

Heller, A., Koch, T., Schmeck, J., van Ackern, K. (1998). Lipid mediators in inflammatory disorders, *Drugs,* vol.55, No. 4, pp.487-96, ISSN 0012-6667.

Hollander, A.P., Corke, K.P., Freemont, A.J., Lewis, C.E. (2001). Expression of hypoxia-inducible factor 1alpha by macrophages in the rheumatoid synovium: implications for targeting of therapeutic genes to the inflamed joint, *Arthritis and Rheumatism,* vol.44, No. 7, pp.1540-4, ISSN 0004-3591.

Ibrahim, S.M., Gold, R. (2005). Genomics, proteomics, metabolomics: what is in a word for multiple sclerosis?, *Current Opinion in Neurology,* vol.18, No. 3, pp.231-5, ISSN 1350-7540.

Isomaki, P., Punnonen, J. (1997). Pro- and anti-inflammatory cytokines in rheumatoid arthritis, *Annals of Medicine,* vol.29, No. 6, pp.499-507, ISSN 0785-3890.

Ivanenkov, Y.A., Balakin, K.V., Tkachenko, S.E. (2008). New Approaches to the Treatment of Inflammatory Disease Focus on Small-Molecule Inhibitors of Signal Transduction Pathways, *Drugs in R&D,* vol.9, No. 6, pp.397-434, ISSN 1174-5886.

Jawed, S., Gaffney, K., Blake, D.R. (1997). Intra-articular pressure profile of the knee joint in a spectrum of inflammatory arthropathies, *Annals of the Rheumatic Diseases,* vol.56, No. 11, pp.686-9, ISSN 0003-4967.

Kallberg, H., Padyukov, L., Plenge, R.M., Ronnelid, J., Gregersen, P.K., van der Helm-van Mil, A., et al. (2007). Gene-Gene and Gene-Environment Interactions Involving HLA-DRB1, PTPN22, and Smoking in Two Subsets of Rheumatoid Arthritis, *The American Journal of Human Genetics,* vol.80, No. 5, pp.867-75, ISSN 0002-9297.

Kleemann, R., Verschuren, L., van Erk, M.J., Nikolsky, Y., Cnubben, N.H.P., Verheij, E.R., et al. (2007). Atherosclerosis and liver inflammation induced by increased dietary cholesterol intake: a combined transcriptomics and metabolomics analysis, *Genome Biology,* vol.8, No. 9, pp.R200, ISSN 1474-760X.

Lamers, R.J.A.N., DeGroot, J., Spies-Faber, E.J., Jellema, R.H., Kraus, V.B., Verzijl, N., et al. (2003). Identification of disease- and nutrient-related metabolic fingerprints in osteoarthritic guinea pigs, *Journal of Nutrition,* vol.133, No. 6, pp.1776-80, ISSN 1096-0007.

Lassmann, H. (2003). Hypoxia-like tissue injury as a component of multiple sclerosis lesions, *Journal of the Neurological Sciences,* vol.206, No. 2, pp.187-91, ISSN 0022-510X.

Lauridsen, M.B., Bliddal, H., Christensen, R., Danneskiold-Samsoe, B., Bennett, R., Keun, H., et al. (2010). (1)H NMR Spectroscopy-Based Interventional Metabolic Phenotyping: A Cohort Study of Rheumatoid Arthritis Patients, *Journal of Proteome Research,* vol.9, No. 9, pp.4545-53, ISSN 1535-3907.

Leo, G.C., Darrow, A.L. (2009). NMR-based metabolomics of urine for the atherosclerotic mouse model using apolipoprotein-E deficient mice, *Magnetic Resonance in Chemistry,* vol.47 Suppl 1, pp.S20-S5, ISSN 1097-458X.

Lin, H.M., Edmunds, S.J., Helsby, N.A., Ferguson, L.R., Rowan, D.D. (2009). Nontargeted Urinary Metabolite Profiling of a Mouse Model of Crohn's Disease, *Journal of Proteome Research,* vol.8, No. 4, pp.2045-57, ISSN 1535-3893.

Loffler, M., Zieker, D., Weinreich, J., Lob, S., Konigsrainer, I., Symons, S., et al. (2011). Wound fluid lactate concentration: a helpful marker for diagnosing soft-tissue infection in diabetic foot ulcers? Preliminary findings, *Diabetic Medicine,* vol.28, No. 2, pp.175-8, ISSN 1464-5491.

Lund-Olesen, K. (1970). Oxygen tension in synovial fluids, *Arthritis and Rheumatism,* vol.13, No. 6, pp.769-76, ISSN 0004-3591.

Lutz, N.W., Viola, A., Malikova, I., Confort-Gouny, S., Ranjeva, J.P., Pelletier, J., Cozzone, P.J. (2007). A branched-chain organic acid linked to multiple sclerosis: First identification by NMR spectroscopy of CSF, *Biochemical and Biophysical Research Communications,* vol.354, No. 1, pp.160-4, ISSN 0006-291X.

MacGregor, A.J., Snieder, H., Rigby, A.S., Koskenvuo, M., Kaprio, J., Aho, K., Silman, A.J. (2000). Characterizing the quantitative genetic contribution to rheumatoid arthritis using data from twins, *Arthritis and Rheumatism,* vol.43, No. 1, pp.30-7, ISSN 0004-3591.

Mapp, P.I., Grootveld, M.C., Blake, D.R. (1995). Hypoxia, Oxidative Stress and Rheumatoid-Arthritis, *British Medical Bulletin,* vol.51, No. 2, pp.419-36, ISSN 0007-1420.

Marchesi, J.R., Holmes, E., Khan, F., Kochhar, S., Scanlan, P., Shanahan, F., Wilson, I.D., Wang, Y.L. (2007). Rapid and noninvasive metabonomic characterization of inflammatory bowel disease, *Journal of Proteome Research,* vol.6, No. 2, pp.546-51, ISSN 1535-3893.

Mayr, M., Madhu, B., Xu, Q. (2007). Proteomics and metabolomics combined in cardiovascular research, *Trends in Cardiovascular Medicine,* vol.17, No. 2, pp.43-8, ISSN 1873-2615.

Metsios, G.S., Stavropoulos-Kalinoglou, A., Panoulas, V.F., Sandoo, A., Toms, T.E., Nevill, A.M., Koutedakis, Y., Kitas, G.D. (2009). Rheumatoid cachexia and cardiovascular disease, *Clinical and Experimental Rheumatology,* vol.27, No. 6, pp.985-8, ISSN 0392-856X.

Montecucco, F., Mach, F. (2009). Common inflammatory mediators orchestrate pathophysiological processes in rheumatoid arthritis and atherosclerosis, *Rheumatology,* vol.48, No. 1, pp.11-22, ISSN 1462-0324.

Munro, R., Capell, H. (1997). Prevalence of low body mass in rheumatoid arthritis: Association with the acute phase response, *Annals of the Rheumatic Diseases,* vol.56, No. 5, pp.326-9, ISSN 0003-4967.

Murata, Y., Ohteki, T., Koyasu, S., Hamuro, J. (2002). IFN-gamma and pro-inflammatory cytokine production by antigen-presenting cells is dictated by intracellular thiol redox status regulated by oxygen tension, *European Journal of Immunology,* vol.32, No. 10, pp.2866-73, ISSN 0014-2980.

Murdoch, C., Muthana, M., Lewis, C.E. (2005). Hypoxia regulates macrophage functions in inflammation, *Journal of Immunology,* vol.175, No. 10, pp.6257-63, ISSN 0022-1767.

Murdoch, T.B., Fu, H., MacFarlane, S., Sydora, B.C., Fedorak, R.N., Slupsky, C.M. (2008). Urinary metabolic profiles of inflammatory bowel disease in interleukin-10 gene-deficient mice, *Analytical Chemistry,* vol.80, No. 14, pp.5524-31, ISSN 0003-2700.

Naughton, D., Whelan, M., Smith, E.C., Williams, R., Blake, D.R., Grootveld, M. (1993). An Investigation of the Abnormal Metabolic Status of Synovial-Fluid from Patients with Rheumatoid-Arthritis by High-Field Proton Nuclear-Magnetic-Resonance Spectroscopy, *FEBS Letters,* vol.317, No. 1-2, pp.135-8, ISSN 0014-5793.

Naughton, D.P., Haywood, R., Blake, D.R., Edmonds, S., Hawkes, G.E., Grootveld, M. (1993). A Comparative-Evaluation of the Metabolic Profiles of Normal and Inflammatory Knee-Joint Synovial-Fluids by High-Resolution Proton Nmr-Spectroscopy, *FEBS Letters,* vol.332, No. 3, pp.221-5, ISSN 0014-5793.

Ng, C.T., Biniecka, M., Kennedy, A., McCormick, J., FitzGerald, O., Bresnihan, B., et al. (2010). Synovial tissue hypoxia and inflammation in vivo, *Annals of the Rheumatic Diseases,* vol.69, No. 7, pp.1389-95, ISSN 0003-4967.

Ngumah, Q.C., Buchthal, S.D., Dacheux, R.F. (2006). Longitudinal non-invasive proton NMR spectroscopy measurement of vitreous lactate in a rabbit model of ocular hypertension, *Experimental Eye Research,* vol.83, No. 2, pp.390-400, ISSN 1096-0007.

Nicoli, F., VionDury, J., ConfortGouny, S., Maillet, S., Gastaut, J.L., Cozzone, P.J. (1996). Cerebrospinal fluid metabolic profiles in multiple sclerosis and degenerative dementias obtained by high resolution proton magnetic resonance spectroscopy, *Comptes Rendus de l Academie des Sciences Serie Iii-Sciences de la Vie-Life Sciences,* vol.319, No. 7, pp.623-31, ISSN 0764-4469.

Nishimoto, N., Sugino, H., Lee, H.M. (2010). DNA microarray analysis of rheumatoid arthritis susceptibility genes identified by genome-wide association studies (vol 12, pg 403, 2010), *Arthritis Research and Therapy,* vol.12, No. 3, ISSN 1478-6362.

Parkes, H.G., Grootveld, M.C., Henderson, E.B., Farrell, A., Blake, D.R. (1991). Oxidative Damage to Synovial-Fluid from the Inflamed Rheumatoid Joint Detected by H-1-Nmr Spectroscopy, *Journal of Pharmaceutical and Biomedical Analysis,* vol.9, No. 1, pp.75-82, ISSN 0731-7085.

Renz, H., von Mutius, E., Brandtzaeg, P., Cookson, W.O., Autenrieth, I.B., Haller, D. (2011). Gene-environment interactions in chronic inflammatory disease, *Nature Immunology,* vol.12, No. 4, pp.273-7, ISSN 1529-2908.

Rieckmann, P., Smith, K.J. (2001). Multiple sclerosis: more than inflammation and demyelination, *Trends in Neurosciences,* vol.24, No. 8, pp.435-7, ISSN 0166-2236.

Ross, R. (1999). Mechanisms of disease - Atherosclerosis - An inflammatory disease, *New England Journal of Medicine,* vol.340, No. 2, pp.115-26, ISSN 0028-4793.

Scannell, G. (1996). Leukocyte responses to hypoxic/ischemic conditions, *New Horizons,* vol.4, No. 2, pp.179-83, ISSN 1063-7389.

Schicho, R., Nazyrova, A., Shaykhutdinov, R., Duggan, G., Vogel, H.J., Storr, M. (2010). Quantitative metabolomic profiling of serum and urine in DSS-induced ulcerative colitis of mice by (1)H NMR spectroscopy, *Journal of Proteome Research,* vol.9, No. 12, pp.6265-73, ISSN 1535-3907.

Serhan, C.N. (2009). Systems approach to inflammation resolution: identification of novel anti-inflammatory and pro-resolving mediators, *Journal of Thrombosis and Haemostasis,* vol.7, pp.44-8, ISSN 1538-7933.

Serkova, N.J., Van Rheen, Z., Tobias, M., Pitzer, J.E., Wilkinson, J.E., Stringer, K.A. (2008). Utility of magnetic resonance imaging and nuclear magnetic resonance-based metabolomics for quantification of inflammatory lung injury, *American Journal Of Physiology-Lung Cellular And Molecular Physiology,* vol.295, No. 1, pp.L152-L61, ISSN 1040-0605.

Simone, I.L., Federico, F., Trojano, M., Tortorella, C., Liguori, M., Giannini, P., Picciola, E., Natile, G., Livrea, P. (1996). High resolution proton MR spectroscopy of cerebrospinal fluid in MS patients. Comparison with biochemical changes in demyelinating plaques, *Journal of the Neurological Sciences,* vol.144, No. 1-2, pp.182-90, ISSN 0022-510X.

Sinclair, A.B., Viant, M.R., Ball, A.K., Burdon, M.A., Walker, E.A., Stewart, P.M., Rauz, S., Young, S.P. (2010). NMR-Based Metabolomic Analysis of Cerebrospinal Fluid and Serum in Neurological Diseases - A Diagnostic Tool?, *NMR in Biomedicine,* vol.23, No. 2, pp.123-32, ISSN 1099-1492.

Sinclair, A.J., Ball, A.K., Burdon, M.A., Clarke, C.E., Stewart, P.M., Cumow, S.J., Rauz, S. (2008). Exploring the pathogenesis of IIH: An inflammatory perspective, *Journal of Neuroimmunology,* vol.201, pp.212-20, ISSN 0165-5728.

Sitkovsky, M., Lukashev, D. (2005). Regulation of immune cells by local. tissue oxygen tension: Hif1 alpha and adenosine receptors, *Nature Reviews Immunology,* vol.5, No. 9, pp.712-21, ISSN 1474-1733.

Stappenbeck, T.S., Cadwell, K., Patel, K.K., Maloney, N.S., Liu, T.C., Ng, A.C.Y., et al. (2010). Virus-Plus-Susceptibility Gene Interaction Determines Crohn's Disease Gene Atg16L1 Phenotypes in Intestine, *Cell,* vol.141, No. 7, pp.1135-64, ISSN 0092-8674.

Strober, W., Fuss, I., Mannon, P. (2007). The fundamental basis of inflammatory bowel disease, *Journal of Clinical Investigation,* vol.117, No. 3, pp.514-21, ISSN 0021-9738.

Summers, G.D., Deighton, C.M., Rennie, M.J., Booth, A.H. (2008). Rheumatoid cachexia: a clinical perspective, *Rheumatology,* vol.47, No. 8, pp.1124-31, ISSN 1462-0324.

Summers, G.D., Metsios, G.S., Stavropoulos-Kalinoglou, A., Kitas, G.D. (2010). Rheumatoid cachexia and cardiovascular disease, *Nature Reviews Rheumatology,* vol.6, No. 8, pp.445-51, ISSN 1759-4790.

Trabold, O., Wagner, S., Wicke, C., Scheuenstuhl, H., Hussain, M.Z., Rosen, N., Seremetiev, A., Becker, H.D., Hunt, T.K. (2003). Lactate and oxygen constitute a fundamental regulatory mechanism in wound healing, *Wound Repair and Regeneration,* vol.11, No. 6, pp.504-9, ISSN 1524-475X.

Trapp, B.D., Bo, L., Mork, S., Chang, A. (1999). Pathogenesis of tissue injury in MS lesions, *Journal of Neuroimmunology,* vol.98, No. 1, pp.49-56, ISSN 0165-5728.

Treuhaft, P.S., McCarty, D.J. (1971). Synovial fluid pH, lactate, oxygen and carbon dioxide partial pressure in various joint diseases, *Arthritis and Rheumatism,* vol.14, No. 4, pp.475-84, ISSN 0004-3591.

Vergadi, E., Chang, M.S., Lee, C., Liang, O.D., Liu, X., Fernandez-Gonzalez, A., Mitsialis, S.A., Kourembanas, S. (2011). Early macrophage recruitment and alternative activation are critical for the later development of hypoxia-induced pulmonary hypertension, *Circulation,* vol.123, No. 18, pp.1986-95, ISSN 0009-7322.

Wang, Z.N., Klipfell, E., Bennett, B.J., Koeth, R., Levison, B.S., Dugar, B., et al. (2011). Gut flora metabolism of phosphatidylcholine promotes cardiovascular disease, *Nature,* vol.472, No. 7341, pp.57-65, ISSN 0028-0836.

Weljie, A.M., Dowlatabadi, R., Miller, B.J., Vogel, H.J., Jirik, F.R. (2007). An inflammatory arthritis-associated metabolite biomarker pattern revealed by H-1 NMR Spectroscopy, *Journal of Proteome Research,* vol.6, No. 9, pp.3456-64, ISSN 1535-3893.

Wheaton, W.W., Chandel, N.S. (2011). Hypoxia. 2. Hypoxia regulates cellular metabolism, *American Journal of Physiology - Cell Physiology,* vol.300, No. 3, pp.C385-C93, ISSN 0363-6143.

White, J.R., Harris, R.A., Lee, S.R., Craigon, M.H., Binley, K., Price, T., Beard, G.L., Mundy, C.R., Naylor, S. (2004). Genetic amplification of the transcriptional response to hypoxia as a novel means of identifying regulators of angiogenesis, *Genomics,* vol.83, No. 1, pp.1-8, ISSN 0888-7543.

Williams, H.R.T., Cox, I.J., Walker, D.G., North, B.V., Patel, V.M., Marshall, S.E., et al. (2009). Characterization of Inflammatory Bowel Disease With Urinary Metabolic Profiling, *American Journal of Gastroenterology,* vol.104, No. 6, pp.1435-44, ISSN 0002-9270.

Young, S.P., Nessim, M., Falciani, F., Trevino, V., Banerjee, S.P., Scott, R.A.H., Murray, P.I., Wallace, G.R. (2009). Metabolomic analysis of human vitreous humor differentiates ocular inflammatory disease, *Molecular Vision,* vol.15, No. 125-29, pp.1210-7, ISSN 1090-0535.

Young, S.P., Wallace, G.R. (2009). Metabolomic analysis of human disease and its application to the eye, *Journal of Ocular Biology, Disease and Informatics,* vol.2, No. 4, pp.235-42, ISSN 1936-8445.

Zhai, G., Wang-Sattler, R., Hart, D.J., Arden, N.K., Hakim, A.J., Illig, T., Spector, T.D. (2010). Serum branched-chain amino acid to histidine ratio: a novel metabolomic biomarker of knee osteoarthritis, *Annals of the Rheumatic Diseases,* vol.69, No. 6, pp.1227-31, ISSN 0003-4967.

12

Clinical Implementation of Metabolomics

Akira Imaizumi et al.*
Amino Acids Basic And Applied Research Group, Frontier Research Laboratories, Institute for Innovation, Ajinomoto, Co., Inc.
Japan

1. Introduction

1.1 Overview

Metabolomics, which is also referred to as metabonomics, metabolic profiling or metabolic fingerprinting, is the comprehensive quantitative measurement of endogenous metabolites within a biological system (Fiehn, 2002; Kaddurah-Daouk et al, 2008; Spratlin et al, 2009). Detection of metabolites is in general carried out in cell extracts, tissue specimens, or various biological fluids including serum, plasma, urine and cerebrospinal fluid (CSF) by liquid chromatography mass spectrometry (LC-MS), gas chromatography–mass spectrometry (GC-MS), capillary electrophoresis–mass spectrometry (CE-MS) or nuclear magnetic resonance spectroscopy (NMR). Metabolomics captures the status of diverse biochemical pathways in a particular situation and can define the metabolic status of an organism (Aranibar et al, 2011; DeFeo et al, 2011; Lu et al, 2008; Roux et al, 2011; Soga, 2007; Yuan et al, 2007). In clinical settings, biomarkers generated from metabolomics have become one of the most essential diagnostic criteria that can be objectively measured and evaluated as indicators of normal or pathological states, as well as a tool to assess responses to therapeutic interventions (Hunter, 2009; Spratlin et al, 2009; van der Greef et al, 2006; Zeisel, 2007). As we describe in this chapter, novel metabolomic markers, for instance, for cancer therapy, glucose intolerance, hepatic steatosis, nephrotic and psychiatric disorders, and their incorporation into clinical decision-making may considerably change future health care.

In order for metabolomics to be successful in clinical settings, it must surpass conventional methods in reliability and predictive capability, and/or should be more informative about disease pathogenesis. Utilizing a systems biology approach in

*Natsumi Nishikata[1], Hiroo Yoshida[2], Junya Yoneda[1], Shunji Takahena[2],
Mitsuo Takahashi[1], Toshihiko Ando[3], Hiroshi Miyano[2], Kenji Nagao[1], Yasushi Noguchi[1],
Nobuhisa Shimba[3,4] and Takeshi Kimura[4]
[1]*Amino Acids Basic And Applied Research Group, Frontier Research Laboratories, Institute for Innovation, Ajinomoto, Co., Inc., Japan*
[2]*Analytical Sciences Group, Fundamental Technology Laboratories, Japan Institute for Innovation, Ajinomoto, Co., Inc.*
[3]*AminoIndex Department, Wellness Business Division, Ajinomoto, Co., Inc., Japan*
[4]*R&D Planning Department, Ajinomoto, Co., Inc., Japan*

biomarker investigation may allow for a deeper understanding of disease associated metabolism. (Jenkins et al, 2004; Kell, 2006). A systems biology approach does not focus on identifying a single target or mechanism of an observed phenotype. Instead it seeks to identify the biological networks or pathways that connect the differing elements of a system (Wheelock et al, 2009). When a shift in equilibrium is observed in a disease, such as altered metabolic fluxes or enzymatic activities, it can be elucidated that those components of the network that are associated with the observed shift are characteristic and potentially descriptive of the disease, and that they accordingly represent potential targets for intervention. Thus, the systems approach in combination with metabolomics, may lead to the discovery of panels of metabolites that more accurately capture the disease status and help acquire information valuable for individualized clinical care (Quinones & Kaddurah-Daouk, 2009). Clinical metabolomics is expected to be a promising technology for personalized medicine and nutrition. A metabolic marker designed to predict individual response including efficacy and side effects during therapeutic intervention for each patient will enable administration of optimal treatments and improve clinical outcome.

1.2 Comprehensive vs. focused metabolomics

The spectrum of biochemicals in a clinical specimen, range from organic acids, amino acids, lipids, nucleic acids and their metabolic intermediates to complex secondary metabolites with signaling functions. Today, however, clinicians in human health care utilize only a very small part of the information contained in the metabolome. Although NMR or MS technology enables a comprehensive (i.e. global) measurement of various small molecules, in many cases, it is simply too difficult to quantify each molecule and understand underlying mechanisms from a global dataset by a single measurement (Steuer, 2006). This has led a number of researchers to look at a focused set of (i.e. local) metabolites such as amino acids or lipids (German et al, 2007; Kimura et al, 2009), where data from multiple measurements such as transcriptomics, proteomics and metabolomics can be effectively integrated to allow more insight into the underlying metabolic alternations, by projecting multiple datasets onto biochemical pathways and analyzing their interactions under a particular physiological state (Caesar et al, 2010; Momin et al, 2011; Noguchi et al, 2008; Zhang et al, 2011).

Recently, there have been reports of trials in integrating different types of 'omics' datasets for the systemic understanding of metabolic phenotypes at multiple levels. Various software packages are available in integrating nonuniform 'omics' datasets (Grimplet et al, 2009; Gruning et al, 2010; Taylor & Singhal, 2009). The link between information and modeling can be achieved by two major types of complementary approaches, a data-oriented exploratory approach, in which data generates information about the structure and relationships between the observed variables in a given system, and a model-based bottom-up approach, in which cybernetic and systems–theoretical knowledge are used to create models that describe mechanisms and dynamics of a system. Formerly, a model-based approach had been used for studying in-vitro cellular or organ systems; however, because of the complexity in modeling whole body systems, recently this approach has been replaced by a data-oriented approach, particularly when dealing with in-vivo 'omics' data in various models including animal models and clinical studies (Dunn et al,

2011). The critical step is the construction of models from the raw dataset of transcriptomics, proteomics, and metabolomics. This may be achieved by using different mathematical techniques ranging from simple Pearson correlations to the use of ordinary differential equations (Wheelock et al, 2009). Through this modeling, fundamental concepts in the understanding of biological systems like robustness, modularity, emergence, etc. are incorporated.

Most studies currently remain focused on local level networks within a set of related genes or protein expressions (Bapat et al, 2010; Kirouac et al, 2010). Yet a combination of different levels of networks can be connected to overview the whole system. A change in the gene regulatory network may have a corresponding effect in the protein–protein interaction network, the metabolic network, etc., which collectively may manifest changes in the pathological phenotype. To understand the whole system, it is critical to integrate knowledge from different datasets. Although some progress has been made in amino acid metabolism, the integration of different types of datasets is still difficult due to differences in dynamic range, scales, or analytical errors, particularly in metabolomic analysis (Ishii et al, 2007; Momin et al, 2011; Noguchi et al, 2008). Therefore, focused-metabolomics, with well managed measurements in terms of accuracy and reproducibility, for lipid, amino acid and glucose metabolism appears to be a realistic approach to illustrate how the phenotype is altered when the metabolic network itself is modified through the alteration of endogenous or environmental factors.

1.3 Generation of multiple metabolite markers

When generating biomarkers from metabolomic analysis, marker identification, verification, and also statistical and experimental evaluations, using bioinformatic techniques of identified candidate markers are required. Recently, various data mining methodologies have been reported for identifying and prioritizing reliable metabolomic markers with high diagnostic capability (Caruana, 2006; Duda, 2001; Gu et al, 2011; Kim et al, 2010; Maeda et al, 2010; Montoliu et al, 2009). In cohort studies, the definite diagnoses of the patients are normally known beforehand. In such trials, "supervised" statistical methods which consider patient classification tend to be more efficient in information utilization and suitable for obtaining targeted metabolite markers.. In contrast, when phenotypes in patients are undetermined, "unsupervised" analysis such as cluster analysis are useful tools for biomarker identification and classification of specimen groups. Moreover, improvement in discriminatory power has been reported when multivariate mathematical models are constructed combining multiple metabolite markers. These approaches include discriminant analysis methods such as linear discriminant analysis, logistic regression analysis, decision trees, the k-nearest neighbor classifier (k-NN), an instance-based learning algorithm, support vector machines or artificial neural networks (Duda, 2001). The Receiver Operating Characteristics (ROC), or the area under the ROC curve (AUC) of multivariate markers is used to represent its discriminatory performance as a trade off between selectivity and sensitivity(Hanley & McNeil, 1982). Obtained metabolomic markers are also required to be experimentally validated using larger datasets from multiple clinical trials and also statistically validated using cross validation, leave-one-out cross validation, and bootstrapping.

2. Practical Issues in the clinical implementation of metabolomics

2.1 Sample stability issues

Enormous information can be obtained by analyzing large numbers of metabolites, and it is utilized for various fields such as health and nutrition. However, the chemical and enzymatic stabilities of most metabolites are unknown. Therefore, inappropriate handling of samples can lead to inaccurate measurements. In this section, blood sampling issues for amino acids analysis as a typical case of sample handling are described. There are mainly four steps in the blood sampling process for amino acids analysis; 1) blood collection, 2) centrifugation, 3) sample storage, and 4) deproteinization. In this section, the crucial points for each step are outlined to highlight the importance of sampling processes in metabolomic studies.

2.1.1 Blood collection

The concentrations of amino acids are known to show circadian rhythms and some of them vary 30% within a day (Forslund et al, 2000). Therefore, it is desirable to collect the blood at a fixed time point. Moreover, since the amino acid concentrations increase after a protein containing meal, blood collecting between 7am and 10am in a fasting state is desirable.

The concentrations of some amino acids are known to be quite different between blood cells and plasma. The differences of essential amino acids are small, but the concentrations of nonessential amino acids can be greater by severalfold in blood cells (Filho et al, 1997). There are also many metabolic enzymes such as arginase in blood cells which will act on the plasma free amino acids (PFAAs). Therefore it is important to verify that haemolysis dose not occur in blood samples. If the blood sample shows heavy haemolysis, it is desirable to take another sample.

If blood samples are left at room temperature after collection until centrifugation, many amino acids are metabolized due to metabolic enzymes from blood cells. In particular, there are many enzymes for metabolizing nonessential amino acids. For instance, glutamine and asparagine are well known to be metabolized to glutamate and aspartate. The concentration change of glutamate at different temperatures is shown in Figure 1. This suggests that it is desirable to cool blood samples after collecting. In another study, we also found that it is essential to cool down the blood samples to 0°C immediately after collecting and that Ice-water is better than the refrigerator or ice because of the faster cooling rate.

However it is not always easy to prepare ice-water in the medical institutions at the time of blood collection. For this reason, we have developed a portable blood tube cooler (CubeCooler™, Figure 2). This cooler is composed of high thermal conductive container (aluminum) and insulator (polyethylene form), which enables the quick cooling of blood samples as well as ice-water and maintains the temperature for 12h (Figure 3). There are many coolers which is commercially available. As far as we have examined, these coolers, however, could not achieve a cooling rate as close to that of ice-water and could not cool blood samples for a long time without differences in temperature arising between tubes inserted in different holes. Thus, the cooler we have developed may be a useful tool not only for amino acid analysis but also for sample management in other metabolomic studies.

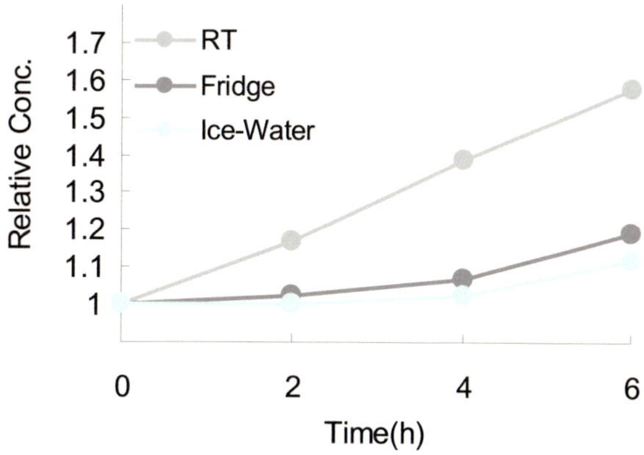

Fig. 1. Effect of cooling on concentration of glutamate in whole blood

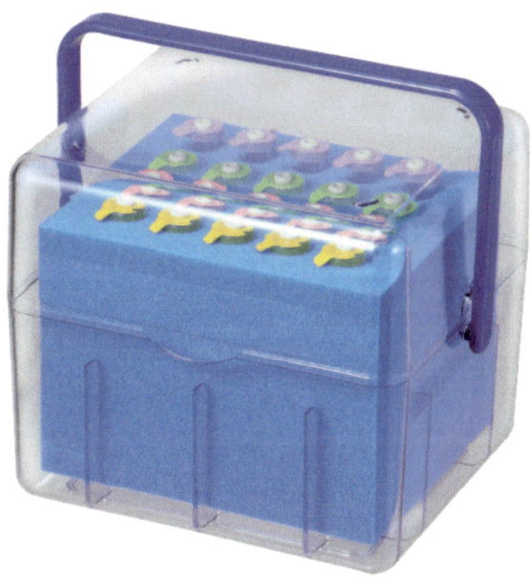

Fig. 2. View of the blood tube cooler (CubeCooler™)

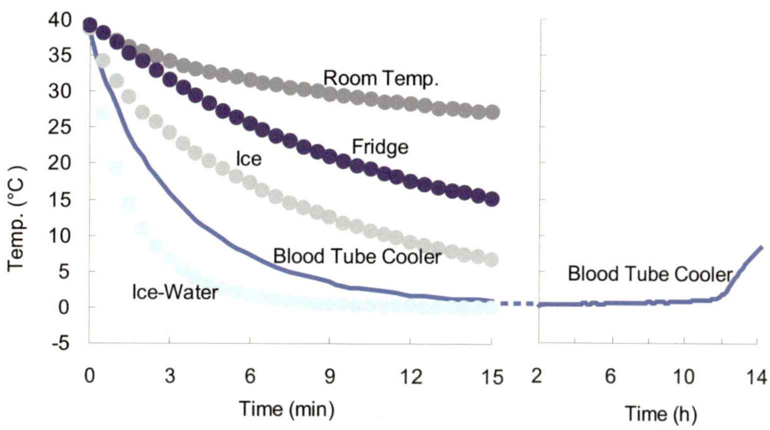

Fig. 3. Cooling rate when the blood tubes are set in various conditions and cooling duration of the blood tube cooler

2.1.2 Centrifugation

It is desirable to store blood samples in ice-water after collection and to separate the plasma from the blood cells within a few hours. As mentioned above, since blood cells contains many amino acids and enzymes, it is important not to contaminate the plasma with platelets. If contamination occurs, the concentrations of some amino acids, such as glutamate, aspartic acid and taurine can be high.

2.1.3 Sample storage

It is necessary to store the plasma in a freezer in case of long term storage. When stored at -20°C, some amino acids, especially glutamate, aspartate and cysteine can gradually decrease. Therefore -80°C freezer should be used for long term storage of plasma samples. When transporting the samples, the samples should be carried in a box filled with dry-ice.

2.1.4 Deproteinization

Since plasma contains proteins such as albumin, deproteinization is necessary before amino acid analysis. When analyzed with amino acid analyzer, plasma is generally mixed with trichloro-acetic acid or sulfo-salicylic acid and the precipitate is centrifuged. Since these reagents are strong acids, it is necessary to rapidly analyze amino acids or store in -80°C freezer so that some amino acids like glutamine are not decomposed due to acid hydrolysis. When analyzing with LC-MS or LC-MS/MS, organic solvents such as methanol and acetonitrile is useful for deproteinization. In this case, the organic solvent may influence the derivatization reaction and separation of amino acids. Since recovery rates for amino acids depend on the procedure of deproteinization, it is desirable to unify the procedure. When analyzing with LC-MS or LC-MS/MS, recovery rates can be calculated by adding stable-isotope-labeled amino acids as internal standards before deproteinization.

2.2 Analytical issues

Nuclear magnetic resonance (Bollard et al, 2001), mass spectrometry (Piraud et al, 2003), gas chromatography mass spectrometry (Thysell et al, 2010), liquid chromatography mass spectrometry (LC-MS) (Lin et al, 2011a), and capillary electrophoresis mass spectrometry (Sugimoto et al, 2010) have been used as primary tools employed for metabolomics.

A clinical metabolomics approach with LC-MS can be broadly classified into comprehensive and targeted analysis. Comprehensive analysis aims to identify and quantify all detectable metabolites in a single run. This analysis offers the advantage of giving much information. In the past, the retention and separation of polar metabolites had been difficult in LC-MS analysis. This was a weakness of LC-MS analysis, and LC-MS was limited to the analysis of hydrophobic metabolites such as lipids. However, the development of column technology enabled the retention and separation of hydrophilic metabolites(Alpert, 1990; Yoshida et al, 2007). This technology has been applied for the research of drug metabolites (Plumb et al, 2003), galactosamine toxicity (Spagou et al, 2011), and renal cell carcinoma diagnosis, staging, and biomarker discovery (Lin et al, 2011a).

In targeted analysis, a selected number of predefined metabolites are quantified. This analysis is sometimes used for quantification of metabolites, which is extracted from comprehensive analysis. Derivatization methods, based on specific reactions to targeted functional groups are major tools in targeted analysis. This method allows for sensitive and selective quantification of endogenous metabolites with amino and carboxyl groups (Tsukamoto et al, 2006; Yang et al, 2006). An advantage of this method is to be able to select a suitable sample preparation for each endogenous metabolite with the same functional group, because of the similar physical and chemical properties. This method is also very important for accurate quantification, because sample stability is different for each endogenous metabolite.

The analysis of amino acids with an amino group has a long history. In 1958, a key application for physiological amino acid analysis was supplanted by ion exchange column chromatography separations on an automated apparatus designed and built by postdoctoral fellow Darrel H. Spackman at the request of his mentor William H. Stein, and Stanford Moore at Rockerfeller University (Moore et al, 1958). This automated system reduced the analytical time from a few weeks to a full day and provided easy to use operation. The present system is used for the study of inborn errors of amino acid metabolism in clinical laboratories (Qu et al, 2001).

Recently, pre-column derivatization reagents for amino acid analyses have been developed, mainly to achieve greater sensitivity and selectivity, and much attention is paid to the design of derivatization reagents for LC-MS (Yang et al, 2006) and LC-MS/MS (Shimbo et al, 2009a; Shimbo et al, 2009b). These reagents have three notable characteristics (Figure 4). First, the reagent must have sufficient hydrophobicity to enable the retention of amino acids. Secondly, is should have a desirable structure which will increases ionization efficiency. Thirdly, it should be designed to provide characteristic and selective cleavage at the bonding site between the reagent moiety and the amino acid in the collision cell of the triple-stage quadrupole mass spectrometer. Using precursor ion scanning, endogenous metabolites with amino groups are can be extracted on ion chromatograms, even in crude biological samples.

3-aminopyridyl-N-hydroxysuccinimidyl carbamate (APDS) reagent is known to provide rapid analysis and separation of amino acids of the same charge to mass ratio on a column (Shimbo et al, 2009b) (Figure 5). This reagent is applied to the modelling of a diagnostic index, "AminoIndex technology", from differences in PFAA profiles between non-cachectic colorectal/breast/lung cancer patients and healthy individuals. (Maeda et al, 2010; Okamoto et al, 2009).

Fig. 4. Typical reaction of amino acids with a derivatizaiton reagent for LC-MS/MS. This reagent has three notable characteristics; 1) sufficient hydrophobicity (benzene ring) 2) increases ionization efficiency (quaternary amine) 3) characteristic and selective cleavage (the reagent moiety and the amino acid).

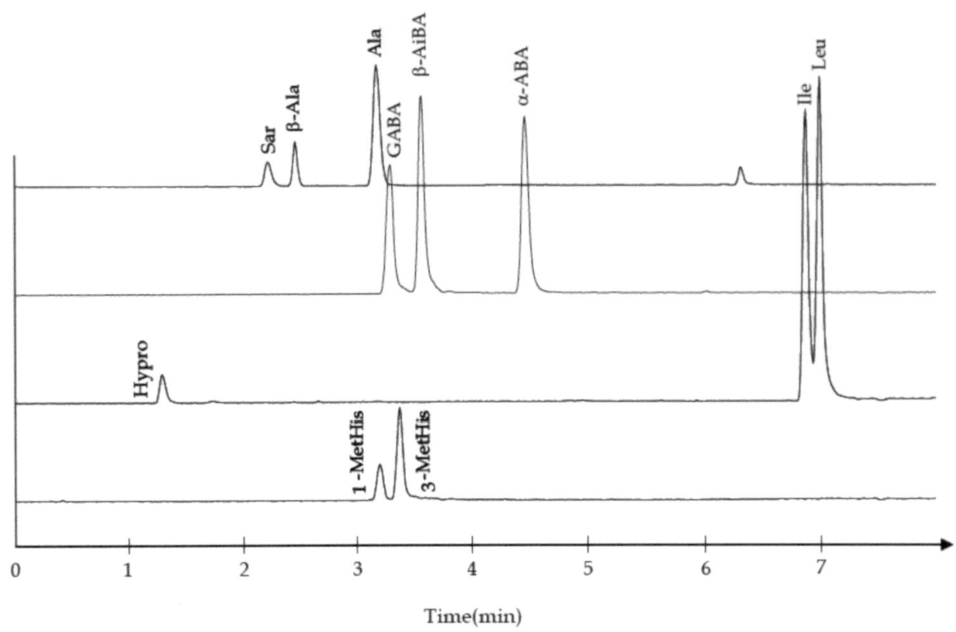

Fig. 5. Typical chromatograms of amino acids which were the same charge to mass ratio on a column.

2.3 Statistical issues

Recently, several applications of metabolome analysis based on computer-aided detection and diagnosis (CAD) has been demonstrated (Duda, 2001; Gu et al, 2011; Kell, 2002; Kim et al, 2010; Montoliu et al, 2009; Righi et al, 2009; Serkova et al, 2011; Taylor et al, 2010). The importance of objective indices for diagnosis based on empirical and statistical knowledge are increasing due to the trend called "Evidence Based Medicine (EBM)". Although CAD follows this trend, the required level of statistical analysis is also increasing and becoming more complex. The requirement of clinical investigation includes not only statistical significance but also feasible and in-depth clinical protocols in which necessary and sufficient conditions need to be satisfied. In this section, multivariate statistical aspects of metabolome analysis focused on the establishment of medical evidence and investigation of biomarkers will be introduced and discussed.

Reproducibility is the most important point of a diagnostics index. It is more complicated to guarantee the statistical reproducibility by multivariate analysis than univariate analysis. Adequate experimental design prior to data collection is therefore crucial for the quality control of the analysis (Hulley, 2006). In general, knowledge obtained from statistical analysis is only capable within the realm in which the data was analyzed and therefore cannot extrapolate beyond the realm. Generally, larger sample size is required in case of multivariate analysis because freedom of variable space is higher than univariate analysis. For example, multivariate analysis of variance (MANOVA) and data simulation are used to determine the appropriate sample size. Additionally, it is sometimes necessary for a data set to be normalized or scaled for unbiased analysis.

The most important point of analysis is algorithm selection. It is well-known as the "no free-lunch theorem", that it is impossible to determine the most suitable algorithm a priori, and that the pros and cons of each algorithm are not always specific, but dependent on each situation. Therefore, preliminary analysis to determine the most felicitous algorithm is necessary in each case. Univariate analysis can be performed to figure the behavior of each metabolite and to select the variable, i.e. dimensionality reduction of variable space, prior to multivariate analysis. It should be noted that the metabolome data are often so connected that there is a potential pitfall of statistical analysis, so-called multicollinearity, where the excess reduction of dimension sometimes can lead to the loss of latent network structure of metabolites. Multivariate analytical methods are applicable for simplification or dimensionality reduction of data to easily figure out visualized images of the "metabolite space" which has huge body of dimensions (metabolites).

Algorithms for multivariate analyses are categorized into two different groups, i.e., unsupervised methods and supervised methods. Unsupervised methods do not require objective variables such as subject status, other observed data, etc., while supervised methods require them for the data set to be analyzed. The examples of multivariate algorithms are listed in Table 1. Unsupervised learning methods are especially useful for investigating the latent structure and decreasing the redundancy of data and therefore they are sometime performed in combination. The advantages of unsupervised methods are that they minimize the loss of information (Maeda et al, 2010). However, whether the results of unsupervised methods can provide the appropriate interpretation or not depends on the setting of parameters or the problem to be analyzed.

Models	Unsupervised learning	Supervised learning	
		Continuous	Discrete
Linear model	Factor analysis	Multiple linear regression (MLR)	Linear discriminant analysis (LDA)
	Principal component analysis (PCA)	Canonical correlation analysis	Canonical discriminant analysis
	Independent component analysis (ICA)	Partial least square regression (PLS)	Partial least square discriminant analysis (PLS-DA)
Nonlinear model	Hierarchical cluster analysis (HCA)	Logistic regression analysis	Naïve Bayes classifier
	K-means cluster analysis	Conditional logistic regression analysis	Support vector machine (SVM)
	Mixture of Gaussians	Generalized linear model (GLM)	Decision trees

Table 1. Algorithm examples for multivariate analysis

On the contrary, supervised methods (Caruana, 2006) themselves contain the objective variables. Therefore the goal of analysis is to find a model (or classifier) in which the error between the model's response and the target traits is minimized to fit the target traits. Target traits can be discrete (e.g., disease vs. healthy, grade of disease) or continuous (e.g., measurement value). Supervised methods are also applicable to discover and predict which metabolites are responsible for the target traits (Maeda et al, 2010; Okamoto et al, 2009; Zhang et al, 2006). However, the generality of the model obtained from those methods can not be always guaranteed because of the potential overfitting or bias of data. Therefore, validation of the obtained model is necessary to establish the usefulness for practical use. Validation methods are categorized into two classes. The first is cross- validation in which single or multiple samples are iteratively left out from the training data set, and the remaining samples are used to evaluate the predictive performance of the model. The other is usage of external validation data set which must not be used for construction of models. Ideally, the latter case in which blinded data set is used is the most appropriate validation. However, it is sometimes difficult to perform the validation test itself.

Various metrics are used as criterion of the performance of diagnosis. In the case of the model in which the object variable contains only two classes (e.g., controls and patients), receiver-operator characteristic (ROC) curve analysis is the most appropriate criteria for evaluating the model because this analysis is independent of both sample size of each group and threshold. As threshold metrics, sensitivity, specificity, positive predictive value (PPV), negative predictive value (NPV), and accuracy are used. Among them, both sensitivity and specificity is independent of sample size and ratio of each group while the others are dependent. Therefore, to determine threshold in terms of PPV, NPV, and accuracy, it is necessary to take into account the "real" distribution of subjects.

3. Examples of clinical implementation of focused metabolomics

3.1 „AminoIndex technology": Example for early cancer diagnosis

Several investigators have reported changes in plasma free amino acid (PFAA) profiles in cancer patients (Cascino et al, 1995; Lai et al, 2005; Lee et al, 2003; Maeda et al, 2010; Naini et al, 1988; Norton et al, 1985; Okamoto et al, 2009; Proenza et al, 2003; Vissers et al, 2005; Zhang & Pang, 1992). Despite evidence of a relationship between PFAA profiles and some types of cancer, few studies have explored the use of PFAA profiles for diagnosis because although PFAA profiles differ significantly between patients, the differences in individual amino acids do not always provide sufficient discrimination abilities by themselves (Cascino et al, 1995; Lai et al, 2005; Naini et al, 1988; Norton et al, 1985; Proenza et al, 2003; Vissers et al, 2005). To address this issue, we have studied using diagnostic indices based on PFAA concentrations that compress multidimensional information from PFAA profiles into a single dimension to maximize the differences between patients and controls.

In previous studies, the alterations in PFAA profiles in cancer patients sometimes seem inconsistent, and some discrepancies existed between our study and those reported (Cascino et al, 1995; Lai et al, 2005; Naini et al, 1988; Norton et al, 1985; Proenza et al, 2003; Vissers et al, 2005). This discrepancy may be due not only to the statistical aspect of data, for example, sample size, the biased distribution of cancer stages, etc., but also to some other factors such as amino acid measurement methods. In contrast to previous studies, we performed analyses using samples in which PFAAs were measured in a unified protocol to guarantee the robustness of analysis in terms of the quality of data (Shimbo et al, 2009a; Shimbo et al, 2009b; Shimbo et al, 2009c).

As a pilot study, we investigated the possibility for early detection of colorectal cancer (CRC) and breast cancer (BC) patients (Okamoto et al, 2009). PFAA profiles were compared between cancer patients (who had CRC or BC) and control subjects. The plasma concentrations of several amino acids in the CRC patients were significantly different from those observed in the controls. The alteration of the PFAA profile in BC differed from that in CRC, with fewer changes observed. Multiple logistic regression analyses with selected variables using each data set resulted in AUC of ROC of0.860 for CRC and 0.906 for BC, respectively when using training data sets. To confirm the performance of the obtained classifier, ROC curves were also generated from the split test data. These reproduced similar diagnostic performances, with AUC of 0.910 for CRC, and 0.865 for BC, respectively.

We then investigated the possibility for early detection of non-small-cell lung cancer (NSCLC) using a larger size of samples (Maeda et al, 2010). 141 NSCLC patients and 423 age-matched, gender-matched healthy controls without apparent cancers were used as the study data set. As a result, fifteen amino acids (Ser, Gly, Ala, Cit, Val, Met, Ile, Leu, Tyr, Phe, His, Trp, Orn, Lys, and Arg) were identified whose profile in plasma were associated with NSCLC. Multiple logistic regression analyses by conditional likelihood methods were performed with variable selection and LOOCV cross-validation using the study data set. The resulting conditional logistic regression model included six amino acids: Ala, Val, Ile, His, Trp, and Orn. The AUC of ROC for the discriminant score was 0.817 in the study data set. It should be noted that conditional logistic (c-logistic) regression analysis can correct the effects of age, gender, and smoking statuses which are potential confounding factors in the discrimination. To verify the robustness of the resulting model, a ROC curve was also generated using the split test data set, which had not been used to construct the model. An AUC of ROC for the discriminant score was 0.812 in the test data set, again demonstrating that the obtained model performed well (Figure 6).

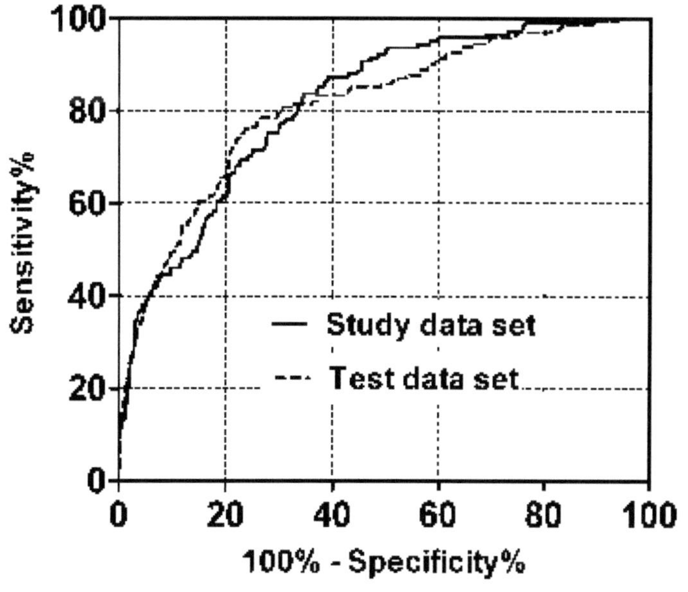

Fig. 6. ROC curves for discriminant scores for the discrimination of NSCLC(Maeda et al, 2010).

It was indicated that the model could discriminate lung cancer patients regardless of cancer stage or histological type. Furthermore, the distribution of the discriminant scores for small-cell lung cancer (SCLC) patients was similar to that for NSCLC patients (Figure 7).

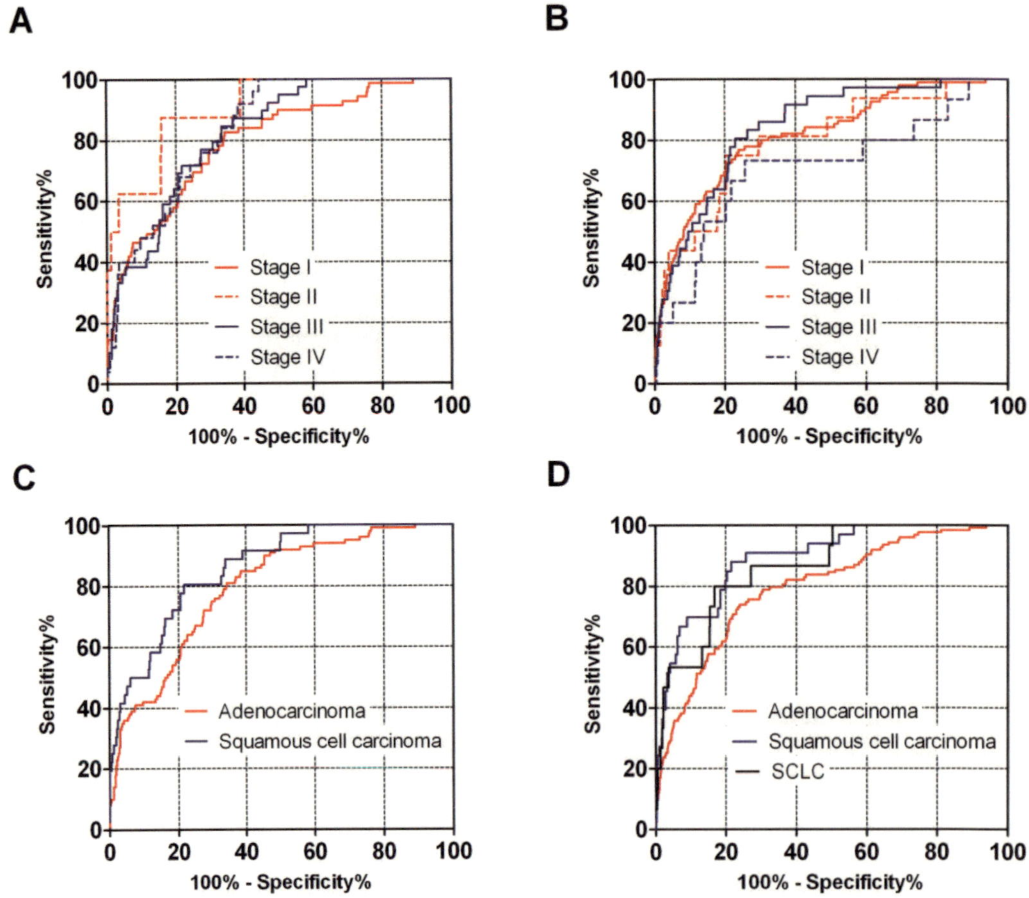

Fig. 7. ROC curves for discriminant scores subgrouped by NSCLC stage and histological type (Maeda et al, 2010). A. ROC curves for cancer stage of study data set. B. ROC curves for cancer stage of test data set. C. ROC curves for histological type of study data set D. ROC curves for histological type of test data set (including SCLC patients).

These studies demonstrated the potential use of PFAA profiling as a focused metabolomics approach for the early detection of patients with various types of cancer. Combining novel analytical techniques and statistical analyses, previously unknown aspects of amino acid metabolism in humans have been revealed. The analysis using considerably larger sample size provided sufficient statistical power to test the robustness of PFAA profiling for cancer diagnosis. We also demonstrated the possibility of detecting cancers, both specifically and broadly, using multivariate analysis to compress the PFAA profile data, even for patients with early stage cancer. Following the further accumulation of data (not shown), AminoIndex® Cancer Screening (AICS) has been commercially released from Ajinomoto Co., Inc., in Japan in April 2011. AICS enables multiple cancer diagnoses simultaneously of gastric, lung, colorectal, prostate and breast cancer.

3.2 „AminoIndex technology“: Example for diagnosis of liver fibrosis

In the clinical pathway of patients with chronic hepatitis C infection, the progression of liver fibrosis leads to cirrhosis and eventually increases the risk of hepatocellular carcinoma (Poynard et al, 2003). The efficacy of current therapy depends on the fibrosis grade, and therefore the detection of fibrosis stage is desirable for determining the clinical settings, i.e., whether treatment is necessary, and what treatment is appropriate (Aspinall & Pockros, 2004; Fried, 2002; Shiffman, 2004). Although fibrosis grading based on biopsy has been considered as a gold standard, there is a high demand for less invasive but effective alternative methods.

In searching for surrogate markers other than biopsy, several methods ranging from the serologic marker-based test (Fibrotest)(Imbert-Bismut et al, 2001) to the ultrasonic-based transient elastography (Fibroscan)(Castera et al, 2005), and others(Lin et al, 2011b) have been suggested. On the other hand, since the liver is an important organ for the metabolism of amino acids, glucose synthesis, fatty acid synthesis, urea synthesis and protein synthesis(Cynober, 2004), it is reasonable to expect any metabolic derangement due to liver failure like liver fibrosis may induce the variation of amino acid metabolism and eventually the variation of PFAA concentration.

In this section we describe the PFAA profiling which was first applied to the diagnosis of liver fibrosis using clinical data(Zhang et al, 2006). The aim of this study was to develop a diagnostics index for the diagnosis of liver fibrosis as a less invasive and effective method using PFAA profiles. The liver specimens were analyzed histologically and graded with the METAVIR scoring system(Metavir., 1994), where F0 means no fibrosis, F1 portal fibrosis without septa, F2 fibrosis with rare septa, F3 portal fibrosis with numerous septa, and F4 cirrhosis. The distribution and variation of the 23 PFAAs of all patients over fibrosis stages is represented in a radar chart, Figure 8.

In the progression of fibrosis from F01 to F4, the decrease of BCAA and inversely the increase of aromatic amino acids, Phe and Tyr, can be observed typically in the profiles of the radar chart. In the non-parametric multi-stage comparison test (Kruskal-Wallis test) , for each amino acid among different fibrosis stages, significant changes in concentration of Phe, Val, Ile, Tyr, Gln, Leu, Met ($p < 0.01$) and ABA (alpha-amino butyric acid, $p < 0.05$) were observed. Dataset including fibrosis stage and PFAA concentrations were analyzed to obtain the diagnostics index for liver fibrosis (AI_fibrosis) in fractional form, (Phe)/(Val) + (Thr+Met+Orn)/(Pro+Gly), which was optimized as a surrogate marker for the liver stages obtained through biopsies. The distribution of molar ratios in two fractional forms over fibrosis stages are shown in Figure 9.

The observation of two molar ratios in the classifier revealed that the former ratio mainly contributed to the F4 discrimination, whereas the latter mainly contributed to discrimination of advanced fibrosis (F3 and F4). For the discriminative power assessment of the surrogate AI_fibrosis as a whole, the area under the curve of receiver operator characteristic curve (ROC AUC) was used. The classifier exhibited high discriminative power for advanced fibrosis (fibrosis stages F3 and F4) from the earlier stages F0-2 and also for cirrhosis (F4) from all other stages, with ROC AUC (95% CI) 0.92 (0.84-1.00) and 0.99 (0.96-1.00), respectively.

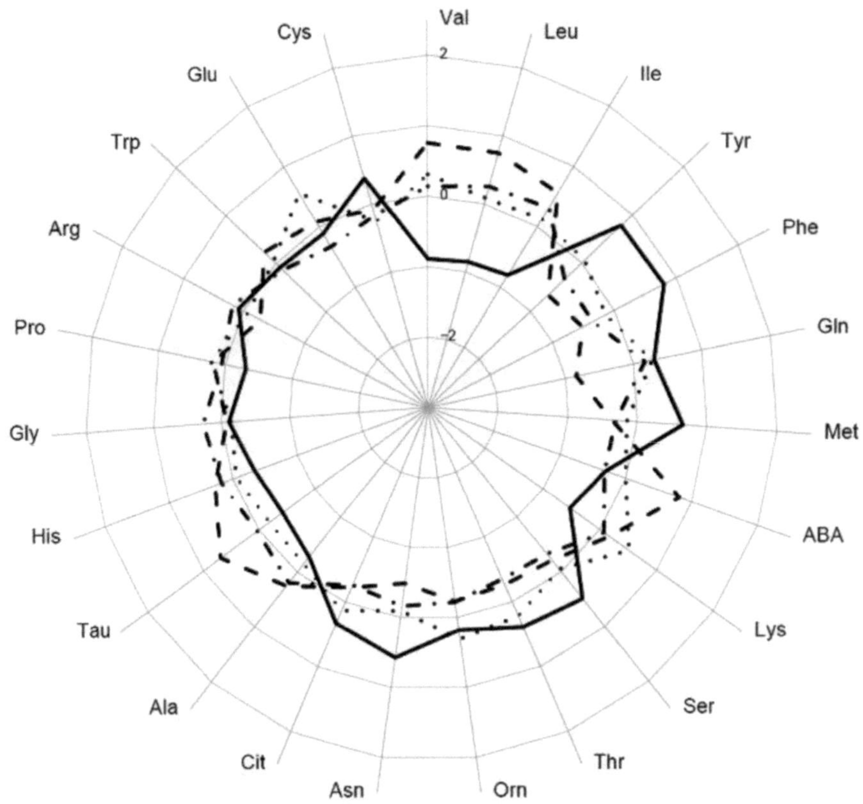

Fig. 8. Radar chart of mean values of PFAAs over fibrosis stages. F01: dashed, F2: dot-dash, F3: dotted, F4: solid. Mean values are scaled in z-score.

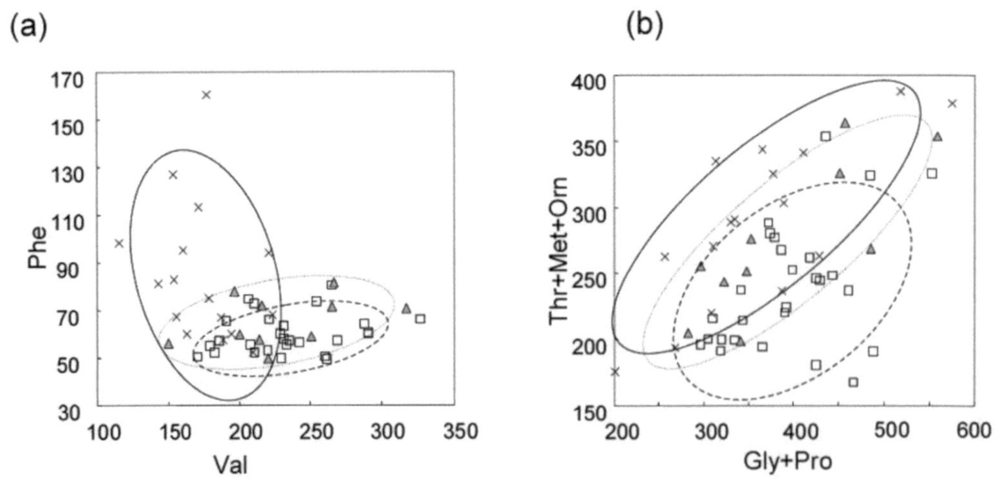

Fig. 9. Molar ratio variation over fibrosis stages. The change in distribution among F0-F2,F3 and F4 stages indicated a stage-dependent trend. Circles are 80% regions of each stage, F0-F2: dashed and square, F3: dotted and triangle, and F4: solid and christcross.

The Fischer's ratio (Val+Leu+Ile)/(Phe+Tyr) was originally created for diagnosis of hepatic encephalopathy (Fischer et al, 1975; Fischer et al, 1976) and has been reported to show good performance in assessing chronic hepatitis (Kano et al, 1991). Therefore a comparison study between the Fischer's ratio and the classifier was undertaken, where the index was generated to have a positive correlation with the degree of fibrosis, showing an inverse pattern to Fischer's ratio. The AI_fibrosis indicated ROC AUC values larger than Fischer's ratio: the ROC AUC values of Fischer's ratio being 0.87 (0.77-0.96) for advanced fibrosis and 0.91 (0.83-0.99) for cirrhosis, respectively. There is a close relationship between the AI_fibrosis and the Fischer's ratio as partially supported by the fact that the ratio Phe/Val correlated well with the inverse of Fischer's ratio ($r = 0.95$) because the BCAAs exhibited good mutual correlation, as did Tyr and Phe. In summary, these results suggest that the AI_fibrosis based on amino acid concentration can be applied to evaluate liver fibrosis as an effective and less invasive method as a surrogate marker for liver biopsy, although future extended validation study is still necessary.

3.3 Lipidomics: A review on the use of lipid metabolomics for clinical use

Lipidomics, a type of focused metabolomics, is the comprehensive measurement of a variety of lipid classes: free fatty acids (FFA), triglycerides (TAGs), cholesterol esters (CEs), lysophosphatidylcholines (LPCs), phosphatidylcholines (PCs), lysophosphatidyl ethanolamines (LPEs), diacylglycerols (DAGs), and sphingomyelins (SMs) and ceramides, generally using LC-MS/MS (Bou Khalil et al, 2010; Bucci, 2011; Dennis, 2009). Several studies have reported the potency of lipidomic analyses for biomarker discoveries in humans in diabetes, non-alcoholic fatty liver disease (NAFLD) (Puri et al, 2009), Alzheimer's disease (Han et al, 2011; Valdes-Gonzalez et al, 2011) and cancers (Hilvo et al, 2011; Min et al, 2011). For instance, Rhee et al reported the LC-MS–based lipid profiling of 189 individuals who developed type 2 diabetes and 189 matched disease-free individuals, with over 12 years of follow up in the Framingham Heart Study (Rhee et al, 2011). They found that lipids of lower carbon number and double bond content were associated with an increased risk of diabetes, whereas lipids of higher carbon number and double bond content were associated with a decreased risk. In addition, Barr et al demonstrated differential serum lipidomics in both NAFLD patients and in a mouse model of NAFLD by ultra performance liquid chromatography-mass spectrometry (UPLC-MS) (Barr et al, 2010). Multivariate statistical analysis of the UPLC-MS datasets revealed metabolic similarities between NAFLD mice and human NAFLD patients in relative serum metabolite levels compared to normal subjects. Lipidomic analysis is also applicable to other biological fluids such as cerebrospinal fluid (CSF), in addition to plasma and serum (Fonteh et al, 2006). For instance, phospholipid profiling in the CSF by nano-HPLC-MS has been reported in Alzheimer's disease (AD) patients, and a statistically significant increase of SMs were observed in CSF from probable AD patients compared to normal subjects (Han et al, 2011).

4. Elucidation of mechanisms underlying metabolomic diagnosis

4.1 Introduction

Many living systems have homeostatic mechanisms to continuously maintain their biological activity. Yet, when a dynamic multi-parametric metabolic response to pathophysiological stimuli is evoked in many disease-associated cells and tissues, it leads to the formation of disease-specific enzymatic metabolite profiles quite different from that of the healthy hosts, and the blood components are significantly influenced as a result.

Blood amino-acid contents are included in such components (referred to as a blood amino-acid profile). It is well known that in the process of feeding, exercising, sleeping, and other activities, the blood amino-acid profile temporarily fluctuates, but within a few hours returns to the normal level through intrinsic homeostatic mechanisms. By contrast, disease-mediated disturbances in the local amino-acid metabolisms may result in formation of a disease-specific change in the blood amino-acid profile. Based on these findings and discussions, we have introduced the AminoIndex® Cancer Screening (AICS) system as a tool for providing new biomarkers to enable the early detection of various cancers.

4.2 Tumor-specific blood amino-acid profile

In order to explain the effects on PFAA profiles by the various tumors, we propose the following simple idea consisting of "three components". As shown in Figure 10, these three components are as follows: a) Metabolic changes in the tumor-bearing organs; b) Metabolic changes in response to the inflammatory reactions; and c) Metabolic changes in various remote healthy organs. In patients with tumors, these three metabolic changes may be evoked simultaneously and their overall effects may be reflected in the tumor-specific blood amino-acid profile. Yet, it is highly unlikely that they contribute evenly to such formation of the tumor-specific blood amino-acid profile during the entire course of the tumor development. It seems more reasonable that these three components contribute individually and differently to the formation of the tumor-specific blood amino-acid profile in the early, the mid and the late (cachexia) stages.

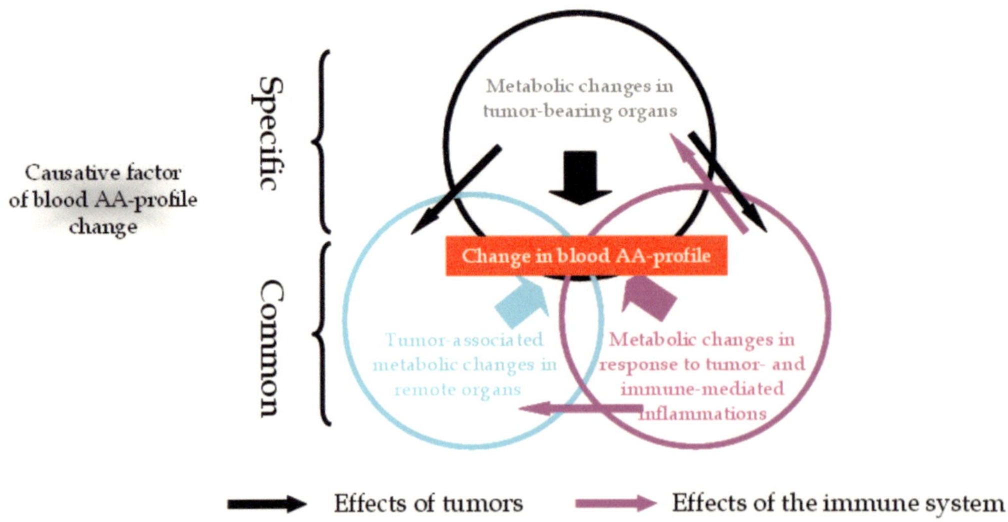

Fig. 10. Scheme for mechanisms underlying tumor-specific metabolic changes

In tumor-bearing hosts, "metabolic changes in tumor-bearing organs", "metabolic changes in response to inflammations" and "metabolic changes in remote organs" are all evoked simultaneously, leading to formation of the tumor-specific amino-acid profile.

4.3 Metabolic changes in the tumor-bearing organs

It has been shown that cancer cells, which can proliferate extraordinarily faster than healthy normal cells, obtain the biological energy required to proliferate by switching to aerobic glycolysis from oxidative phosphorylation (Matthew et al, 2009)(known as "Warburg effect"): several intermediates obtained from the glycolysis pathway might be more easily utilized in nucleotide synthesis and glucose might be used as a carbon source in fatty acid generation (Dang, 2010). At the same time, glutaminolysis was found to be stimulated profoundly, meaning that cancer cell energy generation is highly dependent on glutamine content (Wise & Thompson, 2010). A change in the amino-acid metabolism has been documented even in several noncancerous areas of the tumor-bearing organs. Douvlis has proposed the possibility that many normal tissues show their own specific pattern of the amino-acid absorption but such normal-cell functions may be impaired by amino acids abnormally excreted from neighboring tumor cells (Douvlis, 1999).

4.4 Metabolic changes in response to inflammation

Tumor-associated persistent inflammatory responses are regarded as one of the causative factors for changes in amino-acid metabolism. In addition, inflammation-mediated proliferation of immune competent cells and synthesis of various inflammatory proteins including cytokines and chemokines are also involved (Fox et al, 2005). In almost all of the solid tumors, a variety of inflammatory responses are shown to be evoked in the tumor-surrounding tissues (Mantovani et al, 2008). The tumor-associated inflammation is characterized by a mixture of the anti-tumor inflammatory response, which directs tumor-cell killing, and the tumor-induced inflammation, which stimulates tumor-cell proliferation and promotes neovascularization (Schetter et al, 2010). Therefore, such tumor-associated metabolic changes can be substantially different from those observed in other inflammatory processes seen in pneumonia and colitis. In addition, these tumor-associated inflammations can induce immunocompetent-cell proliferation and antibody production both in tumorous tissues, regional lymph nodes, and the bone marrows, leading to further changes in metabolism (Youn & Gabrilovich, 2010).

4.5 Metabolic changes in various remote normal organs

With an increase in amino-acid demands closely associated with elevated synthesis of nucleotides and proteins in tumor cells, the amounts of amino acids recruited from other tissues and organs are increased by means of enhancing whole body protein turnover, elevating hepatic nonessential amino-acid biosynthesis, and stimulating proteolysis in skeletal muscles accompanied by reduced protein synthesis (Rossi Fanelli et al, 1995). It is well known that in liver and skeletal muscles, tumor-induced negative nitrogen balance can promote intracellular production and extracellular release of glutamine (Medina et al, 1992). In addition, it was reported that in a chronic inflammatory process, the amino-acid metabolism could be influenced even in many remote organs: For instance, absorption of blood cysteine and methionine were both elevated immediately after glutathione synthesis was increased in the liver (Mercier et al, 2002).

5. Future expectations

Although the applications of "AminoIndex technology" are still limited, the foundations for their use for diagnostic purposes are in progress as described above. Studies with clinical data indicate that even with individual variability, the "AminoIndex technology" can be used to separate certain disease and physiological states. We believe that the amino acids are a convenient metabolomic subset to use as a model for the development of metabolomics based diagnostics, and that in the near future, other metabolites could be added to the current analytical platform as practical issues such as stability are solved. At the same time, the universality of the findings must be examined and it should be studied whether the data set we have obtained for the Japanese population is applicable to other populations. We believe that there is a great potential to use metabolome based markers in preliminary diagnostic screening for multiple diseases in which a single measurement of a metabolomic subset can lead to multiple diagnoses. One further advantage of the focused metabolomics multiple metabolite marker approach is that since the biomarkers are generated from a combination of already measured markers, new markers can be generated against any measured target parameter. This means that if a focused metabolomic subset data is obtained at the beginning of a treatment or an experiment, the generation of predictive markers can be attempted with the outcome of the treatment or experiment as the target parameter. We believe this would be of great use in tailor-made medicine and nutrition, as it may be possible to discriminate populations for which certain pharmaceutical or nutritional interventions would be useful or not.

6. References

Alpert, A. J. (1990) Hydrophilic-interaction chromatography for the separation of peptides, nucleic acids and other polar compounds. *J Chromatogr* 499 pp.177-196

Aranibar, N., Borys, M., Mackin, N. A., Ly, V., Abu-Absi, N., Abu-Absi, S., Niemitz, M., Schilling, B., Li, Z. J., Brock, B., Russell, R. J., 2nd, Tymiak, A.&Reily, M. D. (2011) NMR-based metabolomics of mammalian cell and tissue cultures. *J Biomol NMR* 49:(3-4) pp.195-206

Aspinall, R. J.&Pockros, P. J. (2004) The management of side-effects during therapy for hepatitis C. *Aliment Pharmacol Ther* 20:(9) pp.917-29

Bapat, S. A., Krishnan, A., Ghanate, A. D., Kusumbe, A. P.&Kalra, R. S. (2010) Gene expression: protein interaction systems network modeling identifies transformation-associated molecules and pathways in ovarian cancer. *Cancer Res* 70:(12) pp.4809-19

Barr, J., Vazquez-Chantada, M., Alonso, C., Perez-Cormenzana, M., Mayo, R., Galan, A., Caballeria, J., Martin-Duce, A., Tran, A., Wagner, C., Luka, Z., Lu, S. C., Castro, A., Le Marchand-Brustel, Y., Martinez-Chantar, M. L., Veyrie, N., Clement, K., Tordjman, J., Gual, P.&Mato, J. M. (2010) Liquid chromatography-mass spectrometry-based parallel metabolic profiling of human and mouse model serum reveals putative biomarkers associated with the progression of nonalcoholic fatty liver disease. *J Proteome Res* 9:(9) pp.4501-12

Bollard, M. E., Holmes, E., Lindon, J. C., Mitchell, S. C., Branstetter, D., Zhang, W.&Nicholson, J. K. (2001) Investigations into biochemical changes due to diurnal variation and estrus cycle in female rats using high-resolution (1)H NMR spectroscopy of urine and pattern recognition. *Anal Biochem* 295:(2) pp.194-202

Bou Khalil, M., Hou, W., Zhou, H., Elisma, F., Swayne, L. A., Blanchard, A. P., Yao, Z., Bennett, S. A.&Figeys, D. (2010) Lipidomics era: accomplishments and challenges. *Mass Spectrom Rev* 29:(6) pp.877-929

Bucci, M. (2011) Lipidomics: A viral egress. *Nat Chem Biol* 7:(9) pp.577

Caesar, R., Manieri, M., Kelder, T., Boekschoten, M., Evelo, C., Muller, M., Kooistra, T., Cinti, S., Kleemann, R.&Drevon, C. A. (2010) A combined transcriptomics and lipidomics analysis of subcutaneous, epididymal and mesenteric adipose tissue reveals marked functional differences. *PLoS One* 5:(7) pp.e11525

Caruana, R., Niculescu-Mizil, A. (2006) An Empirical Comparison of Supervised Learning Algorithms. In *ICML2006* pp 161-168

Cascino, A., Muscaritoli, M., Cangiano, C., Conversano, L., Laviano, A., Ariemma, S., Meguid, M. M.&Rossi Fanelli, F. (1995) Plasma amino acid imbalance in patients with lung and breast cancer. *Anticancer Res* 15:(2) pp.507-10

Castera, L., Vergniol, J., Foucher, J., Le Bail, B., Chanteloup, E., Haaser, M., Darriet, M., Couzigou, P.&De Ledinghen, V. (2005) Prospective comparison of transient elastography, Fibrotest, APRI, and liver biopsy for the assessment of fibrosis in chronic hepatitis C. *Gastroenterology* 128:(2) pp.343-50

Cynober, L. A. (2004) *Metabolic and therapeutic aspects of amino acids in clinical nutrition*, 2nd edn, CRC Press, 0-8493-1382-1, Roca Raton

Dang, C. V. (2010) Rethinking the Warburg effect with Myc micromanaging glutamine metabolism. *Cancer Res* 70:(3) pp.859-62

DeFeo, E. M., Wu, C. L., McDougal, W. S.&Cheng, L. L. (2011) A decade in prostate cancer: from NMR to metabolomics. *Nat Rev Urol* 8:(6) pp.301-11

Dennis, E. A. (2009) Lipidomics joins the omics evolution. *Proc Natl Acad Sci U S A* 106:(7) pp.2089-90

Douvlis, Z. (1999) Interference of amino acid patterns and tissue-specific amino acids absorption dominance under the influence of tumor cell protein degradation toxins. *Med Hypotheses* 53:(5) pp.450-7

Duda, R. O., Hart, P. E., Stork., D. G. (2001) *Pattern Classification*, 2nd edn, Wiley-Interscience, 0-471-05669-3, New York

Dunn, W. B., Broadhurst, D. I., Atherton, H. J., Goodacre, R.&Griffin, J. L. (2011) Systems level studies of mammalian metabolomes: the roles of mass spectrometry and nuclear magnetic resonance spectroscopy. *Chem Soc Rev* 40:(1) pp.387-426

Fiehn, O. (2002) Metabolomics--the link between genotypes and phenotypes. *Plant Mol Biol* 48:(1-2) pp.155-71

Filho, J. C., Bergstrom, J., Stehle, P.&Furst, P. (1997) Simultaneous measurements of free amino acid patterns of plasma, muscle and erythrocytes in healthy human subjects. *Clin Nutr* 16:(6) pp.299-305

Fischer, J. E., Funovics, J. M., Aguirre, A., James, J. H., Keane, J. M., Wesdorp, R. I., Yoshimura, N.&Westman, T. (1975) The role of plasma amino acids in hepatic encephalopathy. *Surgery* 78:(3) pp.276-90

Fischer, J. E., Rosen, H. M., Ebeid, A. M., James, J. H., Keane, J. M.&Soeters, P. B. (1976) The effect of normalization of plasma amino acids on hepatic encephalopathy in man. *Surgery* 80:(1) pp.77-91

Fonteh, A. N., Harrington, R. J., Huhmer, A. F., Biringer, R. G., Riggins, J. N.&Harrington, M. G. (2006) Identification of disease markers in human cerebrospinal fluid using lipidomic and proteomic methods. *Dis Markers* 22:(1-2) pp.39-64

Forslund, A. H., Hambraeus, L., van Beurden, H., Holmback, U., El-Khoury, A. E., Hjorth, G., Olsson, R., Stridsberg, M., Wide, L., Akerfeldt, T., Regan, M.&Young, V. R. (2000) Inverse relationship between protein intake and plasma free amino acids in healthy men at physical exercise. *Am J Physiol Endocrinol Metab* 278:(5) pp.E857-67

Fox, C. J., Hammerman, P. S.&Thompson, C. B. (2005) Fuel feeds function: energy metabolism and the T-cell response. *Nat Rev Immunol* 5:(11) pp.844-52

Fried, M. W. (2002) Side effects of therapy of hepatitis C and their management. *Hepatology* 36:(5 Suppl 1) pp.S237-44

German, J. B., Gillies, L. A., Smilowitz, J. T., Zivkovic, A. M.&Watkins, S. M. (2007) Lipidomics and lipid profiling in metabolomics. *Curr Opin Lipidol* 18:(1) pp.66-71

Grimplet, J., Cramer, G. R., Dickerson, J. A., Mathiason, K., Van Hemert, J.&Fennell, A. Y. (2009) VitisNet: "Omics" integration through grapevine molecular networks. *PLoS One* 4:(12) pp.e8365

Gruning, N. M., Lehrach, H.&Ralser, M. (2010) Regulatory crosstalk of the metabolic network. *Trends Biochem Sci* 35:(4) pp.220-7

Gu, H., Pan, Z., Xi, B., Asiago, V., Musselman, B.&Raftery, D. (2011) Principal component directed partial least squares analysis for combining nuclear magnetic resonance and mass spectrometry data in metabolomics: application to the detection of breast cancer. *Anal Chim Acta* 686:(1-2) pp.57-63

Han, X., Rozen, S., Boyle, S. H., Hellegers, C., Cheng, H., Burke, J. R., Welsh-Bohmer, K. A., Doraiswamy, P. M.&Kaddurah-Daouk, R. (2011) Metabolomics in early Alzheimer's disease: identification of altered plasma sphingolipidome using shotgun lipidomics. *PLoS One* 6:(7) pp.e21643

Hanley, J. A.&McNeil, B. J. (1982) The meaning and use of the area under a receiver operating characteristic (ROC) curve. *Radiology* 143:(1) pp.29-36

Hilvo, M., Denkert, C., Lehtinen, L., Muller, B., Brockmoller, S., Seppanen-Laakso, T., Budczies, J., Bucher, E., Yetukuri, L., Castillo, S., Berg, E., Nygren, H., Sysi-Aho, M., Griffin, J. L., Fiehn, O., Loibl, S., Richter-Ehrenstein, C., Radke, C., Hyotylainen, T., Kallioniemi, O., Iljin, K.&Oresic, M. (2011) Novel theranostic opportunities offered by characterization of altered membrane lipid metabolism in breast cancer progression. *Cancer Res* 71:(9) pp.3236-45

Hulley, S. B., Cummings, R. B., Browner, W. S., Grady, D. G., Newman, T. B. (2006) *Designing Clinical Research: An Epidemiologic Approach* 3rd edn, Lippincott Williams & Willkins, Inc., Philadelphia

Hunter, P. (2009) Reading the metabolic fine print. The application of metabolomics to diagnostics, drug research and nutrition might be integral to improved health and personalized medicine. *EMBO Rep* 10:(1) pp.20-3

Imbert-Bismut, F., Ratziu, V., Pieroni, L., Charlotte, F., Benhamou, Y.&Poynard, T. (2001) Biochemical markers of liver fibrosis in patients with hepatitis C virus infection: a prospective study. *Lancet* 357:(9262) pp.1069-75

Ishii, N., Nakahigashi, K., Baba, T., Robert, M., Soga, T., Kanai, A., Hirasawa, T., Naba, M., Hirai, K., Hoque, A., Ho, P. Y., Kakazu, Y., Sugawara, K., Igarashi, S., Harada, S., Masuda, T., Sugiyama, N., Togashi, T., Hasegawa, M., Takai, Y., Yugi, K., Arakawa, K., Iwata, N., Toya, Y., Nakayama, Y., Nishioka, T., Shimizu, K., Mori, H.&Tomita, M. (2007) Multiple high-throughput analyses monitor the response of E. coli to perturbations. *Science* 316:(5824) pp.593-7

Jenkins, H., Hardy, N., Beckmann, M., Draper, J., Smith, A. R., Taylor, J., Fiehn, O., Goodacre, R., Bino, R. J., Hall, R., Kopka, J., Lane, G. A., Lange, B. M., Liu, J. R., Mendes, P., Nikolau, B. J., Oliver, S. G., Paton, N. W., Rhee, S., Roessner-Tunali, U., Saito, K., Smedsgaard, J., Sumner, L. W., Wang, T., Walsh, S., Wurtele, E. S.&Kell, D. B. (2004) A proposed framework for the description of plant metabolomics experiments and their results. *Nat Biotechnol* 22:(12) pp.1601-6

Kaddurah-Daouk, R., Kristal, B. S.&Weinshilboum, R. M. (2008) Metabolomics: a global biochemical approach to drug response and disease. *Annu Rev Pharmacol Toxicol* 48 pp.653-83

Kano, T., Nagaki, M., Takahashi, T., Ohnishi, H., Saitoh, K., Kimura, K.&Muto, Y. (1991) Plasma free amino acid pattern in chronic hepatitis as a sensitive and prognostic index. *Gastroenterol Jpn* 26:(3) pp.344-9

Kell, D. B. (2002) Metabolomics and machine learning: explanatory analysis of complex metabolome data using genetic programming to produce simple, robust rules. *Mol Biol Rep* 29:(1-2) pp.237-41

Kell, D. B. (2006) Systems biology, metabolic modelling and metabolomics in drug discovery and development. *Drug Discov Today* 11:(23-24) pp.1085-92

Kim, Y., Koo, I., Jung, B. H., Chung, B. C.&Lee, D. (2010) Multivariate classification of urine metabolome profiles for breast cancer diagnosis. *BMC Bioinformatics* 11 Suppl 2 pp.S4

Kimura, T., Noguchi, Y., Shikata, N.&Takahashi, M. (2009) Plasma amino acid analysis for diagnosis and amino acid-based metabolic networks. *Curr Opin Clin Nutr Metab Care* 12:(1) pp.49-53

Kirouac, D. C., Ito, C., Csaszar, E., Roch, A., Yu, M., Sykes, E. A., Bader, G. D.&Zandstra, P. W. (2010) Dynamic interaction networks in a hierarchically organized tissue. *Mol Syst Biol* 6 pp.417

Lai, H. S., Lee, J. C., Lee, P. H., Wang, S. T.&Chen, W. J. (2005) Plasma free amino acid profile in cancer patients. *Semin Cancer Biol* 15:(4) pp.267-76

Lee, J. C., Chen, M. J., Chang, C. H., Tiai, Y. F., Lin, P. W., Lai, H. S.&Wang, S. T. (2003) Plasma amino acid levels in patients with colorectal cancers and liver cirrhosis with hepatocellular carcinoma. *Hepatogastroenterology* 50:(53) pp.1269-73

Lin, L., Huang, Z., Gao, Y., Yan, X., Xing, J.&Hang, W. (2011a) LC-MS based serum metabonomic analysis for renal cell carcinoma diagnosis, staging, and biomarker discovery. *J Proteome Res* 10:(3) pp.1396-405

Lin, Z. H., Xin, Y. N., Dong, Q. J., Wang, Q., Jiang, X. J., Zhan, S. H., Sun, Y.&Xuan, S. Y. (2011b) Performance of the aspartate aminotransferase-to-platelet ratio index for the staging of hepatitis C-related fibrosis: an updated meta-analysis. *Hepatology* 53:(3) pp.726-36

Lu, X., Zhao, X., Bai, C., Zhao, C., Lu, G.&Xu, G. (2008) LC-MS-based metabonomics analysis. *J Chromatogr B Analyt Technol Biomed Life Sci* 866:(1-2) pp.64-76

Maeda, J., Higashiyama, M., Imaizumi, A., Nakayama, T., Yamamoto, H., Daimon, T., Yamakado, M., Imamura, F.&Kodama, K. (2010) Possibility of multivariate function composed of plasma amino acid profiles as a novel screening index for non-small cell lung cancer: a case control study. *BMC Cancer* 10:(1) pp.690

Mantovani, A., Allavena, P., Sica, A.&Balkwill, F. (2008) Cancer-related inflammation. *Nature* 454:(7203) pp.436-44

Matthew, E. M., Hart, L. S., Astrinidis, A., Navaraj, A., Dolloff, N. G., Dicker, D. T., Henske, E. P.&El-Deiry, W. S. (2009) The p53 target Plk2 interacts with TSC proteins impacting mTOR signaling, tumor growth and chemosensitivity under hypoxic conditions. *Cell Cycle* 8:(24) pp.4168-75

Medina, M. A., Marquez, J.&Nunez de Castro, I. (1992) Interchange of amino acids between tumor and host. *Biochem Med Metab Biol* 48:(1) pp.1-7

Mercier, S., Breuille, D., Mosoni, L., Obled, C.&Patureau Mirand, P. (2002) Chronic inflammation alters protein metabolism in several organs of adult rats. *J Nutr* 132:(7) pp.1921-8

Metavir. (1994) Intraobserver and interobserver variations in liver biopsy interpretation in patients with chronic hepatitis C. The French METAVIR Cooperative Study Group. *Hepatology* 20:(1 Pt 1) pp.15-20

Min, H. K., Lim, S., Chung, B. C.&Moon, M. H. (2011) Shotgun lipidomics for candidate biomarkers of urinary phospholipids in prostate cancer. *Anal Bioanal Chem* 399:(2) pp.823-30

Momin, A. A., Park, H., Portz, B. J., Haynes, C. A., Shaner, R. L., Kelly, S. L., Jordan, I. K.&Merrill, A. H., Jr. (2011) A method for visualization of "omic" datasets for sphingolipid metabolism to predict potentially interesting differences. *J Lipid Res* 52:(6) pp.1073-83

Montoliu, I., Martin, F. P., Collino, S., Rezzi, S.&Kochhar, S. (2009) Multivariate modeling strategy for intercompartmental analysis of tissue and plasma 1H NMR spectrotypes. *J Proteome Res* 8:(5) pp.2397-406

Moore, S., Spackman, D. H.&Stein, W. H. (1958) Automatic recording apparatus for use in the chromatography of amino acids. *Fed Proc* 17:(4) pp.1107-15

Naini, A. B., Dickerson, J. W.&Brown, M. M. (1988) Preoperative and postoperative levels of plasma protein and amino acid in esophageal and lung cancer patients. *Cancer* 62:(2) pp.355-60

Noguchi, Y., Shikata, N., Furuhata, Y., Kimura, T.&Takahashi, M. (2008) Characterization of dietary protein-dependent amino acid metabolism by linking free amino acids with transcriptional profiles through analysis of correlation. *Physiol Genomics* 34:(3) pp.315-26

Norton, J. A., Gorschboth, C. M., Wesley, R. A., Burt, M. E.&Brennan, M. F. (1985) Fasting plasma amino acid levels in cancer patients. *Cancer* 56:(5) pp.1181-6

Okamoto, N., Miyagi, Y., Chiba, A., Akaike, M., Shiozawa, M., Imaizumi, A., Yamamoto, H., Ando, T., Yamakado, M.&Tochikubo, O. (2009) Diagnostic modeling with differences in plasma amino acid profiles between non-cachectic colorectal/breast cancer patients and healthy individuals. *Int J Med Med Sci* 1:(1) pp.1-8

Piraud, M., Vianey-Saban, C., Petritis, K., Elfakir, C., Steghens, J. P., Morla, A.&Bouchu, D. (2003) ESI-MS/MS analysis of underivatised amino acids: a new tool for the diagnosis of inherited disorders of amino acid metabolism. Fragmentation study of 79 molecules of biological interest in positive and negative ionisation mode. *Rapid Commun Mass Spectrom* 17:(12) pp.1297-311

Plumb, R. S., Stumpf, C. L., Granger, J. H., Castro-Perez, J., Haselden, J. N.&Dear, G. J. (2003) Use of liquid chromatography/time-of-flight mass spectrometry and multivariate statistical analysis shows promise for the detection of drug metabolites in biological fluids. *Rapid Commun Mass Spectrom* 17:(23) pp.2632-8

Poynard, T., Yuen, M. F., Ratziu, V.&Lai, C. L. (2003) Viral hepatitis C. *Lancet* 362:(9401) pp.2095-100

Proenza, A. M., Oliver, J., Palou, A.&Roca, P. (2003) Breast and lung cancer are associated with a decrease in blood cell amino acid content. *J Nutr Biochem* 14:(3) pp.133-8

Puri, P., Wiest, M. M., Cheung, O., Mirshahi, F., Sargeant, C., Min, H. K., Contos, M. J., Sterling, R. K., Fuchs, M., Zhou, H., Watkins, S. M.&Sanyal, A. J. (2009) The plasma lipidomic signature of nonalcoholic steatohepatitis. *Hepatology* 50:(6) pp.1827-38

Qu, Y., Slocum, R. H., Fu, J., Rasmussen, W. E., Rector, H. D., Miller, J. B.&Coldwell, J. G. (2001) Quantitative amino acid analysis using a Beckman system gold HPLC 126AA analyzer. *Clin Chim Acta* 312:(1-2) pp.153-62

Quinones, M. P.&Kaddurah-Daouk, R. (2009) Metabolomics tools for identifying biomarkers for neuropsychiatric diseases. *Neurobiol Dis* 35:(2) pp.165-76

Rhee, E. P., Cheng, S., Larson, M. G., Walford, G. A., Lewis, G. D., McCabe, E., Yang, E., Farrell, L., Fox, C. S., O'Donnell, C. J., Carr, S. A., Vasan, R. S., Florez, J. C., Clish, C. B., Wang, T. J.&Gerszten, R. E. (2011) Lipid profiling identifies a triacylglycerol signature of insulin resistance and improves diabetes prediction in humans. *J Clin Invest* 121:(4) pp.1402-11

Righi, V., Durante, C., Cocchi, M., Calabrese, C., Di Febo, G., Lecce, F., Pisi, A., Tugnoli, V., Mucci, A.&Schenetti, L. (2009) Discrimination of healthy and neoplastic human colon tissues by ex vivo HR-MAS NMR spectroscopy and chemometric analyses. *J Proteome Res* 8:(4) pp.1859-69

Rossi Fanelli, F., Cangiano, C., Muscaritoli, M., Conversano, L., Torelli, G. F.&Cascino, A. (1995) Tumor-induced changes in host metabolism: a possible marker of neoplastic disease. *Nutrition* 11:(5 Suppl) pp.595-600

Roux, A., Lison, D., Junot, C.&Heilier, J. F. (2011) Applications of liquid chromatography coupled to mass spectrometry-based metabolomics in clinical chemistry and toxicology: A review. *Clin Biochem* 44:(1) pp.119-35

Schetter, A. J., Heegaard, N. H.&Harris, C. C. (2010) Inflammation and cancer: interweaving microRNA, free radical, cytokine and p53 pathways. *Carcinogenesis* 31:(1) pp.37-49

Serkova, N. J., Standiford, T. J.&Stringer, K. A. (2011) The Emerging Field of Quantitative Blood Metabolomics for Biomarker Discovery in Critical Illnesses. *Am J Respir Crit Care Med*

Shiffman, M. L. (2004) Management of patients with chronic hepatitis C virus infection and previous nonresponse. *Rev Gastroenterol Disord* 4 Suppl 1 pp.S22-30

Shimbo, K., Kubo, S., Harada, Y., Oonuki, T., Yokokura, T., Yoshida, H., Amao, M., Nakamura, M., Kageyama, N., Yamazaki, J., Ozawa, S. I., Hirayama, K., Ando, T., Miura, J.&Miyano, H. (2009a) Automated precolumn derivatization system for analyzing physiological amino acids by liquid chromatography/mass spectrometry. *Biomed Chromatogr* 24:(7) pp.683-91

Shimbo, K., Oonuki, T., Yahashi, A., Hirayama, K.&Miyano, H. (2009b) Precolumn derivatization reagents for high-speed analysis of amines and amino acids in biological fluid using liquid chromatography/electrospray ionization tandem mass spectrometry. *Rapid Commun Mass Spectrom* 23:(10) pp.1483-92

Shimbo, K., Yahashi, A., Hirayama, K., Nakazawa, M.&Miyano, H. (2009c) Multifunctional and highly sensitive precolumn reagents for amino acids in liquid chromatography/tandem mass spectrometry. *Anal Chem* 81:(13) pp.5172-9

Soga, T. (2007) Capillary electrophoresis-mass spectrometry for metabolomics. *Methods Mol Biol* 358 pp.129-37

Spagou, K., Wilson, I. D., Masson, P., Theodoridis, G., Raikos, N., Coen, M., Holmes, E., Lindon, J. C., Plumb, R. S., Nicholson, J. K.&Want, E. J. (2011) HILIC-UPLC-MS for exploratory urinary metabolic profiling in toxicological studies. *Anal Chem* 83:(1) pp.382-90

Spratlin, J. L., Serkova, N. J.&Eckhardt, S. G. (2009) Clinical applications of metabolomics in oncology: a review. *Clin Cancer Res* 15:(2) pp.431-40

Steuer, R. (2006) Review: on the analysis and interpretation of correlations in metabolomic data. *Brief Bioinform* 7:(2) pp.151-8

Sugimoto, M., Wong, D. T., Hirayama, A., Soga, T.&Tomita, M. (2010) Capillary electrophoresis mass spectrometry-based saliva metabolomics identified oral, breast and pancreatic cancer-specific profiles. *Metabolomics* 6:(1) pp.78-95

Taylor, N. S., Weber, R. J., White, T. A.&Viant, M. R. (2010) Discriminating between different acute chemical toxicities via changes in the daphnid metabolome. *Toxicol Sci* 118:(1) pp.307-17

Taylor, R.&Singhal, M. (2009) Biological Network Inference and analysis using SEBINI and CABIN. *Methods Mol Biol* 541 pp.551-76

Thysell, E., Surowiec, I., Hornberg, E., Crnalic, S., Widmark, A., Johansson, A. I., Stattin, P., Bergh, A., Moritz, T., Antti, H.&Wikstrom, P. (2010) Metabolomic characterization of human prostate cancer bone metastases reveals increased levels of cholesterol. *PLoS One* 5:(12) pp.e14175

Tsukamoto, Y., Santa, T., Yoshida, H., Miyano, H., Fukushima, T., Hirayama, K., Imai, K.&Funatsu, T. (2006) Synthesis of the isotope-labeled derivatization reagent for carboxylic acids, 7-(N,N-dimethylaminosulfonyl)-4-(aminoethyl)piperazino-2,1,3-benzoxadiazol e (d6) [DBD-PZ-NH2 (D)], and its application to the quantification and the determination of relative amount of fatty acids in rat plasma samples by high-performance liquid chromatography/mass spectrometry. *Biomed Chromatogr* 20:(4) pp.358-64

Valdes-Gonzalez, T., Goto-Inoue, N., Hirano, W., Ishiyama, H., Hayasaka, T., Setou, M.&Taki, T. (2011) New approach for glyco- and lipidomics--molecular scanning of human brain gangliosides by TLC-Blot and MALDI-QIT-TOF MS. *J Neurochem* 116:(5) pp.678-83

van der Greef, J., Hankemeier, T.&McBurney, R. N. (2006) Metabolomics-based systems biology and personalized medicine: moving towards n = 1 clinical trials? *Pharmacogenomics* 7:(7) pp.1087-94

Vissers, Y. L., Dejong, C. H., Luiking, Y. C., Fearon, K. C., von Meyenfeldt, M. F.&Deutz, N. E. (2005) Plasma arginine concentrations are reduced in cancer patients: evidence for arginine deficiency? *Am J Clin Nutr* 81:(5) pp.1142-6

Wheelock, C. E., Wheelock, A. M., Kawashima, S., Diez, D., Kanehisa, M., van Erk, M., Kleemann, R., Haeggstrom, J. Z.&Goto, S. (2009) Systems biology approaches and pathway tools for investigating cardiovascular disease. *Mol Biosyst* 5:(6) pp.588-602

Wise, D. R.&Thompson, C. B. (2010) Glutamine addiction: a new therapeutic target in cancer. *Trends Biochem Sci* 35:(8) pp.427-33

Yang, W. C., Mirzaei, H., Liu, X.&Regnier, F. E. (2006) Enhancement of amino acid detection and quantification by electrospray ionization mass spectrometry. *Anal Chem* 78:(13) pp.4702-8

Yoshida, H., Mizukoshi, T., Hirayama, K.&Miyano, H. (2007) Comprehensive analytical method for the determination of hydrophilic metabolites by high-performance liquid chromatography and mass spectrometry. *J Agric Food Chem* 55:(3) pp.551-60

Youn, J. I.&Gabrilovich, D. I. (2010) The biology of myeloid-derived suppressor cells: the blessing and the curse of morphological and functional heterogeneity. *Eur J Immunol* 40:(11) pp.2969-75

Yuan, K., Kong, H., Guan, Y., Yang, J.&Xu, G. (2007) A GC-based metabonomics investigation of type 2 diabetes by organic acids metabolic profile. *J Chromatogr B Analyt Technol Biomed Life Sci* 850:(1-2) pp.236-40

Zeisel, S. H. (2007) Nutrigenomics and metabolomics will change clinical nutrition and public health practice: insights from studies on dietary requirements for choline. *Am J Clin Nutr* 86:(3) pp.542-8

Zhang, G. F., Sadhukhan, S., Tochtrop, G. P.&Brunengraber, H. (2011) Metabolomics, pathway regulation, and pathway discovery. *J Biol Chem* 286:(27) pp.23631-5

Zhang, P. C.&Pang, C. P. (1992) Plasma amino acid patterns in cancer. *Clin Chem* 38:(6) pp.1198-9

Zhang, Q., Takahashi, M., Noguchi, Y., Sugimoto, T., Kimura, T., Okumura, A., Ishikawa, T.&Kakumu, S. (2006) Plasma amino acid profiles applied for diagnosis of advanced liver fibrosis in patients with chronic hepatitis C infection. *Hepatol Res* 34:(3) pp.170-7

Part 6

Improving Analytics

13

Improvement in the Number of Analytic Features Detected by Non-Targeted Metabolomic Analysis: Influence of the Chromatographic System and the Ionization Technique

R. Pandher[1], E. Naegele[2], S.M. Fischer[3] and F.I Raynaud[1]
[1]*The Institute of Cancer Research, Pharmacokinetics and Metabolomics,*
[2]*Agilent Technologies Research and Development, Waldbronn,*
[3]*Agilent Technologies, Metabolomics Laboratory, Santa Clara, CA,*
[1]*UK*
[2]*Germany*
[3]*USA*

1. Introduction

The development and optimisation of genomic, transcriptomic and proteomic technologies have significantly contributed to the assessment of biological systems and increased our understanding of gene function and regulation (Kitano 2002, Brown et al., 1999, Pandey and Mann 2000). In addition, metabolic fingerprinting or metabolomics complement these approaches by measuring low molecular weight chemicals in biological samples (Nicholson and Lindon, 2008). The elucidation of the links between genetic regulation, kinetic activities of enzymes and metabolic reactions is key to understanding homeostatic regulation of living organisms and the effects of food, diurnal variations disease and drugs (Nicholson et al., 2003, van der Greef et al., 2003, Plumb et al., 2003). Mapping of these various interactions is likely to result in applications in disciplines such as agriculture and medicine (Lee et al., 2007, Borodina and Nielson 2007, Wishart 200, Ducruix et al., 2006). Several analytic tools have been applied to profile the metabolome.

LC-MS studies are a more recent introduction to the field of metabolomics compared with the more established techniques of GC-MS and NMR. LC-MS can be used for the analysis of metabolites with a wide range of molecular weights than those detectable by GC-MS including polar and non-volatile compounds. With LC-MS, many different chromatographic phases and thus separation techniques are available when compared with GC-MS (Dunn, 2008).

Targeted metabolomic studies allow the identification and quantification of defined sets of metabolites and are performed using triple Quadrupole mass spectrometers which provide sensitivity and selectivity. Non-targeted global metabolomic studies are carried out on instruments with good mass accuracy such as time of flight and orbitrap mass analysers. In metabolomic profiling, comparison of biological samples collected under different conditions is performed by multivariate statistical analysis in order to to identify significant differences between the groups. In metabolomics a "feature" is a molecular ion (*m/z*) coupled to a retention time (RT) that is generated following data processing, where feature finding is performed in conjunction with noise reduction and alignment of data. A drift in mass accuracy or retention time will affect the experimental results by creating additional numbers of novel features. An increase in variability of peak area or height may mask differences between experimental groups. The number of features, and their intensities in a number of replicate analysis of a given sample can define the robustness of an analytical run prior to complex statistical analysis of the data. . The acceptance criteria for reproducibility and repeatability differ between laboratories. In the studies presented in this chapter a coefficient of variation (CV) threshold of 25% was set which is in line with similar metabolomic studies published in the literature (Crews et al., 2009; Lai et al., 2010).

Plasma represents an important biofluid and global metabolite profiles have been derived from a variety of LCMS methods (Sabatine et al., 2005, Want et al., 2006, Bruce et al., 2008 and Zelena et al., 2009). We have previously shown that 2 different QTOF instruments produced the same number of reproducible features from tissue culture media extracts (Pandher et al., 2009). The goal of this study as to develop an efficient methodology using the Agilent LC Infinity system and Jet Stream Technology for metabolomic reverse phase LC-MS approaches. Here we describe the number of features obtained in human plasma extracts with a conventional rapid resolution chromatographic system and a QTOF mass spectrometer equipped with an electrospray ionization source in positive ionization mode. The improvement in the number of features and reproducibility following chromatographic separation with the Agilent 1290 Infinity LC system at various flow rates is also presented together with the peak capacity of a selected number of analytes. In addition, the impact of further optimization of the analytical conditions (temperature, flow rate) with the Jet Stream ionization technology on the number and reproducibility of the ions detected is presented.

2. Materials and method

Water (LC-MS grade), acetonitrile (LC-MS grade) and formic acid (Aristar grade) were all purchased from Fisher Scientific (Loughborough, UK). Leucine enkephalin was purchased from Sigma (Poole, UK). The external standards creatine (CAS no: 57-00-1), carnitine (CAS no: 541-14-0), colchicine (CAS no: 64-86-8), hydrocortisone (CAS no: 50-23-7), phenylalanine (CAS no: 673-06-3), tryptophan (CAS: 73-22-3) and hippuric acid (CAS no: 495-69-2) were purchased from Sigma (Poole, UK). Standard stock solutions of 1 mM were prepared in water or DMSO as appropriate and in human plasma. Human plasma was collected in heparinised tubes from healthy donors and centrifuged at 1500 x g for 15 minutes at 4°C. Plasma was then stored at -80°C until analysis.

3. Sample extraction

Plasma samples (200 µl) were extracted with 4 volumes of acetonitrile, using a 96 well protein precipitation plate (Whatman, Maidstone, UK). The plate was vortexed for 1 min before a vacuum was applied. The filtered samples were collected in a 96 deep well plate and plasma extracts were pooled and aliquoted out for further analysis.

3.1 Sample analysis

For optimization of the analytical conditions, triplicate samples containing the spiked analytical standards were injected. For reproducibility studies, 3 replicates of unspiked and spiked plasma samples were analysed.

3.2 Liquid chromatographic separation

The HPLC systems used were the conventional Agilent 1200 and the Agilent 1290 Infinity LC system. Most of the analytic separation was achieved on a Waters Acquity column HSS T3 C_{18} (100 mm × 2.1 mm, I.D 1.8μm particles) and a 150 mm column was also tested.

Different chromatographic conditions were evaluated:

A mobile phase of 100% 0.1% formic acid was run isocratic for 0.5 minutes followed by a linear gradient ending in 100% acetonitrile over 7.5 minutes or 5.5 minutes or 3.5 minutes, followed by 100% acetonitrile over 2 minutes. After returning to the original conditions, the system was left to equilibrate for 3 minutes prior to the next injection. Different flow rates were evaluated (0.4 ml/min on the 1200 HPLC and 0.4, 0.6, 0.8 and 1 ml/min on the 1290 Infinity UHPLC). The same gradients were used on the 150 mm column. Columns were previously equilibrated with the injection of 5 plasma extracts.

3.3 Mass spectrometric analysis

Mass spectrometry was performed in positive ionization mode on a QTOF (6530, Agilent). Two different sources were evaluated: the classical electrospray ionization source and the Jet Stream technology. With the ESI source, parameters were set with a capillary voltage of 4 kV in positive ionisation mode. The fragmentor voltage was 140 V and skimmer was 65 V. The gas temperature was 250°C, drying gas 10 l/min and nebulizer 40 psig. Nitrogen was used as a drying gas. MS spectra were acquired in full scan analysis over an m/z range of 70-1000 using extended dynamic range and a scan rate of 1.4 spectra/second. To maintain mass accuracy during the run time, a reference mass solution containing reference ions 121.0508 and 922.0097 was used.

With the Jet Stream technology; parameters were set with a capillary voltage of 4 kV in positive ionisation mode. The fragmentor voltage was 140 V and skimmer was 65 V. The gas temperature was 250°C, drying gas 6 l/min and nebulizer 60 psig. Nitrogen was used as a drying gas. The sheath gas temperature was tested and optimized from 200°C to 400°C by increments of 50°C and the sheath gas flow rate was 11l/min. Total ion spectra were acquired in full scan analysis over an m/z range of 70-1000 using extended dynamic range 2GHz and an acquisition rate of 2Hz.

Sampling rates of 2, 4 and 6Hz were tested with the Jet Stream technology at 0.6 ml/min and 0.8 ml/min with the sheath gas at 200°C in order to evaluate the number of sampling points collected across each peak and the reproducibility of the analysis.

4. Data processing

Sample features were extracted with the molecular feature extractor (MassHunter Workstation Software (version B.01.03)). Data were processed using the following conditions: restrict retention time to 0.20 - 8.5 min, restrict m/z to 100-800, absolute height threshold: 25000 or 2500, mass tolerance: 0.05, peaks with height: > 100 counts, isotope grouping: peak spacing tolerance: 0.0025 m/z, plus 7.0 ppm, isotope model: common organic model, mass filters: filter mass list: 20 ppm.

The list of features consisting of retention times (RT) and molecular masses was then analysed using GeneSpring MS Analysis Platform (v1.2, Agilent Technologies, Inc., Santa Clara, CA) where they were aligned and normalized.

Data were then imported into Excel spreadsheets and mean, SD and CV of all features was calculated.

5. Results and dicussion

There are a number of experimental variables, related to chromatography or mass spectrometry, that can impact the reproducibility of metabolomic profile data. This in turn can compromise the validity of the data's biological relevance and applicability. The importance of several variables, including flow rate, column length, mass spectrometric conditions were all evaluated in terms of the number of features found.

Following triplicate separation of human plasma on the conventional 1200 LC system at 0.4 ml/min with a 7.5 min gradient followed by 2 min isocratic on a 10 cm column with the ESI source, 1324 total features were detected out of which 795 (60%) showed less than 25% CV Table 1).

Method	System pressure	Total no. of features	No. of features <25% CV	% of features with <25% CV	% change in features with <25% relative to conventional system
LC1200 0.4ml/min	450	1324	795	60	n/a
LC1290 0.4ml/min	270	1714	925	54	+16%
LC1290 0.6ml/min	390	2559	1149	45	+44%
LC1290 0.8ml/min	502	2263	1074	47	+35%
LC1290 1.0ml/min	605	1805	305	17	-38%

[a]Chromatography performed on a reverse phase Waters Acquity T3 column with a 7.5min 0.1% formic acid: acetonitrile gradient. Mass spectrometry analysis performed on 6530 QTOF in ESI mode. Data was extracted using MassHunter Qualitative software package and GeneSpringMS and then exported to Excel where statistics were performed.

Table 1. Effect of flow rate on the number and reproducibility of features present in technical replicates of human plasma extracts conventional 1200 LC and novel 1290 Infinity LC in ESI mode [a] (n=3).

Under similar conditions, the novel 1290 LC system generated 16% higher and reproducible features compared with the conventional system without any change in analytic conditions. Increasing the flow rate to 0.6 ml/min with the same gradient increased the number of reproducible features to 1149 allowing a 44% improvement when compared with the 1200 LC system. When the flow rate was increased from 0.4 to 0.6 ml/min many additional features were detected that were not previously observed following separation by the 1200 system. Careful examination of the features showed that they were mainly ions that had not previously eluted from the column at 0.4 ml/min. Further increase in flow rate showed that fewer reproducible features were detected. The pressure in the system with the column installed was significantly lower in the 1290 system with a back pressure of at 132 bar at 0.4 ml/min on the 1290 versus 450 bars on the 1200 respectively. At 1 ml/min, the pressure on the 1290 system was 605 bars only. It is possible that the different composition of the pistons and their independent operation together with the novel mixing technology used in the 1290 Infinity LC system can explain the decreased pressures compared with the conventional 1200 (data not shown)

The length of the gradient was then shortened to 5.5 minutes and 3.5 minutes respectively but this resulted in a significant decrease in reproducibility (Table 2). In fact, we noted that the isocratic segment of the gradient had to be extended in order to avoid carry-over from previous samples which defeated the purpose of a shorter analytic run (data not shown).

Method[b]	Total no. of features	No. of features <25% CV	% of features with <25% CV	% change in features with <25% relative to conventional system
LC1200 0.4ml/min *7.5min gradient*	1324	795	60	n/a
LC1290 0.6ml/min *7.5 min gradient*	2559	1149	45	+44%
LC1290 0.6ml/min *5.5 min gradient*	1989	1086	55	+37%
LC1290 0.6ml/min *3.5 min gradient*	1889	1017	54	+28%

[a]Chromatography performed on a reverse phase Waters Acquity T3 column with a 7.5min 0.1% formic acid: acetonitrile gradient. Mass spectrometry analysis performed on 6530 QTOF in ESI mode. Data was extracted using MassHunter Qualitative software package and GeneSpringMS and then exported to Excel where statistics were performed.

Each gradient was preceded by 0.5min of 100% A (0.1% formic acid in water) and followed by 2min of 100% B (0.1 formic acid in acetonitrile).

Table 2. The effect of gradient duration on the number and reproducibility of features present in technical replicates of human plasma extracts using conventional 1200 and novel 1290 Infinity LC systems in ESI mode[a] (n=3).

Following triplicate analysis of human plasma on the conventional 1200 LC system using the 150 mm column with our previously described gradient, there was no significant improvement in total or reproducible number of features when compared with the 100 mm column regardless of the flow rate (data not shown). Our conclusion for the data from the ESI source was that a flow rate of 0.6 ml/min was optimal with the 100 mm column with the original 7.5 minute gradient.

We then proceeded to evaluate the effect of the Jet Stream technology on the number of features detected and their repeatability. Incremental temperatures of 50°C of heated nitrogen sheath gas; from 200°C to 400°C were applied and evaluated. At 0.6 ml/min with a sheath gas of 200°C, both the total and reproducible features were more than doubled when compared to the equivalent result with the ESI source. Overall, 50% of features showed less 25% CV over triplicate analysis (Table 3). This represents a 173% increase in reproducible features when compared with the conventional 1200 LC system and the ESI source.

Method	Total no. of features	No. of features <25% CV	% of features with <25% CV	% change in features with <25% relative to conventional system
ESI LC1200 0.4ml/min	1324	795	60	n/a
JS LC1290 0.6ml/min *200°C*	4357	2176	50	+173%
JS LC1290 0.8ml/min *200°C*	4312	2512	58	+215%
JS LC1290 0.6ml/min *250°C*	4396	2294	52	+189%
JS LC1290 0.8ml/min *250°C*	4810	2708	56	+241%
JS LC1290 0.6ml/min *300°C*	4463	2565	57	+223%
JS LC1290 0.8ml/min *300°C*	5130	2869	56	+261%
JS LC1290 0.6ml/min *350°C*	4693	2762	59	+247%
JS LC1290 0.8ml/min *350°C*	5257	2994	57	+277%
JS LC1290 0.6ml/min *400°C*	4919	2707	55	+241%
JS LC1290 0.8ml/min *400°C*	5095	3310	65	+316%

[a]Chromatography performed on a reverse phase Waters Acquity T3 column with a 7.5min 0.1% formic acid: acetonitrile gradient. Mass spectrometry analysis performed on 6530 QTOF in ESI mode. Data was extracted using MassHunter Qualitative software package and GeneSpringMS and then exported to Excel where statistics were performed.

Table 3. The effect of Jet Stream (JS) sheath gas and flow rates on number and reproducibility of features present in technical replicates of human plasma extracts the novel 1290 Infinity LC system coupled to a 6530 QTOF using Jet Stream technology [a] (n=3)

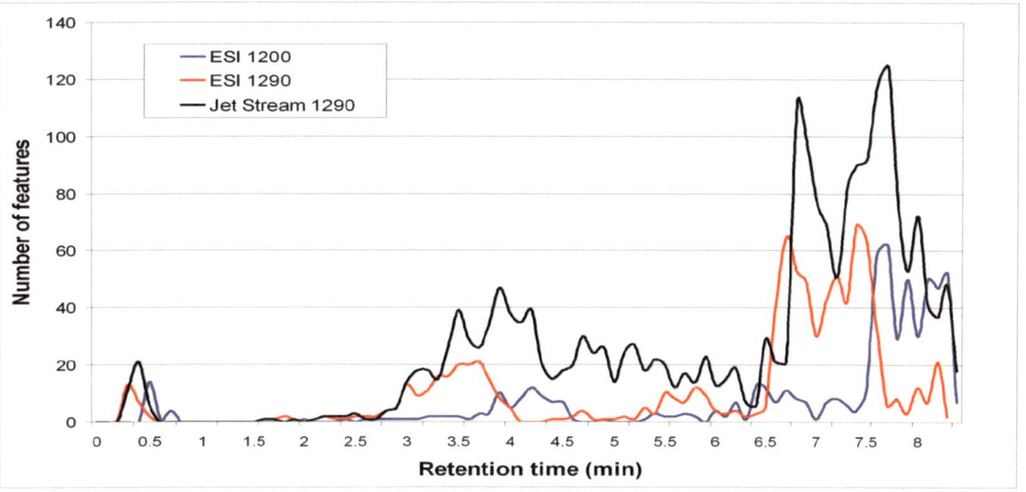

Fig. 1. Number of reproducible features against retention time. Number of reproducible features generated following data processing using the conventional 1200 LC system with ESI, the novel 1290 Infinity LC system with ESI and the novel LC system with Jet Stream technology.

The increased number of features ionized by the Jet Stream technology when compared with ESI is illustrated in Figure 1 which shows overlayed total ion chromatograms. In contrast to our results with the ESI source, increasing the flow rate to 0.8 ml/min further increased the number of reproducible features when compared with 0.6 ml/min (Table 3). Increasing the temperature of the sheath gas by increments of 50°C at both 0.6 and 0.8 ml/min gradually increased the number of features. At 400°C, a striking 5095 features were detected; 3310 of which showed less than 25% CV.

We were concerned that this significant increase in features with increase in temperature could be the result of thermal degradation of ions. . To address this, a Venn diagram derived from GeneSpring MS shows the ions present at at both temperatures, at 200°C or 400°C only. (Figure 2).

The heat plot in Figure 3 demonstrates that all the 1491 ions present at 400° C were weak whereas more than half of the 888 were stronger in intensity, suggesting that the increase in temperature fragmented the ions and that thermal degradation occurred. Further examination of the features at 200°C in Jet Stream and ESI at 0.6 ml/min, showed that >2000 features were specific to Jet Stream alone. By analysing the most intense features we found that they were not split features due to errors in the automatic processing software. We were concerned that some of these features may be detectable in ESI at a lower threshold. Therefore, we proceeded to lower the threshold to 2500 in both Jet Stream and ESI and found that 10599 and 15424 total features and 2792 and 4163 reproducible features were detected respectively in the two systems with 26% reproducibility obtained using both systems. Comparison of these features showed that a proportion was only present in Jet Stream and the remainder were detectable at low intensity and not reproducible.

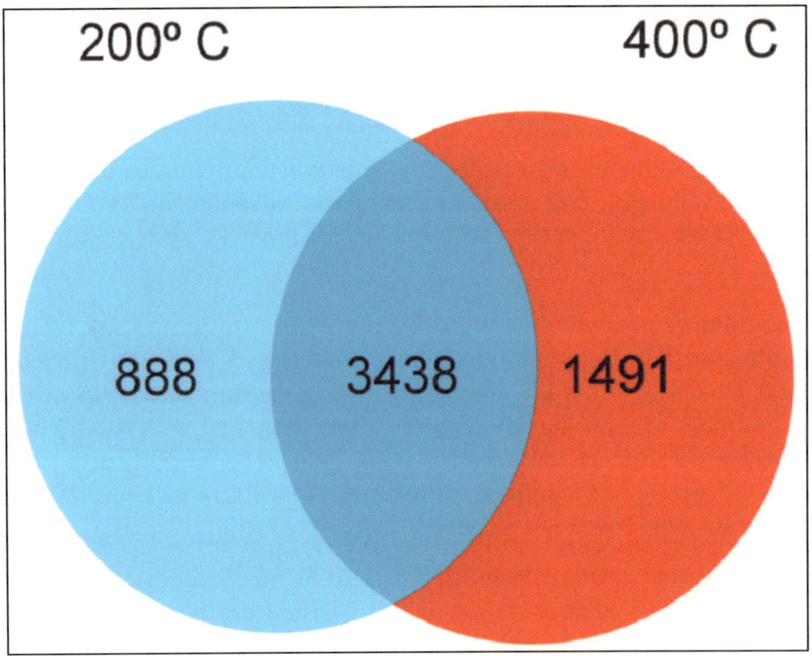

Fig. 2. Comparison of features observed with Jet Stream and ESI. Venn diagram showing features found in human plasma extracts using LC 1290 coupled to 6530 QTOF coupled with Jet Stream technology using sheath gas temperature of either 200°C or 400°C at 0.8 ml/min. The features present at 200°C were 888, 1491 were present only at 400°C and 3438 were present at both temperatures.

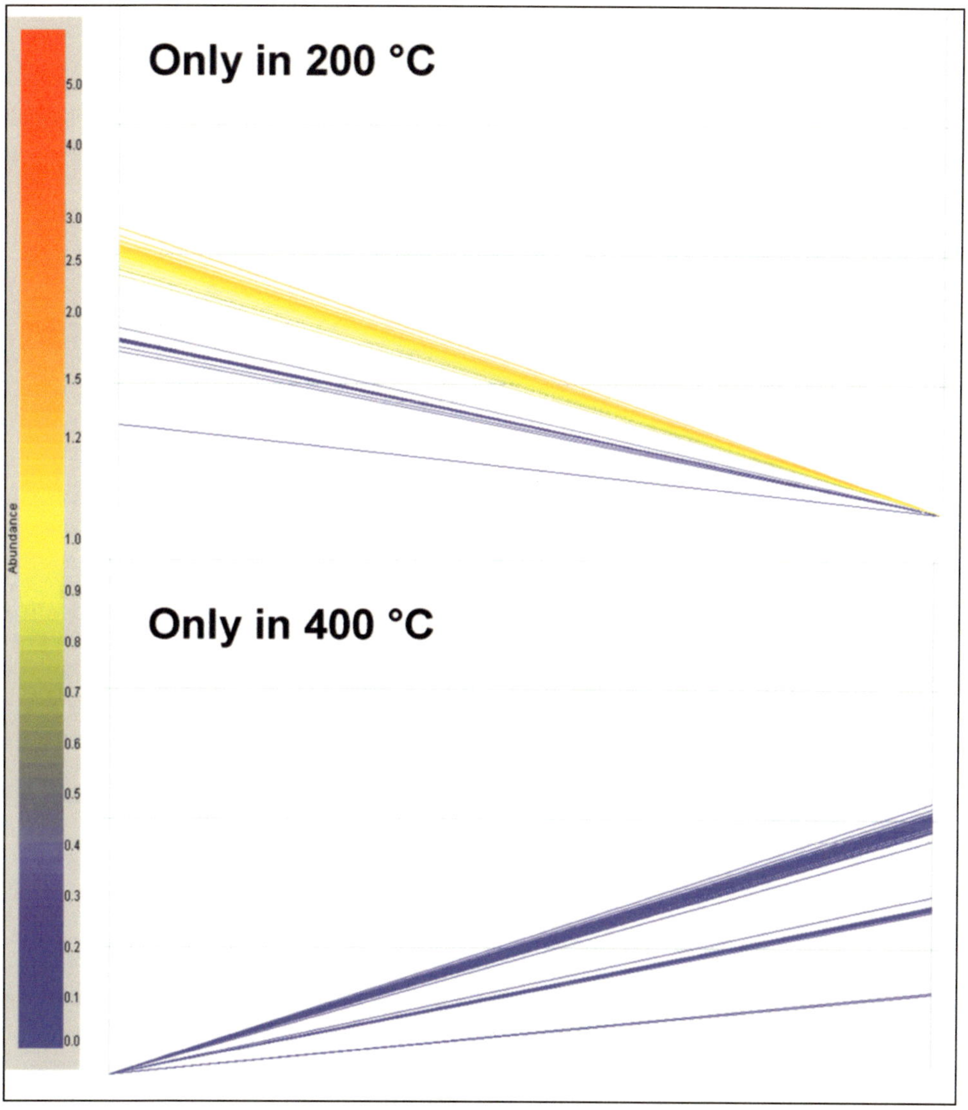

Fig. 3. Abundance of features observed exclusively at 200°C and 400°C. Heat plot depicting the abundance of features present in human plasma extracts using LC1290 coupled to 6530 QTOF coupled with Jet Stream technology at either 200°C or 400°C at 0.8 ml/min. Each line represents one feature found exclusively at either temperature, with red representing those features present in a low high intensity and in blue those present in low intensity.

In summary, the Jet Stream technology increased the overall number of features when compared with the ESI but thermal degradation occurred above 200°C, which is therefore the optimal temperature to use under the conditions studied.

Our data demonstrates the advantage of the new LC system which allowed operation at higher flow rates with low back pressure and very reproducible analysis. The increase in flow rate resulted in a predictable increase in peak capacity (Figure 4).

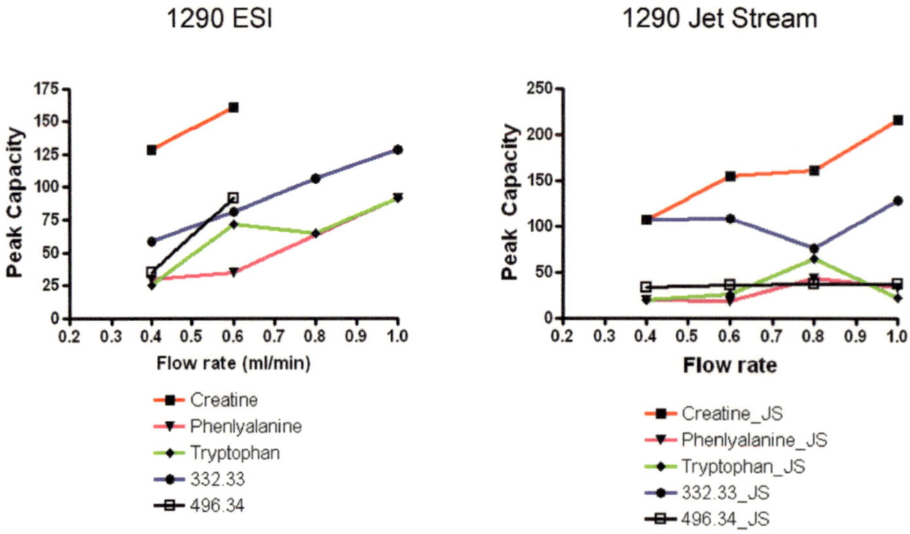

Fig. 4. Peak capacity versus flow rate determined for creatine (m/z 132.07), phenylalanine (166.08), tryptophan (m/z 204.10) and glycerophosphocholine (m/z 496.34) and ion m/z 332.33 using the conventional 1200 LC system and novel 1290 LC system coupled to ESI and Jet Stream technology. Chromatographic separation was carried out on a 100mmx2.1mm ID reverse phase Waters Acquity T3 column with a 7.5min 0.1% formic acid: acetonitrile gradient followed by 2 min isocratic.

However, saturation of desolvation may have occurred at flow rates of 0.8 ml/min and above, as mentioned previously, and features started to disappear. For example we could no longer detect creatine (*m/z* 132.07) and glycerophosphocholine (*m/z* 496.34) with ESI above a flow rate of 0.6 ml/min. The Jet Stream technology detected these features at higher flow rate when compared with the ESI and showed the expected linear increase in peak capacity with flow rate.

We were concerned however, that at higher flow rates, less data points were collected across chromatographic peaks and proceeded to test various scan rates (2, 4 and 6Hz) at 0.6 and 0.8 ml/min (Table 4a). For a limited number of compounds the number of points across peaks for early and late eluting metabolites (0.5-0.8 min, n=2, and 6-6.5 min, n=4) were measured). Decreasing the scan rate increased the number of points across the peaks but decreased the sensitivity of the analysis resulting in a significant decrease in the total and reproducible number of features (Table 4b).

For example, at 0.8 ml/min, at 2Hz; only 4566 features were detected and 2056 with less than 25%CV as opposed to 3716 and 1294 at 4Hz and 3233 and 893 at 6Hz. At 2Hz, there were an average number of 5 points across chromatographic peaks versus 15 at 4Hz, and in late eluting peaks an average of 10 points were monitored across peaks at 2Hz and 25 at 4Hz. Interestingly, the overall percentage of reproducible features was similar at all acquisition rates suggesting that the loss of sensitivity observed at 4 or 6GHz occurs equally across total features whether they are variable or not. The number of reproducible features before 2 min at 0.6 ml/min was: 28 at 2Hz versus 40 and 41 at 6Hz (data not shown). The lower number of reproducible features at 2Hz when compared to 4 and 6 suggests that early polar metabolites being defined with less than 10 points across peaks are less reproducibly detected.

m/z,RT	2GHz 200°C 0.6mlmin	4GHz 200°C 0.6mlmin	6GHz 200°C 0.6mlmin	*m/z*,RT	2GHz 200°C 0.8mlmin	4GHz 200°C 0.8mlmin	6GHz 200°C 0.8mlmin
116.07_0.47	8 (7.5)	22 (5.4)	38 (6.7)	*116.07_0.36*	6 (4.5)	17 (5.5)	22 (5.2)
182.08_0.50	8 (3.8)	9 (1.3)	11 (2.7)	*182.08_0.38*	7 (4.0)	7 (1.6)	9 (3.1)
312.36_6.20	11 (0.4)	27 (0.4)	33 (0.2)	*312.36_5.95*	10 (0.5)	15 (1.0)	30 (0.8)
496.34_6.00	10 (2.4)	23 (2.5)	27 (2.5)	*496.34_5.70*	8 (1.6)	19 (2.0)	24 (2.6)
520.34_5.80	9 (1.5)	19 (1.0)	25 (1.3)	*520.34_5.50*	7 (1.6)	13 (1.60)	23 (0.9)
522.34_6.30	10 (1.0)	29 (1.0)	37 (0.8)	*522.34_7.20*	10 (0.8)	28 (0.8)	32 (0.5)

[a]Chromatography performed on a reverse phase Waters Acquity T3 column with a 7.5min 0.1% formic acid: acetonitrile gradient. Mass spectrometry analysis performed on 6530 QTOF in ESI mode. Chromatograms were then analysed using MassHunter Qualitative software.

Table 4a. Evaluation of number of points across chromatographic peaks and mass accuracy of selected metabolites using 2Hz, 4Hz and 6Hz scan rates at 0.6ml/min and 0.8ml/min [a]

Method	Total no. of features	No. of features <25% CV	% of features with <25% CV
1290 Infinity LC 0.6ml/min 200°C 2HZ	4566	2056	45
1290 Infinity LC 0.6ml/min 200°C 4HZ	3058	1410	46
1290 Infinity LC 0.6ml/min 200°C 6HZ	2231	1073	48
1290 Infinity LC 0.8ml/min 200°C 2HZ	4460	2602	26
1290 Infinity LC 0.8ml/min 200°C 4HZ	3716	1294	35
1290 Infinity LC 0.8ml/min 200°C 6HZ	3233	893	28

[a]Chromatography performed on a reverse phase Waters Acquity T3 column with a 7.5min 0.1% formic acid: acetonitrile gradient. Mass spectrometry analysis performed on 6530 QTOF in ESI mode. Chromatograms were then analysed using MassHunter Qualitative software.

Table 4b. The effect of scan rate on number and reproducibility of features present in technical replicates of human plasma extracts the novel 1290 Infinity LC system coupled to a 6530 QTOF using Jet Stream technology [a] (n=3)

This study clearly demonstrates the considerable challenges associated with reproducibly and sensitively acquiring metabolomic data. Increasing the flow rates eluted more non-polar metabolites off the column but eventually at the detriment of polar metabolites that became undersampled. Whatever choice is made in analytic conditions cannot be optimal for all metabolites. It has to be noted that the number of features do not correspond to the number of metabolites. For example other studies have described up to 23 features for a given metabolite which further complicates matters (these may include multiply charged ions and in source fragmentation ions) (Evans 2009).

In conclusion, our study describes the much improved effect of the 1290 LC system together with the Jet Stream technology on the number of features detected compared with the 1200 LC system and ESI. It is clear that this increased number of features corresponds both to an increased number of metabolites eluting from the column at higher flow rates and an additional number of species being reproducibly ionized by the Jet Stream technology when compared with the ESI source. Our preferred analytic conditions use the 100 mm analytic column, a 7.5 min gradient (total run time 10 min plus 3 min equilibration) with a flow rate of 0.6 ml/min and an acquisition rate of 2Hz.

6. References

Brown, P. O.; Botstein, D. Nat Genet 1999, 21, 33-37.

Bruce S.J., Jonsson P., Clorarec C., Trygg J., Marklund S.L., Moritz T. Anal. Biochem., 2008, 372, 237-249.

Crews, B., Wikoff, W.R., Patti, G.J., Woo, H.K., Kalisiak, E., Heideker, J., and Siuzdak, G. Anal Chem 2009, 81, 8538-8544.

Dunn, W.B. Phys Biol 2008, 5, 100-111

Evans, A.M., DeHaven C.D., Barrett T., Mitchell M. and Milgram E. J Anal Chem 2009 81 (16), pp 6656–6667

Kitano, H. Nature 2002, 420, 206-210.

Lai, L., Michopoulos, F., Gika, H., Theodoridis, G., Wilkinson, R.W., Odedra, R., Wingate, J., Bonner, R., Tate, S., and Wilson, I.D. Mol Biosyst 2010 6, 108-120.

Lee, S. H.; Woo, H. M.; Jung, B. H.; Lee, J.; Kwon, O. S.; Pyo, H. S.; Choi, M. H.; Chung, B. C. Anal Chem 2007, 79, 6102-6110.

Lenz, E. M.; Bright, J.; Knight, R.; Wilson, I. D.; Major, H. Analyst 2004, 129, 535-541.

Nicholson, J. K.; Lindon, J. C. Nature 2008, 455, 1054-1056.

Nicholson, J. K.; Wilson, I. D. Nat Rev Drug Discov 2003, 2, 668-676.

Pandey, A.; Mann, M. Nature 2000, 405, 837-846.

Pandher, R., Ducruix, C., Eccles, S.A., and Raynaud, F.I. J Chromatogr B Analyt Technol Biomed Life Sci. 2009, 877, 1352-1358.

Plumb, R.; Granger, J.; Stumpf, C.; Wilson, I. D.; Evans, J. A.; Lenz, E. M. Analyst 2003, 128, 819-823.

Quinones, M.P., and Kaddurah-Daouk, R. Neurobiol Dis 2009 35, 165- 176.

Sabatine M.S., Liu, E., Morrow, D.A., Heller E, McCarroll R., Wiegand R., Berriz G.F., Roth F.P., Gerszten R.E., . Circulation. 2005, 112. 3868-3875.

Want E.J., O'Maille G., Smith C.A., Brabdon T.R., Uritboonthai W., Qui C., Trauger S.A., Siuzdak, G. Anal. Chem., 2006, 78, 743-752.

van der Greef, J.; Stroobant, P.; van der Heijden, R. Curr Opin Chem Biol 2004, 8, 559-565.

Zelena E., Dunn W, D. Broadhurst, S. Francis-McIntyreK. Carroll, P.Begley, S. O'Hagan, J.D. Knowles, A. Halsall, HUSERMET Consortium, I.D. Wilson,D. Kell, D. Anal. Chem., 2009, 81, 1357-1364.

Part 7

Metabolomics in Safety Assessments

14

Challenges for Metabolomics as a Tool in Safety Assessments

George G. Harrigan and Bruce Chassy
Regulatory Product Characterization and Safety Center, Monsanto Company, St. Louis, USA

1. Introduction

Agriculture's ability to supply an abundance of nutritious foods and feeds to nourish the world's growing population faces serious challenges (Foresight, 2011). In order to meet these challenges, plant breeders will be required to continuously improve agricultural productivity as well as enhance food and feed quality. In recent years, the development of methods for the direct introduction of new traits to produce transgenic varieties - also known as GM crops - has proven to be a powerful tool in the hands of breeders. In most countries, however, GM crops are subjected to rigorous pre-market regulatory assessments that require numerous laboratory and field studies and which consume time and resources (Kalaitzandonakes et al., 2007).

Comprehensive compositional analyses represent a key component of the pre-market safety evaluations of GM crops (Harrigan, et al., 2010). These analyses typically include the measurement of levels of key nutrients such as protein, storage oil, fiber, amino acids, fatty acids, vitamins, as well as crop-specific metabolites such as gossypol and cyclopropenoid fatty acids in cotton or isoflavones in soybean. The Organization of Economic Cooperation and Development (OECD) has produced a series of consensus documents that identify key analytes in a number of major crop varieties (http://www.oecd.org). These documents carefully review the composition and uses for each crop and identify those components that contribute to nutritional or functional food or feed value as well as components that might confer health-beneficial, health-protective, or harmful effects (e.g. allergens, anti-nutrients, and potential toxicants). The large-scale compositional studies performed as part of regulatory assessments must follow internationally accepted guidelines. These are outlined in detail by Codex Alimentarius (Codex Alimentarius, 2008) and OECD. In most cases, these studies are typically conducted under Good Laboratories Practice (GLP), a practice that places a high premium on documentation and reconstructability of data, method validation and personnel training, and a requirement for professionally staffed Quality Assurance Units.

The fact that different crops produce foods or feeds with differing compositions, along with the fact that human and animal diets vary greatly in their consumption of these crops, means that each crop plays a unique role in diet and health. Most plant foods in the human diet make significant contributions to the total intake of just a few macro- and micronutrients (Senti and Rizek, 1974; Chassy, 2010). It is therefore important to assure that no changes have occurred that would lower the dietary intake of an essential nutrient; on the other hand, large changes in the content of one or more nutrients in a crop that supplies an infrequently consumed food, one which is consumed in small amounts in the diet, or one which is not an important source of that nutrient in the diet, are of no health consequence and will have no adverse effect on health (Chassy, 2010).

The identification and analysis of a key set of relevant metabolites is often referred to a "targeted" compositional analysis. Analyses utilize quantitative assays and the overall approach allows the generation of data that is easily interpretable from a nutrition and food/feed safety aspect. Furthermore, since the small molecule metabolite pool in seed is of low abundance relative to macromolecular components, measurement of macronutrients approximates the total seed biomass. For example, the small molecule metabolite pool in corn grain is only ~5% of the total biomass (corn is dominated by starch, fiber, protein, and fat). Anti-nutrient components in grain such as phytic acid and raffinose (which represent much of the small molecule metabolite pool) are measured in regulatory assessments. Other small molecules metabolites can be included if they are an intended endpoint of compositional or nutritional modification. Otherwise analytical measurement of the metabolites that constitute this pool, mainly ubiquitous free amino acids, sugars, and organic acids), is of little value owing to the extreme sensitivity of metabolite levels to environmental influences and the negligible contribution they make to safety and nutritional content (Herman et al., 2009; Skogerson et al., 2010, Harrigan et al., 2007).

In fact, levels of all crop compositional components are influenced markedly by environment (Harrison and Harrigan, 2011; Harrigan, et al., 2010; Zhou et al., 2011a, 2011b). To illustrate, as far back as 1983, it was noted that "The concentration of the isoflavones vary from [soybean] variety to variety, and there are also differences when the same variety is grown in different locations" (Eldridge and Kwolek, 1983). Given the extensive scientific literature on isoflavone variability, it was unsurprising that Gutierrez-Gonzalez et al. (2009) recently concluded that "The range of values of isoflavones is overwhelming, even for homozygous genotypes growing in the same year and location, which greatly complicates genetic studies." This is true for almost all crop compositional components as evidenced by challenges in enhancing nutritional quality in staple crops through conventional approaches. Figure 1 illustrates the type of variability than can be observed for metabolites such as isoflavones.

The use of multiple geographically separate sites is required in regulatory assessments to allow compositional studies across a wide range of environmental conditions. Indeed, information on compositional variation in conventional crops with respect to their responsiveness to environmental factors is necessary to provide context to evaluations of new GM crops. Studies incorporating four to five replicated field sites utilizing randomized complete block designs with three blocks per comparator are typical in regulatory assessments, although the European Food Safety Authority (EFSA) has recently mandated a minimum of eight replicated sites utilizing randomized complete block designs with four blocks (EFSA, 2011).

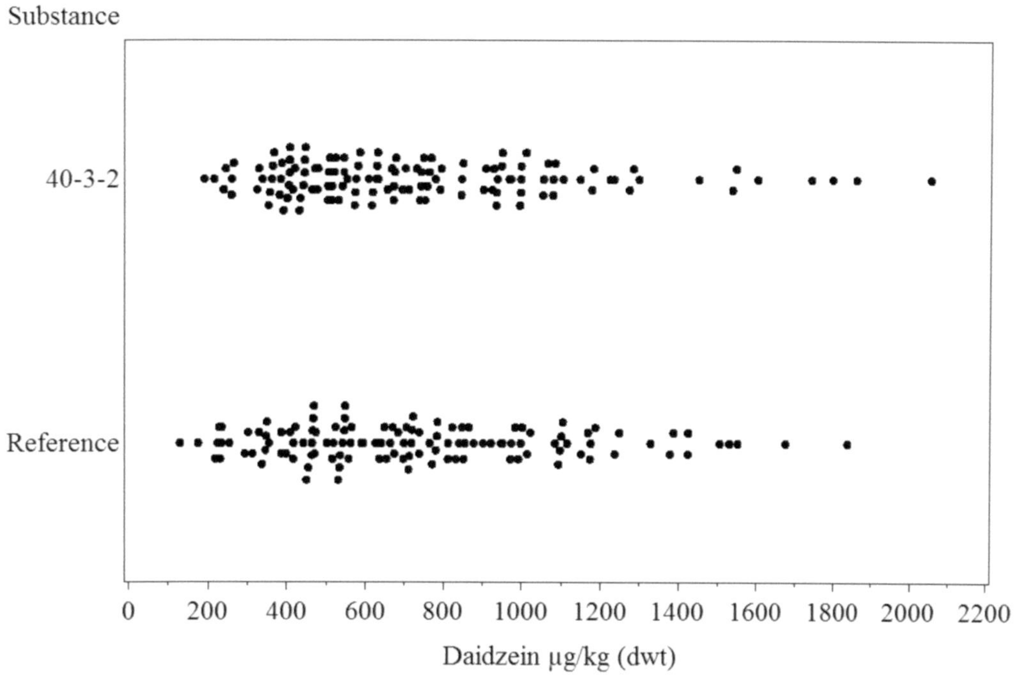

Fig. 1. An overview of variability in isoflavone levels. Datapoints show daidzein values from an analysis of GM (40-3-2) and conventional reference comparators from a total of nine (2001-2009) growing seasons. A total of 112 unique GM varieties were assessed (Zhou et al., 2011b). This type of information presents context to any GM-non-GM pairwise comparison and would be a required component of any metabolomic assessments.

Results to date from these large-scale compositional studies have generally demonstrated that the effect of transgene insertion is significantly less than the impact of environmental or germplasm variation on conventional crops (Harrigan et al., 2010). This has allowed some to question the relevance and design of compositional assessments. One review, for example, suggests that "the current complexity and resource requirements for compositional studies on transgenic crops containing input traits are not justified by a commensurate understanding of safety" (Herman et al., 2009).

Despite continued confirmation that conventional breeding and environmental variation contribute to compositional variability more so than transgene insertion (Ricroch et al., 2011), and the resource-intensiveness of the large-scale studies currently required for regulatory assessments, there remains some interest in the application of profiling technologies to compare GM and conventional crops. These are often posited in terms of "gap-filling" (Heineman et al., 2011) or "case-by-case" (Davies, 2010) evaluations. It is also perceived by many (e.g., Kok et al., 2008) that measurement of "primarily the low-molecular weight molecules" is more relevant to safety than proteomic or transcriptomic profiling due to a closer relationship to "the plant phenotype and nutritional and toxicological characteristics". This potential advantage of metabolic profiling could be extended as an improvement over, for example, measurements of gross levels of protein, fat, and fibers, key nutritional but essentially "safe and inert" components of food. It is noteworthy that Kok et al. (2008) define metabolomics as the "generation of profiles of secondary metabolites" whereas most metabolic profiling experiments to date have focused on primary metabolites. It has also been suggested that untargeted profiling techniques are unbiased while "targeted" compositional analysis is biased. Finally, advocates of metabolomic profiling have suggested that such an approach can detect potentially deleterious totally novel metabolites that would have been missed by "targeted" analysis, although it should be noted that many profiling technologies require standards of known identity to accurately identify and measure specific metabolites thus limiting this potential advantage. In addition, in examples where a new traditionally bred plant variety has caused toxic effects, this has been attributable to increased levels of well-known toxicants (Chassy, 2010).

Profiling technologies have confirmed on a case-by-case basis the compositional "equivalence" of GM crops to their conventional near-isogenic comparators (Ricroch et al., 2011). Profiling technologies are, however, unlikely to provide immediately interpretable data in safety assessments that would provide added value to, or otherwise enhance, rigorously quantitative assessments of known nutrients and anti-nutrients that comprise foodstuffs. In the case of metabolic profiling, this can be directly attributable to i) the intrinsically safe nature of food itself, ii) inconsistencies in metabolite coverage versus quantitative capabilities afforded by different data acquisition technologies, iii) the ubiquitous and innocuous nature of small molecule metabolites identified in profiling as well as extreme variability in metabolite levels even within homozygous genotypes, and iv) the "chasm" between the large number of data generated in profiling experiments and the ability to interpret them in a way that is meaningful to nutrition and food safety. We now expand on these observations and further emphasize that a clear distinction between "substantial equivalence" and food safety should be promoted.

2. Key challenges for the omics

1. *Domesticated crops have been selected to serve human needs and have an extensive history of safe consumption. Extensive information on levels of nutrients and crop-specific antinutrients is available. These can be measured through highly quantitative assays to provide interpretable data of direct relevance to food nutrition and safety.*

Of over 250,000 plant species, only 7000 are considered as foodstuffs (Khoshbahkt and Hammer, 2008), and even fewer, 150, supply over 90% of all plant food. Three major crops, *i.e.* maize, wheat, and rice, supply over 66%. Choices made in crop domestication and breeding have enabled food and feed qualities that serve human needs. Numerous path changes between wild and domesticated plants are known and include *e.g.* loss of spontaneous shattering of seed head on ripening and greater uniformity of seed ripening and germination, both of which facilitate human harvest. A key path change however is reduction and even loss of toxic compounds in the edible parts of domesticated crops. In other words, human selection has resulted in crop phenotypes and compositions that distinguish domesticated varieties from their natural counterparts, are more suited to human diets and needs, and are safer and more nutritious. Interestingly, one review (Jones, 1998) asked "Why are so many food plants cyanogenic?" and concluded that "Cyanogenesis by plants is not only a surprisingly effective chemical defence against casual herbivores, but it is also easily overcome by careful pre-ingestion food processing, this latter skill being almost exclusive to humans." In other words, because "cyanogenic plants are surprisingly well protected from herbivory and yet can be readily detoxified by food processing, ... early farmers fortuitously chose these plants above all others for cultivation."

Of course, many modern foodstuffs are still associated with "ancestral" secondary metabolites that may confer nutritional or safety concerns at elevated levels. Classic examples include glycoalkaloids in potato (NIEHS, 1998), β-N-oxalyl-L-α,β-diaminopropionic acid (ODAP) in *Lathyrus sativus* (Bell, 2003), psoralens in celery (Beier and Oertli, 1983), and gossypol in cotton (Sunlkumar et al., 2006). Targeted measurement of these components as opposed to broad-based compositional screening is recommended by Herman et al. (2009); in other words, compositional assessments should focus on molecules explicitly associated with safety concerns. This is consistent with the observation that in the very few examples where a new plant variety has caused toxic effects it has been attributable to well-know toxicants associated with conventionally bred crops and not to a hitherto undetected metabolite (Chassy, 2010).

It is noteworthy that such targeted assessments could easily facilitate a partnership with omics researchers conducting semi-targeted profiling on pathways associated with toxic metabolites to support both early development and commercialization of nutritionally enhanced products. Such a partnership could, at least in principle, mitigate the current regulatory burden imposed on new GM crops (Graff et al., 2009; Potrykus, 2010) and promote the application of omics within modern agricultural biotechnology.

2. *Information on compositional variation in conventional crops with respect to their responsiveness to environmental and genetic factors is necessary to provide context to evaluations of new GM crops. The need to assess natural variation is also true for metabolomics yet little information on the impact of conventional breeding on metabolite profiles is available. The inconsistent coverage of metabolites offered through different data acquisition platforms may provide challenges in establishing a coherent literature in this area.*

Ironically, as mentioned earlier, continued confirmation that conventional breeding, environment, and germplasm contribute to compositional variation more than transgene insertion has coincided with increased interest in the use of 'omics technologies. This paradox is compounded by the fact that results from these technologies have only further highlighted the equivalence of GM crops to their conventional counterparts and reaffirmed the substantial effect of environment and germplasm on compositional and biochemical variability (see Ricroch et al., 2011). Although there are complexities in the interpretation of data generated through modern profiling technologies (Broadhurst and Kell, 2006; Lay et al., 2006) including the fact that the data is not quantitative and there is no standardized framework for comparisons, the lack of variation between GM crops and their conventional comparators at the transcriptomic, proteomic, and metabolomic level has, nonetheless, been independently corroborated. These profiling evaluations extend to a wide range of plants including wheat (Baker et al., 2006; Gregersen et al., 2005; Ioset et al., 2007), potato (Catchpole et al., 2005; Defernez et al., 2004; Lehesranta et al., 2005), soybean (Cheng et al., 2008), rice (Dubouzet et al., 2007; Wakasa et al., 2006), tomato (Le Gall et al., 2003), tobacco, *Arabidopsis* (Kristensen et al., 2005), and *Gerbera* (Ainasoja et al., 2008).

As with the compositional studies reported above, results from many of the 'omics studies emphasize the need to understand natural variation in levels of endogenous metabolites in providing biological context to pair-wise differences in any recorded profiles (see Figure 1). Levels of compositional components are sensitive to environmental conditions. This has been established for, for example, protein and oil in key crops (Panthee et al., 2005; Lam et al., 2010). Protein levels in soybean seed generally average ~40% dry weight (dwt), with values reported in the USDA soybean germplasm collection, for example, ranging from 34.1 to 56.8% dwt (Wilson, 2004). In a recent meta-analysis of environmental effects on soybean composition, Rotundo and Westgate (2009) observed that water stress, temperature, and/or nitrogen supply all affected protein levels measured in mature seed.

Variability is even greater for lower abundance small molecule metabolites. Vitamin E (α-tocopherol) is typically only a minor component in soybean but is known to be important in maintaining oxidative stability of soybean oil. Levels in soybean seed are affected by environment and germplasm. For example, Britz et al. (2008) showed a greater than 2-fold variation in levels across three locations in the U.S. over a period of four years. Levels in soybean seed harvested from six different locations in Eastern Canada over a single year ranged from 0.87 to 3.32 mg/100g dwt (Seguin et al. 2009). Seguin et al. (2010) point out that environmental factors associated with variability in vitamin E levels include drought, temperature, and even crop management systems. The "overwhelming variability" of isoflavones was mentioned in the introduction (see Figure 1). As will be discussed later, this "overwhelming variability" can be considered to apply to levels of small molecule metabolites in harvested seed and grain of most crops.

Encouragingly, many comparative profiling studies on GM and non-GM crops have been designed to include at least one element of genotypic or environmental variability. This is exemplified in the following two examples, both of which reaffirm the need to provide biological context to pairwise-differences between two comparators.

In Baker et al. (2006) NMR-based metabolic profiles of three GM wheat varieties and the corresponding parents were generated. The incorporated transgenes encoded high-molecular weight subunits of the storage protein, glutenin. The wheat varieties were grown at two different sites over three different growing seasons (1999 -2001). Differences between the GM and parental lines were within the same range as the differences between the control lines grown on different sites and in different years. Analogous to the approach adopted in targeted compositional analyses adopting OECD recommendations, this study emphasized the importance of data from multiple years and multiple sites and that environmental variation influences metabolome composition.

In Catchpole et al. (2005) two GM potato varieties modified in fructan chemistry were grown over two different seasons (2001, 2003). Metabolic profiles of the GM and five conventional crops were generated using flow-injection MS (FIE-MS), GC-MS, and LC-MS. These demonstrated that differences between the GM and conventional potatoes were due to the intended metabolic changes, but aside from these targeted changes, the GM crops were "substantially equivalent to traditional cultivars". A major finding recognized by the authors was the large variation in the metabolic profiles of the conventional crops and, as such, the study emphasized the importance of understanding genotypic variability in assessments of compositional changes.

An often overlooked aspect of the Catchpole et al. (2005) paper is their demonstration that levels of glycoalkaloids (α-chaconine and α-solanine) were normal in the GM potatoes, a result that is easily interpretable from a food and feed perspective. Our understanding of nutrients and anti-nutrients forms the basis of attempt to modify crops through conventional breeding or agricultural biotechnology. It has allowed crops to be developed by conventional breeding that are deliberately non-equivalent to their parental progenitors in a wide range of nutritional (and agronomic) characteristics.. As Rischer et al. (2006) point out "For centuries, conventional plant breeding programs have produced new traits, higher yields and improved quality. However, little attention has been paid to metabolic changes occurring in successive generations. The issue has gained importance only recently in the context of defining thresholds for safety assessments of GM crops." It is not immediately obvious why these hitherto neglected metabolites should now be at the center of such attention. Indeed, there are few studies on small molecule metabolite changes in crops where macro-molecular composition has been deliberately changed through conventional breeding (e.g. high oil and high protein corn, high oil soybean).

Catchpole et al. (2005) in their demonstration of the compositional equivalence of GM potatoes to conventional lines also remark on the large metabolite variation in conventional potato as follows; "These significant differences [between conventional cultivars] were never sought as desired traits in traditional breeding programs, and overall composition has not given cause for public safety concerns". Overall, however, experimental designs that will both account for natural variation and have enough power to identify differences that can be attributed to transgene insertion will offer opportunities to maximize the value of omics technologies as tools in plant breeding and the development of new crops.

3. *Metabolomics offers opportunities to generate data on large numbers of metabolites. Most of these metabolites will be low in abundance and levels will be highly variable. They are also more likely to include central (and hence ubiquitous) metabolites such as sugars, organic acids, and free amino acids; metabolites that are not immediately associated with safety or nutritional relevance.*

Compositional assessments of new foodstuffs generally focus on the article of commerce, most typically harvested seed or grain. This material is generally characterized by high levels of starch, protein, fat, and fibers, with the small metabolite pool being low in abundance. For example, approximately 95-98% of maize grain is comprised the aforementioned materials; the small metabolite pool in grain, is of low abundance (~2-5% of grain biomass) and its levels are highly dependent on changes in the macromolecular pool. Soybean seed is comprised 40% protein, 20% fat, and 15% fiber. The residual 15% is comprised mainly of sugars (e.g. sucrose, raffinose, stachyose, glucose, galactose, fructose) of which the principal two, raffinose and stachyose, are measured in regulatory assessments. The fact that the small molecule metabolite pool in seed or grain is of low abundance and influenced by levels of the major macromolecular nutrients accounts for its extensive variability (Skogerson et al., 2010; Harrigan et al., 2007).

Skogerson et al. (2010) sought to assess genetic and environmental impacts on the metabolite composition of corn grain. Their data acquisition technology (gas chromatography-mass spectrometry) measured 119 identified metabolites including free amino acids, free fatty acids, sugars, organic acids, and other small molecules in a range of corn hybrids derived from 48 inbred lines crossed against two different tester lines (from the C103 and Iodent heterotic groups) and grown at three locations in Iowa (Table 1). Different metabolic phenotypes were clearly associated with the two distinct tester populations. Overall, grain from the C103 lines contained higher levels of free fatty acids and organic acids, whereas grain from the Iodent lines were associated with higher levels of amino acids and carbohydrates. In addition, the fold-range of genotype mean values [composed of six samples each (two tester crosses per inbred × three field sites)] for identified metabolites ranged from 1.5- to 93-fold with sugars and polyols being particularly variable. Interestingly, some grain metabolites showed a non-normal distribution over the entire corn population, which could, at least in part, be attributed to large differences in metabolite values within specific inbred crosses relative to other inbred sets.

Metabolite class	No. of analytes	Affected by Tester[a]	Affected by Location[b]
free amino acids	26	14	2
sterols, amines, and others	17	6	1
organic acids	17	6	0
free fatty acids and related metabolites	17	5	0
sugar alcohols	18	5	0
mono-, di-, and trisaccharides	16	1	0
sugar acids	8	0	1

[a]This indicates a statistically significant difference ($p<0.0001$) between hybrids derived from a cross with one tester (C103 heterotic group) versus another tester (Iodents heterotic group) [b]This indicates a statistically significant difference ($p<0.0001$) across the three locations in this study

Table 1. Variation in Metabolites due to Genotype or Environmental Variation

In an analogous report on the same samples, Harrigan et al. (2007) concluded that, given such variability, measurement of the small metabolite pool, was unlikely to prove useful to a comparative assessment of GM crops unless a given metabolite was an intended nutritional or toxicological endpoint. In fact, it is not immediately obvious how the data generated in Skogerson et al. (2011) could be used to determine which hybrids in this study were the safest.

In its report in 2004 the US National Research Council made pointed remarks about this disconnect as summarized in the following quotes. ".. severe imbalances between highly advanced analytical technologies and limited ability to interpret the results and predict health effects that result from the consumption of food that is genetically modified" and "….inherent difficulties, however, in identifying all of the constituents detected in profiling methods or understanding the activity and potential biological consequence of all genes in an organism severely limit the usefulness of these methods for predictive purposes.." Unable to bridge this gap, many profiling proponents make an assumption of safety on the non-GM comparator and consider statistical differences to equate with unintended effects. This tendency is described later.

4. *Another challenge in establishing a coherent literature on the impact of conventional and other approaches to breeding on natural variability in metabolite as well as determining a framework to establish nutritional meaning from metabolite analysis is the differential coverage of metabolites offered through the numerous data acquisition platforms available to omics researchers.*

As described in numerous articles on metabolomics, (*e.g.* Goodacre et al., 2004; Rischer and Oksman-Caldentey, 2006; Kusano et al., 2011) the large physico-chemical diversity of small molecule metabolites renders comprehensive metabolomic profiling through a single data acquisition technology impossible. A range of technologies associated with different detection capabilities (metabolite coverage and sensitivity), precision, resolution, throughput and reproducibility are now extensively deployed by the research community. Nuclear magnetic resonance spectroscopy (NMR), gas-chromatography mass spectrometry (GC-MS), liquid-chromatography (LC)-MS utilizing different ionization modes, Fourier-transform MS, and capillary electrophoresis (CE)- MS have all been applied in comparative assessments of GM and non-GM crops. MS approaches predominate over NMR analyses given their greater sensitivity and coverage; however this advantage does come at the expense of quantitation (i.e. MS would need an internal standard for every metabolite to be quantitated) and with a large number of unidentified MS signals in any metabolite profile. Whilst it has been suggested that untargeted profiling techniques are unbiased, it is clear that selection of a specific data acquisition technology *is* a bias and that this type of analytical bias would need to be justified by pre-specified experimental hypotheses. This justification would be critical in a Regulatory environment.

Recognizing inherent limitations for any given data acquisition technology Kusano et al. (2011) applied a multi-platform approach to an evaluation of transgenic tomato. These authors used a combination of GC-MS, LC-MS, and CE-MS with each technology covering distinct metabolite classes. Free amino acids, sugars and organic acids were covered by GC-MS, larger molecules (e.g. flavonoids) by LC-MS whereas CE-MS measured specific cations and anions. Overall, the data generated 175 unique identified metabolites but a total of 1460 with "no or imprecise metabolite annotation." Of the identified metabolites, only 56 were observed in at least two platforms. A total of 261 peaks showed no correlation with experimental factors (transgene, cultivar, tissue type) and had to be removed from statistical analyses.

It is worth pointing out that two studies that assessed the metabolic profiles of grain from GM maize containing the *Cry1ab* gene and that utilized the same data acquisition platform (NMR) differed in their conclusions on the impact of transgene insertion on levels of free amino acids (Manetti et al., 2006; Piccioni et al., 2009). Manetti et al. (2006) reported that the GM crop included higher levels of sugars (glucose, sucrose, meliobiose), GABA, glutamine, and succinate and decreased levels of alanine, asparagine, and choline. Piccioni et al. (2009) reported lower levels of all amino acids, lower sugars, and lower succinate (and other organic acids). Piccioni et al. (2009) were also able to report on metabolites observed in their NMR profiles but absent in those of Manetti et al. (2006). Key design differences between the two studies include different parental lines, different growth conditions, and sample extraction protocols.

Levandi et al. (2008) utilized CE-MS to compare levels of 27 metabolites in three different GM maize lines also containing the Cry1ab gene. Some of these metabolites (e.g., certain free amino acids, choline, GABA) were also recorded in the NMR platforms of Mannetti et al

(2006) and Piccioni et al. (2009). No consistent association of these metabolites with the GM trait when assessed over all three GM lines was observed, a conclusion in line with the combined results of Mannetti et al. (2006) and Piccioni et al. (2009). Several of the metabolites reported by Levandi et al. (2006) are more typically associated with other taxonomic groupings, for example, graveolin (*Ruta graveolus*, Rutaceae) and lunarine, (*Lunaria annua*, Brassicaceae). The assignment of peaks to metabolites not typically associated with a genus or family would almost certainly require extensive validation in a Regulatory environment.

Leon et al. (2009) utilized FT-MS on the same samples assessed by Levandi et al. (2008). This allowed coverage of 5500 mass signals of which approximately 1000 could be assigned an elemental composition. Those elemental compositions could be associated (through MasstTRIX) with specific metabolic pathways (KEGG); these associations are referred to as "isomeric hits". This approach would identify any differences in GM and non-GM metabolic profiles, especially where an elemental composition could be assigned, to be tentatively associated with biochemical differences. Overall, it was shown that a greater number of isomeric hits in pathways such as arachidonic acid metabolism, free amino acid metabolism, purine metabolism, and folate biosynthesis were associated with the GM samples. A list of 33 possible compounds that could distinguish the GM and non-GM varieties was generated, of which 12 could be confirmed in an orthogonal assay (CE-MS). The authors then indicated that only four of these could be considered as potential GM "biomarkers"; L-carnitine, apigenidin, 5, 6-dihydroxyindole, and one unidentified metabolite. There is little further literature on levels of these metabolites in maize and the association of these metabolites as GM biomarkers is almost certainly premature. Further, the interpretability of the Levandi et al. (2009) approach is not at all clear; there are fewer isomeric hits associated with inositol phosphate metabolism, yet levels of phytic acid have been well-established to be near-identical in GM and non-GM maize. The association of isomeric hits for bile acid biosynthesis, which is not typically associated with plant metabolism, is also difficult to interpret.

In summary, different metabolic profiling platforms applied to similar biological questions will yield non-overlapping solutions. This is due to differential metabolite coverage (even within similar data acquisition technologies) and is compounded both by the number of unidentified signals observed in current metabolite profiles and, in some cases, "identification" of metabolites not previously known to be biosynthetically associated with the plant species or genus in question.

3. Equating statistical equivalence with biosafety

Predetermined criteria would need to be established for any study protocol, data acquisition steps or statistical analyses utilized in a safety assessment. As alluded to earlier, sampling from multiple replicated field sites would be required. Discussions on the number of replicates required to generate meaningful results from omics experiments are available as well as on the potential for bias and over-fitting (Broadhurst and Kell, 2006; Goodacre et al., 2007). Here we focus on the routine misinterpretation of "statistical significance" (Goodman, 2008) and the tendency to associate statistically significant differences between GM and a non-GM comparator as, minimally, an unintended effect, and often to imply the statistical difference raises a question about the safety of a newly evaluated crop.

As indicated earlier, compositional assessments of GM crops involve direct comparisons of levels of key nutrients and anti-nutrients in the new crop variety to those of a near-isogenic conventional comparator. Statistical evaluations of the compositional data have typically utilized classical frequentist significance testing. There are, however, several features of significance hypothesis testing that impact its application to compositional comparisons between crops with different agronomic qualities (Lecoutre, et al., 2001). Berger (1985), for example, stated, "We know from the beginning that the point null hypothesis is almost certainly not exactly true, and that this will always be confirmed by a large enough sample. What we are really interested in determining is whether or not the null hypothesis is approximately true." There are many factors that impact crop composition, *including agronomic traits we seek to modify through plant breeding*, (e.g. Scott et al., 2006; Uribelarrea et al., 2004; Dornbos and Mullen, 1992; Hymowitz et al., 1972; Wilcox and Shibles, 2001; Yin and Vyn, 2005) and any compositional changes that accompany enhanced agronomic quality may confound interpretation of results generated through significance testing.

Statistical significance is used only as a first step in comparative assessments. The interpretation of statistical significance from a p-value, *the probability of an observed result or a more extreme result occurring if the null hypothesis were true*, does not imply biological significance (Goodman, 2008). Statistically significant differences do not imply large differences between GM and conventional comparators or that these comparators can be easily distinguished from a biological perspective. In fact, the power of the experimental designs (multiple highly replicated field trials) adopted in current compositional assessments allows statistical significance to be assigned even where there are very small difference in mean values of a given component but where the distribution of component values overlap extensively. As such, significance approaches must be accompanied with further data analysis encompassing discussion of magnitudes of differences, assessments of component ranges, and the sensitivity of component values to environmental factors such as location. This is consistent with the recommendation by Codex Alimentarius (2008, Ch. 44) that "The statistical significance of any observed differences should be assessed in the context of the range of natural variations for that parameter to determine its biological significance." It is further consistent with observations of high variability in crop composition recorded in the scientific literature. The current scientific consensus is that, in most if not all cases, statistically significant differences between GM and near-isogenic conventional controls represent modest and nutritionally meaningless differences in magnitude. For example, a recent review of studies on GM crop composition showed that over 99% of all nutrient and antinutrients comparisons, where significant differences at the 5% level (α=0.05) in mean values were observed, had a relative magnitude difference less than 20%. These differences are considerably less than the range of values attributable to germplasm and environmental factors (Harrigan et al., 2010).

Most metabolic profiling experiments utilize significance testing and Rischer and Oksman-Caldentey (2006) refer to unintended effects as "effects which represent a statistically significant difference (e.g. in chemical composition of the GM plant compared with a suitable non-GM plant)" although they acknowledge that such differences would have to be evaluated in the context of natural variability. One review that endorses the use of omics in safety assessments suggests that "the amount of variation from genetic engineering should be small (~3%)." (Heineman et al., 2011). Whilst this particular number is unrealistic since it falls well within the natural variability of metabolite levels and is even less than typical experimental error, setting a universal threshold for relative magnitude of differences as a trigger for further safety assessments of GM crops has been considered. In 2000, the Nordic Council of Ministers recommended that if a component in a GM crop differed from the conventional control by ±20% in relative magnitude, additional analyses of the GM crop were warranted (cited in Hothorn and Oberdoerfer, 2006). This concept was refined to account for the nutritional relevance of a component and the experimental precision of its measurement (Hothorn and Oberdoerfer, 2006). Threshold ranges for GM components were suggested as follows; 0.833-1.20 of the conventional control for "nutritionally very relevant" components (minerals, vitamins, anti-nutrients, bioactives, essential amino acids, and fatty acids), 0.769-1.30 for "relevant" (non-essential amino and fatty acids), and 0.667-1.50 for components of "less relevance" (proximates, fiber). Suggestions for the use of limits and triggers of this kind have been criticized for their failure to fully account for the role and contributions of the specific crop in the human diet; and with GM crops in particular since they are often not eaten as such but are used as a source of macronutrients such as oil, starch and protein (Chassy, 2008; Chassy, 2010). As noted previously, most plant foods in the human diet make significant contributions to the total intake of just a few macro- and micronutrients and therefore even large compositional changes in a single crop plant might produce little impact on the nutritional value of the overall diet. Chassy (2010) has observed that composition cannot be viewed in isolation since the composition of the diet is far more important than the composition of a single variety of a single crop. Strictly numerical approaches have not been adopted in compositional studies and there is no reason they would be relevant to profiling experiments.

At least one profiling study has attempted to apply statistical equivalence testing but again falls prey to the dubious association of equivalence with safety. Kusano et al. (2011) compared a GM-tomato (a miraculin protein expressor) to not only to the parental line but to a panel of conventional reference varieties. The statistical design (described by the authors as a proof-of-safety test) involved comparing the difference between test and control and the determining whether these differences fell within equivalence limits established by the reference varieties. However such a design makes more of a statement about the selection of the reference substances and the control to which the GM-trait is introgressed, and not about the effect of transgene insertion; the same test-to-control differences can be equivalent or non-equivalent contingent on whether a limited or diverse range of genotypes is available. The overall conclusion from the study however was that "miraculin over-expressors are remarkably similar to the control line".

In summary, there are no defined data analyses strategies currently being consistently applied to profiling data that would facilitate interpretability of data.

4. Conclusion

There are clearly divergent views about the utility of 'omics sciences in food safety assessments. This paper has discussed some of the reasons metabolic profiling technologies are, however, unlikely to provide immediately interpretable data in safety assessments that would otherwise enhance rigorously quantitative assessments of known nutrients and anti-nutrients that comprise foodstuffs. Indeed, it is not clear to the present authors that any new types of data are in fact necessary to judge GM or other foods as safe. We are also unaware of any "gaps" in our compositional knowledge that might compromise safety and in fact, our current understanding of plant anti-nutrients and toxicants, allows GM solutions to enhancing food safety (e.g. Sunilkumar et al., 2006). The last 25 years of research on GM plants and 15 years of commercial experience planting GM crops without harm or incident suggest that no difference in safety that would require further analysis exists between GM and crops bred by other strategies. All breeding induces genetic changes and these changes give rise to transcriptomic, proteomic and metabolomic alterations.

We consider that metabolic profiling could increase its value in food safety science as well as in the development of nutritionally enhanced crops as follows;

1. *Improved compositional analysis*. One potential target for future research could be to develop metabolic screening methods that afford a comprehensive compositional assessment in a single suite of determinations rapidly and at lower cost than traditional targeted analysis. It is known that the metabolites in a cell form a large, complex and interconnected network; one possible approach would be elucidation of key metabolic compound whose determination might provide insight into the global concentrations of numerous other metabolites. If such a validated analytical method could be developed it would great aid research and development and would be particularly valuable in assessments of nutritionally enhanced crops where changes in a specific pathway are sought. However, metabolomic technologies are not able to supply this kind of analysis and data.
2. *Detection of novel toxicants*. Targeted analysis is inherently incapable of assessing levels of metabolites that are not selected (targeted) for analysis. Proponents of metabolic profiling have argued that profiling might detect the emergence of previously unknown novel toxicants presumably created by the breeding process. However, the abundance of a few macro-components (protein, fiber, carbohydrate, lipids) and numerous minor metabolites leaves little compositional "space" for novel toxicants. If wholly new molecules were created by the spontaneous evolution of a new pathway or pathways necessary for its biosynthesis, the chances that sufficient quantities would be present to exert an adverse effect are small indeed. Perhaps this is why such effects have not yet been observed by science or why coherent hypotheses as to how a novel toxicant would be generated by a specific breeding process appear to be sparse in the literature.
3. *Detection of unintended effects*. Proponents of metabolic profiling often suggest that a profile itself may be an indicator that unintended changes had occurred. Methods to draw safety conclusions based on differences in metabolic profiles do not yet exist, and certainly as we have discussed above, no reason to assume that differences in profiles imply a safety concern; in fact, by any objective measure, there is no such technique as metabolomic profiling. What we have today is a series of distinct and emerging powerful scanning techniques each of which surveys a slightly different molecular landscape with variable degrees of resolution. Clearly, the number of metabolites present in crops is very large and the power of targeted metabolic profiling will become increasingly useful in analyzing the chemical complexity of prospective commercial releases as they progress through initial research and development phases.

Metabolomics is an expanding and exciting field of research. The rapidly expanding scope of the metabolomic profiling technologies tempts us to test their applicability to a wide array of analytical challenges. We have, on the other hand, a long history of safe experience with plant breeding. We know that many unintended changes take place in plant breeding, however, these are almost without exception innocuous. There is no reason to believe that GM breeding should require any new or different data set than other forms of breeding.

It seems clear to the present authors that there is no role for metabolic profiling in food safety assessment. We agree that modern targeted metabolic profiling technologies can rapidly identify pathway perturbations and, if judiciously applied and interpreted, might enhance food safety science, although traditional analytical methods can still be used to assess if changes in pathways and metabolite pools have occurred. If incorporated into the early selection stages of a prospective new trait targeted metabolic profiling may greatly aid in the selection of metabolites that need to be considered during the compositional phase of a risk assessment. To quote Larkin and Harrigan (2007) "However, it should be self-evident that GM crops ought not to be considered a single monolithic class that is either good or bad for the economy, agriculture or the environment. Each novel crop should be considered on its own merits and demerits. If we ever get to that point we will have achieved something positive out of the GM controversy." It is our hope that colleagues will take this as a challenge to further metabolic profiling in the advancement of food safety and nutritional enhancement of crops.

5. Acknowledgements

Figure 1 was prepared by Jay Harrison of the Statistics Technology Center, Monsanto Company.

6. References

Ainasoja, M. M., Pohjala, L. L., Tammela, P.S. M., Somervuo, P. J., Vuorela, P. M. & Teeri, T. H. 2008. Comparison of transgenic Gerbera hybrida lines and traditional varieties shows no differences in cytotoxicity or metabolic fingerprints. Transgenic Res, 17, 793-803.

Baker, J. M. Hawkins, N. D. Ward, J. L., Lovegrove, A., Napier, J. A., Shewry, P. R., & Beale, M. H. 2006. A metabolomic study of substantial equivalence of field-grown genetically modified wheat. Plant Biotechnol J, 4, 381-92.

Beier, R. C. & Oertli, E. H. 1983. Psoralen and other linear furocoumarins as phytoalexins in celery. Phytochem, 22, 2595-97.

Bell, A. E. 2003. Nonprotein amino acids of plants: Significance in medicine, nutrition, and agriculture. J Agric Food Chem, 51, 2854–65.

Berger, J.O. 1985. Statistical Decision Theory and Bayesian Analysis. 2nd edition: Springer-Verlag.

Britz, S. J., Kremer, D. F., & Kenworthy, W. 2008. Tocopherols in soybean seeds; Genetic variation and environmental effects in field-grown crops. J Am Oil Chem Soc, 85, 931-36.

Broadhurst, D. I. & Kell, D. B. 2006. Statistical strategies for avoiding false discoveries in metabolomics and related experiments. Metabolomics, 2, 171-96.

Catchpole, G. S., Beckmann, M., Enot, D. P., Mondhe, M., Zywicki, B., Taylor, J., Hardy, N., Smith, A., King, R. D., Kell, D. B., Fiehn, O. & Draper, J. 2005. Hierarchical metabolomics demonstrates substantial compositional similarity between genetically modified and conventional potato crops. Proc Natl Acad Sci U S A, 102, 14458-62.

Chassy, B., Egnin, M., Gao, Y., Glenn, K., Kleter, G. A., Nestel, P., Newell-Mcgloughlin, M., Phipps, R. H. & Shillito, R. 2008. Recent developments in the safety and nutritional assessment of nutritionally improved foods and feeds. Comp Rev Food Sci Food Safety, 7, 50-113.

Chassy, B. 2010. Can 'omics inform a food safety assessment? Reg Toxicol Pharmacol,58, S62-70.

Cheng, K. C., Beaulieu, J., Iquira, E., Belzile, F. J., Fortin, M. G., & Stromvik, M. V. 2008. Effect of transgenes on global gene expression in soybean is within the natural range of variation of conventional cultivars. J Agric Food Chem, 56, 3057–67.

Codex Alimentarius. 2008. Guideline for the conduct of food safety assessment of foods derived from recombinant-DNA plants. CAC/GL 45-2003.

Davies, H. 2010. A role for "omics" technologies in food safety assessment. Food Control, 21, 1601-10.

Defernez, M., Gunning, Y. M., Parr, A. J., Shepherd, L. V., Davies, H. V. & Colquhoun, I. J. 2004. NMR and HPLC-UV profiling of potatoes with genetic modifications to metabolic pathways. J Agric Food Chem, 52, 6075-85.

Dornbos, D. & Mullen, R. 1992. Soybean seed protein and oil contents and fatty acid composition adjustments by drought and temperature. J Am Oil Chem Soc, 69, 228-31.

Dubouzet, J. G., Ishihara, A., Matsuda, F., Miyagawa, H., Iwata, H. & Wakasa, K. 2007. Integrated metabolomic and transcriptomic analyses of high-tryptophan rice expressing a mutant anthranilate synthase alpha subunit. J Exp Bot 58, 3309-21.

Eldridge, A. C. & Kwolek W. F. 1983. Soybean isoflavones: effect of environment and variety on composition. J Agric Food Chem, 31, 394-6.

EFSA. 2011. EFSA Panel on Genetically Modified Organisms (GMO); Scientific opinion on guidance for risk assessment of food and feed from genetically modified plants. EFSA Journal, 9, 2150 [37pp]. Available online:
www.efsa.europa.eu/efsajournal.htm

Foresight. The Future of Food and Farming (2011) Final Project Report. The Government Office for Science, London.

Goodacre, R., Vaidyanathan, S., Dunn, W. R., Harrigan, G. G. & Kell, D. B. 2004. Metabolomics by numbers-Acquiring and understanding global metabolite data. Trends Biotechnol, 22, 245-52.

Goodacre, R., Broadhurst, D., Smilde, A. K., Kristal, B.S., Baker, D. J., Beger, R., Bessant, C., Connor, S., Capuani , G., Craig, A., Ebbels, T., Kell, D. B., Manetti, C., Newton, J., Paternostro, G., Somorjai, R., Sjostrom, M., Trygg, J. & Wulfert, F. 2007. Proposed minimum reporting standards for data analysis in metabolomics. Metabolomics, 3, 231-41.

Goodman, S. 2008. A dirty dozen: Twelve p-value misconceptions. Sem Hematol, 45, 135-40.

Graff, G. D., Zilberman, D. & Bennett, A. B. 2009. The contraction of agbiotech product quality innovation. Nat Biotechnol, 27, 702-4.

Gregersen, P. L., Brinch-Pedersen, H. & Holm, P. B. 2005. A microarray-based comparative analysis of gene expression profiles during grain development in transgenic and wild type wheat. Transgenic Res, 14, 887-905.

Gutierrez-Gonzalez, J. J., Wu, X., Zhang, J., Lee, J. D., Ellersieck, M., Shannon, J. G., Yu, O., Nguyen, H. T. & Sleper, D.A. 2009. Genetic control of soybean seed isoflavone content: importance of statistical model and epistasis in complex traits. Theor Appl Genet, 119, 1069-83.

Harrigan, G. G. & Harrison, J. M. 2011. Assessing compositional variability through graphical analysis and Bayesian statistical approaches; case studies on transgenic crops. Biotechnol Genet Eng Rev, 28, 1-18.

Harrigan, G. G., Stork, L. G., Riordan, S. G., Reynolds, T. L., Ridley, W. P., Masucci, J. D., Macisaac, S., Halls, S. C., Orth, R., Smith, R. G., Wen, L., Brown, W. E., Welsch, M., Riley, R., Mcfarland, D., Pandravada, A. & Glenn, K. C. 2007. Impact of genetics and environment on nutritional and metabolite components of maize grain. J Agric Food Chem, 55, 6177-85.

Harrigan, G. G., Lundry, D., Drury, S., Berman, K., Riordan, S. G., Nemeth, M. A., Ridley, W. P. & Glenn, K. C. 2010. Natural variation in crop composition and the impact of transgenesis. Nat Biotechnol, 28, 402-4.

Heineman, J. A., Kurenbach, B. & Quist, D. 2011. Molecular profiling - a tool for addressing emerging gaps in the comparative risk assessment of GMOs. Environ Int, 37, 1285-93.

Herman, R. A., Chassy, B. M. & Parrott, W. 2009. Compositional assessment of transgenic crops: an idea whose time has passed. Trends Biotechnol, 27, 555-7.

Hothorn, L. A. & Oberdoerfer, R. 2006. Statistical analysis used in the nutritional assessment of novel food using the proof of safety. Reg Toxicol Pharmacol, 44, 125-135.

Hymowitz, T., Collins, F. I., Panczner, J. & Walker, W. M. 1972. Relationship between the content of oil, protein, and sugar in soybean seed. Agron J, 64, 613-16.

Ioset, J. R., Urbaniak, B., Ndjoko-Ioset, K., Wirth, J., Martin, F., Gruissem, W., Hostettmann, K. & Sautter, C. 2007. Flavonoid profiling among wild type and related GM wheat varieties. Plant Mol Biol, 65, 645-54.

Kok, E. J., Keijer, J., Kleter, G. A. & Kuiper, H. A. 2008. Comparative safety assessment of plant-derived foods. Regul Toxicol Pharmacol, 50, 98-113.

Kristensen, C. Morant M., Olsen C. E., Ekstrøm, C. T., Galbraith, D. W., Møller, B. L., & Bak, S. 2005. Metabolic engineering of dhurrin in transgenic Arabidopsis plants with marginal inadvertent effects on the metabolome and transcriptome. Proc Natl Acad Sci U S A, 102, 1779-84.

Jones, D.A. 1998. Why are so many food plants cyanogenic? Phytochem, 47, 155-62.

Kalaitzandonakes, N., Alston, J.M. & Bradford, K. J. 2007. Compliance costs for regulatory approval of new biotech crops. Nat Biotechnol, 25, 509-11.

Khoshbahkt, K. & Hammer, K. 2008. How many plant species are cultivated? Genet Resour Crop Evol, 55, 925-28

Kusano, M., Redestig, H., Hirai, T., Oikawa, A., Matsuda, F., Fukushima, A., Arita, M., Watanabe, S., Yano, M., Hiwasa-Tanase, K., Ezura, H. & Saito, K. 2011. Covering chemical diversity of genetically-modified tomatoes using metabolomics for objective substantial equivalence assessment. PLoS, 6, e16989.

Lam, H-M., Xu, X., Liu, X., Chen, W., Yang, G., Wong F. L., Li, M.-W., He, W., Qin, N., Wang, B., Li, J., Jian, M., Wang, J., Shao, G., Wang, J., Sun , S. S.-M., & Zhang, G. 2010. Resequencing of 31 wild and cultivated soybean genomes identifies patterns of genetic diversity and selection. Nature Gen, 42, 1053-59.

Larkin, P. & Harrigan, G. G. 2007. Opportunities and surprises in crops modified by transgenic technology: metabolic engineering of benzylisoquinoline alkaloid, gossypol and lysine biosynthetic pathways. Metabolomics, 3, 371-82.

Lay Jr., J.O, Borgmann, S., Liyanage, R. & Wilkins, C.L. 2006. Problems with the "omics". Trends Anal Chem, 25, 1046-56.

Le Gall, G., Colquhoun, I. J., Davis, A. L., Collins, G. J. & Verhoeyen, M. E. 2003. Metabolite profiling of tomato (*Lycopersicon esculentum*) using ^{1}H NMR spectroscopy as a tool to detect potential unintended effects following a genetic modification. J Agric Food Chem, 51, 2447-56.

Lecoutre, B., Lecoutre, M.-P. & J. Poitevineau, J. 2001. Uses, abuses and misuses of significance tests in the scientific community: Won't the Bayesian choice be unavoidable? Int Stat Rev, 69, 399-417.

Lehesranta, S. J., Davies, H. V., Shepherd, L. V., Nunan, N., Mcnicol, J. W., Auriola, S., Koistinen, K. M., Suomalainen, S., Kokko, H. I. & Karenlampi, S. O. 2005. Comparison of tuber proteomes of potato varieties, landraces, and genetically modified lines. Plant Physiol, 138, 1690-9.

Leon, C., Rodriguez-Meizoso, I., Lucio, M., Garcia-Canas, V., Ibanez, E., Schmitt-Kopplin, P. & Cifuentes, A. 2009. Metabolomics of transgenic maize combining Fourier transform-ion cyclotron resonance-mass spectrometry, capillary electrophoresis-mass spectrometry and pressurized liquid extraction. J Chromatog A, 1216, 7314-23.

Levandi, T., Leon, C., Kaljurand, M., Garcia-Canas, V. & Cifuentes, A. 2008. Capillary electrophoresis time-of-flight mass spectrometry for comparative metabolomics of transgenic versus conventional maize. Anal Chem, 80, 6329-35.

Manetti, C., Bianchetti, C., Casciani, L., Castro, C., Di Cocco, M. E., Miccheli, A., Motto, M. & Conti, F. 2006. A metabonomic study of transgenic maize (*Zea mays*) seeds revealed variations in osmolytes and branched amino acids. J Exp Bot, 57, 2613-25.

NIEHS. 1998. National Institute of Environmental Health Sciences; α-Chaconine [20562-03-2] and α-solanine [20562-02-1]. Review of toxicological literature. http://ntp-server.niehs.nih.gov/index.cfm?objectid=6F5E930D-F1F6-975E-7037ACA48ABB25F4

NRC. 2004. National Research Council: Safety of Genetically Engineered Foods. Approaches to Assessing Unintended Health Effects. The National Academies Press, Washington, D.C.

Panthee, D. R., Pantalone, V. R., West, D. R., Saxton A. M., & Sams, C. E. 2005. Quantitative trait loci for seed protein and oil concentration and seed size in soybean. Crop Sci, 45, 2015-22.

Piccioni, F., Capitani, D., Zolla, L. & Mannina, L. 2009. NMR metabolic profiling of transgenic maize with the Cry1A(b) gene. J Agric Food Chem, 57, 6041-49.

Potrykus, I. 2010. Regulation must be revolutionized. Nature, 466, 561

Ricroch, A. E., Berge, J. B. & Kuntz, M. 2011. Evaluation of genetically engineered crops using transcriptomic, proteomic, and metabolomic profiling techniques. Plant Physiol, 155, 1752-61.

Rischer, H. & Oksman-Caldentey, K.-M. 2006. Unintended effects in genetically modified crops: revealed by metabolomics? Trends Biotechnol, 24, 102-4.

Rotundo, J.L. & Westgate, M. E. 2009. Meta-analysis of environmental effects on soybean seed composition. Field Crops Res, 110, 147-56.

Scott, M. P., Edwards, J. W., Bell, C. P., Schussler, J. R. & Smith, J. S. 2006. Grain composition and amino acid content in maize cultivars representing 80 Years of commercial maize varieties. Maydica, 51, 417-23.

Seguin, P., Turcotte, P., Tremblay, G., Pageau, D., & Liu, W. 2009. Tocopherols concentration and stability in early maturing soybean genotypes. Agron. J, 101, 1153-59.

Seguin, P., Tremblay, G., Pageau, D. & Liu, W. 2010. Soybean tocopherol concentrations are affected by crop management. J Agric Food Chem, 58, 5495-5501.

Senti, F.R. & Rizek, R. R.1974. An overview of GRAS regulations and the effect from the viewpoint of nutrition, In Hanson C. H. (ed) CSSA Special Publication. Vol. 5. SSA, Madison, WI.

Skogerson, K., Harrigan, G. G., Reynolds, T. L., Halls, S. C., Ruebelt, M., Iandolino, A., Pandravada, A., Glenn, K. C. & Fiehn, O. 2010. Impact of genetics and environment on the metabolite composition of maize grain. J Agric Food Chem, 58, 3600-10.

Sunilkumar, G., Campbell, L. M., Puckhaber, L., Stipanovic, R. D. & Rathore, K. S. 2006. Engineering cottonseed for use in human nutrition by tissue-specific reduction of toxic gossypol. Proc Natl Acad Sci U S A, 103, 18054-59.

Uribelarrea, M., Below, F. E. & Moose, S.P. 2004. Grain composition and productivity of maize hybrids derived from the Illinois protein strains in response to variable nitrogen supply. Crop Sci, 44, 1593-1600.

Yin, X.H. & Vyn, T. J. 2005. Relationships of isoflavone, oil, and protein in seed with yield of soybean. Agron J, 97, 1314-21.

Wakasa, K., Hasegawa, H., Nemoto, H., Matsuda, F., Miyazawa, H., Tozawa, Y., Morino, K., Komatsu, A., Yamada, T., Terakawa, T. & Miyagawa, H. 2006. High-level tryptophan accumulation in seeds of transgenic rice and its limited effects on agronomic traits and seed metabolite profile. J Exp Bot, 57, 3069-78 .

Wilcox, J.R. & Shibles, R. M. 2001. Interrelationships among seed quality attributes in soybean. Crop Sci, 41, 11-4.

Wilson, R. F. 2004. Seed composition, In H.R. Boerma and J.E. Specht (ed.) Soybeans: Improvement, production and uses, 3rd ed. ASA, CSSA, SSA, Madison, WI.

Zhou, J., Berman, K. H., Breeze, M. L., Nemeth, M. A., Oliveira, W. S., Braga, D. V., Berger, G. U. & Harrigan, G. G. 2011a. Compositional variability in conventional and glyphosate-tolerant soybean varieties grown in different regions in Brazil. J Agric Food Chem, 59, online

Zhou, J., Harrigan, G. G., Berman, K. H., Webb, E. G., Klusmeyer, T. H. & Nemeth, M. A. 2011b. Stability of the compositional equivalence of grain from insect-protected corn and seed from herbicide-tolerant soybean over multiple seasons, locations and breeding germplasms. J Agric Food Chem, 59, 8822-28.

15

Metabolomics Approach for Hazard Identification in Human Health Assessment of Environmental Chemicals

Suryanarayana V. Vulimiri[1], Brian Pachkowski[2], Ambuja S. Bale[1] and Babasaheb Sonawane[1]
[1]*U.S. Environmental Protection Agency, Office of Research and Development,*
[2]*Oak Ridge Institute for Science and Education Postdoctoral Fellow,*
National Center for Environmental Assessment, Washington DC,
USA

1. Introduction

Exposure to xenobiotics induces complex biochemical responses in mammalian cells resulting in several perturbations in cellular toxicity pathways. Within the context of systems biology, such biochemical perturbations can be studied individually using "omics" approaches such as toxicogenomics, transcriptomics, proteomics and metabolomics (Heijne et al., 2005). The objective of this chapter is to examine how the metabolomics approach can be used in identifying the risk posed by environmental chemicals to human health using selective examples of organ toxicity. Metabolomics is a medium-to-high throughput technique employing predominantly mass spectrometry (MS) and nuclear magnetic resonance (NMR) technology (Roux et al., 2011) for the identification and characterization of endogenous metabolites of low molecular weight (<1800 Da) arising from different biochemical pathways either as primary or secondary metabolites (Idle & Gonzalez, 2007). The sum total of all small metabolites is referred to as the "metabolome". Metabolomics has also been applied to the identification of low molecular weight, exogenous metabolites of xenobiotics (Roux et al., 2011; Rubino et al., 2009). With these capabilities, metabolomics represents a relatively quick and informative approach for assessing the physiological response to environmental chemicals.

2. Human health risk assessment

Chemicals in the environment could pose potential risks to human health. In order to inform the assessment of risks from chemical exposures, the U.S. National Research Council (NRC) published a report entitled, "*Risk Assessment in the Federal Government: Managing the Process,*" more commonly known as the "Red Book" (NRC, 1983), which has been widely accepted and endorsed by the U.S. Environmental Protection Agency (U.S. EPA) and other federal agencies. This risk assessment process consists of four steps: hazard identification, dose-response assessment, exposure assessment, and risk characterization. The focus herein is on hazard identification, which has been defined as "identification of the contaminants that are suspected to pose health hazards, quantification of the concentrations at which they are present in the environment, a description of the specific forms of toxicity (i.e. neurotoxicity, carcinogenicity, etc.) that can be caused by the contaminants of concern, and an evaluation of the conditions under which these forms of toxicity might be expressed in exposed humans" (NRC, 1994).

For human health assessment of chemicals, non-cancer or cancer risk values are derived based on the selection of a critical endpoint of toxicity or several endpoints (e.g. biochemical, pathological, physiological, and behavioral abnormalities) of adverse health outcomes. Uncertainty factors are applied to the lowest dose associated with the critical health outcome(s) in order to derive the resulting exposure level for non-cancer toxicity. These uncertainty factors attempt to account for exposure duration, pharmacokinetic, and pharmacodynamic data gaps associated with inter- and intra-species extrapolation.

The U.S. EPA and the International Agency for Research on Cancer (IARC) evaluate the evidence for carcinogenesis in humans from epidemiological, experimental animal, and mechanistic data to determine the qualitative cancer classification for humans. In addition, the U.S. EPA evaluates exposure-response relationships and develops quantitative cancer risk values based on the observed tumors that correspond to a unit exposure (U.S. EPA, 2005). Uncertainties with cancer risk values are presented and are generally associated with the mode of action (MOA) for carcinogenicity.

One of the major concerns with cancer risk assessment is false-positive animal tumor findings. Having an understanding of the mechanism(s) leading to carcinogenicity would help in developing a better perspective of whether a carcinogen in experimental animals is likely to be a carcinogen in humans. For example, correlating a metabolomic profile of a suspected carcinogen between human exposures (environmental or occupational) and experimental animal exposure studies would be highly useful. If similar biochemical markers were to appear across the human and animal metabolomic profiles, that information would help in informing similarities or differences in interspecies mechanisms. Further, if the chemical was demonstrated to be a carcinogen in animals through a traditional two-year animal bioassay, but there was inconclusive epidemiological evidence, the similarity in metabolomic data could be used along with other mechanistic data (e.g. mutagenicity/genotoxicity assays, cell proliferation findings, oxidative stress, epigenetics, etc.) to support or refute human carcinogenicity. In this regard, metabolomics information could be used to support mechanistic data to augment the animal and human findings.

3. The potential of "omics" data to inform mode of action of environmental chemicals

In developing a human health evaluation for environmental chemical hazard identification, it is ideal to have information on the key mechanistic events leading to an adverse health outcome. In this regard, mode of action (MOA) is an important part of hazard identification. MOA can be defined as "a sequence of key events and processes, starting with interaction of an agent with a cell, proceeding through operational and anatomical changes, and resulting in cancer formation" (U.S. EPA, 2005). A 'key event' is defined as "an empirically observable precursor which is by itself a necessary component of the MOA or is a biologically-based measurable marker for such a component" (U.S.

EPA, 2005). *In vitro* systems (e.g. cell cultures) and *in vivo* models (e.g. experimental animals or human population studies) have identified several of the early key events such as oxidative stress, inflammation, genotoxicity, and cytotoxicity that occur from toxicant exposure. Since metabolomics measures biological response at the molecular level, this approach can identify the metabolites associated with the sequence of 'key events' and the processes inherent to the mechanism(s) of xenobiotic toxicity. The metabolomics approach could generate several mode(s) of action hypotheses using a nontargeted approach. The individual MOA hypothesis thus generated could be tested in targeted approaches (e.g. measuring glutathione reduction from oxidative stress) using more conventional assays. Metabolomics data could be used to further inform the mode(s) of action in experimental animals associated with carcinogenicity or with non-cancer health outcomes, which may help to confirm the relevancy of the observations in experimental animals to humans.

4. Metabolomics approach in investigating environmental chemical exposure

Environmental chemicals act through multiple toxicity pathways via a multitude of mechanisms (Guyton et al., 2009). To date, very limited toxicogenomics information has been applied to the field of risk assessment (Boverhof & Zacharewski, 2006; Mortensen & Euling, 2011). There is a paucity of relevant metabolomic information for application to human health risk assessment of environmental chemicals. To date, the available literature suggests an informative role of metabolomics in understanding the mode of action of environmental xenobiotics (Vulimiri et al., 2011; Vulimiri et al., 2009).

In general, human data are relatively sparse for many environmental chemicals with respect to both non-cancer and cancer health outcomes. Most human data are occupational, where exposure levels are generally higher than those encountered in the environment. In many cases, there are not any environmental or occupational human exposures that could be used to corroborate the animal data. In these cases, animal data are generally used to develop non-cancer and cancer risk values for human health assessments thereby raising issues of uncertainty associated with interspecies extrapolation. Metabolomic data could be used to fill in such data gaps. For example, *in vitro* metabolomic assays with human cells may be developed to compare with animal metabolomic profiles to determine if there are potentially similar mechanisms of toxicity or for identifying toxicity pathways. As a result, this characterization of biochemical mechanisms of toxicity would inform hazard identification for use in the human health risk assessment process.

5. Ability of metabolomics to differentiate gender, phenotypic, and genetic differences, and organ-specific effects

Since metabolomics analyzes endogenous and exogenous (xenobiotic-derived) low molecular weight metabolites, this approach has been applied to the differentiation of metabolic profiles between phenotypes and genotypes. As briefly discussed below, metabolomics has the ability to inform gender, genetic, and phenotypic differences as well as organ-specific effects. For understanding the toxicity of environmental agents, the utility of such information would clarify toxicodynamic uncertainty associated with the extrapolation between species as well as within species (i.e. human) variability.

5.1 Gender differences

Metabolomics approach can differentiate between gender-dependent differences in the metabolic profiles of untreated or control experimental animals. In a study aimed at the identification of novel biomarkers of effect using chemicals with diverse modes of action, one research group (van Ravenzwaay et al., 2007) pooled the metabolomic data (i.e. plasma profiles) from eleven individual experiments involving 670 male and female control Wistar rats over a span of one year. After principal component analysis (PCA), the authors found that the metabolic profiles were clustered into separate groups for males and females suggesting a stable metabolome of the control rats during the study period. This observation highlights the ability of metabolomics to potentially identify unique gender responses to chemical exposure.

5.2 Genetic differences and phenotypic effects

The metabolomics approach can be further used for studying the relationship between the genotype and phenotype of the organism. Genetic polymorphisms in human genes are known to modify exposure to environmental health hazards and are a source of uncertainty when assessing risk from environmental chemicals (Ginsberg et al., 2009; Kelada et al., 2003). Genetic differences have been shown to reflect changes in the metabolite profiles of individuals. In a human population, genetically determined variants (e.g. those associated with fatty acid metabolism) in metabolic phenotype (metabotype) have been identified by simultaneously detecting single-nucleotide polymorphisms (SNPs) in a genome-wide association study (GWAS) and endogenous serum metabolites (Gieger et al., 2008).

This is also evident in the field of functional genomics where a change in phenotype is observed due to gene-related alterations (reverse-genetics) such as deletions or insertions leading to silent mutations as in yeasts (Raamsdonk et al., 2001). Specific gene mutations can also be evaluated using metabolic footprinting (Szeto et al., 2010). Such metabolomic data could provide basic information regarding gene product function, particularly in the context of environmental exposure.

Also, the genotypes of animals, as in genetically manipulated animal models (e.g. gene knockout and transgenic mice), have been used effectively for understanding the metabolism of toxicants mediated by cytochrome P450 (CYP) isozymes, in order to further elucidate mechanisms of toxicity. For example, MS-based approaches were able to distinguish between the metabolic profiles for *Cyp2e1*-null, CYP1A2-humanized, and wild-type mice after exposure to the ubiquitous dietary carcinogen 2-amino-1-methyl-6-phenylimidazo[4,5-b]pyridine (Chen et al., 2007) or the hepatotoxic agent acetaminophen (Chen et al., 2008). As a result, the metabolomics approach could be used to identify mechanistic changes stemming from genetic differences.

5.3 Organ-specific effects

Metabolomics approach has been utilized in identifying specific profiles that are altered in different organs in response to toxicant injury. The following represents a brief discussion of three select organ toxicities associated with exposure to a given chemical or mixture of chemicals. The extent of the metabolomic datasets varies for each type of organ toxicity; however, discussion focuses on how metabolomic investigations have contributed to some understanding of mechanisms of toxicity. Some of these metabolic changes are specific but many are nonspecific for the selected organ toxicities.

5.3.1 Hepatotoxicity (Carbon tetrachloride)

Carbon tetrachloride (CCl_4) is a well-studied hepatotoxicant and carcinogen that has been used as a fumigant (Manibusan et al., 2007). The mechanisms underlying CCl_4 toxic responses are fairly well understood. In rats, CCl_4 exposure results in mild to severe centrilobular necrosis of the liver with elevation of serum alanine aminotransferase (ALT) and aspartate aminotransferase (AST) activities suggestive of hepatotoxicity. In addition, CCl_4 induces lipid peroxidation, oxidative stress, and regenerative proliferation, eventually leading to hepatocarcinogenesis. The following brief discussion focuses on how metabolomics has been used to identify metabolites associated with these known mechanisms of toxic response.

Metabolomic analyses have identified several aldehydes (e.g. formaldehyde, acetaldehyde, propanal, butanal, pentanal, hexanal and malondialdehyde) excreted in urine (De Zwart et al., 1997) or in exhaled breath (Gee et al., 1981) from rodents dosed with CCl_4. Similarly, short-chain hydrocarbons (e.g. pentane and ethane) generated during lipid peroxidation have been reported in the breath of humans suggestive of CCl_4-induced oxidative stress (Hwang & Kim, 2007). These reactive aldehydes arising from lipid peroxidation have been shown to be quenched by reduced glutathione (GSH), causing GSH depletion which leads to cirrhosis in the liver of rats (Cabre et al., 2000). Further, GSH depletion causes a perturbation of prooxidant-antioxidant balance eventually leading to oxidative stress (Ichi et al., 2009). Metabolomic studies have provided support that oxidative stress induced by CCl_4 generates reactive oxygen species (ROS) capable of causing DNA damage as shown by an increase in 8-oxodeoxyguanosine (8-oxodG) and malondialdehyde-deoxyguanosine (M_1G) adducts (Beddowes et al., 2003). Also, lipid peroxidation products, such as 4-hydroxynonenal, can form etheno adducts, which are capable of inducing mutations in critical genes through base-pair substitutions (Barbin, 2000; Chung et al., 1999).

Further, CCl_4-induced oxidative stress inhibits energy metabolism as observed by changes in levels of Krebs cycle intermediates (i.e. citric acid, 2-oxoglutarate, and succinate), which has an overall effect on glutathione metabolism eventually leading to hepatotoxicity (Huang et al., 2008; Kim et al., 2008; Lin et al., 2009; Robertson et al., 2000). These earlier cellular perturbations further affect Ca^{+2} homeostasis, cause protein and phospholipid degradation, and potentially lead to cytotoxicity (Manibusan et al., 2007), apoptosis and necrosis of the liver tissue (Sun et al., 2001). Furthermore, membrane phospholipid degradation leads to the release of high levels of *o*-phosphoethanolamine, a well-known biomarker of cytotoxicity that can be detected using a metabolomic approach (Wang et al., 2009).

CCl_4 toxicity also leads to an inflammatory response characterized by the release of arachidonic acid, which is the precursor for a cascade of events leading to prostaglandin and leukotriene synthesis involving cyclooxygenases and lipoxygenases, respectively (Basu, 2003), and other newer inflammatory biomarkers, such as resolvins, protectins, and maresins (Liu et al., 2009; Serhan, 2009). Following cytotoxicity, liver tissue undergoes regenerative proliferation and activation of the urea cycle characterized by polyamine biosynthesis, the latter being a hallmark of cellular proliferation and differentiation (Heby, 1981). Detection of increased levels of urea cycle metabolites such as putrescine, ornithine, spermine, and spermidine (Pegg et al., 1981) in CCl_4-exposed rats by metabolomics approach gives an early indication of regenerative cell proliferation, which has been detected at a later stage of disease progression using conventional toxicity assays.

Overall, metabolomics studies have identified important biomarkers for CCl_4-induced toxicity including: aldehydes (lipid peroxidation); GSH/GSSG ratio (oxidative stress); 8-oxodG, M_1G, and etheno adducts (genotoxicity); *o*-phosphoethanolamine (cytotoxicity); arachidonic acid and prostaglandins (inflammation); and polyamines (regenerative cell proliferation), which are also observed in conventional assays (Vulimiri et al., 2011). This brief discussion further supports the use of metabolomics for investigating toxicity associated with environmentally related chemicals.

5.3.2 Neurotoxicity (Methylmercury)

Methylmercury is a chemical that was originally used as a fungicide for agricultural purposes in the early 20th century. In one of the earliest studies of methylmercury toxicity, occupational exposure to this chemical resulted in detriments in neurological functions in factory workers and changes in brain morphology with specificity to the cerebellar area. Additionally, in one of the most noted disasters in the 20th century, methylmercury exposure through consumption of fish was determined to be the causative agent for producing neurological deficits in fishermen and neurodevelopmental abnormalities in children in Minamata Bay, Japan (Ekino et al., 2007).

Traditional approaches in hazard characterization have concluded that the most sensitive (critical) effect observed from methylmercury exposure is neurobehavioral changes in children [as reviewed in (U.S. EPA, 2005) and Agency for Toxic Substances and Disease Registry (ATSDR, 1999)]. An uncertainty that was highlighted in both of these human health assessment evaluations was that of pharmacodynamic variability in the human population. One approach to address pharmacodynamic variability is to have an understanding of the key mechanism(s) that results in the sensitive effect. There have been several reviews on the neurological mechanisms associated with methylmercury (Ekino et al., 2007; Myers et al., 2009). One of the primary mechanisms associated with methylmercury neurotoxicity in both the adult and developing brain is cytotoxicity. Methylmercury has been demonstrated to induce oxidative stress in the nervous system. Additionally, *in vitro* studies have reported that methylmercury treatment results in apoptosis in glial cells, cerebellar granule cells, and in neuronal cell lines such as differentiated PC12 cells and neuroblastoma cells. Methylmercury exposure in rats has also resulted in apoptosis in the cerebellum as well as loss of astrocytes in monkeys. The cell death pathway for methylmercury has been highly characterized as reviewed recently (Ceccatelli et al., 2010). In summary, several studies have demonstrated that methylmercury induces caspase-dependent apoptosis in primary neuronal cell cultures and cell lines.

Metabolomic analysis for methylmercury was conducted using an *in vitro* methodology proposed by the European Commission Joint Research Centre in 2008 (van Vliet et al., 2008). Dynamic reaggregating neuronal cultures prepared from 16-day old rat fetal telencephalon were used for this analysis. The identified putative biomarkers were γ-aminobutyric acid (GABA), choline, glutamine, creatine, and spermine. The authors were able to speculate potential mechanisms of methylmercury from results of traditional biochemical assays (Eskes et al., 2002; Monnet-Tschudi et al., 1996) where the enzymatic activities of neuronal enzymes including glutamine synthesis, choline acetyltransferase, and glutamic acid decarboxylase were reported to be significantly decreased. In essence, the decreased levels of GABA and choline in the *in vitro* metabolomic study may help to explain these decreases in enzymatic activity which then leads to neuron-specific toxicity. This information correlates to the observed increases in apoptotic cell death associated with methylmercury. Similarly, the increased creatine levels that were reported from methylmercury treatment were correlated to gliosis (proliferation of astrocytes in the central nervous system) by the authors. Creatine is generally linked to brain osmoregulation (Bothwell et al., 2001) and increased creatine levels would lead to increased activity of glial cells that could result in gliosis.

Thus, a very limited metabolomic dataset for methylmercury adds to the mechanistic profile of this compound and helps establish a temporality of the key events leading to cytotoxicity in the brain. In the case of using the metabolomic data for methylmercury, the observed changes in the neuronal metabolites provide supportive, early evidence of later stage events leading to brain cytotoxicity. Metabolomic data from *in vivo* studies or from incidental human exposure would demonstrate any pharmacodynamic differences and if so, how such differences could be quantified through the available data rather than the default use of uncertainty factors.

5.3.3 Pulmonary toxicity (Cigarette smoke)

Cigarette smoking is one of the major etiological reasons for pulmonary toxicity and lung cancer. Cigarette smoke contain at least 4000 chemicals from different chemical classes, of which at least 60 are well-established carcinogens in experimental animals (Hecht, 2006). Cigarette smoke has been shown to induce oxidative stress and an inflammatory response in lungs, and chronic exposure is known to cause cancer. While cigarette smoke is most deleterious to the smoker, environmental or passive exposure to smoke (i.e. second-hand smoke) can lead to adverse health effects to bystanders (U.S. EPA, 1992). Using the metabolomics approach in A549 human lung epithelial cells it has been shown that several biochemical pathways are altered by either the whole smoke (WS) or its component phases i.e. wet total particulate matter (WTPM) or gas/vapor phase (GVP) (Vulimiri et al., 2009). Exposures to the different phases lead to unique biochemical alterations in A549 cells. For example, WTPM exposure resulted in changes in the Krebs cycle and urea cycle metabolites, whereas GVP exposure resulted in alterations in hexosamines and in lipid metabolism. These investigators found that exposure to cigarette smoke was associated with perturbations in the metabolism of glutathione, amino acids, lipids, and nucleotides. Alterations were also seen in the urea cycle, Krebs cycle, and the production of polyamines and cellular cofactors.

Oxidative stress is an important consequence associated with cigarette smoke. The oxidants produced or found in cigarette smoke can react with and damage cellular macromolecules including proteins, DNA, and lipid membranes. Additionally, these oxidants can act as signaling molecules that can influence cell proliferation (Faux et al., 2009). However, cells contain antioxidant defenses, such as GSH to prevent oxidative damage (Rahman & MacNee, 1999). The metabolomics approach has indeed found changes in metabolites associated with these effects of cigarette smoke. Predominant changes in metabolites from exposure to WS included glutathione and γ-glutamylglutamine, which showed 51 and 13-fold increases compared to control cells. The increased levels of metabolites within the glutathione pathway (i.e. glutathione and γ-glutamylglutamine) suggest a protective response against oxidative stress, which can result from WS. These data correlate with human microarray data that demonstrated an antioxidant response to cigarette smoke through the induction of genes associated with glutathione metabolism (Spira et al., 2004). Further, Vulimiri and colleagues (Vulimiri et al., 2009) observed increased levels (16.4-fold) of *o*-phosphoethanolamine, a marker of phospholipid degradation that may indicate cell membrane damage. Additional observations included increased arachidonic acid levels that may suggest an inflammatory response and markers (e.g. putrescine) of cell proliferation.

Conversely, there was a statistically significant decrease in glutathione levels from exposure to WTPM and GVP when compared to controls. Predominant metabolite changes in these phases were acetylcarnitine (4.5-fold increase) and palmitoleate (5.2-fold increase) for WTPM and GVP, respectively, both of which indicated changes in lipid metabolism (Vulimiri et al., 2009). In summary, these authors (Vulimiri et al., 2009) demonstrated how metabolomics could differentiate metabolic responses caused by exposure to complex mixtures (i.e. cigarette smoke) and also provided empirical data for metabolic changes for known markers of toxicity (e.g. decreased GSH, membrane damage) associated with exposure to cigarette smoke.

6. Integrating "omics" data into risk assessment

The field of toxicology is advancing with several of the high throughput screening techniques taking major roles in understanding toxicity pathways. More specifically toxicogenomics and proteomics have contributed to identifying the mechanisms of toxicity pathways (Burchiel et al., 2001; Heijne et al., 2005). Integration of data from different "omics" approaches would help clarify the mode of action of xenobiotics especially at low dose levels, which are relevant to environmental exposures. Thus, combining "omic" techniques such as toxicogenomics, transcriptomics, proteomics, and metabolomics promises to provide a robust understanding of biological responses to xenobiotics. Under this paradigm, effects on genes and their downstream products, namely proteins and the metabolites produced by proteins, can be assessed together. Researchers have utilized diverse "omics" approaches to understand the mode(s) of action or mechanism of different toxicity pathways in various experimental models. Some studies have indeed shown a good correlation between the genomics and metabolomics approaches. For example, metabolomics approach has demonstrated alterations in the pathways associated with lipid peroxidation, DNA damage and repair, and cell proliferation in rats exposed to CCl_4 which was consistent with the expression of transcripts associated with steatosis/fibrosis that were specific to cell injury and regeneration (Chung et al., 2005a; Chung et al., 2005b; Fountoulakis et al., 2002). Additionally, metabolomics has identified biochemical alterations in pathways associated with oxidative stress, inflammation, cell proliferation, and cytotoxicity in human lung epithelial cells exposed to cigarette smoke (Vulimiri et al., 2009). These observations are supported by data from genomic (Harvey et al., 2007; Meng et al.,

2006; Spira et al., 2004) and proteomic (Carter et al., 2011; Kelsen et al., 2008; Zhang et al., 2008) investigations that also detected markers associated with the pathways affected by exposure to cigarette smoke.

Conversely, the integration of "omics" technologies has also demonstrated some limitations. For example, a study by Steiling and co-workers found a discrepancy between genomic and proteomic data. Specifically, gel electrophoresis followed by LC-MS analysis identified 41 proteins whose expression would not have been detected by microarray analysis (Steiling et al., 2009). Such data highlight the importance of assessing more downstream markers (i.e. proteins or metabolites) that may provide a more accurate understanding of the biological responses to chemical exposures.

7. Shift in the risk assessment paradigm from apical endpoints to biochemical perturbations

Historically, human health risk assessment relied on identifying apical endpoints, defined as "observable outcomes in a whole organism, such as a clinical sign or pathologic state, indicative of a disease state that can result from exposure to a toxicant", such as birth defects, neurologic deficits, and tumor number that are obtained from high-dose animal bioassays (Krewski et al., 2010). However, the mechanistic relevance of these data can be tenuous considering the need for interspecies extrapolation and that environmental exposures may be orders of magnitude lower. Recently, the NRC report on "*Toxicity Testing in the 21st Century: A Vision and Strategy*" suggested a fundamental change in the risk assessment paradigm where the reliance on apical endpoints of toxicity would be replaced in favor of *in vitro* toxicity testing for identifying perturbations in biochemical pathways. Such an approach would employ cell lines, particularly human-based lines, and high throughput screening assays (e.g. genomics, proteomics) with computational toxicology methods. As a result, this approach would also lead to an overall increase in the efficiency of toxicity testing and decrease in animal usage [as reviewed in (Krewski et al., 2010)]. As pointed out by the NRC committee, the use of these low-dose, human-based *in vitro* systems would negate concerns associated with high-dose, animal data. Since metabolomics can identify early perturbations in biochemical pathways, this technology is poised to become an important element of this proposed risk assessment paradigm.

8. Advantages of metabolomics approach for environmental chemical assessment

Metabolomics approach offers several advantages for understanding the mechanisms of toxicity of environmental chemicals and informing human health assessments. One advantage of the metabolomics approach is the relatively non-invasive (e.g. urine samples) nature of this technique. In this context, samples from humans subjected to an incidental environmental exposure can be easily collected, analyzed, and correlated to metabolomic profiles from animal studies in order to identify critical effects from the exposure. Aside from quantitative differences, metabolic pathways are evolutionally conserved across different species; metabolomics data can be qualitatively extrapolated or interpreted at the molecular level among and between species. Unlike genomics and proteomics, metabolomics databases offer information on the structural, physicochemical, pharmacological and spectral profiles as well as biological functions of metabolites (Go, 2010). From a technical standpoint, sample preparation is relatively minimal for metabolomic analyses compared to genomics and proteomics approaches. Also, since metabolomics allows for the *a priori* selection of metabolites, the investigator has better control over the number of standards that need to be procured or synthesized for analysis.

A survey of the available literature already demonstrates the value of using metabolomics for elucidating the mode of action of environmental chemicals. Since a number of matrices are amenable to metabolomic analysis, various systems including *in vitro* and *in vivo* models can be employed to identify metabolic perturbations associated with exposure to environmental chemicals. Some of the mammalian *in vitro* systems involved exposing A549 human lung carcinoma cells to cigarette smoke (Vulimiri et al., 2009), human embryonic stem cells to various teratogens (West et al., 2010), and reaggregating neuronal cultures to methylmercury (van Vliet et al., 2008). From *in vivo* studies, intact organs from exposed experimental animals have been evaluated to include the analysis of intact lung after intratracheal instillation of silica dust (Hu et al., 2008) as well as liver and lung after exposure to 1-nitronaphthalene (Azmi et al., 2005). Also, metabolomics has been able to demonstrate different metabolic responses between mixtures of environmental chemicals. For example, in assessing the inflammatory response, Schmelzer et al (2006) found unique lipid profiles between rats exposed to 1-nitronaphthalene and those exposed to a mixture of 1-nitronaphthalene and ozone. As highlighted earlier in this chapter, metabolomics approach has been shown to differentiate the metabolic changes attributed to the complex mixtures of whole smoke versus its two constituent physical phases (Vulimiri et al., 2009).

9. Limitations of metabolomics research

Regardless of whether a chemical being investigated is a pharmaceutical or environmental pollutant, a critique of metabolomics has been the detection of metabolites that appear to be changed independent of the toxicant, its mode of action, and target tissue. Such common metabolites were previously termed the "usual suspects" by Robertson and often include 2-oxoglutarate, citrate, succinate, and trimethyl amine/trimethyl amine oxide (Robertson, 2005). From a data interpretation standpoint, these "usual suspects" may confound the identification of metabolites that have true value as specific markers of organ toxicity by skewing results from pattern separation analyses (Connor et al., 2004; Robertson, 2005). Connor et al (2004) found that urinary metabolite changes, including many of the "usual suspect" metabolites, can be associated with systemic or secondary toxicity effects, namely reductions in food intake and body weight. Additionally, the authors suggest that such metabolites may still inform mechanisms of toxicity when put into the context of diet and weight change. Conversely, the authors did identify exposures where common metabolites were specific for toxicant exposure. For example, succinate and 2-oxoglutarate were specifically associated with exposure to the hepatotoxicant α-naphthylisothiocyanate, rather than resulting from secondary toxicity (Connor et al., 2004). Research by other investigators has identified metabolites as novel biomarkers associated with specific diseases and for specific toxicities. For example, 5-oxoproline has been demonstrated to be a specific marker of bromobenzene-induced hepatotoxicity (Waters et al., 2006).

10. Future directions/research needs

A goal of this chapter has been to highlight how metabolomics approach can be used to better understand the toxicity of environmental chemicals, with a particular focus on hazard identification and mode of action. However, due to its relative infancy compared to conventional toxicity assays and other "omic" technologies, metabolomic information for environmental chemicals is sparse. As interest in metabolomics increases and as this approach becomes more accessible to researchers focused on environmental chemicals, this database will grow. A contemporary issue regarding xenobiotics is the influence of the microbiome on immunity, metabolism, and human health [reviewed in (Han et al., 2010)]. This microbiome involves the sum of all the microorganisms (e.g. bacteria) that internally and externally reside on an individual or animal. Of these microorganisms, the gut microbiota has been shown to influence the metabolism of several xenobiotics. Accordingly, the microbiome is likely to be important in the toxicity of environmental chemicals and pertinent to human health assessments (Betts, 2011). Thus, the influence of the microbiome is an important aspect of chemical toxicity that should be further studied using metabolomics.

11. Conclusions

Metabolomics is an emerging medium-to-high throughput technique which measures the endogenous biochemicals affecting different metabolic pathways and can be useful in characterizing the hazards of environmental chemicals. Identifying metabolic perturbations caused in mammalian cell systems following chemical exposure helps in elucidating the predominant toxicity pathways. Some of the advantages of using metabolomic data for hazard identification—one of the key steps in risk assessment—include the ability to inform gender, genetic, and organ-specific effects in a relatively expedient manner. As briefly discussed, metabolomics can identify early biochemical perturbations associated with toxicity in the hepatic, nervous, and pulmonary systems caused by selected environmental chemicals. As surveyed, various research systems using metabolomics demonstrate how metabolomic data could be used for hazard identification and mode of action characterization for environmental chemicals. Overall, metabolomics represents an opportunity to develop a better understanding of the toxicity of environmental chemicals and could further impact the human health assessment of these chemicals.

12. Disclaimer

The views expressed in this chapter are those of the authors and do not necessarily represent the views or policies of the U.S. Environmental Protection Agency.

13. References

ATSDR (1999). Toxicological profile for mercury, Agency for Toxic Substances and Disease Registry, Atlanta, GA, Available from: http://www.atsdr.cdc.gov/toxprofiles/tp46.pdf

Azmi, J.; Connelly, J.; Holmes, E.; Nicholson, J.K.; Shore, R.F. & Griffin, J.L. (2005). Characterization of the biochemical effects of 1-nitronaphthalene in rats using global metabolic profiling by NMR spectroscopy and pattern recognition, *Biomarkers,* Vol. 10, No. 6, pp. 401-16.

Barbin, A. (2000). Etheno-adduct-forming chemicals: from mutagenicity testing to tumor mutation spectra, *Mutat Res,* Vol. 462, No. 2-3, pp. 55-69.

Basu, S. (2003). Carbon tetrachloride-induced lipid peroxidation: eicosanoid formation and their regulation by antioxidant nutrients, *Toxicology,* Vol. 189, No. 1-2, pp. 113-27.

Beddowes, E.J.; Faux, S.P. & Chipman, J.K. (2003). Chloroform, carbon tetrachloride and glutathione depletion induce secondary genotoxicity in liver cells via oxidative stress, *Toxicology,* Vol. 187, No. 2-3, pp. 101-15.

Betts, K.S. (2011). A study in balance: how microbiomes are changing the shape of environmental health, *Environ Health Perspect,* Vol. 119, No. 8, pp. a340-6.

Bothwell, J.H.; Rae, C.; Dixon, R.M.; Styles, P. & Bhakoo, K.K. (2001). Hypo-osmotic swelling-activated release of organic osmolytes in brain slices: implications for brain oedema in vivo, *J Neurochem,* Vol. 77, No. 6, pp. 1632-40.

Boverhof, D.R. & Zacharewski, T.R. (2006). Toxicogenomics in risk assessment: applications and needs, *Toxicol Sci,* Vol. 89, No. 2, pp. 352-60.

Burchiel, S.W.; Knall, C.M.; Davis, J.W., 2nd; Paules, R.S.; Boggs, S.E. & Afshari, C.A. (2001). Analysis of genetic and epigenetic mechanisms of toxicity: potential roles of toxicogenomics and proteomics in toxicology, *Toxicol Sci,* Vol. 59, No. 2, pp. 193-5.

Cabre, M.; Camps, J.; Paternain, J.L.; Ferre, N. & Joven, J. (2000). Time-course of changes in hepatic lipid peroxidation and glutathione metabolism in rats with carbon tetrachloride-induced cirrhosis, *Clin Exp Pharmacol Physiol,* Vol. 27, No. 9, pp. 694-9.

Carter, C.A.; Misra, M. & Pelech, S. (2011). Proteomic Analyses of Lung Lysates from Short-Term Exposure of Fischer 344 Rats to Cigarette Smoke, *J Proteome Res,* Vol. (In Press), No.

Ceccatelli, S.; Dare, E. & Moors, M. (2010). Methylmercury-induced neurotoxicity and apoptosis, *Chem Biol Interact,* Vol. 188, No. 2, pp. 301-8.

Chen, C.; Krausz, K.W.; Idle, J.R. & Gonzalez, F.J. (2008). Identification of novel toxicity-associated metabolites by metabolomics and mass isotopomer analysis of acetaminophen metabolism in wild-type and Cyp2e1-null mice, *J Biol Chem,* Vol. 283, No. 8, pp. 4543-59.

Chen, C.; Ma, X.; Malfatti, M.A.; Krausz, K.W.; Kimura, S.; Felton, J.S.; Idle, J.R. & Gonzalez, F.J. (2007). A comprehensive investigation of 2-amino-1-methyl-6-phenylimidazo[4,5-b]pyridine (PhIP) metabolism in the mouse using a multivariate data analysis approach, *Chem Res Toxicol,* Vol. 20, No. 3, pp. 531-42.

Chung, F.L.; Nath, R.G.; Nagao, M.; Nishikawa, A.; Zhou, G.D. & Randerath, K. (1999). Endogenous formation and significance of 1,N2-propanodeoxyguanosine adducts, *Mutat Res,* Vol. 424, No. 1-2, pp. 71-81.

Chung, H.; Hong, D.P.; Jung, J.Y.; Kim, H.J.; Jang, K.S.; Sheen, Y.Y.; Ahn, J.I.; Lee, Y.S. & Kong, G. (2005a). Comprehensive analysis of differential gene expression profiles on carbon tetrachloride-induced rat liver injury and regeneration, *Toxicol Appl Pharmacol,* Vol. 206, No. 1, pp. 27-42.

Chung, H.; Hong, D.P.; Kim, H.J.; Jang, K.S.; Shin, D.M.; Ahn, J.I.; Lee, Y.S. & Kong, G. (2005b). Differential gene expression profiles in the steatosis/fibrosis model of rat liver by chronic administration of carbon tetrachloride, *Toxicol Appl Pharmacol,* Vol. 208, No. 3, pp. 242-54.

Connor, S.C.; Wu, W.; Sweatman, B.C.; Manini, J.; Haselden, J.N.; Crowther, D.J. & Waterfield, C.J. (2004). Effects of feeding and body weight loss on the 1H-NMR-based urine metabolic profiles of male Wistar Han rats: implications for biomarker discovery, *Biomarkers,* Vol. 9, No. 2, pp. 156-79.

De Zwart, L.L.; Venhorst, J.; Groot, M.; Commandeur, J.N.; Hermanns, R.C.; Meerman, J.H.; Van Baar, B.L. & Vermeulen, N.P. (1997). Simultaneous determination of eight lipid peroxidation degradation products in urine of rats treated with carbon tetrachloride using gas chromatography with electron-capture detection, *J Chromatogr B Biomed Sci Appl,* Vol. 694, No. 2, pp. 277-87.

Ekino, S.; Susa, M.; Ninomiya, T.; Imamura, K. & Kitamura, T. (2007). Minamata disease revisited: an update on the acute and chronic manifestations of methyl mercury poisoning, *J Neurol Sci,* Vol. 262, No. 1-2, pp. 131-44.

Eskes, C.; Honegger, P.; Juillerat-Jeanneret, L. & Monnet-Tschudi, F. (2002). Microglial reaction induced by noncytotoxic methylmercury treatment leads to neuroprotection via interactions with astrocytes and IL-6 release, *Glia,* Vol. 37, No. 1, pp. 43-52.

Faux, S.P.; Tai, T.; Thorne, D.; Xu, Y.; Breheny, D. & Gaca, M. (2009). The role of oxidative stress in the biological responses of lung epithelial cells to cigarette smoke, *Biomarkers,* Vol. 14 Suppl 1, No. 90-6.

Fountoulakis, M.; de Vera, M.C.; Crameri, F.; Boess, F.; Gasser, R.; Albertini, S. & Suter, L. (2002). Modulation of gene and protein expression by carbon tetrachloride in the rat liver, *Toxicol Appl Pharmacol,* Vol. 183, No. 1, pp. 71-80.

Gee, D.L.; Bechtold, M.M. & Tappel, A.L. (1981). Carbon tetrachloride-induced lipid peroxidation: simultaneous in vivo measurements of pentane and chloroform exhaled by the rat, *Toxicol Lett,* Vol. 8, No. 6, pp. 299-306.

Gieger, C.; Geistlinger, L.; Altmaier, E.; Hrabe de Angelis, M.; Kronenberg, F.; Meitinger, T.; Mewes, H.W.; Wichmann, H.E.; Weinberger, K.M.; Adamski, J.; Illig, T. & Suhre, K. (2008). Genetics meets metabolomics: a genome-wide association study of metabolite profiles in human serum, *PLoS Genet,* Vol. 4, No. 11, pp. e1000282.

Ginsberg, G.; Smolenski, S.; Neafsey, P.; Hattis, D.; Walker, K.; Guyton, K.Z.; Johns, D.O. & Sonawane, B. (2009). The influence of genetic polymorphisms on population variability in six xenobiotic-metabolizing enzymes, *J Toxicol Environ Health B Crit Rev,* Vol. 12, No. 5-6, pp. 307-33.

Go, E.P. (2010). Database resources in metabolomics: an overview, *J Neuroimmune Pharmacol,* Vol. 5, No. 1, pp. 18-30.

Guyton, K.Z.; Kyle, A.D.; Aubrecht, J.; Cogliano, V.J.; Eastmond, D.A.; Jackson, M.; Keshava, N.; Sandy, M.S.; Sonawane, B.; Zhang, L.; Waters, M.D. & Smith, M.T. (2009). Improving prediction of chemical carcinogenicity by considering multiple mechanisms and applying toxicogenomic approaches, *Mutat Res,* Vol. 681, No. 2-3, pp. 230-40.

Han, J.; Antunes, L.C.; Finlay, B.B. & Borchers, C.H. (2010). Metabolomics: towards understanding host-microbe interactions, *Future Microbiol,* Vol. 5, No. 2, pp. 153-61.

Harvey, B.G.; Heguy, A.; Leopold, P.L.; Carolan, B.J.; Ferris, B. & Crystal, R.G. (2007). Modification of gene expression of the small airway epithelium in response to cigarette smoking, *J Mol Med (Berl),* Vol. 85, No. 1, pp. 39-53.

Heby, O. (1981). Role of polyamines in the control of cell proliferation and differentiation, *Differentiation,* Vol. 19, No. 1, pp. 1-20.

Hecht, S.S. (2006). Cigarette smoking: cancer risks, carcinogens, and mechanisms, *Langenbecks Arch Surg,* Vol. 391, No. 6, pp. 603-13.

Heijne, W.H.; Kienhuis, A.S.; van Ommen, B.; Stierum, R.H. & Groten, J.P. (2005). Systems toxicology: applications of toxicogenomics, transcriptomics, proteomics and metabolomics in toxicology, *Expert Rev Proteomics,* Vol. 2, No. 5, pp. 767-80.

Hu, J.Z.; Rommereim, D.N.; Minard, K.R.; Woodstock, A.; Harrer, B.J.; Wind, R.A.; Phipps, R.P. & Sime, P.J. (2008). Metabolomics in lung inflammation:a high-resolution (1)h NMR study of mice exposedto silica dust, *Toxicol Mech Methods,* Vol. 18, No. 5, pp. 385-98.

Huang, X.; Shao, L.; Gong, Y.; Mao, Y.; Liu, C.; Qu, H. & Cheng, Y. (2008). A metabonomic characterization of CCl4-induced acute liver failure using partial least square regression based on the GC/MS metabolic profiles of plasma in mice, *J Chromatogr B Analyt Technol Biomed Life Sci,* Vol. 870, No. 2, pp. 178-85.

Hwang, E.S. & Kim, G.H. (2007). Biomarkers for oxidative stress status of DNA, lipids, and proteins in vitro and in vivo cancer research, *Toxicology,* Vol. 229, No. 1-2, pp. 1-10.

Ichi, I.; Kamikawa, C.; Nakagawa, T.; Kobayashi, K.; Kataoka, R.; Nagata, E.; Kitamura, Y.; Nakazaki, C.; Matsura, T. & Kojo, S. (2009). Neutral sphingomyelinase-induced ceramide accumulation by oxidative stress during carbon tetrachloride intoxication, *Toxicology,* Vol. 261, No. 1-2, pp. 33-40.

Idle, J.R. & Gonzalez, F.J. (2007). Metabolomics, *Cell Metab,* Vol. 6, No. 5, pp. 348-51.

Kelada, S.N.; Eaton, D.L.; Wang, S.S.; Rothman, N.R. & Khoury, M.J. (2003). The role of genetic polymorphisms in environmental health, *Environ Health Perspect,* Vol. 111, No. 8, pp. 1055-64.

Kelsen, S.G.; Duan, X.; Ji, R.; Perez, O.; Liu, C. & Merali, S. (2008). Cigarette smoke induces an unfolded protein response in the human lung: a proteomic approach, *Am J Respir Cell Mol Biol,* Vol. 38, No. 5, pp. 541-50.

Kim, K.-B.; Chung, M.W.; Um, S.Y.; Oh, J.S.; Kim, S.H.; Na, M.A.; Oh, H.Y.; Cho, W.-S. & Choi, K.H. (2008). Metabolomics and biomarker discovery: NMR spectral data of urine and hepatotoxicity by carbon tetrachloride, acetaminophen, and D-galactosamine in rats, *Metabolomics,* Vol. 4, No. 377-392.

Krewski, D.; Acosta, D., Jr.; Andersen, M.; Anderson, H.; Bailar, J.C., 3rd; Boekelheide, K.; Brent, R.; Charnley, G.; Cheung, V.G.; Green, S., Jr.; Kelsey, K.T.; Kerkvliet, N.I.; Li, A.A.; McCray, L.; Meyer, O.; Patterson, R.D.; Pennie, W.; Scala, R.A.; Solomon, G.M.; Stephens, M.; Yager, J. & Zeise, L. (2010). Toxicity testing in the 21st century: a vision and a strategy, *J Toxicol Environ Health B Crit Rev,* Vol. 13, No. 2-4, pp. 51-138.

Lin, Y.; Si, D.; Zhang, Z. & Liu, C. (2009). An integrated metabonomic method for profiling of metabolic changes in carbon tetrachloride induced rat urine, *Toxicology,* Vol. 256, No. 3, pp. 191-200.

Liu, J.Y.; Tsai, H.J.; Hwang, S.H.; Jones, P.D.; Morisseau, C. & Hammock, B.D. (2009). Pharmacokinetic optimization of four soluble epoxide hydrolase inhibitors for use in a murine model of inflammation, *Br J Pharmacol,* Vol. 156, No. 2, pp. 284-96.

Manibusan, M.K.; Odin, M. & Eastmond, D.A. (2007). Postulated carbon tetrachloride mode of action: a review, *J Environ Sci Health Part C,* Vol. 25, No. 3, pp. 185-209.

Meng, Q.R.; Gideon, K.M.; Harbo, S.J.; Renne, R.A.; Lee, M.K.; Brys, A.M. & Jones, R. (2006). Gene expression profiling in lung tissues from mice exposed to cigarette smoke, lipopolysaccharide, or smoke plus lipopolysaccharide by inhalation, *Inhal Toxicol,* Vol. 18, No. 8, pp. 555-68.

Monnet-Tschudi, F.; Zurich, M.G. & Honegger, P. (1996). Comparison of the developmental effects of two mercury compounds on glial cells and neurons in aggregate cultures of rat telencephalon, *Brain Res*, Vol. 741, No. 1-2, pp. 52-9.

Mortensen, H.M. & Euling, S.Y. (2011). Integrating mechanistic and polymorphism data to characterize human genetic susceptibility for environmental chemical risk assessment in the 21st century, *Toxicol Appl Pharmacol*, Vol. No. (In Press).

Myers, G.J.; Thurston, S.W.; Pearson, A.T.; Davidson, P.W.; Cox, C.; Shamlaye, C.F.; Cernichiari, E. & Clarkson, T.W. (2009). Postnatal exposure to methyl mercury from fish consumption: a review and new data from the Seychelles Child Development Study, *Neurotoxicology*, Vol. 30, No. 3, pp. 338-49.

NRC (1983). *Risk assessment in the federal government: Managing the process*, National Academies Press, Washington, DC. Available from: http://www.nap.edu/openbook.php?isbn=0309033497

NRC (1994). *Science and judgment in risk assessment*, National Academies Press, Washington, DC. Available from: http://www.nap.edu/catalog/2125.html

Pegg, A.E.; Matsui, I.; Seely, J.E.; Pritchard, M.L. & Poso, H. (1981). Formation of putrescine in rat liver, *Med Biol*, Vol. 59, No. 5-6, pp. 327-33.

Raamsdonk, L.M.; Teusink, B.; Broadhurst, D.; Zhang, N.; Hayes, A.; Walsh, M.C.; Berden, J.A.; Brindle, K.M.; Kell, D.B.; Rowland, J.J.; Westerhoff, H.V.; van Dam, K. & Oliver, S.G. (2001). A functional genomics strategy that uses metabolome data to reveal the phenotype of silent mutations, *Nat Biotechnol*, Vol. 19, No. 1, pp. 45-50.

Rahman, I. & MacNee, W. (1999). Lung glutathione and oxidative stress: implications in cigarette smoke-induced airway disease, *Am J Physiol*, Vol. 277, No. 6 Pt 1, pp. L1067-88.

Robertson, D.G. (2005). Metabonomics in toxicology: a review, *Toxicol Sci*, Vol. 85, No. 2, pp. 809-22.

Robertson, D.G.; Reily, M.D.; Sigler, R.E.; Wells, D.F.; Paterson, D.A. & Braden, T.K. (2000). Metabonomics: evaluation of nuclear magnetic resonance (NMR) and pattern recognition technology for rapid in vivo screening of liver and kidney toxicants, *Toxicol Sci*, Vol. 57, No. 2, pp. 326-37.

Roux, A.; Lison, D.; Junot, C. & Heilier, J.F. (2011). Applications of liquid chromatography coupled to mass spectrometry-based metabolomics in clinical chemistry and toxicology: A review, *Clin Biochem*, Vol. 44, No. 1, pp. 119-35.

Rubino, F.M.; Pitton, M.; Di Fabio, D. & Colombi, A. (2009). Toward an "omic" physiopathology of reactive chemicals: thirty years of mass spectrometric study of the protein adducts with endogenous and xenobiotic compounds, *Mass Spectrom Rev*, Vol. 28, No. 5, pp. 725-84.

Schmelzer, K.R.; Wheelock, A.M.; Dettmer, K.; Morin, D. & Hammock, B.D. (2006). The role of inflammatory mediators in the synergistic toxicity of ozone and 1-nitronaphthalene in rat airways, *Environ Health Perspect*, Vol. 114, No. 9, pp. 1354-60.

Serhan, C.N. (2009). Systems approach to inflammation resolution: identification of novel anti-inflammatory and pro-resolving mediators, *J Thromb Haemost*, Vol. 7 Suppl 1, No. 44-8.

Spira, A.; Beane, J.; Pinto-Plata, V.; Kadar, A.; Liu, G.; Shah, V.; Celli, B. & Brody, J.S. (2004). Gene expression profiling of human lung tissue from smokers with severe emphysema, *Am J Respir Cell Mol Biol*, Vol. 31, No. 6, pp. 601-10.

Steiling, K.; Kadar, A.Y.; Bergerat, A.; Flanigon, J.; Sridhar, S.; Shah, V.; Ahmad, Q.R.; Brody, J.S.; Lenburg, M.E.; Steffen, M. & Spira, A. (2009). Comparison of proteomic and transcriptomic profiles in the bronchial airway epithelium of current and never smokers, *PLoS One,* Vol. 4, No. 4, pp. e5043.

Sun, F.; Hamagawa, E.; Tsutsui, C.; Ono, Y.; Ogiri, Y. & Kojo, S. (2001). Evaluation of oxidative stress during apoptosis and necrosis caused by carbon tetrachloride in rat liver, *Biochim Biophys Acta,* Vol. 1535, No. 2, pp. 186-91.

Szeto, S.S.; Reinke, S.N.; Sykes, B.D. & Lemire, B.D. (2010). Mutations in the Saccharomyces cerevisiae succinate dehydrogenase result in distinct metabolic phenotypes revealed through (1)H NMR-based metabolic footprinting, *J Proteome Res,* Vol. 9, No. 12, pp. 6729-39.

U.S. EPA (2005). Guidelines for carcinogenic risk assessment, Risk Assessment Forum, U.S. Environmental Proection Agency, Washington, DC, EPA/630/P-03/001F. Available from:
http://www.epa.gov/raf/publications/pdfs/CANCER_GUIDELINES_FINAL_3-25-05.PDF

U.S. EPA (1992). Respiratory health effects of passive smoking: lung cancer and other disorders, U.S. Environmental Protection Agency, Washington, DC, EPA/600/6-90/006F. Available from: http://www.epa.gov/ncea/ets/pdfs/acknowl.pdf

U.S. EPA (2001). Water Quality Criterion for the Protection of Human Health: Methylmercury, U.S. Environmental Protection Agency, Washington, DC, EPA-823-R-01-001. Available from:
http://water.epa.gov/scitech/swguidance/standards/criteria/aqlife/pollutants/methylmercury/upload/2009_01_15_criteria_methylmercury_mercury-criterion.pdf

van Ravenzwaay, B.; Cunha, G.C.; Leibold, E.; Looser, R.; Mellert, W.; Prokoudine, A.; Walk, T. & Wiemer, J. (2007). The use of metabolomics for the discovery of new biomarkers of effect, *Toxicol Lett,* Vol. 172, No. 1-2, pp. 21-8.

van Vliet, E.; Morath, S.; Eskes, C.; Linge, J.; Rappsilber, J.; Honegger, P.; Hartung, T. & Coecke, S. (2008). A novel in vitro metabolomics approach for neurotoxicity testing, proof of principle for methyl mercury chloride and caffeine, *Neurotoxicology,* Vol. 29, No. 1, pp. 1-12.

Vulimiri, S.V.; Berger, A. & Sonawane, B. (2011). The potential of metabolomic approaches for investigating mode(s) of action of xenobiotics: case study with carbon tetrachloride, *Mutat Res,* Vol. 722, No. 2, pp. 147-53.

Vulimiri, S.V.; Misra, M.; Hamm, J.T.; Mitchell, M. & Berger, A. (2009). Effects of mainstream cigarette smoke on the global metabolome of human lung epithelial cells, *Chem Res Toxicol,* Vol. 22, No. 3, pp. 492-503.

Wang, C.; Yang, J. & Nie, J. (2009). Plasma phospholipid metabolic profiling and biomarkers of rats following radiation exposure based on liquid chromatography-mass spectrometry technique, *Biomed Chromatogr,* Vol. 23, No. 10, pp. 1079-85.

Waters, N.J.; Waterfield, C.J.; Farrant, R.D.; Holmes, E. & Nicholson, J.K. (2006). Integrated metabonomic analysis of bromobenzene-induced hepatotoxicity: novel induction of 5-oxoprolinosis, *J Proteome Res,* Vol. 5, No. 6, pp. 1448-59.

West, P.R.; Weir, A.M.; Smith, A.M.; Donley, E.L. & Cezar, G.G. (2010). Predicting human developmental toxicity of pharmaceuticals using human embryonic stem cells and metabolomics, *Toxicol Appl Pharmacol,* Vol. 247, No. 1, pp. 18-27.

Zhang, S.; Xu, N.; Nie, J.; Dong, L.; Li, J. & Tong, J. (2008). Proteomic alteration in lung tissue of rats exposed to cigarette smoke, *Toxicol Lett,* Vol. 178, No. 3, pp. 191-6.